Alison Davies

Chaetanthera and Oriastrum

Alison Davies

Chaetanthera and Oriastrum

A systematic revision of Chaetanthera Ruiz and Pav. and the reinstatement of Oriastrum Poepp. and Endl. (Asteraceae: Mutisieae)

Südwestdeutscher Verlag für
Hochschulschriften

Imprint
Any brand names and product names mentioned in this book are subject to trademark, brand or patent protection and are trademarks or registered trademarks of their respective holders. The use of brand names, product names, common names, trade names, product descriptions etc. even without a particular marking in this work is in no way to be construed to mean that such names may be regarded as unrestricted in respect of trademark and brand protection legislation and could thus be used by anyone.

Publisher:
Südwestdeutscher Verlag für Hochschulschriften
is a trademark of
Dodo Books Indian Ocean Ltd., member of the OmniScriptum S.R.L Publishing group
str. A.Russo 15, of. 61, Chisinau-2068, Republic of Moldova Europe
Printed at: see last page
ISBN: 978-3-8381-1910-6

Zugl. / Approved by: Muenchen, LMU, Diss., 2010

Copyright © Alison Davies
Copyright © 2010 Dodo Books Indian Ocean Ltd., member of the OmniScriptum S.R.L Publishing group

Contents

Effective publication according to ICBN (Vienna 2006).. 5
Acknowledgements 5
Abstract... 6

PART A: Systematics of *Chaetanthera* & *Oriastrum*

I   Introduction.. 9
    1   Philosophy of taxonomic research 9
    2   Goals of taxonomic research... 10
    3   Application to *Chaetanthera* & *Oriastrum* 10
        3.1   Historical background 10
        3.2   Novel sources of variation 11
        3.3   Re-defining taxonomic relationships 13
II   Materials, methods & measurements 14
    1   Materials 14
    2   Methods 14
        2.1   Capitula & floral dissection... 14
        2.2   Involucral bract anatomy studies 14
        2.3   Achene & pappus studies 14
        2.4   Pollen studies.. 15
        2.5   Chromosome studies... 16
        2.6   Images & illustrations 16
        2.7   Geographical data 16
    3   Measurements 16
III   Variation of characters in *Chaetanthera* & *Oriastrum* 18
    1   General remarks 18
    2   Morphological & anatomical variation... 18
        2.1   Life cycle & habit 18
        2.2   Leaves 22
        2.3   Indumentum.. 24
        2.4   Involucral bracts 24
            2.4.1   Arrangement.. 24
            2.4.2   Morphology of the inner involucral bracts................................. 25
            2.4.3   Anatomy of the inner involucral bracts...................................... 26
        2.5   Pappus 27
        2.6   Floret variation: ray dimorphism & showiness 28
        2.7   Anthers & pollen 32
        2.8   Styles & stigmas 35
        2.9   Achenes 37
            2.9.1   Achene shape... 37
            2.9.2   Carpopodium 37
            2.9.3   Achene hairs 38
            2.9.4   Achene testa epidermis 40
    3   Cytological variation 43
    4   Genetic variation 47
IV   Form, function & habitat 49
    1   Introduction.. 49
    2   Homoplasy 49
    3   Habitats & stress... 50
    4   Form & function... 50

	4.1	Rosettes, succulence & sclerophylly in *Chaetanthera*	50
	4.2	Cushions, stem buds & scleroid bracts in *Oriastrum*	51
	4.3	Fragile pappus & indehiscent achenes: dispersal brakes	52
	4.4	Xeromorphic adaptations in pollen, pericarp & testa epidermis	53
	4.5	Competitive ability	54
	4.6	Reproductive strategy	54 |//

V The biogeography of *Chaetanthera* & *Oriastrum* 56
  1 The distribution of *Chaetanthera* 56
    1.1 Species distributions 56
      1.1.1 Species endemic to the Altiplano 56
      1.1.2 Species endemic to Chile 56
      1.1.3 Species from Chile & Argentina 57
    1.2 Species hotspots 57
      1.2.1 The diversity hotspot "Coquimbo" (29° - 30°S) 58
      1.2.2 The diversity hotspot "Santiago" (33° - 34°S) 60
    1.3 Trichome diversity: evidence of species radiation 61
    1.4 Lamina morphology & genetic affinity: evidence of species radiations 63
      1.4.1 Linear-leaved group 63
      1.4.2 Dentate-ciliate leaved group 66
  2 The distribution of *Oriastrum* 68
    2.1 Species distributions 68
      2.1.1 Species endemic to either Argentina or Chile or Peru 70
      2.1.2 Species from Argentina & Chile 70
      2.1.3 Species from Argentina, Chile, Bolivia & Peru 70
    2.2 Species hotspots 70
    2.3 Species radiations 71
      2.3.1 Santiago 71
      2.3.2 Altiplano group 71
      2.3.3 Altoandino group 72
      2.3.4 Andino group 72

VI "The variety of all things forms a pleasure" 77
  1 Introduction 77
  2 *C. glabrata*: polymorphism & the El Niño effect 78
    2.1 Introduction 78
    2.2 Variation & distribution 78
    2.3 El Niño Southern Oscillation (ENSO) 81
    2.4 Results: observed phenotypic plasticity 81
    2.5 Conclusion 84
  3 *C. linearis* & *C. albiflora*: active hybridisation 85
    3.1 Introduction 85
    3.2 Variation & distribution 85
    3.3 Results: hybrid zones 86
    3.4 Conclusions 89
  4 Perennial, scapose *Chaetanthera*: polymorphism or hybridisation 89
    4.1 Introduction 89
    4.2 Variation & distribution 90
    4.3 Analysing variation: are the traditional characters truly diagnostic? 91
      4.3.1 Results 91
      4.3.2 Summary 93
    4.4 Analysing variation: principle components analysis 93
      4.4.1 Methods 93
      4.4.2 Results 95

	4.5	Analysing variation: HYWIN	98
	4.5.1	Results of the HYWIN analysis	99
	4.5.2	Summary	100
	4.6	Extinction or botanist's pride?	103
VII	Discussion & conclusions of systematic analysis		104
1	Uniting characters - grouping species		104
2	Evolutionary adaptation to environmental stress		108
3	Current dynamic change in *Chaetanthera*		109
4	Conclusions		112

PART B: Taxonomy of *Chaetanthera* & *Oriastrum*

VIII	*Chaetanthera* & *Oriastrum*: an historic review		115
1	Introduction		115
2	Chronological changes in *Chaetanthera* & *Oriastrum*		115
3	Nomenclatural notes for *Chaetanthera* & *Oriastrum*		118
	3.1	*Chaetanthera ciliata*: lectotype of *Chaetanthera* Ruiz & Pav.	118
	3.2	*Oriastrum pusillum*: generitype of *Oriastrum* Poepp. & Endl.	119
	3.3	David Don: nomenclatural types & nomenclatural obscurity	119
		3.3.1 Don's nomenclatural types	119
		3.3.2 Nomenclatural obscurity of Don names	120
	3.4	Rudolfo Philippi: lectotypes & nomina nuda	120
IX	*Chaetanthera* Ruíz & Pav.		122
1	Diagnosis		122
2	Subgeneric division of *Chaetanthera*		123
	2.1	*Chaetanthera* Ruíz & Pav. subgenus *Chaetanthera*	123
	2.2	*Chaetanthera* subgenus *Tylloma* (D.Don) Less.	124
3	Key to the species		125
4	Species descriptions		129
	4.1	*Chaetanthera* Ruíz & Pav. subgenus *Chaetanthera*	129
		1. *Chaetanthera albiflora* (Phil.) A.M.R. Davies	129
		2. *Chaetanthera chilensis* (Willd.) DC s.l.	133
		3. *Chaetanthera ciliata* Ruiz & Pav.	139
		4. *Chaetanthera depauperata* (Hook. & Arn.) A.M.R. Davies	142
		5. *Chaetanthera elegans* s.l. Phil.	145
		6. *Chaetanthera glandulosa* J. Rémy var. *glandulosa*	149
		7. *Chaetanthera glandulosa* var. *gracilis* A.M.R. Davies	152
		8. *Chaetanthera incana* Poepp. ex Less.	154
		9. *Chaetanthera linearis* Poepp. ex Less.	157
		10. *Chaetanthera linearis* x *albiflora*	160
		11. *Chaetanthera microphylla* (Cass.) Hook. & Arn.	161
		12. *Chaetanthera moenchioides* Less.	164
		13. *Chaetanthera multicaulis* DC.	169
		14. *Chaetanthera perpusilla* (Wedd.) Anderb. & S. E. Freire.	172
		15. *Chaetanthera peruviana* A. Gray	174
		16. *Chaetanthera ramosissima* D. Don ex Taylor & Phillips	177
		17. *Chaetanthera* x *serrata* Ruíz & Pav.	180
		18. *Chaetanthera taltalensis* (Cabrera) A.M.R. Davies	183
	4.2	*Chaetanthera* subgenus *Tylloma* (D.Don) Less.	186
		19. *Chaetanthera euphrasioides* (DC.) F. Meigen	186
		20. *Chaetanthera flabellata* D. Don ex Taylor & Phillips	189

		21.	*Chaetanthera flabellifolia* Cabrera	191
		22.	*Chaetanthera frayjorgensis* A.M.R. Davies	193
		23.	*Chaetanthera glabrata* (DC) F. Meigen	196
		24.	*Chaetanthera kalinae* A.M.R.Davies	200
		25.	*Chaetanthera limbata* (D. Don) Less.	202
		26.	*Chaetanthera philippii* B. L Rob.	204
		27.	*Chaetanthera pubescens* A.M.R. Davies	207
		28.	*Chaetanthera renifolia* (J. Rémy) Cabrera	210
		29.	*Chaetanthera schroederi* G. F. Grandjot & K. Grandjot	212
		30.	*Chaetanthera spathulifolia* Cabrera	215
		31.	*Chaetanthera splendens* (J. Rémy) B. L Rob.	218
		32.	*Chaetanthera villosa* D. Don ex Taylor & Phillips	220

X	*Oriastrum* Poepp. & Endl.			223
1	Diagnosis			223
2	Subgeneric division of *Oriastrum*			224
	2.1	*Oriastrum* Poepp. & Endl. subgenus *Oriastrum*		224
	2.2	*Oriastrum* Poepp. & Endl. subgenus *Egania* (J.Rémy) A.M.R. Davies		225
3	Key to the species			226
4	Species descriptions			228
	4.1	*Oriastrum* subgenus *Oriastrum*		228
		1.	*Oriastrum chilense* (J. Rémy) Wedd.	228
		2.	*Oriastrum gnaphalioides* (J. Rémy) Wedd.	231
		3.	*Oriastrum lycopodioides* (J. Rémy) Wedd.	235
		4.	*Oriastrum pusillum* Poepp. & Endl.	238
		5.	*Oriastrum werdermannii* A.M.R.Davies	241
	4.2	*Oriastrum* subgenus *Egania*		243
		6.	*Oriastrum abbreviatum* (Cabrera) A.M.R. Davies	243
		7.	*Oriastrum acerosum* (J. Rémy) Phil.	246
		8.	*Oriastrum achenohirsutum* (Tombesi) A.M.R. Davies	249
		9.	*Oriastrum apiculatum* (J. Rémy) A.M.R. Davies	251
		10.	*Oriastrum cochlearifolium* A. Gray.	254
		11.	*Oriastrum dioicum* (J. Rémy) Phil.	256
		12.	*Oriastrum famatinae* A.M.R. Davies	259
		13.	*Oriastrum polymallum* Phil.	261
		14.	*Oriastrum pulvinatum* Phil.	265
		15.	*Oriastrum revolutum* (Phil.) A.M.R. Davies	268
		16.	*Oriastrum stuebelii* (Hieron.) A.M.R. Davies var. *stuebelii*	272
		17.	*Oriastrum stuebelii* var. *cryptum* A.M.R.Davies	275
		18.	*Oriastrum tarapacensis* A.M.R. Davies	278
		19.	*Oriastrum tontalensis* A.M.R. Davies	280

XI	Summary of taxonomic revision	283
1	Changes in the nomenclature of *Chaetanthera* & *Oriastrum*	283
2	Updates in the descriptive taxonomy of *Chaetanthera* & *Oriastrum* species	284
3	Additional information concerning *Chaetanthera* & *Oriastrum*	284
XII	References	286
XIII	Appendix 1	293

Effective publication according to ICBN (Vienna 2006)

The original thesis from which this publication is derivbed was successfully defended on the 28th April 2010 (http://edoc.ub.uni-muenchen.de/11538/) under the auspices of the Faculty of Biology at the Ludwig-Maximilians-Universität München. As required by Article 30.5 of the International Code of Botanical Nomenclature (Vienna code 2006) I declare that this printed publication, first distributed in November 2010 subsequent to its successful defense, is regarded by me to be effectively published under that code.

Acknowledgements

This is dedicated to Ric, Tim, Isabel and Nicolas.

The completion of this work is due in great part to numerous people who have contributed both directly and indirectly. Thank you. Especial thanks goes to my husband Dr. Ric Davies who has provided unwavering support and encouragement throughout. I am deeply indebted to my supervisor, Jürke Grau, who made this research possible. Thank you for your support and guidance, and for your compassionate understanding of wider issues. The research for this study was funded by part-time employment on digital archiving projects coordinated via the Botanische Staatssammlung Munchen (INFOCOMP, 2000 – 2003; API-Projekt, 2005). Appreciative thanks go to the many friends and colleagues from both the Botanische Staatssammlung and the Botanical Institute who have provided scientific and social support over the years. Philomena Bodensteiner, Peter Döbbeler, Tini Ehrhart, Matthias Erben and Lilo Klingenberg deserve a special mention. Much of the evidence supporting the final conclusions in the thesis was acquired through laboratory research and S.E.M. imaging. This would not have been possible without the patient schooling and friendly assistance of Mrs. E. Vosyka, Dr. E. Facher and Mr E. Marksteiner.

Travel grants were received from the DAAD (the German Academic Exchange Service) to visit Chile twice, in 2001 and 2002. My stay in Chile and the field excursions were made far easier thanks to Sunke Nef & family, Diettrich & Arnulf Becker, and Melica Muñoz-Schick in Santiago, and Roberto Rodriguez, Señora Gleisner & family, and Marcelo Baeza & family in Concepción. Special thanks are extended to Mary Kalin Arroyo, an active botanist and ecologist in Chile, who has an ongoing scientific interest in *Chaetanthera*.

Permission to use digital images was kindly granted by Gustavo Aldunate, Stephan Beck, Michal Belov (www.Chileflora.com), Mauricio Bonifacino (MVFA, Uruguay), Michael Dillon, María Terese Eyzaguirre (Fundación R.A. Philippi), Jürke Grau, Michel Grenon, Stephan Halloy, Irene Till-Bottraud (Station Alpin Joseph Fourier, Grenoble) and Mauricio Zuñiga.

Finally, a huge "Thank you" to my parents Jean & Iain Robertson and my parents-in-law Elizabeth & Peter (†).

"...a traveller should be a botanist, for in all views plants form the chief embellishment."

DARWIN, 'Darwin's Journal of a Voyage round the World', p. 599 (1896)

Abstract

Chaetanthera Ruiz & Pav. (30 species, 1 variety, 2 hybrid forms) and *Oriastrum* Poepp. & Endl. (18 species, 1 variety) are among the most species-rich Astereaceae genera of the Chilean Flora. Formerly combined under one name, the two genera have been extensively revised. *Chaetanthera* is found mainly in Chile, with one Peruvian species and several scattered populations of other species in Andean Argentina. *Oriastrum* inhabits the higher elevations of the Andes, spread over Chile, Argentina, Bolivia and Peru. Systematic studies focussing on morphological and anatomical variation of characters taken from habit, involucral bracts, and achenes, combined with palynological and genetic (nr DNA) information are used to circumscribe *Chaetanthera* with two subgenera – *Chaetanthera* subgenus *Chaetanthera* and *Chaetanthera* subgenus *Tylloma* (D.Don) Less., and reinstate *Oriastrum* with two subgenera – *Oriastrum* subgenus *Oriastrum* and *Oriastrum* subgenus *Egania* (J.Rémy) A.M.R. Davies.

Character variation is discussed in the context of form, function and habitat, with emphasis on the evolutionary adaptiveness of character traits seen in the two allied genera. *Chaetanthera* appears to show primary adaptation to cold and several secondary adaptations to arid conditions, typical of modern Chilean landscapes. *Oriastrum* taxa appear well-adapted to the cold, high elevations of the Andes, and show secondary developments trending towards an insular syndrome.

The collated bio-geographical information of the taxa is considered in terms of endemism, hotspots and species radiations. *Chaetanthera* taxa have 2 loci of diversity hotspots in Chile – in Coquimbo and in Santiago. Trichome diversity and capitula morphology trends are used as evidence of species radiations in *Chaetanthera*. *Oriastrum* taxa are notable for parallel radiations of morphologically similar species within particular Andean zones: i.e., Altoandino or Altiplano.

Case studies concerning three groups of *Chaetanthera* taxa are presented. The first case highlights the effect of the El Niño on the polymorphic *C. glabrata* along the Chilean Pacific coast. The second case deals with current active hybridisation between *C. linearis* and *C. albiflora* in the semi-arid Andean foothills. In the last example, incipient speciation and polymorphism between *C. chilensis* and *C. elegans* in southern Central Chile is discussed. Various statistical techniques for the analysis of hybridisation events are applied.

All taxa are keyed out and described. Novel taxa are described and imaged or illustrated. Nomenclatural issues and lectotypification of 15 *Chaetanthera* names and 6 *Oriastrum* names are effected. *Chaetanthera* is described here with one novel species (*C. pubescens* A.M.R. Davies), one novel variety (*C. glandulosa* var. *microphylla* A.M.R. Davies), a new name (*C. frayjorgensis* A.M.R. Davies), and three new combinations: *C. albiflora* (Phil.) A.M.R. Davies, *C. depauperata* (Hook. & Arn.) A.M.R. Davies, *C. taltalensis* (Cabrera) A.M.R. Davies. *Oriastrum* is described here with four new species and one new variety: *O. werdermannii* A.M.R. Davies, *O. famatinae* A.M.R. Davies, *O. tarapacensis* A.M.R. Davies, *O. tontalensis* A.M.R. Davies and *O. stuebelii* var. *cryptum* A.M.R. Davies respectively. Five novel combinations are presented: *O. abbreviatum* (Cabrera) A.M.R. Davies, *O. achenohirsutum* (Tombesi) A.M.R. Davies, *O. apiculatum* (J.Rémy) A.M.R. Davies, *O. revolutum* (Phil.) A.M.R. Davies and *O. stuebelii* (Hieron.) A.M.R. Davies var. *stuebelii*.

Part A:

Systematics of *Chaetanthera* & *Oriastrum*

I Introduction

1 Philosophy of taxonomic research

The craft of taxonomy is a technology using available sources of science and art to communicate facts and truth (after SMITH 1994). Taxonomy is neither a science in the Popperian sense (POPPER 1934) because, despite using scientific techniques, the end result is not falsifiable data; nor is taxonomy an art because it deals in natural tangible events, not imaginative interpretation. Consider the following short logic problem, to which a solution is given at the end of the introduction.

Schnapsmann ist wieder mal am Kritzeln. Er malt – völlig überflüssigerweise – zehn Bäumchen auf die Rückseite des Bierfilzes.
„Bierling" sagt er, „wie lassen sich diese zehn Bäume in funf Reihen zu je vier Bäumen pflanzen?"
Bierling ahnt schon, was auf ihn zukommt, und besorgt sich zunächst einen größeren Vorrat von Bierfilzen zum Kritzeln. Gerade als er anfangen will, fällt Schnapsmann ein, die Bedingungen noch etwas zu erschweren.
„Bierling" sagt er, „es gibt mehrere Lösungen. Bei keiner dürfen Baumreihen parallel laufen. Und kein Baum darf mehr als zwei Baumreihen angehören."
Da diese Aufgabe ziemlich schwierig ist, bestellt er sich zunächst noch ein Bier – grosszügigerweise auch eines für sein freund – , lehnt sich genüßlich zurück und beobachtet, wie Bierling an das Problem herangeht.

WERNECK [1]

This rather light-hearted puzzle has a serious side in reflecting how taxonomy requires logical deduction but is not fully deterministic. The "trees" are analogous to taxonomic entities, and in themselves need to be clearly defined. The questions, "how is it possible to arrange them in groups?" and "what connections can be found or deduced between the groups?" are absolutely central to a monographic revision. Of course, it is possible for there to be more than one solution.

The principles of post-Hennigian plant taxonomy and systematics (HENNIG 1950, 1966) support the development of a meaningful solution, and not merely a catalogue of species. That is,

---

[1] **English translation of Werneck's logic problem.** Schnapsmann is scribbling again! He draws, quite unnecessarily, ten trees on the reverse of a beer mat. "Bierling", he asks, "how can these ten trees be planted in five rows of four trees?" Bierling, suspecting what is to come, procures himself a large supply of beer mats for scribbling on. Just as Bierling is about to start, it occurs to Schnapsmann to make the conditions harder. "Bierling", he says, "there is more than one solution. The rows of trees must never run in parallel. And no tree may be in more than two rows." As the task is quite difficult, Schanpsmann orders himself a beer, and generously also one for his friend, leans back comfortably and watches Bierling tackle the problem.

that the proposed network or hierarchy identifies monophyletic groups [of species] by defining specific characters unambiguously and plotting their distribution. The value of such strict application of monophyly and taxonomic concept is under debate (e.g., EBACH et al. 2006, HÖRANDL 2007, ZANDER 2007). HÖRANDL (2007) proposed that modern evolutionary classifications should use cladograms as a tool for recognising relationships and processes, but not primarily for the definition of taxa. However, systematics does not have to be defined as "phylogenetics with applications in classification" (ZANDER 2007), but has greater potential predictive value in observing how taxa evolve and adapt, i.e., understanding the diversity aspect of plants and their evolution.

2 Goals of taxonomic research

There are two main goals of taxonomic research. These were eloquently formulated by CAMP (1940), later re-used by TURNER at the 1985 Compositae Generic Limits Symposium, and are summarised here. To achieve the first goal, a revision may follow established lines such as weeding out synonymy, and listing and describing new material found since the last monographic treatment. This defines the species, or units of evolution (FUNK 1985). The second goal of a revision may be not so much aimed at cataloguing known species, as aimed at the study of the origin, evolution and dispersal of a group of plants. A proponent of the latter school of thought "...knows that the plants in his hands, in themselves, do not constitute an orthogenetic series but are only the ends of a much-branched and often tangled system of descents. [He] is likely to have a vastly different concept as to what constitutes a genus from the one who is merely cataloguing the species of a group..." (CAMP 1940).

Both goals require a good understanding of the taxonomically important characters and their variation. Naming a plant using obvious characters not necessarily important in relationships is quite different to using whatever characters one can to detect relationships (STEVENS 2001).

3 Application to *Chaetanthera* & *Oriastrum*

3.1 Historical background

The two South American Mutisieae (Compositae) genera *Chaetanthera* Ruiz & Pav. (1794) and *Oriastrum* Poepp. & Endl. (1842) have an intricately linked history, spanning over 150 years. The 42 species from these two genera have often been grouped together, one genus subsuming the other into a single large polymorphic, paraphyletic genus, most recently by CABRERA (1937). This apparently great infra-generic diversity within a paraphyletic *Chaetanthera* (sensu Cabrera) is also attractive to scientists wishing to conduct diversity studies and breeding system research (ARROYO et al. 2007a, b; 2006a, b).

Chaetanthera and *Oriastrum* were most recently united by CABRERA (1937) in his comprehensive monographic revision of *Chaetanthera*. He commented in his introduction:

I Introduction

> "...*pude observar que la revisión de las especies chilenas, publicada por Reiche en su Flora de Chile [1904], es notablemente insuficiente e para un perfecto conocimiento de este género [**Chaetanthera**]. Los caracteres diferenciales dados para las especies afines a **Ch. serrata** y para ciertas formas incluidas en **Oriastrum** son muy poco apreciables...*"
>
> CABRERA 1937, p.87 [2]

He highlights two of several areas in *Chaetanthera* and *Oriastrum* that unfortunately remained problematic despite the extensive cross-referencing to type material in the Museo de Historia Natural de Santiago (Chile) and the synonymisation of over 270 names. The majority of Cabrera's illustrations remain the best depictions of the species. However, poorly studied character variation between species played a major role in hazy species concepts. This was certainly largely an artefact of the small sample sizes available to Cabrera, especially in the high Andean taxa. For example, Cabrera recorded eleven collections of *Oriastrum acerosum* (J. Rémy) Phil. (= *Chaetanthera acerosa* (J. Rémy) Benth. & Hook. f.). This study included thirty-eight collections. Similarly, those taxa that have since been shown to be highly polymorphic, such as *Chaetanthera chilensis*, or *Chaetanthera glabrata*, were also clarified with larger samples.

Taxonomic revisions concerning these two genera to date (e.g., REMY in GAY 1849, BENTHAM 1873, REICHE 1904, CABRERA 1937) have been more concerned with describing the taxa, selecting characters that easily identify species. Variation in habit, life form, the leaf shape and form, the imbricate initially foliaceous then reduced to entirely membranous involucral bracts, and the achene indumentum have all been used in the past to define species and groups of species. Latterly, Cabrera used the presence of foliaceous outer involucral bracts to unite the species. These taxa were considered distinct from the genera *Brachyclados* D.Don – a dwarf shrub with biseriate imbricate non-foliaceous coriaceous involucral bracts, and the uni-specific *Pachylaena* D. Don ex Hook. & Arn. (KATINAS 2008) – an acaulescent, fleshy-leaved herb with condensed heads, multiseriate imbricate non-foliaceous involucral bracts and plumose pappus. Taxonomic inertia in *Chaetanthera* sensu Cabrera was sustained by retaining a ponderous sub-generic hierarchy. This reiterated the traditional so-called "artificial" groupings based on easily definable macro-characters.

3.2 Novel sources of variation

The starting point in this study was to consider the traditional characters used to define the species and the groups of species. These were typically sourced from macro-morphological features such as habit and leaf shape. It became apparent that although simple microscopy had intimated variation in characters at a micro-morphological level, these had often been poorly and inconsistently observed. Following the technological advances in electron microscopy, the potential value of micro-morphological characters in defining and understanding taxonomic relationships

---

[2] **English translation.** "During my studies I observed that the revision of the Chilean species published by Reiche in his Flora of Chile was insufficient for the understanding of this genus. The differentiating characters given for the species affiliated with *Chaetanthera serrata* and for certain taxa included in *Oriastrum* are of little value."

within the Compositae (Asteraceae) was of great interest in the 1980's. While SCOTT (1985) assessed the value of micro-characters as generic markers in the Eupatiorieae and concluded that there was often a high degree of correlation between the micro-characters and others, SUNDBERG (1985) concluded that there was little correspondence between occurrence of micro-morphological characters as generic markers in the Astereae and the traditionally accepted generic limits. However, there is no doubt that where taxonomic variation exists the technology provided a powerful tool to observe it better. Since then, numerous studies have used micro-morphological variation to great effect in understanding variation and evolution of groups of taxa (e.g., FREIRE & KATINAS 1995).

Two micro-morphological aspects – pollen structure and achene hair variation – have been the focus of four key studies that have led to the re-assessment of generic limits between *Chaetanthera* and *Oriastrum*. These are PARRA & MARTICORENA (1972), HANSEN (1991), DAVIES & FACHER (2001) and TELLERÍA & KATINAS (2004).

The first reliable modern indication that the differences between *Chaetanthera* and *Oriastrum* could be more than intuitive came from the pollen investigations of Chilean Mutisieae conducted by PARRA and MARTICORENA in 1972. This revealed two subtypes of pollen amongst the taxa in *Chaetanthera* and *Oriastrum* – the "*Chaetanthera* type" and the "*Acerosa* type". The latter type was observed in all *Oriastrum* taxa studied. Further pollen work by TELLERÍA & KATINAS (2004) supported the early work published by PARRA & MARTICORENA and detailed structural differences between the pollen of *Chaetanthera* and *Oriastrum*. It also confirmed pollen observations made during this study.

In 1991, HANSEN reported 2 types of achene twin hairs (so-called 'Zwillingshaare') in *Chaetanthera* sensu Cabrera, observed while revising the *Gerbera*-complex. He recognised the significance of this variation and wrote:

"*Cabrera divided* **Chaetanthera** *into 6 [7] subgenera, but a re-evaluation using cladistics may be useful. It seems illogical to consider all the c. 40 species [of Chaetanthera] as one genus while splitting the* **Gerbera**-*complex in many genera*"

HANSEN 1991, p.32.

This achene work was followed up by DAVIES & FACHER (2001) and showed that *Chaetanthera* and *Oriastrum* taxa could be divided and grouped according to their achene hair type, or absence thereof. As a result, doubt was cast on the subgeneric structure as a natural taxonomic construct. However, the apparent correlation between geographical location (especially altitude) and achene hair type could not be disregarded.

No modern systematic revision is complete without a consideration of the genetic diversity within a group. Molecular work (led by Kalin Arroyo at the Universidad de Chile, Santiago) concerning these two genera was conducted parallel to this taxonomic revision, and was published by HERSHKOVITZ et al. (2006). Collaborative exchanges were an important aspect of both research projects. The phylogenetic analysis on the nrDNA results showed a clear divergence between *Chaetanthera* and *Oriastrum* and complement the macro- and micro-morphological variation studied here.

3.3 Re-defining taxonomic relationships

Observations at the micro-morphological level have increasingly implied that *Chaetanthera* sensu Cabrera was not simply a single, large, polymorphic genus, but actually an unnatural paraphyletic entity. CRONQUIST (1985) was of the opinion that in order to have conceptually useful genera (in the Compositae) we must accept that many genera are inherently ill-defined and unresponsive to efforts at precision. But when a group of taxa become conspicuous as a result of how difficult they are to handle, it can be a strong indicator that a novel or more detailed approach is needed to better understand them. Clearly then, *Chaetanthera* and *Oriastrum* were ideal subjects for more intensive study. As the Chilean botanist MARTICORENA was moved to write on learning about the impending revision in 2001:

> *"A ver si aparece un botanico valiente que rehabilite algunos de los generos originales que furon unidos en un solo genero por Cabrera. Siempre he pensado que probablemente alguno de ellos deberia rehabilitado".*
>
> MARTICORENA pers. comm.[3]

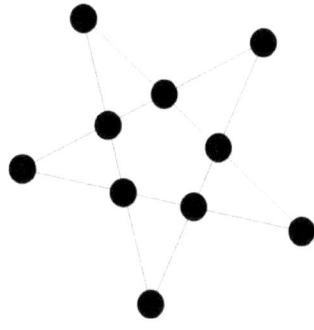

Fig. 1: A solution to Schnapsmann's problem.

---

[3] **English translation.** "It would be a courageous botanist indeed to relocate some of the genera united to a single one by Cabrera. I always thought that some of them should be moved."

II   Materials, methods & measurements

1   Materials

The revision is principally based on the study of herbarium material. Over 2000 loans and collections from B, BM, CONC, E, F, G, GH, K, LP, LPB, M, MA, MO, MSB, NY, P, SGO and W were studied. Additionally, images were gratefully received from OXF, SI, US and USM. Material from the private herbaria of C. Ehrhart, J. Grau and M.T. Kalin Arroyo was also specially loaned. Images from the internet collection of JAL were also invaluable for understanding some species. Material collected by the author on two field trips to Central Chile in November 2001 and February-March 2002 is deposited in MSB.

Seeds from field trips to Chile (2001, 2002), and also received from C. Ehrhart (2002), and M.T. Kalin Arroyo (2002 – 2003) were sown in the experimental greenhouses in the Botanische Garten München-Nymphenburg.

2   Methods

2.1   Capitula & floral dissection

For the detailed examination of capitula and floral parts, the material was briefly boiled (1 minute) in tap water before dissection. No surfactant was necessary. Excessive boiling resulted in loss of structural definition. The material was observed under a stereo microscope and sketched.

2.2   Involucral bract anatomy studies

Inner involucral bracts were removed from herbarium material and dehydrated using 70% - 90% alcohol for 2 days before embedding in stubs of histo-resin and dried for 2 – 3 days. The stubs were sliced in 9 µm slices using a microtome. The slices were stained with Safranin, washed with H_2O[dist.], then stained a second time with Astra-blau, washed with 20 – 30 % alcohol and stored in H_2O[dist.]. The slices were mounted on gelatine-smeared slides and allowed to dry somewhat before having a slide cover glued on top. The slides were observed using a Leica DMRBE light microscope and and the images digitally stored for further observation and data collection.

2.3   Achene & pappus studies

Pappus was simply mounted in gelatine and stored as a permanent slide for light microscope observations. For the testa epidermis studies mature achenes were selected and allowed to soak in 10% KOH for 6 – 12 hours. They were then washed with H_2O [dist.] and set aside in Glycerol water. Prior to using the freeze-microtome the achenes were bisected transversely near the broadest part and the outer carpel coat and cotyledons removed. The testa epidermis, including the endosperm, was then sliced transversely using the freeze-microtome. Slices were stained with Cotton Blue for an hour then washed with Lactic Acid and mounted.

For observations under the S.E.M. achenes and pappus from both ray and disk florets were carefully extracted from the herbarium material and placed directly on carbon-coated stubs. Persistent pappus setae were removed from the achenes to reduce charging problems in the vacuum. The stubs were sputtered with platinum using a Bal-tec sputterer. The material was observed with a high vacuum using LEO 483VP scanning electron microscope (S.E.M.) at both low and high magnifications where appropriate, and the images digitally stored for further observation and data collection.

2.4 Pollen studies

Mounting pollen (fresh or herbarium) directly onto S.E.M. carbon stubs without any treatment resulted in sticky clumped material. Therefore both acetolysis and critical point drying techniques were tested.

The acetolysis procedure used here was based on that described by Erdtman (1960). Collect max anther/min plant material. Add 2-3 ml 10% KOH. Boil in water bath for 10 minutes, stirring constantly. Sieve, using 100µm sieve, and wash with H_2O [dist.]. Centrifuge at 3000 r.p.m. for 3 minutes, and decant. Add 5-7 ml glacial acetic acid, centrifuge and decant. In fume cupboard, prepare 9:1 Acetic acid anhydride: H_2SO_4. Add to sediment. Place in lukewarm water bath and bring to boil, centrifuge and decant. Add 5-7 ml glacial acetic acid, centrifuge and decant. Add 5-7 ml H_2O [dist.] and 1-2 drops 10% KOH, centrifuge and decant. Add 5-7 ml H_2O [dist.], centrifuge and decant. Rinse Eppendorfer corolla tubes with 50% C_2H_5OH. Add 1 ml 70% C_2H_5OH to sediment and pipette into Eppendorfer tubes and store.

Material mounted on S.E.M. carbon stubs at this stage and allowed to air-dry resulted in severely collapsed pollen grains coated in a milky-opaque layer. Several probes which had very few pollen grains in them were observed under the light microscope at this stage.

Critical Point Drying (CPD) employs the properties of liquids (e.g. CO_2) to change from a liquid phase to a gaseous phase at a critical point of temperature and pressure. The technique relies on the rapid removal of water from the specimen by phase change. An alternative would be to use sublimation by freeze-drying. The procedure was carried out as follows: Etch identification numbers onto CPD tubes. Remove 70% C_2H_5OH from Eppendorfer tubes without disturbing the sediment. Add 100% acetone. Mix and put into CPD tubes, plugging tubes with cotton wool. Top up with acetone, ensuring no air bubbles. Place in pressure chamber cartridge and fill with acetone. Close chamber and fill with CO_2. After 10 minutes empty and refill with CO_2. Repeat 3 times in total. Heat up to 37°C (80 bars); maximum 40°C (90 bar). Leave for 5-10 minutes (at critical point). Very slowly reduce pressure. As it drops below 50 bars switch off heater. Allow pressure to bleed off until at room pressure. This will take 1-2 hours. Store samples in desiccator until ready to mount, sputter and observe under the S.E.M. The images were digitally stored.

2.5 Chromosome studies

Fresh root tips were collected and immediately immersed in chilled 8-hydroxychinolin (C_9H_7NO). Dirt and gravel was cleaned off, and root tips stored for at least 4 hours at 4 °C. Then they were either placed in 3:1 EtOH (glacial Acetic Acid) and stored before hydrolysing or hydrolysed immediately. The root tips were hydrolysed by placing them in 0.5 HCl and warming for 10 minutes at 60°C. Then they were placed in H_2O [dist.] and a chromosome squash was prepared. The chromosomes were stained with Orcein and observed with a light microscope, using phase contrast for the best results. The images were digitally stored for further observation.

2.6 Images & illustrations

All illustrations were done by the author using textured paper ("Runzelkorn") and a combination of black pencil and ink. The originals were scanned at 600 dpi and exported as TIFF files. Habit illustrations were taken from photos and slides where possible. Images have become increasingly available over the Internet. Permission to use images was kindly granted by Michal Belov (www.Chileflora.com), Mauricio Bonifacino (MVFA, Uruguay), Michael Dillon, Tini Ehrhart, María Terese Eyzaguirre (Fundación R.A. Philippi), Mark Gardener, Jürke Grau, Michel Grenon, Iréne Till-Bottraud (Station Alpin Joseph Fourier, Grenoble) and Mauricio Zuñiga.

2.7 Geographical data

Localities for specimens in Argentina, Bolivia, Chile and Peru were confirmed and pinpointed using a combination of atlases and maps available at the library in the Institut für Systematische Botanik, München, and from the web sites listed in the references. The Concepción database of herbarium specimens (CONC) also has the majority of localities cross-referenced. Bolivian localities were mostly pinpointed using the maps provided at http://patepluma.virtualave.net/patemaps.htm#cp. However, it seems this link is no longer available.

3 Measurements

Measurements for the species descriptions were made from sketches of the rehydrated material. All the data given for size (length, width, or height) are always taken at the broadest point of the plant organ. Some measurements are not self-evident. These are illustrated in Fig. 2. Height and width, particularly of the capitula but also of the floral parts, can be deceptive in herbarium material as a result of pressing and dessication. The approximate ratio of measurements made on dry material to fresh material was calculated to be about 0.7. Measurements given in the descriptions are unadjusted for this factor and taken from herbarium material unless otherwise stated.

The leaves can be described as linear or indistinctly petiolate. For linear leaves, the total length and the width at the broadest point are given. For indistinctly petiolate leaves, where the leaf is divided into a petiole-like component and a lamina, the total length of the leaf is given, followed by the length and width of the lamina.

II Materials, methods & measurements

Capitula details often include shape, sometimes height and disk diameter where known from fresh examples. The imbricate, multiseriate involucral bracts are described as belonging to one of three series, outer, middle and inner. The outer and middle series have a clear leaf-like component, and are distinguished as involucral bracts by having basal membranous or hyaline alae. Due to the nature of variation in the dimensions of the involucral bracts in these series total length is the only measurement given. The measurements for the inner series of involucral bracts (example in Fig. 2) are given as the total length and maximum width. In the case of *Oriastrum* taxa whose inner involucral bracts have a well-defined apical appendage, the length of the appendage is given as well as the total length of the involucral bract.

The ray floret corolla measurements are divided into total corolla length, the length of the tube, the width of the outer lip (ligule) and the length of the bifid inner lip.

The achene dimensions are given as maximum width and total length not including a carpopodium. Pappus features such as colour, length, number of series, attachment, setae width, barb length, attachment and frequency are given in the descriptions.

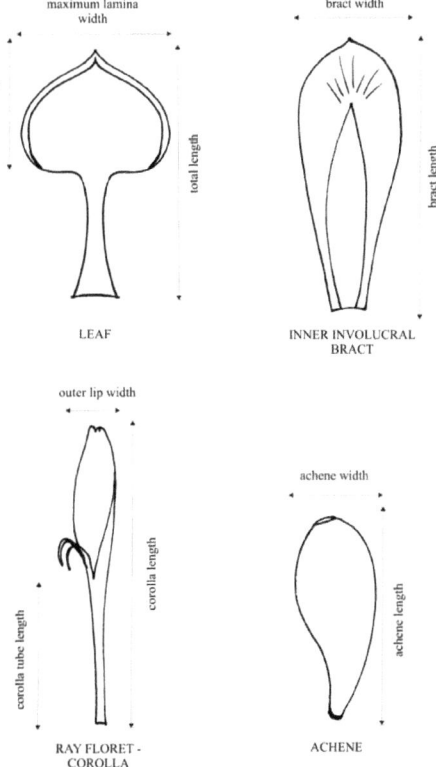

Fig. 2: Definition of the measurements used in descriptions.

III Variation of characters in *Chaetanthera* & *Oriastrum*

1 General remarks

The variation in the characters is important at two levels of taxonomic hierarchy: between species and between groups of species. That is, characters that define species and those that define groups of species. Examples of within-species variation, particularly polymorphism, are discussed in Chapter VI. The characters are individually outlined, with specific examples from both *Chaetanthera* and *Oriastrum* where appropriate.

Some characters, such as life cycle and leaf outline variation, have often been considered to be taxonomically important characters in the past. They are presented in the context of the overall conclusions of this revision. Other characters, such as involucral bract anatomy, pappus diversity, testa epidermis structure, and carpopodium presence are novel from this research. Pollen diversity and achene hair variation has already been published elsewhere (PARRA & MARTICORENA, 1972; HANSEN, 1991; DAVIES & FACHER, 2001; TELLERÍA & KATINAS, 2004), although the pollen variation was independently analysed for this research. The images for both pollen and achene hair sections are original. The nrDNA variation, published by HERSHKOVITZ et al. (2006), is briefly outlined.

The taxonomic implications of the variation are then discussed in the context of generic and sub-generic limits of *Chaetanthera* and *Oriastrum*, i.e. characters defining groups of species.

2 Morphological & anatomical variation

2.1 Life cycle & habit

Perennials and annuals are found in both *Chaetanthera* and *Oriastrum*. *Chaetanthera* annuals are generally small (< 50 cm height) and occasionally dwarf in stature (< 10 cm height/ length). They have a short, erect stem, generally no taller than 1-2 cm, crowned with a loose whorl of leaves that often wither during the growing season. This is shown in Fig. 3A; an aerial view of a juvenile plant of the annual *C. linearis*. The flowering stems or branches originate from a compacted node above the stem rosette and may be decumbent to erect, with sparsely spaced opposite to alternate leaves which terminate in a dense whorl or rosette below the capitula (Fig. 3B).

The similarity of the leaves to the outer involucral bracts often makes it difficult to determine where the cauline leaves end and the bracts start. Occasionally, the change is marked by a single, disproportionately small leaf or scale. Identification of this feature is most problematic in the taxa with densely leafy stems.

The leaves of the perennial species can either be arranged rosette-like as tightly whorled leaves around a shortly extended stem, as in *C. villosa* (Fig. 3C), or as loosely rosulate clusters below ascending – erect, monocephalous scapes, as in *C. chilensis* (Fig. 3D). These are termed pseudo-rosettes in this study.

In *Chaetanthera*, the seasonal longevity of the rosette is associated with the life cycle. The annual species have ephemeral stem rosettes. These are present in the post-germination development and juvenile stages, but are withered by the flowering stage, i.e. stem rosettes play no role in reproduction and dispersal. The perennials have seasonally persistent annual rosettes with either more or less sessile (*Chaetanthera* subgenus *Tylloma*) or scapose (*Chaetanthera* subgenus *Chaetanthera*) capitula. There are two exceptions to this. The high elevation annual *C. renifolia* has loosely whorled leaves up the stem (no evidence of rosettes) and sessile capitula. The mid-elevation perennial *C. glandulosa* has not been observed to have rosettes of any kind. This localised Chilean endemic is poorly known, and has an array of adaptations to aridity unique within *Chaetanthera*.

The leaf rosette in *Chaetanthera* has been duplicated by the foliaceous multi-seriate involucral bracts. An aerial view of the above-ground habit of the high elevation perennials *C. villosa* and *C. spathulifolia* would show clusters of dwarf rosettes with sessile capitula. *C. philippi*, also a high elevation perennial, has branched flowering stems with few leaves and no apparent rosettes or rosulate leaves. However, the function of the leaves is taken over by the foliaceous involucral bracts, which give the impression of a rosette. The annuals have ephemeral stem rosettes and extended branching flowering stems. This is seen, for example, in the high elevation annuals *C. flabellifolia* and *C. kalinae* and in the lower elevation annuals like *C. glabrata* or *C. flabellata*. The rhizomatous *C. chilensis* and related taxa have annual rosettes supporting monocephalous scapes.

Stem woodiness in *Chaetanthera* is more prevalent than the previous literature records. *C. glandulosa* (Fig. 4A), with its aerial erect branching woody stems is always described as the only woody sub-shrub in the genus. However, closer analysis shows that several species, e.g. *C. philippii*, *C. chilensis*, *C. villosa*, also have woody stems and woody roots (Fig. 4). The woody stems can be slender, branching woody rhizomes, as seen in *C. chilensis* (Fig. 4B). *C. philippii* (Fig. 4C) has creeping to ascendant woody stems that are often buried in the typically unstable substrate (scree) where it is commonly found. Deep-lying, branching woody rhizomes, as in *C. villosa* (Fig. 4D) are also found. The extent of the rhizomes depends on the local nature of the substrate in which the plants are growing. Longer rhizomes are found in plants growing in looser substrates such as scree, volcanic rubble and humus-rich soils at lower elevations. *C. x serrata* has stolons that creep along the surface of the ground.

The herbaceous dwarf *Oriastrum* species can be divided into two natural groups according to the presence or absence of clusters of more-or-less subterranean perennating buds. Due to the type of substrate on which many of these species are commonly found, i.e., scree or landslips, the buds may appear to be subterranean at any given point in time. There are two principal growth forms seen in *Oriastrum* subgenus *Oriastrum* and *Oriastrum* subgenus *Egania*. The first form is a loosely-branched spreading habit (Fig. 5A, 5B). The leaves are often decussate. *O. werdermannii* or *O. dioicum* are typical examples. The second form is a compact dwarf habit with short branching stems, as shown in Fig. 5C. The leaves are densely whorled or decussate on the stem. Typical species include *O. polymallum* (Fig. 5C, 5D) and *O. pusillum*.

Figure 3: Life forms of *Chaetanthera*. **A.** Juvenile habit of the annual *C. linearis*. Photo©A.M.R. Davies. **B.** Adult habit of the annual *C. kalinae*. Photo©C. Ehrhart. **C.** Adult habit of the perennial *C. villosa*. Photo©J. Grau. **D.** Adult habit of the perennial *C. chilensis* Photo©M. Bonifacino.

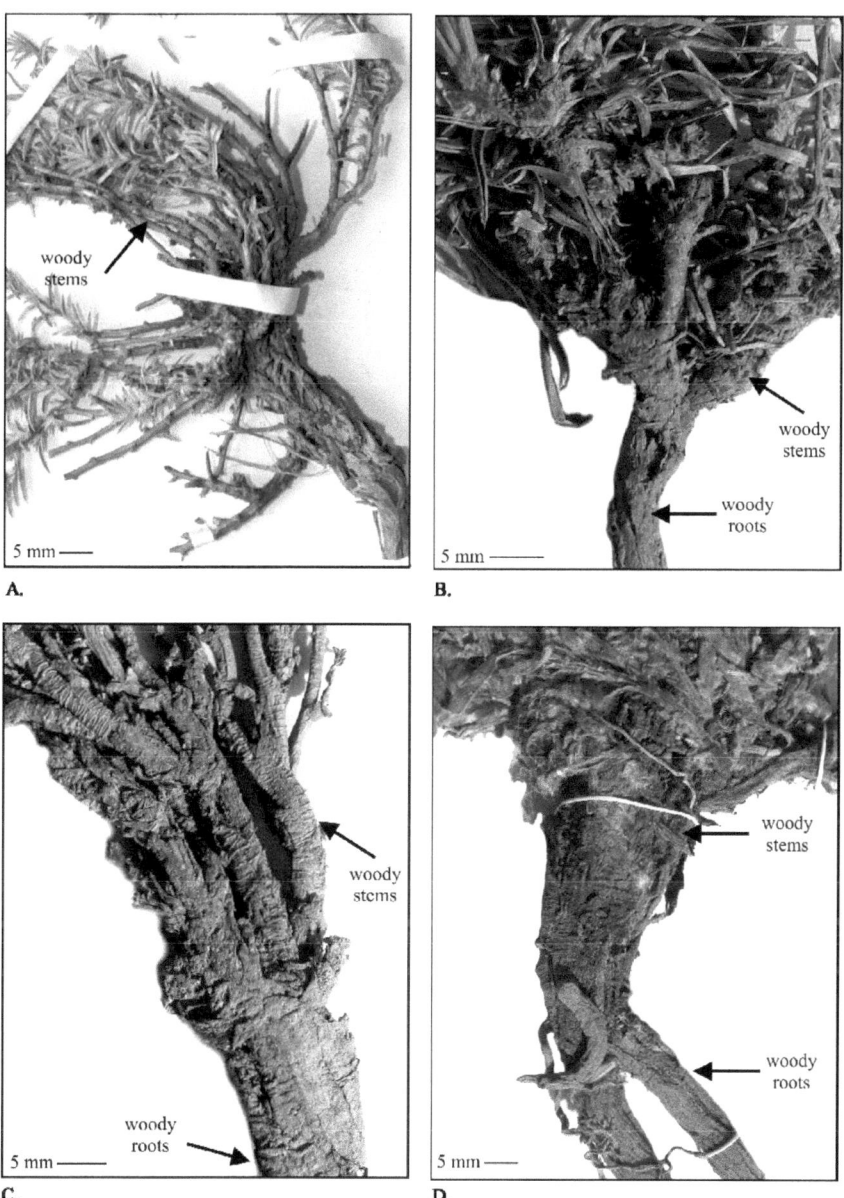

Figure 4: Perennial habit. **A.** Woody stems of *C. glandulosa*. **B.** Woody rhizomes and roots of *C. chilensis*. **C.** Woody stems and roots of *C. philippii*. **D.** Thick woody roots of *C. villosa*.

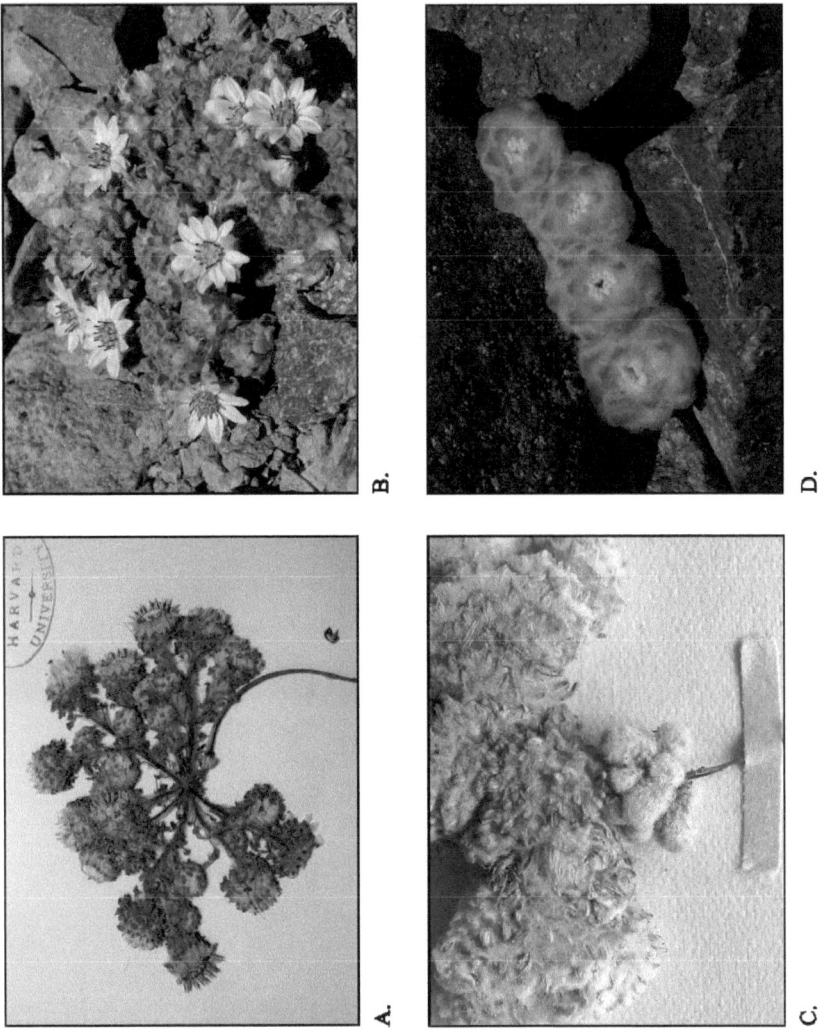

Figure 5: Life forms of *Oriastrum* species. **A.** *O. werdermannii*. Werdermann 507 (GH). **B.** *O. gnaphalioides* ©Irène Till-Bottraud. **C.** *O. polymallum*. Werdermann 250 (M). **D.** *O. polymallum*. ©Irène Till-Bottraud.

2.2 Leaves

Variation in *Chaetanthera* and *Oriastrum* leaves is illustrated in Fig. 6. The frequency and arrangement of the leaves varies from densely whorled, imbricate clusters forming a pseudo-rosette, to remotely decussate or alternately arranged leaves. They can be strap-like or involute or conduplicate. The leaf shape can be linear, as *C. linearis* or *C. glandulosa* (Fig. 6A, 6B), truncate

(*C. euphrasioides*), spathulate (Fig. 6C - *C. kalinae*) or oblong-obovate or oblanceolate as those of *C. villosa* (Fig. 6D). Linear-lanceolate leaves are found only in a few species, such as *C. elegans* (Fig. 6E).

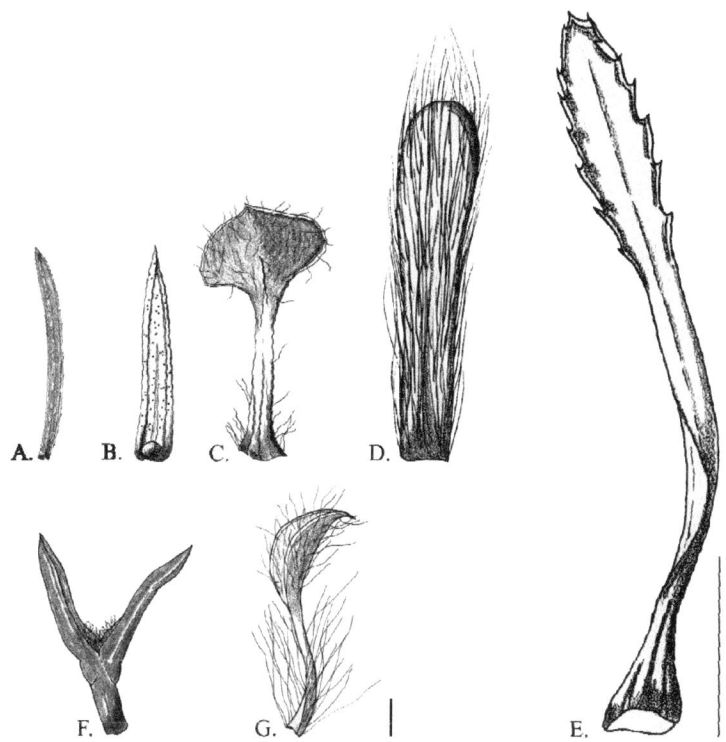

Figure 6: Leaf diversity. Illustrations showing selected leaf outline diversity. **A.** *C. linearis*. **B.** *C. glandulosa*. **C.** *C. kalinae*. **D.** *C. villosa*. **E.** *C. elegans*. Scale bar (A – E) = 10 mm. **F.** *O. chilense*. **G.** *O. dioicum*. Scale bar (F, G) = 1 mm.

Some species are defined by their flabellate (*C. flabellata*) or reniform (*C. renifolia*) leaves. The leaves can be sessile as in *C. villosa*, or petiolate – described as indistinctly petiolate here because really they are longly attenuate, as in *C. kalinae*. The leaf texture can vary from herbaceous (*C. elegans*) to somewhat fleshy or carnose (*C. glabrata*). The leaf colour varies from glaucous to yellow-green, grey-green, or bright green. The leaf margins can be entire and limbate (*C. glabrata*) or dentate (*C. elegans*), spinose to ciliate (*C. ciliata*) or crenate serrate (*C. splendens*). The leaf tips are acute to obtuse, sometimes mucronate.

Oriastrum leaves can be like those of *O. dioicum*, which has linear leaves. The apices vary from acute to aciculate or obtuse. The leaves may be basally dilated and connate. A stem leaf pair is illustrated in Fig. 6F. *O. chilense* (Fig. 6G) has typical leaves for *Oriastrum* subgenus *Oriastrum*; indistinctly petiolate with spathulate to orbicular laminas.

2.3 Indumentum

The indumentum can be restricted to the leaf axils of principally glabrous plants, or may be present on all vegetative plant parts, viz. leaves, and stems. The dorsal surfaces of the ray florets are often sericeous, but this is not a taxonomically valuable character. The leaf surfaces always seem smooth, with any rugulose effect considered to be a consequence of desiccated carnose leaves. Although glabrous leaves are common (*C. glabrata*, *C. linearis*), more interesting are the variations in the types of indumentum. The leaves can be shortly pubescent, like those of *O. acerosum*, to densely silvery villous e.g., *C. villosa*, *C. chilensis*, or even floccose (*O. gnaphalioides*, *C. taltalensis*). Alternatively, the leaves can have sessile (*C. glandulosa*) or shortly pedicellate glands (*C. frayjorgensis*). In some species, several types of trichome can be observed in combination, for example in *C. limbata*, or *C. pubescens*. A unique trichome type for *Chaetanthera* has been observed on *C. pubescens*. This is a simple 2-3 celled clavate trichome. *C. schroederi* has been observed to have long lanate hairs composed of several cells stacked together. A further novel trichome has been observed on the leaves of *C. ciliata* and *C. peruviana*. It is a compound twin-celled trichome with a reduced basal inflatable cell (ca 50 μm L) and a long filamentous cell (ca 250 μm L)

2.4 Involucral bracts

The variation among the species is so distinctive that in most cases the species can be identified simply by observing the form and sequence of the involucral bracts.

2.4.1 Arrangement

The arrangement of the involucral bracts applies to both *Chaetanthera* and *Oriastrum*. The involucral bracts are imbricate, in several series; that is arranged contiguously in several whorls. The involucral bracts can be loosely categorised into outer, middle and inner involucral bracts, even though the progressive reduction of the laminar component is continuous. Fig. 7 shows a typical sequence of involucral bracts, illustrated from *C. kalinae*. Outer involucral bracts initially have a dominant lamina component, identical to the leaf lamina in outline and size, but the ratio of lamina to membranous margin decreases along the series. The absolute length of the involucral bracts also decreases along the series. The middle involucral bracts have alate margins that are 2/3 to entirely membranous, with a reduced or vestigial lamina component. The length of the involucral bracts increases from their initial minimum. The inner involucral bract series are entirely membranous and can be either translucent or with striate maculate (darkly coloured) apical regions as seen in *Chaetanthera*, or with distinctly delineated coloured apical regions, as seen in *Oriastrum*. In some species the size of the capitula and number of involucral bracts are so reduced that not all series apply. The lamina component can be appressed to the capitula or reflexed. This feature is often lost in dried herbarium specimens, but can be seen in re-hydrated material.

III Variation of characters in *Chaetanthera* & *Oriastrum*

2.4.2 Morphology of the inner involucral bracts

The inner involucral bracts of *Chaetanthera* species (e.g., *C. linearis*, *C. glabrata*) are transversely differentiated into a central mesophyllous zone and two marginal membranous alae. The alae are only 1 – 2 cells wide (Fig. 8A), and relatively long, compared to the central zone. *C. villosa* has rather thicker, poorly differentiated alae, and is generally more mesophyllous. In contrast, *C. glandulosa* has very little mesophyllous structure in the inner involucral bracts.

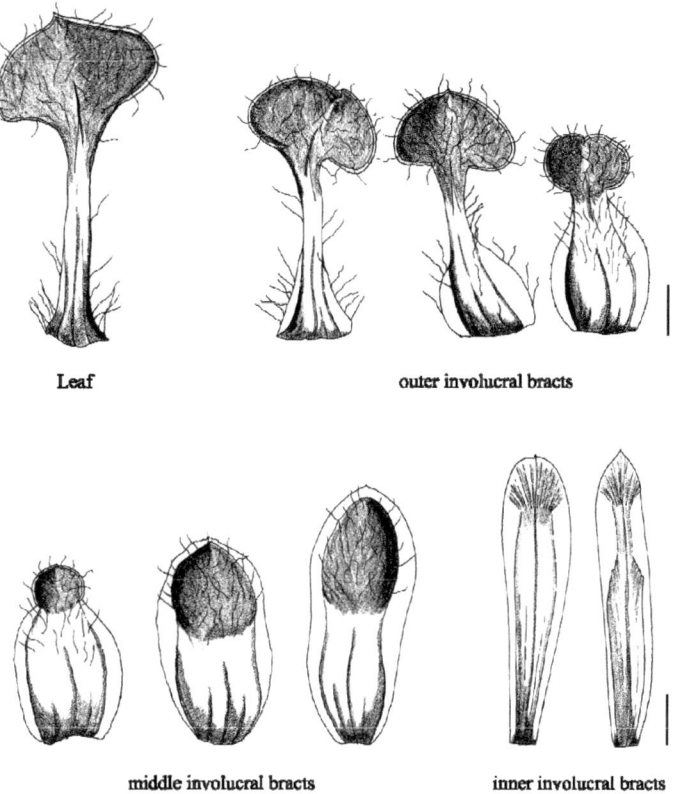

Figure 7: Involucral bract series. Leaf and involucral bract illustration (*C. kalinae*) showing the gradual changes in the imbricate series from leaf (left) to outer, middle and inner involucral bracts (right). Scale bar = 2 mm.

The inner involucral bracts of *Oriastrum* species are vertically differentiated into a generally thick (ca 5 – 10 cells), transversely gradually tapering lower region, and distinctly delineated apices

(Fig. 8B). The apices are often darkly maculate (*O. dioicum*) but can be translucent (*O. pulvinatum*), or even coloured (*O. pusillum*).

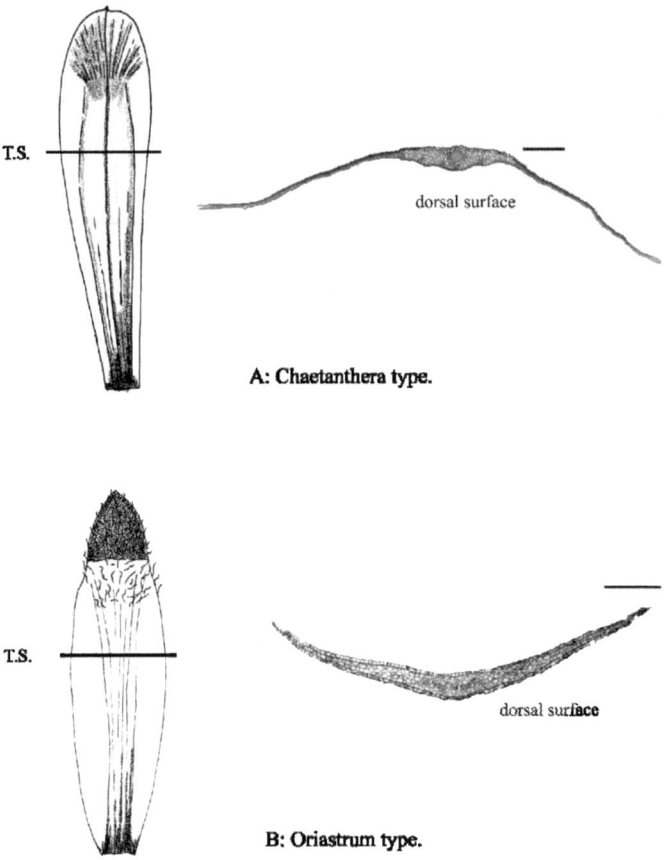

Figure 8: Inner involucral bracts. Illustrations of the typical inner involucral bract outlines (dorsal view) and macro-structure of *Chaetanthera* (A, left) and *Oriastrum* (B, left). Images of a transverse section (T.S.) taken through the cellular structure of the lower tissues of *Chaetanthera* (A, right) and *Oriastrum* (B, right). Scale bar = 200 μm.

2.4.3 Anatomy of the inner involucral bracts

The anatomy of the inner involucral bracts of *Chaetanthera* and *Oriastrum* can be divided into three types according to the distribution of parenchymatous and sclerenchymatous cells in the apices and bases. Fig. 9 shows schematic illustrations of the three main types, A, B and C, of inner involucral bract sections summarized into apical and basal regions for each bract type (1 = apical, 2

= basal). In type A, the apices are be packed with woody, sclerenchymatous cells (Fig. 9A1), while the bases have distinct clusters of sclerenchymatous cells around the vascular bundles (Fig. 9A2). This is typical for species in *Oriastrum* like *O. pusillum* and *O. acerosum*. Type B bracts have clusters of sclerenchymatous cells around the vascular bundles throughout the bract (Fig. 9B1, 9B2). This is typical *Chaetanthera* inner involucral bract anatomy (*C. linearis*). Type C is similar to type B (Fig. 9C1, 9C2) but also has a continuous sclerenchymatous adaxial surface. It is considered a subgroup of type B, and was observed in *C. chilensis*.

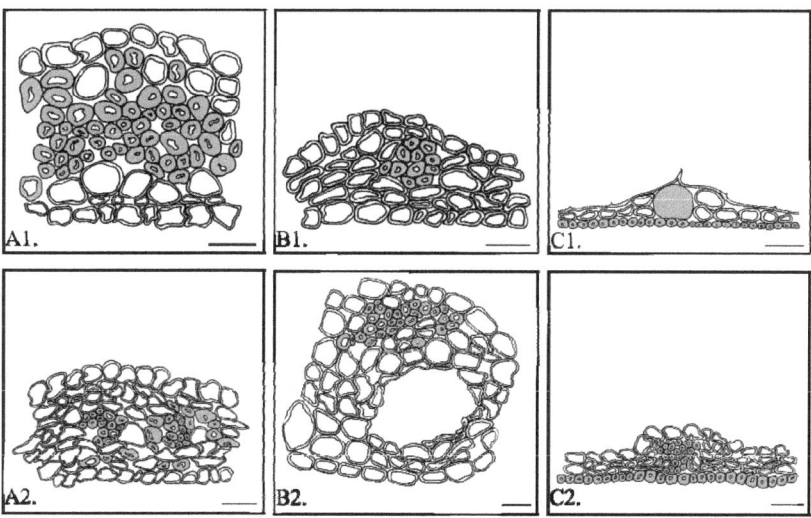

Figure 9: Inner involucral bract anatomy. Drawings taken from light microscope images of transverse sections (T.S.) through apical (1) and basal (2) regions of the inner involucral bracts. Shaded areas indicate sclerenchymatous cells. **A.** Oriastrum type. Scale bar = 25 µm. **B.** Chaetanthera type. Scale bar = 20 µm. **C.** Chaetanthera subtype "chilensis". Scale bar = 25 µm.

2.5 Pappus

The pappus of *Chaetanthera* and *Oriastrum* is composed of white barbellate setae arranged in 1 – 2 (or sometimes more) monomorphic series (Fig. 10A). Pappus setae length varies from 2.5 – 14 mm. Although often included in descriptions, it is a function of the general plant habit (large capitula: long pappus) and not a taxonomically useful indicator. In some species the base of the pappus setae is fused – united by a multicellular wall (Fig. 10B, C). This is especially true for some annual and perennial species in *Chaetanthera* subgenus *Chaetanthera*. The pappus appears to dehisce at maturity in many high altitude taxa, e.g., *C. villosa, O. acerosum,* fig. 10D). In several species the setae can be longly ciliate to plumose, commonly so at the setae base (Fig. 10E), but are never truly plumose as those of *Pachylaena*. When present, the basal cilia present occupy the lower 10 – 20 (30) % of the setae and are at least twice as long as the remaining barbs.

The pappus setae cell width is a fairly constant 5 – 10 µm wide with only few exceptions. *O. dioicum, C. flabellata, C. peruviana* and *C. elegans* have cells about 15µm wide, and *O. pulvinatum* and *O. acerosum* have cells 24µm wide. The pappus setae width is measured as the number of cells. The taxa can be divided into three groups based on the number of cells wide the setae are. Narrow [1 – 2 (3)], typical [(4)5 – 7 (8)], and wide [>10]. *O. cochlearifolium* has up to 12 cells at the base of the setae. *C. limbata* can have between 10 and 17 cells at the base and *O. chilense* can have up to 14 cells at the base.

The barbs are often laxly alternately arranged in a single plane with one parallel arrangement of barbs along the setae and some having a tetrahedral cross-section with two parallel rows of barbs. Barb length varies from short (50µm), typical (94 – 140µm), to long (190 – 230µm). For comparison, the barb frequency, barb adhesion and barb inflexion were observed in the mid-region of the setae because variation in the presence of cilia in the basal region and the tendency of the pappus to end in dense clusters of shorter barbs. The barb frequency along a 100µm stretch of setae was constant within a species but variable between species. The species observed were split into two groups with regard to barb frequency: infrequent (3 – 4 / 5 – 6), or frequent ((7) 8 – 15). Barb adhesion refers to the proportion of the margin cell forming the barb that is not exerted from the main plane of the setae. The adhesion of the barbs was classified as shortly (<40%) adhered (Fig. 10E), medium (50%) adhered (Fig. 10F), or longly (>60%) adhered (Fig. 10H). The free part of the barb may be lax and spreading (Fig. 10H) or appressed (Fig. 10F). Commonly, the pappus setae were continuously barbed, i.e., every marginal cell was barbellate. In three exceptions (*O. chilense, O. pusillum* and *O. lycopodioides*) barbs were often separated by 2 – 3 non-barbellate cells.

Chaetanthera species generally have indehiscent setae ca. 7 cells wide. Fig. 10F (*C. linearis*) shows the midsection of a typical pappus setae of *Chaetanthera* species. The barbs are appressed, frequent, and medium to longly adhered (50 - 60% of barb is fixed to setae). High altitude species such as *C. villosa* or *C. spathulifolia* have indehiscent setae whose barbs are longly ciliate to sub-plumose. The group of species most closely related to *C. glabrata* are the only taxa to have basal cilia on the pappus setae. Some *Chaetanthera* species have tetra-branchial setae, e.g., *C. linearis* and *C. microphylla* (Fig. 10G). The low montane Central Andean species *C. euphrasioides* has infrequent, longly adhered, spreading, barbs, a more typical feature of *Oriastrum* species.

Oriastrum species tend to have pappus ca. 4 cells wide with dehiscent, ciliate setae and comparatively infrequent, shortly to longly adhered, spreading barbs (Fig. 10H). *Oriastrum* taxa are diverse in their mode of barb adhesion.

2.6 Floret variation: ray dimorphism & showiness

Chaetanthera species are monoecious. The ray florets are functionally female, with greatly reduced sterile anthers (inconspicuous staminodes). They are radiate bilabiate and zygomorphic (after KATINAS et al. 2008), with a shortly 3-dentate ligule and a bifid inner lip, arranged in a single series, and yellow (e.g., *C. linearis, C. glabrata*) or white (*C. albiflora, C. philippii*). *C. ciliata* has been observed as having a cerise form (Hershkovitz, pers. comm.; F. Lira photostream 11.06.2008) and *C. microphylla* usually has brick-red rays.

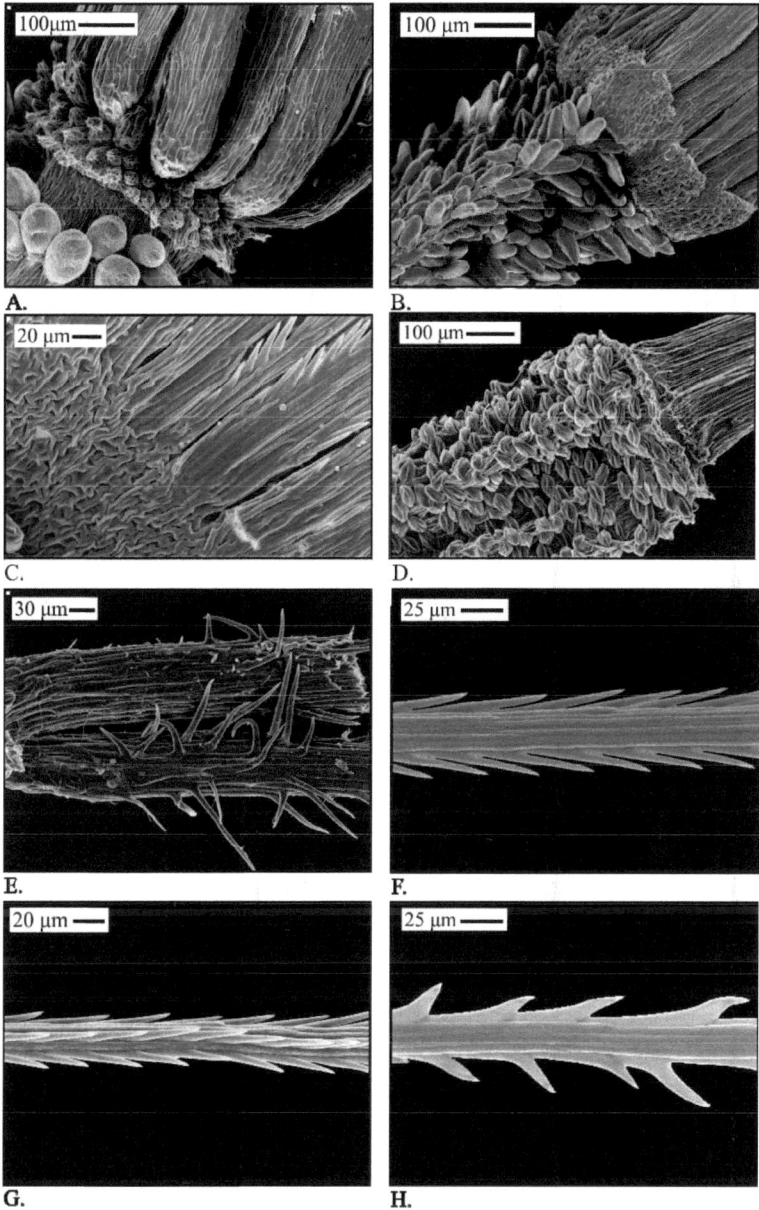

Figure 10: Pappus diversity. **A.** *C. philippii*. **B.** *C. chilensis*. **C.** *C. linearis*. **D.** *O. acerosum*. **E.** *C. philippii*. **F.** *C. linearis*. **G.** *C. microphylla*. **H.** *O. acerosum*.

Species such as *C. serrata* and *C. philippii* have distinct pink to red dorsal stripes on the exerted ray ligules. With the exception of *C. perpusilla* and *C. ramosissima*, the ray florets are conspicuously exerted beyond the involucrum. The disk florets are non-radiate, bilabiate and slightly zygomorphic. They are always yellow. The typical form of disk and ray florets is illustrated in Fig. 11 (E, F).

Oriastrum species may be monoecious or gynodioecious. The pistillate ray florets from bisexual capitula (Fig. 11A) are radiate bilabiate and zygomorphic, and have irregularly formed 3-dentate ligules, and a bifid inner lip. They are usually cream-coloured or white, often flushed with pink. *O. abbreviatum* has yellow rays and some herbarium sheets report yellow rays for *O. cochlearifolium*. The disk florets are are non-radiate bilabiate and slightly zygomorphic. They are always yellow, and bisexual with fertile anthers (Fig. 11B). Gynodioecious species are found only in a closely related group of species in *Oriastrum* subgenus *Egania* and include *O. dioicum, O. polymallum, O. pulvinatum, O. revolutum* and *O. stuebelii*. Some individuals in these species have capitula entirely composed of pistillate ray florets, which are somewhat dimorphic (see Fig. 11C, D).

Figure 11: Floret variation. Ray and disk florets from the gynodioecious *O. dioicum* (A - D), and the monoecious *C. kalinae* (E - F). **A.** Ray floret from bisexual capitula. **B.** Disk floret from bisexual capitula. **C & D.** Dimorphic ray florets from female capitula. **E.** Ray floret from bisexual capitula. **F.** Disk floret from bisexual capitula. Scale bar = 2 mm.

The comparative length of the ray floret ligule (short or long) has been used as a character defining the difference between *Chaetanthera* and *Oriastrum* species in previous publications (CABRERA, 1937; TELLERÍA & KATINAS, 2004). While it certainly holds true that the ray florets of *Oriastrum* species (3.0 – 13.0 mm) are often shorter than those of many *Chaetanthera* species (4.0

– 26.0 mm), more than half the *Chaetanthera* species have ray florets whose mean length lies within the normal range of *Oriastrum* ray floret lengths.

Measurements were made of the ray floret length, ray floret tube length, ray ligule width, disk floret length and disk floret tube length for all species as part of the species descriptions. When mean ray length (RL) is plotted against mean disk length (DL) (see Fig. 12, data in Table 21a, Appendix 1) the same relationship is seen between the disk and ray florets for both *Chaetanthera* and *Oriastrum* species. *C. taltalensis* has smaller florets overall but have the same ray length to disk length proportion as *C. kalinae*, and the trend is continued in *C. villosa*, even though the rays of *C. villosa* (24.0 mm) are six times longer than those of *C. taltalensis* (3.8 mm). In *Oriastrum*, *O. famatinae* (RL = 4.4 mm) and *O. acerosum* (RL = 8.0 mm) have the same ray length to disk length proportion, despite the difference in size. There are exceptions to this relationship in both genera. For example, *C. elegans* has longer rays than expected. *O. polymallum* has distinctly shorter ray florets while *O. werdermannii* has longer ray florets than expected.

The ray and disk floret variation was analyzed in the context of the altitude zones in which the species are generally found. An index of ray floret length to disk floret length (RL:DL) was generated. The means of the RL:DL indices for the species in each altitude zone were tested using Students paired T-test assuming a two-tailed distribution for equal variance (homoscedastic t-Test). Tests for significance within *Oriastrum*, over 2 altitude zones (Andino – Altoandino 2700 – 3500 m; Andino superior > 3500 m) showed no significant differences in the floret proportions ($p = 0.29$). Only a weak significance was detected between the Mediterranean-matorral and Andean species within *Chaetanthera* ($p < 0.01$).

However, when all the species in both genera were attributed either to a lowland Mediterranean-matorral zone (<1500 m.a.s.l.) or an upland Andean zone (>1800 m.a.s.l.), there was a significant difference (Students T-test, $p < 0.0001$) between the RL:DL ratios. I conclude that the lowland species have more showy capitula i.e., tend to have longer ray florets in relation to the disk florets (n = 15, RL:DL = 1.75 ± 0.09) than the upland species (n=31, RL:DL = 1.39 ± 0.04). This renders the use of ray floret ligule measurements as a generic identifier redundant.

Within the lowland species the following had extreme ray:disk ratios: *C. elegans* (2.3), *C. microphylla* (1.2) and *O. werdermannii* (2.3). Within the Andean species the following had extreme ray:disk ratios: *C. renifolia* (1.7), *C. peruviana* (1.7) and *O. polymallum* (0.6). All these species also show peculiarities in aspects such as locality, breeding system and other morphological and anatomical features. Although there are some disjunctions of size between selected taxa, in the majority of species of *Oriastrum* and *Chaetanthera*, ray floret length and disk floret length reflect the overall aspect of the capitula, while the ratio between them appears, to be a function of altitude. The possible correlates between altitude zone, habitat and breeding system are not discussed here.

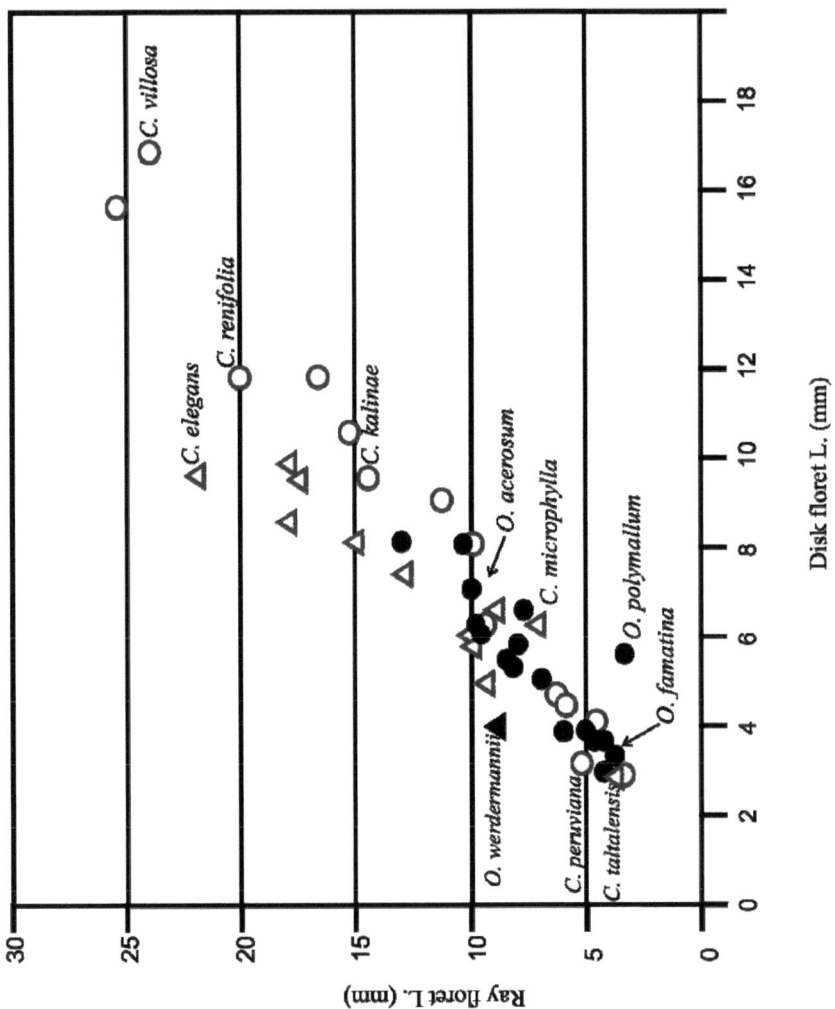

Figure 12: Ray floret length (RL) and disk floret length (DL) variation. Graph plotting RL against DL for *Oriastrum* (black shapes) and *Chaetanthera* (red shapes) for lowland (<1500 m) [▲] and upland (>1800 m) [●] species.

2.7   Anthers & pollen

The form of the apical anther appendages have been used in other Mutisieae genera to define taxon limits, e.g. *Ianthopappus* (ROQUE & HIND 2001). The *Gochnatia* complex (FREIRE et al. 2002) (now considered as tribe Gochnatieae after PANERO & FUNK 2002), is apparently uniquely defined by the having a combination of apiculate anther appendages and smooth style branches. In *Chaetanthera* and *Oriastrum* the anther apices were observed to be invariably longly acute and smooth. The basal caudae of the anthers of *Chaetanthera* and *Oriastrum* are ciliate. This was the feature that gave its name to the genus – "chaeta" from the Greek for setae or long hair; "anthera" from the Latin for anther (RUIZ & PAVON 1794). It is typical of many Mutiseae, and the closely related genera *Brachyclados* and *Pachylaena* were also observed with similar basal ciliate caudae.

Pollen studies conducted by the author tally with those results published in the recent, definitive study by TELLERÍA & KATINAS (2004). They described two main pollen types found among the species studied: type I (Acerosa type compact tectum) and type II (Chilensis type, tectum with internal radial cavities). Type I is circumscribed by taxa in the genus *Oriastrum*, while type II is found in all *Chaetanthera* taxa. A subtype of type II ectosexine structure is compact and not columellate. Some species are reportedly variable for these ectosexine subtypes.

All taxa in *Chaetanthera* and *Oriastrum* share an imperforate tectum with microspines and a tricolporate aperture with a marked mesoaperture. Two types of combined equatorial (Eq.) and polar (Pol.) shape are distinguishable. Pollen grains can either have a subrectangular (Eq.) and subcircular (Pol.) shape, as seen in *O. apiculatum* (Fig. 13) or an elliptic (Eq.) and subtriangular (Pol.) shape, as seen in *Chaetanthera* species (e.g. *C. ciliata, C. glabrata* and *C. linearis* (Fig. 13.).

The exine can be thick (11 - 20µm) as in *C. splendens* (Fig. 14A) or thin (5-7µm) as in *O. cochlearifolium* (Fig. 14B). The ectosexine is compact or finely columellate, the endosexine has stout ramified columellae and the nexine is either dumb-bell shaped (Fig. 14A) or elliptical (Fig. 14B). The columellae are clearly illustrated for *C. spathulifolia* in Fig. 14D. The "nipple" at the poles of *C. splendens* is the apoclpial field, indicating that the grain is so-called parasyncolpate. This is, according to TELLERÍA & KATINAS, incidental within *Chaetanthera*, occuring in some pollen grains of *C. philippii, C. limbata, C. linearis, C. splendens* and *C.* "valdiviana". (N.B. *C. valdiviana* is now a synonym of *C. serrata*). This was observed in *C. glabrata* during this study.

The inverse S.E.M image (Fig. 14C) of *C. glabrata* shows the details of the aperture, showing ecto-, meso-, and endoaperture.

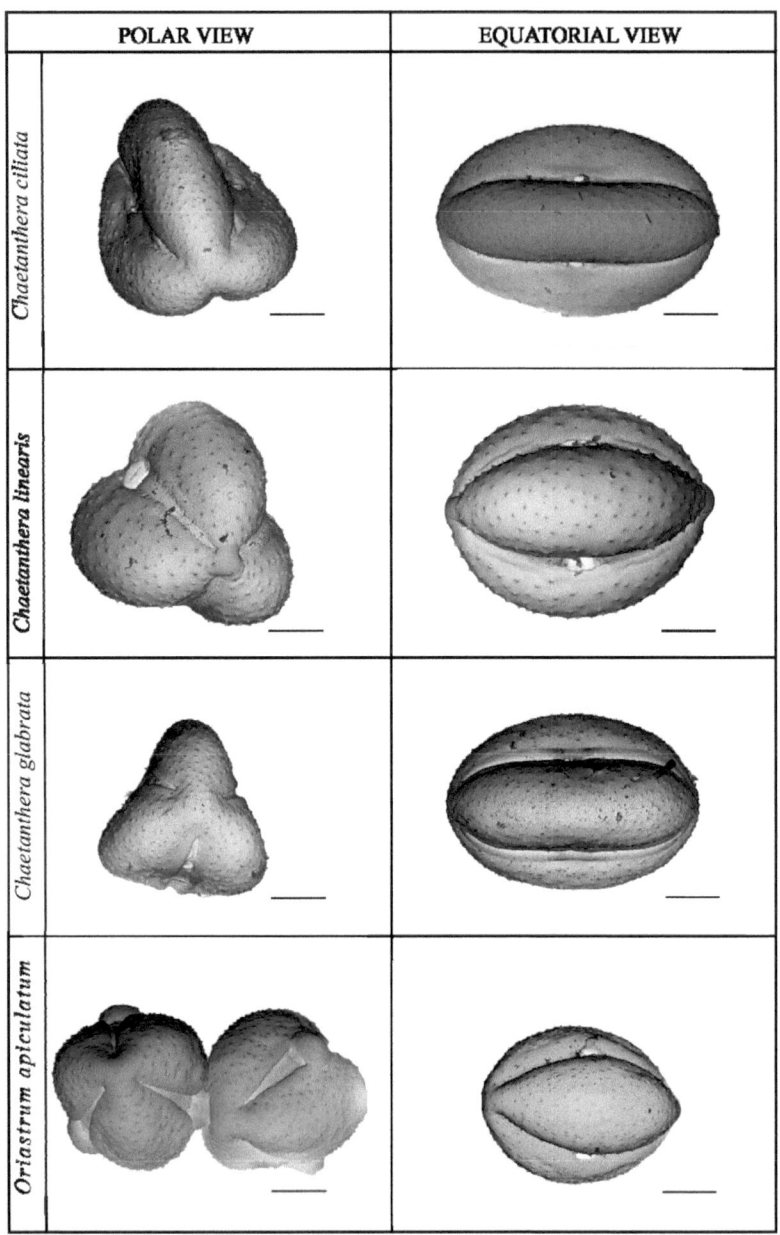

Figure 13: External pollen structure. S.E.M. Images (inverted) of the polar and equatorial views of *C. ciliata*, *C. glabrata*, *C. linearis* and *O. apiculatum*.

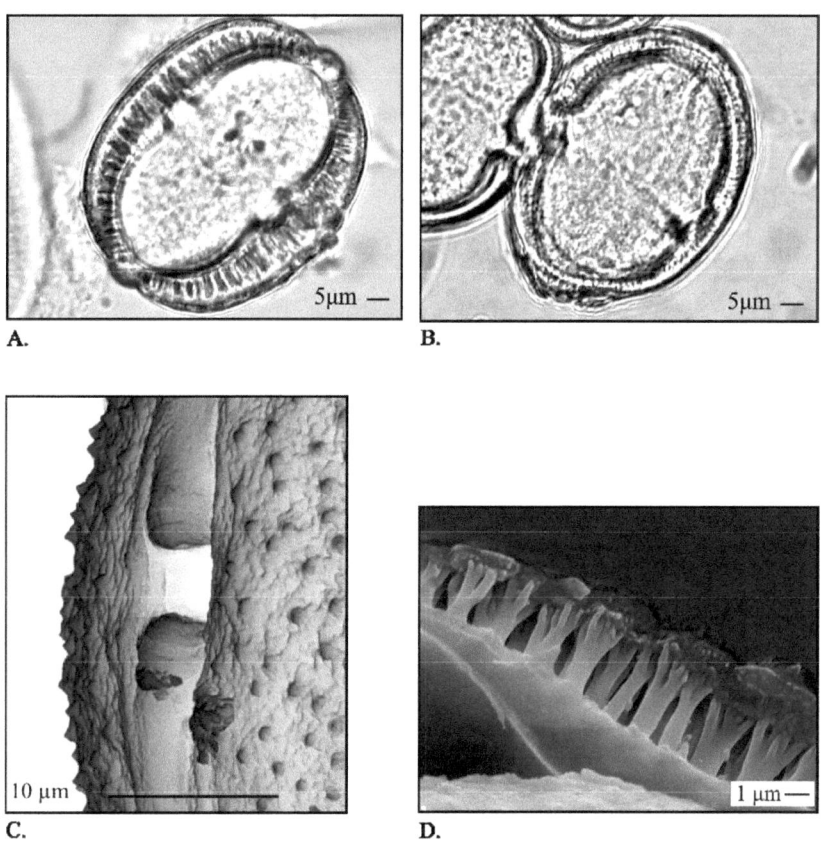

Figure 14: Internal pollen structure. **A.** *C. splendens*. **B.** *O. cochlearifolium*. **C.** *C. glabrata*; details of the aperture. **D.** *C. spathulifolia*; exine structure. A, B: polarized LM images. C, D: S.E.M. images.

2.8 Styles & stigmas

The functional stigma arms of *Chaetanthera* are bifid; long (on average 5% of the total style length) in relation to the style and oblong or truncate at the apices. Fig. 15 shows some of the recorded stigma lobe and stigmatic hair variation. *C. linearis* (Fig. 15A, B), *C. ciliata* (Fig. 15C) and *C. moenchioides* (Fig. 15D) all have flattened, smooth, triangular hairs on the stigma apices. There is little or no variation in the form of the hairs on the stigmatic surfaces between *Chaetanthera* species.

The functional stigma arms of *Oriastrum* species are bilobate; short in comparison to the style (on average 2.5% of the total style length), and are rounded or blunt at the apex. The hairs on the stigma apices are long, rounded, with a lightly echinate surface. Fig. 15E – H shows two examples: *O. dioicum* and *O apiculatum*.

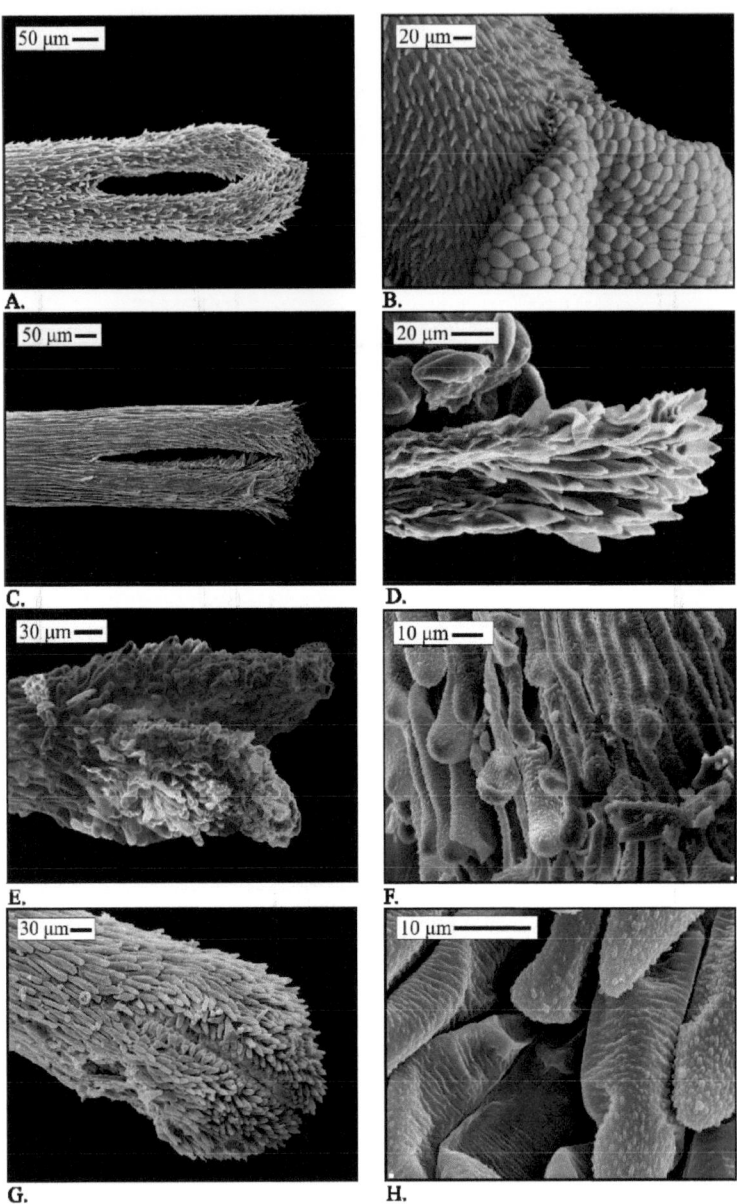

Figure 15: Stigma lobe and stigmatic hair variation. **A.** *C. linearis*. **B.** *C. linearis* (detail of stigmatic hairs). **C.** *C. ciliata*. **D.** *C. moenchioides* (detail of stigmatic hairs). **E.** *O. dioicum*. **F.** *O. dioicum* (detail of stigmatic hairs). **G.** *O. apiculatum*, **H.** *O. apiculatum* (detail of stigmatic hairs). A, C - H = S.E.M. images, B = T.E.M. image.

2.9 Achenes

The achenes of *Chaetanthera* and *Oriastrum* show variation in the shape, the carpopodium, the achene hairs or pericarp indumentum, and the cell structure of the testa epidermis.

2.9.1 Achene shape

The achenes in *Oriastrum* and *Chaetanthera* s.str. are variable in size and shape depending on maturity, position in the capitula and viability. The shape here refers to the viable, mature achenes. *Chaetanthera* achenes are generally turbinate – sometimes sinuously so. The high altitude taxa, such as *C. villosa* and *C. spathulifolia* have large achenes (5 – 7 mm long) that are asymmetrically obovate, and laterally flattened (Fig. 16A). More typically the achenes are narrowly turbinate and vary from 2 – 4 mm long (Fig. 16B). The achenes of species related to *C. glabrata* occasionally appear shortly rostrate. *Oriastrum* achenes are small (on average 2 mm long) often distinctly angled, and obovate to fatly pyriform. A typical example are the achenes of *O. pusillum*, as shown in Fig. 16C. Mature achenes of the taxa in *Oriastrum* subgenus *Egania* can be very difficult to find. However, they are either small and angled turbinate, as in *O. revolutum* (not illustrated), or large (to 5 mm long), asymmetrically obovate and laterally flattened, as seen in *O. acerosum* (Fig. 16D).

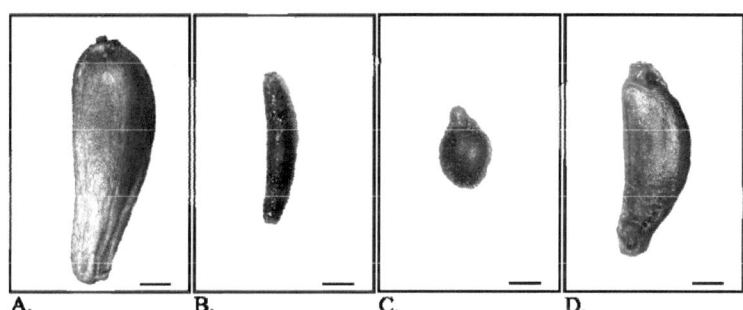

Figure 16: Achene shape. **A.** *C. villosa*, **B.** *C. glabrata*, **C.** *O. pusillum*, **D.** *O. acerosum*. Scale bar = 1 mm.

2.9.2 Carpopodium

The carpopodium, or abscission zone, is found between the receptacle and the base of the achene. MUKHERJEE & NORDENSTAM (2004) defined 14 types of carpopodium in the Asteraceae, although their sample included only one Mutisieaen representative (*Gerbera jamesonii*) that reportedly had no carpopodium (op. cit., p. 39). The carpopodium type was concluded to be diacritical at the species level. In this study two types can be identified. These are illustrated in Fig. 17.

The first type is a poorly differentiated to completely absent abscission zone (Fig. 17A, C, E, G). The second type has a carpopodium which is more or less symmetric in a complete ring, composed of rows of thick-walled carpopodial cells (Fig. 17B, D, F, H). The perennial species in both genera often have a reduced or poor carpopodium, e.g. *C. glandulosa* – 17B, *C. spathulifolia* – 17G, *O. dioicum* – 17E. *O. apiculatum* (not illustrated) has a small but symmetrical ring-shaped carpopodium, possibly a pseudo-carpopodium, consisting of pericarp cells rather than specialized carpopodial cells. The annual species in *Chaetanthera* have a clearly defined carpopodium, e.g. *C. glabrata* – 17F, *C. ciliata* – 17H. The high elevation, annual species in *Oriastrum*, e.g. *O. pusilla* – 17A, *O. lycopodioides* – 17C have poor to non-existent carpopodia.

The carpopodium structure seems constant within a species, and in some cases between closely related species. There seems to be some linkage between life-cycle, geographical elevation and carpopodium presence/ absence.

2.9.3 Achene hairs

The myxogenic phenomenon seen on *Chaetanthera* and *Oriastrum* achenes was first described by A. Gray (1861) in a note accompanying the description of "*Oriastrum cochlearifolium*". The *Chaetanthera* [s.l.] pericarp is pellucid or hyaline and often highly hydrophilic (GRAU 1980). ANDERBERG & FREIRE (1990) postulated that the trichomes themselves were myxogenic, in contrast to *Anthemis*, where the mucilage comes from the achene epidermis cells directly (OBERPRIELER et al. 2001). Observations have shown that the twin-celled trichomes forming the pericarp indumentum on many *Chaetanthera* achenes, and on *Oriastrum* subgenus *Oriastrum* achenes do not retain their form after wetting, losing their integrity. On wetting *C. taltalensis* achenes, the twin hairs swell about 30% in length then burst, releasing a mass that extends in coiled strings up to 4-5 mm from the achene. It retains its string-like form for some time (10-15 minutes) before gelling into an amorphous mass. Although it dries solid, ike glue, it becomes gel-like on re-wetting. Those achenes with no hairs do not exhibit myxogenic properties.

The achene hair variation described here was first published by DAVIES & FACHER (2001). Three basic achene hair types have been identified: "Zwillingshaare" (after HESS 1938) that can be either lanceolate or spherical, filiform hairs and single-celled papillae. The "Zwillingshaare" can be further described as conforming to the non-glandular, basic-basic or basic-rounded subtypes of FREIRE & KATINAS (1995), with an anticlinal primary division.

Chaetanthera species typically have large ((80) 90 – 120 (130)µm long) obovate-elliptic to lanceolate "Zwillingshaare". The species e.g., *C. incana* (Fig. 18A), with these sorts of hairs generally have hairs on both the disk and ray achenes, and are very seldom glabrescent. The "Zwillingshaare" are more conical or rounded in one or two annual species, i.e., *C. flabellata* (Fig. 18B) and *C. euphrasioides* (Fig. 18C).

High elevation perennial taxa such as *C. villosa* or *C. spathulifolia* (Fig. 18D) can have tiny spherical twin hairs that vary from 8 – 16µm, which are sparsely distributed over the achene surface. Often the hairs are present in an incipient stage and they are frequently poorly preserved, so achenes may often appear to be glabrous.

III Variation of characters in *Chaetanthera* & *Oriastrum*

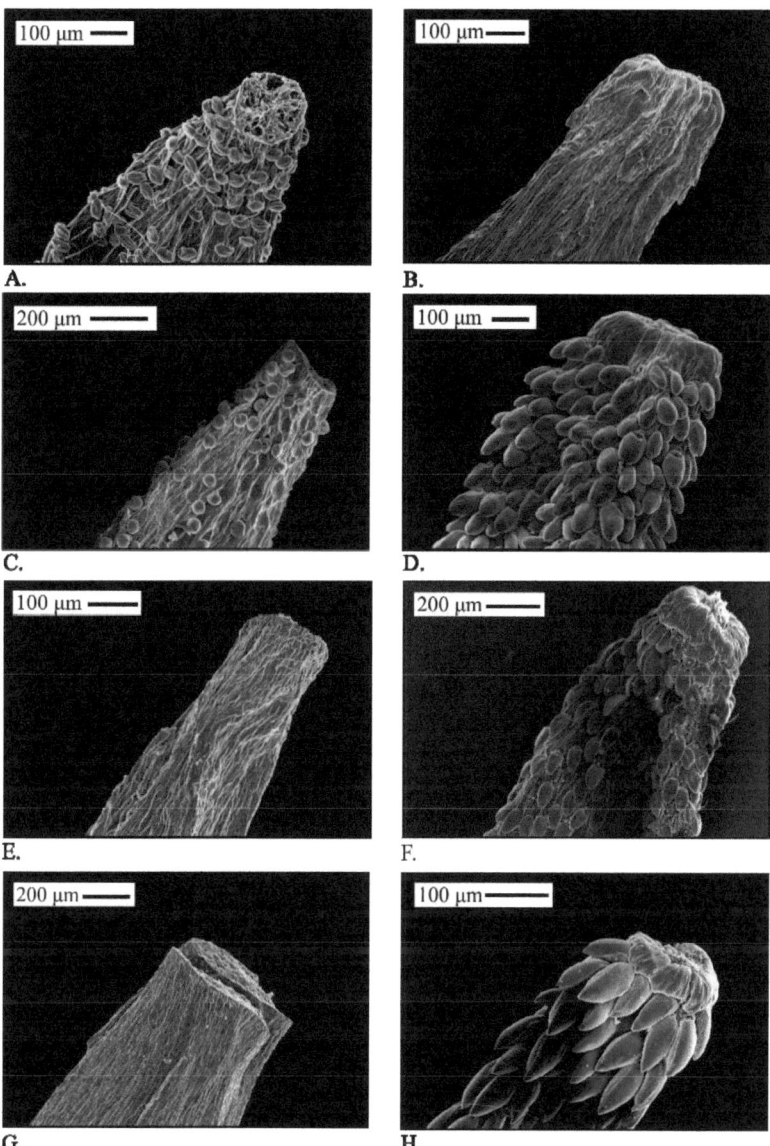

Figure 17: Carpopodium variation. **A.** *O. pusillum*, **B.** *C. glandulosa*, **C.** *O. lycopodioides*, **D.** *C. chilensis*, **E.** *O. dioicum*, **F.** *C. glabrata*, **G.** *C. spathulifoila*, **H.** *C. ciliata*.

These species are excellent for finding examples of the ontogeny of 4-celled twin hairs. Different collections of *C. philippii* (e.g. *Werdermann 934* NY; *Johnston 5987* GH) have been observed as having achenes with large obovate "Zwillingshaare" and achenes with tiny, poorly preserved "Zwillingshaare". The collections are otherwise morphologically comparable. There is considerable infrageneric variation in the achene hair types found in *Oriastrum*. None of the species have the large inflated lanceolate twin hairs that are typical of *Chaetanthera*. Species from *Oriastrum* subgenus *Oriastrum* have inflated spherical to ampulliform "Zwillingshaare" of varying sizes. Larger (35 – 50μm in length) spherical hairs are found on *O. pusillum* (Fig. 18E) and *O. chilense* achenes. Somewhat smaller (20 – 35μm) spherical hairs can be found on *O. lycopodioides* specimens, while tiny (10 – 20μm) ampulliform to nearly deltoid Zwillingshaare are seen on *O. gnaphaliodes* specimens (Fig. 18F). These are comparable to the non-glandular, basic-rounded subtype of FREIRE & KATINAS (1995).

Species from *Oriastrum* subgenus *Egania* do not have inflated twin hairs. Some plants within populations of *O. acerosum* or *O. achenohirsutum* can have achenes with long simple filiform hairs (Fig. 18G). They are built of two vertically stacked dispropotionately sized cells (periclinal division) ca. 450μm long. Oblate, flattened hairs, (the single-celled papillae of DAVIES & FACHER 2001), were observed in two species, *O. pulvinatum* and *O. revolutum* (Fig. 18H). They densely cover the achenes and measure 25 – 45μm in length. The ontogeny of these hairs was not observed, although further observations indicate that these are composed of 2 cells only. These seem to be raised basal cells, originating from an anticlinally divided epidermal mother cell, but arrested before they develop on to the next division that would classify them as typical twin hairs of HESS (1938) and ROTH (1977). A few species are characteristically glabrous, i.e. *O. polymallum*. However, it is more common to find some sort of achene indumentum, even if they are not present on both ray and disk achenes, or even on different collections of the same species. No correlation to disk and ray distribution of achene hairs, or the lack thereof, was observed, and the loss of achene hairs is, in this case, not taxonomically relevant. This study concurs with HANSEN (1991) and FREIRE & KATINAS (1995) that the glabrous state is a parallelism.

2.9.4 Achene testa epidermis

The testa epidermis is the cell layer separating the pericarp from the testa itself. Two *Chaetanthera* species – *C. microphylla* and *C. ramosissima* – were studied by GRAU (1980) as part of a wider survey of the Mutisieae testa epidermis variation. They showed a distinctive U-shaped thickening of the testa, typical of a "Mutisia – Gochnatia" group. Two aspects of the testa epidermis were considered here: the surface view and the transverse section through the testa epidermis, testa and endosperm. Four types of surface ornamentation were observed: narrow-oblong cells in parallel (Fig. 19A); sinuous, pseudo-dendritic cells (Fig. 19B); rounded-oblong cells in parallel (Fig. 19C, D); and tightly sinuous tesselated cells (19E). Four types of sclerenchymatous thickenings were observed from the transverse sections. Thin irregularly O-shaped walls were observed in *O. acerosum* (Fig. 19A), *O. revolutum* and *O. polymallum*. Multiple, U-shaped ribs in the cells were observed in *O. chilense*, *O. pusillum* (Fig. 19B), *O. gnaphalioides* and *O. lycopodioides*. Thickened U-shaped cell walls (Fig. 19C) were observed in *C. chilensis*, *C. glandulosa*, *C. kalinae*, *C. moenchioides*, and *C. euphrasioides*. Thickened O-shaped walls with thinner tops, sometimes with lateral villi (Fig. 19D) were observed in *C. glabrata*, *C. philippii*, *C. renifolia* and *C. villosa*.

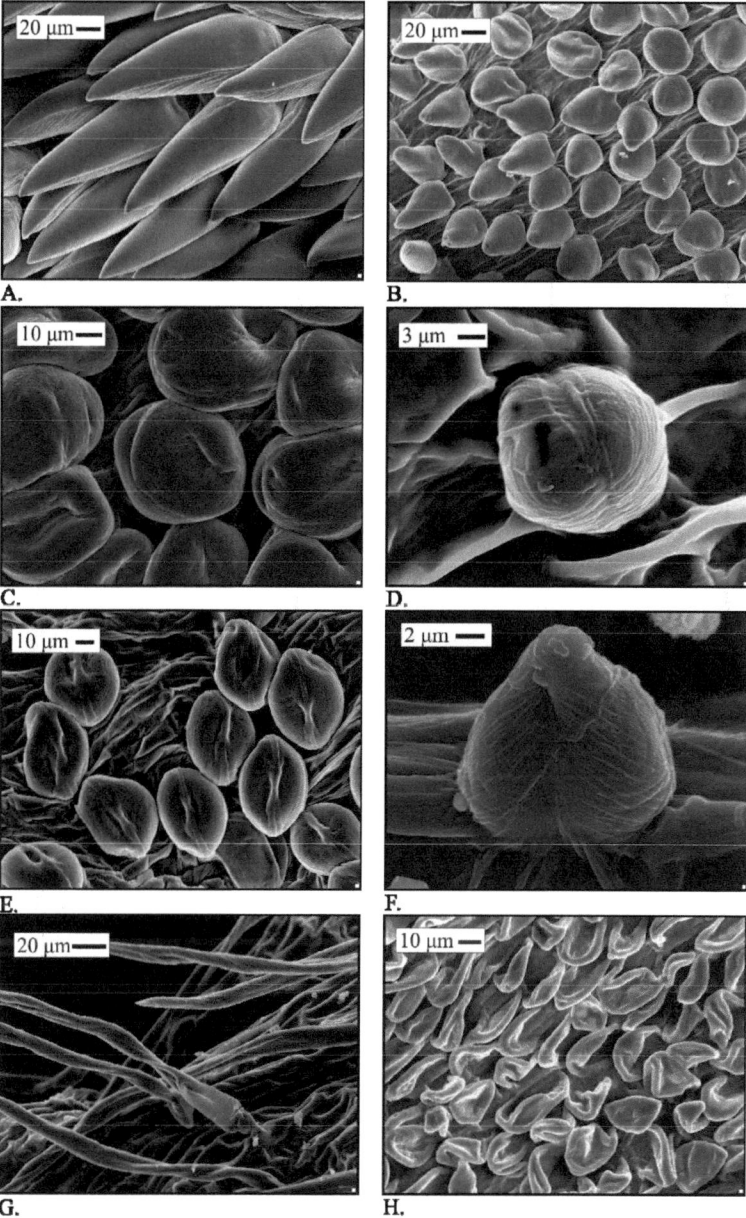

Figure 18: Achene hair variation. **A.** *C. incana*. **B.** *C. flabellata*. **C.** *C. euphrasioides*. **D.** *C. spathulifolia*. **E.** *O. pusillum*. **F.** *O. gnaphalioides*. **G.** *O. acerosum*. **H.** *O. revolutum*.

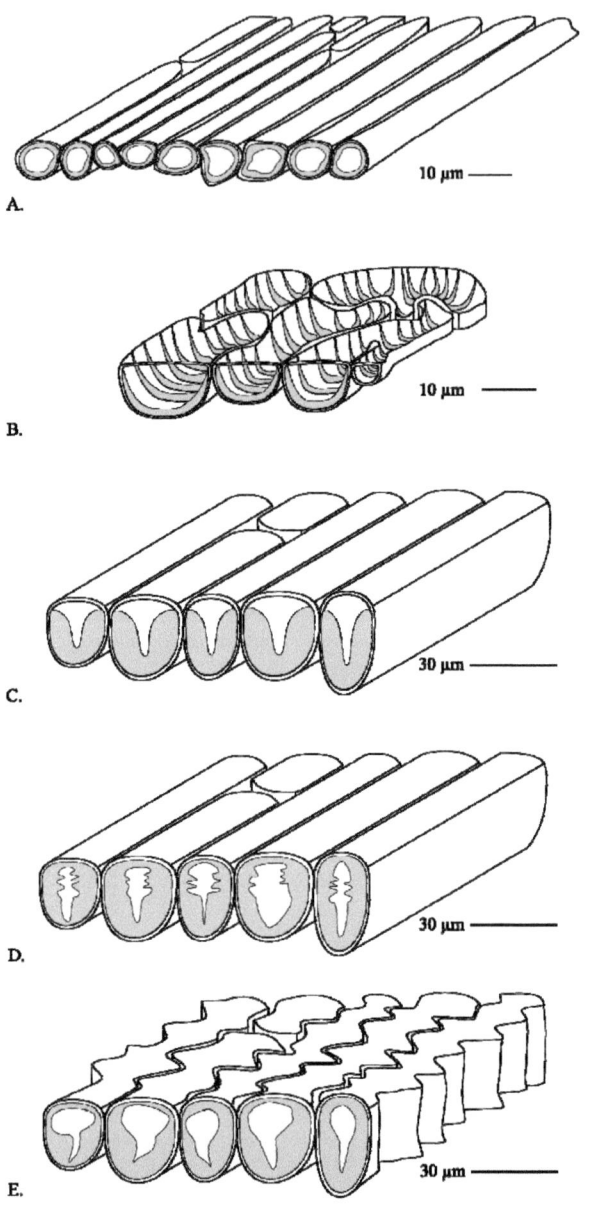

Figure 19: Testa epidermis variation. **A.** Egania type, **B.** Oriastrum type, **C & D.** Chaetanthera type, **E.** *C. renifolia*.

The "Egania" type is represented by narrow-oblong cells in parallel with thin, irregularly O-shaped walls. The "Oriastrum" type is typified by the presence of multiple U-shaped ribs. Three of four observed species had sinuous, pseudo-dendritic cells; *O. gnaphalioides* had ribs but parallel cells. The "Chaetanthera" type (= Grau's Mutisia-Gochnatia type, 1980, op.cit.) is typified by having U to O thickened walls. Excepting *C. renifolia*, which has tightly sinuous tesselated cells, all species observed had parallel cells.

The testa epidermis cells were up to 10 µm (rarely 15µm) high for the thin walled and ribbed cells. The testa epidermis cells ranged from 20 – 50µm high for the thick U-walled and O-walled cells. The testa itself varied in thickness 5 – 10 (15µm) and appears as an indeterminate cell mass under a light microscope. Rhomboid or oblong oxalate crystals were observed in *C. moenchioides, C. chilensis, C. renifolia, C. glabrata, C. kalinae* and *C. philippii*. The crystals were not present in the other species studied (*C. glandulosa, C. villosa, O. pusillum, O. acerosum, O. polymallum*). The endosperm was 1 (seldom 2) cells thick (3 – 7 µm).

3 Cytological variation

Six *Chaetanthera* species (*C. albiflora, C. frayjorgensis, C. glabrata, C. linearis, C. moenchioides,* and *C. villosa*) were successfully cultivated in Munich, and sporophytic counts recorded.

Some of these were re-counts (*C. albiflora, C. glabrata,* and *C. linearis*) of numbers published by GRAU (1986). The haploid count n = 14 for *C. ramosissima* (= *C. tenella*) by POWELL ET AL. (1974) has not yet been confirmed. BAEZA & SCHRADER published short notes on the karyotype analyses of *C. microphylla* (2005a) and *C. chilensis* and *C. ciliata* (2005b) using FISH (flourescent in-situ hybridisation). BAEZA & TORRES DIAZ published the novel chromosome count of *Oriastrum dioicum* [= *Chaetanthera pentacaenoides*] (2n = 20) in 2006 and BAEZA et al (2007) confirmed a count of 2n = 2x = 26 for *C. moenchioides*.

Chaetanthera taxa have variable sporophytic chromosome counts of xn = 22, 24, 26, 28 and 38. The diploid number 2n = 22 is the basic chromosome complement seen in *C. linearis* and *C. villosa* (Fig. 20a, 20b).

Both species were observed to have very similar karyotypes. There are eight (numbered 1-8 in decreasing total length) pairs of long metacentric (sometimes abbreviated to 'm') chromosomes. Number 9 is more-or-less metacentric, but notable for often having visible NORs (Nucleolar Organising Region). Numbers 10 and 11 are subtelocentric. *C. ciliata* (2n = 22) is documented as having an entirely metacentric chromosome complement. *C. chilensis* (2n = 22) has one subtelocentric chromosome pair (number 8) in an otherwise metacentric chromosome complement (BAEZA & SCHRAEDER 2005b). *C. albiflora*, with 2n = 24, has an NOR on chromosome pair 7. The distribution of metacentric and subtelocentric or acrocentric chromosomes in its karyotype is varied. It is not illustrated here. *C. microphylla*, also with 2n = 24, is recorded as having an asymmetrical complement (8m + 4st) (BAEZA & SCHRAEDER 2005a).

C. glabrata (Fig. 21) and *C. frayjorgensis* have 2n = 4x - 6 = 38. The karyotype analysis revealed that all chromosomes appear to be present in fours, except numbers 1, 4 and 10, which are present only as pairs. This could be an example of descending dysploidy whereby an ancestral tetraploid (4n = 44) has lost three chromosome pairs, resulting in a complement of 38. Loss of chromosomes is said to be tolerated particularly well by polyploids because of internal duplication of chromosomes (GRIFFITHS et al., 1996).

Despite several field records of hybrids between *C. linearis* (2n = 22) and *C. albiflora* (2n = 24), and the successful cultivation of such a hybrid, the chromosome number was not recorded. It would be interesting to see how the plants circumvent the problem of combining homeologous sets of chromosomes.

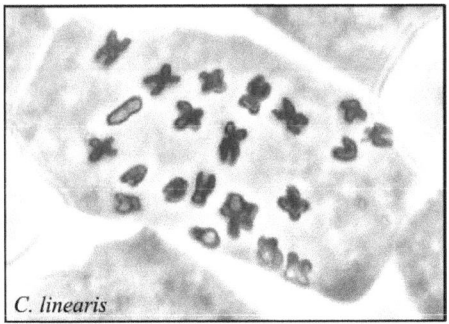 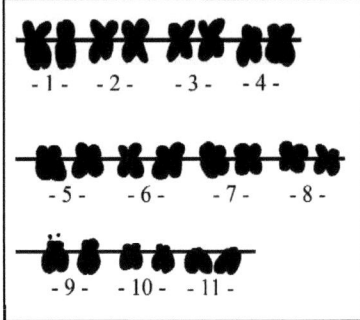

Figure 20a: Chromsome diversity. Polarised Light Microscope images of arrested metaphase root-tip squashes (left), and karyotype diagrams (right) of *C. linearis*.

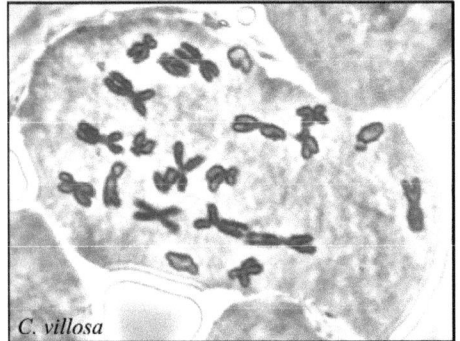 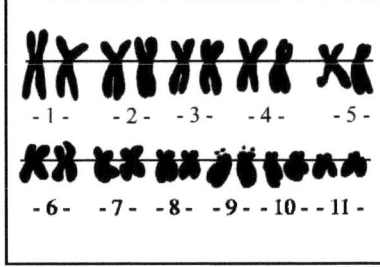

Figure 20b: Chromsome diversity. Polarised Light Microscope images of arrested metaphase root-tip squashes (left), and karyotype diagrams (right) of *C. villosa*.

Table 1 summarizes the novel and published chromosome counts of *Chaetanthera* and *Oriastrum* species. In 10 different *Chaetanthera* species haploid counts vary from n = 11, 12, 13 to n = 14. These counts support the theory that *Chaetanthera* is very polymorphic in its cyto-evolutionary mechanisms (BAEZA & SCHRAEDER 2005b). It is not unusual to see examples of dysploid reduction during cyto-evolution. This same mechanism has resulted in diverse chromosome numbers, genome organisation and chromosome structure in *Boronia* L (SHAN et al. 2003.), for example.

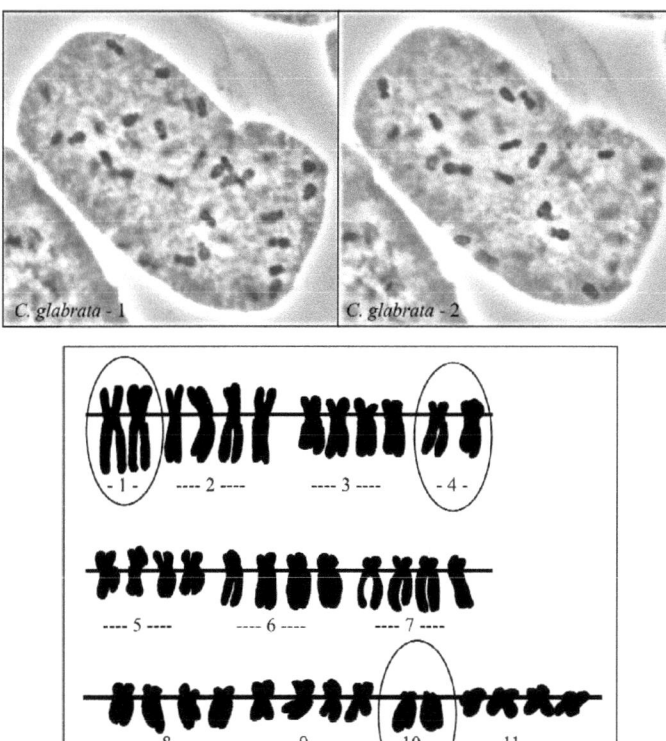

Figure 21: Chromsome diversity. Polarised Light Microscope (P.LM.) images of arrested metaphase II root-tip squashes and karyotype diagrams of *C. glabrata*. The two P.LM. images are taken at different depths of view in the sample to reveal chromosomes at different positions. The circled chromosome pairs on the diagram are the non-duplicated pairs.

Species	Chromosome count	Collection locality of material analysed	Source
C. albiflora	2n = 24	Chile. IV Región de Coquimbo, Limarí, Parque Nacional Fray Jorge, E.Bayer (M)	Grau 1986
	2n = 24	Chile. IV Región de Coquimbo, Limarí, PN de Fray Jorge, Bosque Fray Jorge, erste Hänge im Landesinneren nach den Steilkurven, 11.2002, Ehrhart 6.40 (M)	Davies & Vosyka, new count
C. chilensis	2n = 22	Chile. VIII Región de Bio-Bío, Quillón, road between Quillón & Cabrero, Canchilla, 110 m, 01.2003, C. Baeza 4204 (CONC).	Baeza & Schrader 2005b
C. ciliata	2n = 22	Chile. VIII Región de Bio-Bío, Quillón, Road between Quillón and Cabrero, Canchilla, 110 m, 01.2003, C. Baeza 4205 (CONC).	Baeza & Schrader 2005b
C. frayjorgensis	2n = 38	Chile. IV Región de Coquimbo, 800 m antes del desvio hacia Ovalle a lado del Servicentro, 11.2001, Lopez s.n. (M)	Davies & Vosyka, new count
C. glabrata	2n = 38	Chile. III Región de Atacama, Huasco, Huasco- Carrizal Bajo, 9km nördlich Huasco, Felshänge, 50 m, 03.12.2002 Ehrhart 6.142 (M)	Davies & Vosyka, new count
	2n = 28	Chile. Región Metropolitana de Santiago, östl. Las Condes, Straße nach Farellones, 890 m, 11.1980, Grau 2450 (M)	Grau 1986
	2n = 38	Chile. Región Metropolitana de Santiago, Farellones, 2100 m, 02.03.2002, Kalin Arroyo s.n. (CONC)	Davies & Vosyka, new count
C. incana	2n = 22	Baeza, Ruiz & Negritto in Gayana Botanica 65 (2): 237 – 240, 2009.	
C. linearis	2n = 22	Chile. Región Metropolitana de Santiago, Straße nach Farellones 1500m, J.Grau s.n. (M)	Grau 1986
	2n = 22	Chile. Región Metropolitana de Santiago, Straße nach Farellones, Kurve 20, 1500m, 02.2002, Davies 2002/001 (M)	Davies & Vosyka, new count
	2n = 22	Chile. Región Metropolitana de Santiago, Chacabuco, La Dormida, Hellwig s.n. (M)	Grau 1986
C. moenchioides	2n = 26	Chile. Región Metropolitana de Santiago, Cuesta La Dormida, 1600 m, 02.1998, Grau s.n. (M)	Davies & Vosyka, new count
		Chile. VIII Región de Bio-Bío, Antuco, entre Villa Peluca y bifurcación a Rayenco, 710 m, 12.2002, C. Baeza 4198 (CONC)	Baeza et al. 2007
C. microphylla	2n = 24	Chile. VIII Región de Bio-Bío, Comuna de Quillón, Puente El Roble, 64 m, 12.2002, C. Baeza 4177 (CONC)	Baeza & Schrader, 2005a
C. tenella	n = 14	No information available.	Powell et al. 1974
C. villosa	2n = 22	Chile. IX Región de La Araucanía, Prov. Malleco, Volcán Lonquimay, 1500 m, 03.2002, Davies & Grau 2002/058 (M)	Davies & Vosyka, new count
Oriastrum dioicum (C. pentacaenoides in lit.)	2n = 20	Chile. Región Metropolitana de Santiago, Valle Nevado, Cerro Tres Puntas, 3600 m, 01.2003, C. Torres s.n. (CONC)	Baeza & Torres Diaz 2006

Table 1: Summary of the novel and published chromosome counts of *Chaetanthera* and *Oriastrum* species.

4 Genetic variation

Chaetanthera has been genetically sampled in a few studies (i.e. KARIS ET AL. 1992; BREMER 1994; H.-J. KIM 2004,), but these studies were aimed at detecting tribal relationships and only included a couple of species at most. Some taxa included in these studies were verifiably misidentified, e.g. in BREMER (1994), and the species selection has never taken the wide subgeneric variation into account. Most recently, PANERO & FUNK (2008) re-analysed the generic placement of the anomalous genera in the Mutisiinae using 10 chloroplast loci, with the result that the herbaceous shrub *Brachyclados* D.Don has been relocated closer to the scapose *Trichocline* Cass., and the acaulescent *Pachylaena* D.Don ex Hook. & Arn. closer to *Mutisia* Lf. while *Chaetanthera* s.l. (represented by *O. dioicum*) remains part of an unresolved clade uniting all these groups. *C. linearis* is in the process of being sequenced (PANERO, 2009 pers. comm.)

The first comprehensive, reliable DNA analysis and subsequent phylogenetic reconstruction of *Chaetanthera* and *Oriastrum* species was by HERSHKOVITZ ET AL. (2006). Their paper is summarized briefly below.

Chloroplast (cp) DNA ndhF sequences and nuclear ribosomal (nr) DNA ITS sequences were taken from 27 *Chaetanthera* taxa and 14 *Oriastrum* taxa. The cpDNA and the nrDNA data gave different relations among the species but this was considered to be a sampling artefact in the ndhF sequences. The clearest phylogenetic model was produced with the ITS sequences, giving a "between species" maximum parsimony bootstrap consensus tree. This model was then scaled using a penalized likelihood estimate of divergence dates of *Chaetanthera* [and *Oriastrum*]. This is adapted for Fig. 22. The morphological species names are identified to the right of the branches generated by the ITS data analysis. The supra-specific hierarchies of *Chaetanthera* and *Oriastrum* are aligned to the right of the species list.

At the species level some taxa have genetic variation that is not reflected in the current morphological concept. For example, several samples representing *O. gnaphalioides* (*Arroyo* et al. *25079, 25086, 02-154, 25127*) show quite divergent nrDNA sequences. These are all combined under the one specific epithet in the morphological-taxonomic work, albeit recognised as a polymorphic entity that is arguably susceptible to altitudinal variations in water relations. Similarly, the closely related, but distinct species of *C. albiflora, C. linearis* and *C. microphylla* are nested together; the former with several genetically different collections, the latter two taxa apparently genetically indistinguishable.

Above the species level the nrDNA shows two significant divergences. The first split divides the sampled taxa into "Clade A" – a grouping analogous to *Chaetanthera*, and "Clade B" – a grouping analogous to *Oriastrum*. These two groups are then further subdivided into clades that are analogous to *Chaetanthera* subgenus *Chaetanthera*, *Chaetanthera* subgenus *Tylloma*, *Oriastrum* subgenus *Oriastrum*, and *Oriastrum* subgenus *Egania*.

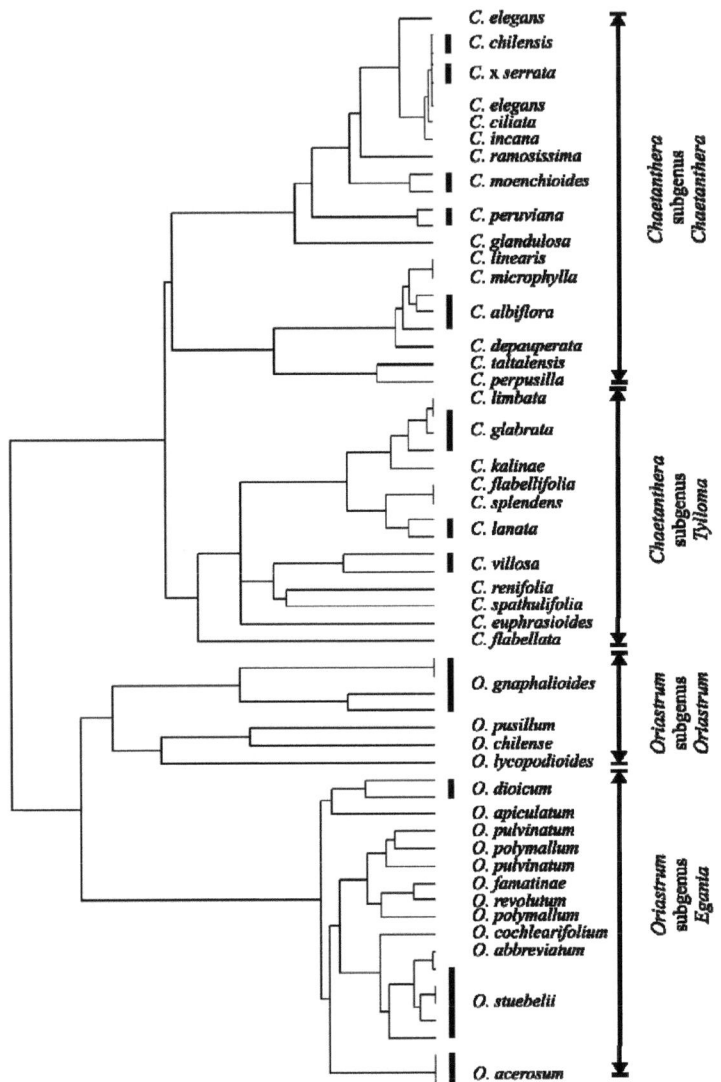

Figure 22: Species phylogeny based on nrDNA (ITS) data, with accepted names. Derived from HERSHKOVITZ et al. (2006).

IV   Form, function & habitat

1   Introduction

The Asteraceae are famously plastic in their ability to adapt to widely differing habitats (e.g. FUNK et al. 2005). A shared life form can be interpreted as having the same kind of morphological and/or physiological adaptation to a certain ecological factor. "Form, function and habitat" is the mantra of an evolutionary systematist. Only once the link (function) between the form and habitat is understood, can the pattern of evolution and selective or adaptive advantage be established. The publication of a *Chaetanthera* (including *Oriastrum*) species phylogeny based on nrDNA prior to the completion of this thesis means a morphological cladistic analysis would be largely redundant. ZANDER (2007) considers that a valuable classification should reflect functional evolution, but at the same time he derides the simplistic mapping of morphology onto a molecular tree. Therefore, rather than forcing a phylogenetic reconstruction of morphological taxa in genetically predetermined clades, the characters important in the systematics of *Chaetanthera* and *Oriastrum* are considered in the context of a functional evolutionary pattern – combining genealogy (phylogeny) and ecological adaptiveness of evolutionary divergence (after MAYR & BOCK 2002 fide VAN WYCK 2007). This allows for the circumscription of a group of species of shared evolutionary history, but who have since diverged to occupy different niches.

Form and function of characters in *Chaetanthera* (including *Oriastrum*) was discussed with respect to the pollen diversity by TELLERIA & KATINAS (2004). Phylogeny and habitat change was discussed in the light of the genetic diversity in *Chaetanthera* (including *Oriastrum*) by HERSHKOVITZ et al. (2006).

2   Homoplasy

Traditional characters such as life cycle (annual: perennial) and leaf shape (linear, sessile: spathulate, indistinctly petiolate) are the pit-falls of cladistics analyses, often creating artificial clades as a result of homoplasy. Homoplasy, or similarity not as a direct result of common ancestry, is also typical of micro-morphological characters (SCOTT 1985), often rendering them less than useful for a cladistic analysis and species phylogeny. For example, the presence of minute spherical achene twin hairs in several species of *Chaetanthera*, which otherwise has large, lanceolate achene twin hairs. There are also *Oriastrum* species with somewhat larger spherical twin hairs. While the spherical twin hairs in both genera are homologous, the taxa in these two groups are defined by more than the single character. Thus, the presence of this achene hair type is an example of homoplasy. Interestingly, the related high elevation acaulous perennial herb *Pachylaena atriplicifolia* has minute spherical achene twin hairs, while *Brachyclados lycioides* has larger spherical twin hairs and large, lanceolate twin hairs on its achenes (DAVIES & FACHER 2001). In combination, morphological and genetic data can often highlight homoplasy rather than resolve it (Coreopsideae, KIMBALL & CRAWFORD 2004; *Cicer*, DAVIES et al. 2007).

3 Habitats & stress

The habitats that *Chaetanthera* and *Oriastrum* typically occupy are found in the lower realms of the Central Chilean Province and the high elevations of the Puna and Altoandean Provinces (after CABRERA & WILLINK 1980). The climate of the Central Chilean province is characterised by a pronounced summer drought, while the climate of the high elevation Puna and Altoandean provinces is characterised by dry summers, winter precipitation in the form of snow, low temperatures and high diurnal insolation. Generally, aridity becomes less of a stress factor with increasing elevation because of the increased precipitation and reduced evapo-transpiration and reduced temperatures (KOERNER 1999, MARTORELL & EXCURRA 2002). Of special interest is the significance of cloud and fog belts as moisture providers in montane environments and along the northern coastal areas of Chile (see e.g. DILLON 1997).

Stress and disturbance characterize the type of life strategy a plant might adopt (after GRIMES 1977), where stress (e.g. water limitation) limits the biomass production and disturbance consists of the mechanisms of biomass destruction (e.g. extreme weather, predation). Strategies are evolved to tolerate or avoid stress and disturbance, whereby plant structures vary in such a way so as to confer functional advantages. The adaptive value of plant functional traits is contributed to by processes of plasticity, ecological sorting and adaptive evolution (ACKERLEY, 2003). The stress factors exerted on *Chaetanthera* and *Oriastrum* species can be broadly defined as aridity and frigidity. One commonly adopted cold resistance mechanism utilised by high Andean hemicryptophytes is freezing tolerance by using thermal refuges (i.e. between rocks) as insulation (SQUEO et al. 1996). This is seen in high elevation *Chaetanthera* and in *Oriastrum*.

4 Form & function

4.1 Rosettes, succulence & sclerophylly in *Chaetanthera*

The rosette form is attributed with conferring diverse advantages such as protection of the apical meristems against subfreezing temperatures, optimization of light interception, and the harvesting of precipitation, in particular nebulous drizzle and fog (MARTORELL & EXCURRA 2002). Succulence is an attribute usually associated with increasing aridity, being a form of water storage. Sclerophylly defines the material properties of evergreen leaves of plants growing in xeric habitats that contend with periodic drought or moisture limitation (BALSAMO et al. 2003).

The initial presence of a stem rosette in the juvenile forms of the annual species is likely an adaptation to optimising limited water resources during a critical phase of growth. The occurrence of the adult rosette form in *Chaetanthera* is decoupled from the aridity of the habitat in which the rosette-forming species are found, i.e. high elevation rosettes (e.g. *C. villosa*): low elevation loss of rosettes (e.g. *C. glabrata*). Instead, the rosettes, and in particular the rosette-like involucres, more likely play a role in the protection of the apical meristems against subfreezing temperatures, and optimization of light interception, as suggested for other high altitude taxa (see KOERNER 1999).

HANSEN (1991) proposed two stem morphoclines in the Mutisieae s.str. One line of reduction was postulated to follow reduction in the number of [flowering] branches and numbers of capitula until attaining stem monocephaly. A second path of reduction was proposed for stem/branch length reduction leading to rosulate leaves and scaposity. The reduction from multiple flowering branches to monocephalous stems has happened once, within *Chaetanthera* subgenus *Chaetanthera*. This is seen in the perennial scapose group comprising *C. chilensis* and associated taxa. The genus *Chaetanthera* shows stem reduction to the stage of rosulate leaves and indistinct stem. The senescent loss of stem rosettes during the seasonal development of the annuals is considered to be apomorphic. Within the succulent-leaved *Chaetanthera* subgenus *Tylloma* there is a small cluster of high elevation taxa comprising *C. villosa/ C.spathulifolia/ C. renifolia* whose flowering stems are reduced to ± sessile capitula.

The foliaceous imbricate involucral bracts imply that *Chaetanthera* is primarily adapted to high elevation cold environments. The ability to form stem rosettes is common to all species (no data for *C. glandulosa*). Generally, the stem rosettes of the annuals are advantageous for short-term water capture at a critical growth phase, after which resources are relocated and the stem rosettes senesce. This is probably a secondary adaptation to aridity. In the rosulate perennials with sessile capitula the stem rosettes assume a secondary insulation role. In the rosulate scapose perennials the stem rosettes persist for one growing season.

Leaf succulence is typical of all taxa in *Chaetanthera* subgenus *Tylloma* at all elevations. It also occurs in one species in *Chaetanthera* subgenus *Chaetanthera* (*C. albiflora*) that is found along the northern coast of Chile.

The only *Chaetanthera* species to exhibit sclerophylly is *C. glandulosa*. It is found in the xeric *Austrocedrus* woodlands at mid-elevations (2000 m). It not only has small, linear sclerophyllous leaves, but also a densely glandular indumentum. Woodiness in other *Chaetanthera* species is not associated with aridity or xeric habitats. The short woody stems and deep rhizomes of, for example, *C. villosa* and *C. philippii*, are part of an over-wintering strategy, and an advantageous adaptation to frost tolerance (CARLQUIST ref. 2008).

4.2   Cushions, stem buds & scleroid bracts in *Oriastrum*

Most *Oriastrum* species have densely arranged leaves on branched stems, often with short internodes, resulting in a loosely or densely formed cushion. Sometimes the stems of species like *O. chilense* or *O. achenohirsutum* can be condensed or shortly extended depending on the substrate. Because of the presence – as in *Chaetanthera* – of foliaceous outer involucral bracts, the impression can be of a small rosette. These are not homologous with the rosette development in *Chaetanthera*. The cushion habit, characterised by a high density of branches and leaves, with short internodes giving a compact architecture is one of the growth forms best adapted to the alpine habitat. Because of their low stature and compact form, cushions attenuate the effect of extreme environmental conditions by being efficient heat traps (after ARROYO et al. 2003). Typical cushion species of the high Andes of central Chile such as *Azorella monantha* (Apiaceae - CAVIERES et al. 2005), *Laretia acaulis* (ALLIENDE & HOFFMAN 1983), or *Oreopolus glacialis* (Rubiaceae - BADANO et al. 2002) are significant members of the Andean ecosystems in which they grow, promoting species diversity (e.g. BADANO & CAVIERES 2006). *Oriastrum* species are dwarf plants, forming cushions to maximum of 15 cm diameter. They typically grow away from the dominant cushion-forming vegetation. *O. abbreviatum*, itself a cushion-forming species, is the only taxon recorded as growing epiphytically on cushions. High elevation species such as *O. stuebelii* can be found amongst *Lepidophyllum* tussocks, *O. revolutum* amongst *Stipa frigida*. Other dwarf cushion species form characteristic vegetation associations e.g. the *Nototricho auricomae – Chaetantheretum sphaeroidalis* (= *O. polymallum*) association of NAVARRO (1993) found at high elevations (> 4800 m) in SW Bolivia, or the *Nassauvia pungens-Chaetanthera lycopodioides* (= *O. lycopodioides*) association of the Upper Andean Zone in Central Chile (HOFFMAN et al. 1997).

Additional to the cushion forming habit, the perennials of *Oriastrum* subgenus *Egania* are hemicryptophytes: that is they have herbaceous stems with densely hairy bud clusters at soil level. These bud clusters are an adaptation for cold tolerance (KOERNER 1999).

The coloration of the bract apices was postulated to be involved in pollinator displays by TELLERÍA & KATINAS (2004). However, when the capitula are fully open, the bracts are almost always hidden by the exerted ray florets, somewhat negating any possible open display functions. Typical of the Asteraceae, the capitula of *O. apiculatum* and *O. pusillum* were observed to close in low light conditions and towards dusk, when the dark or maculate inner involucral bracts apices became visible. Additionally, the apices of the bracts are, with only one or two exceptions (i.e. *O. werdermannii*), densely covered with a white indumentum. Anatomically, the apices are distinctively packed with sclereid cells. It is possible that the apices play an important protective role for the capitula, not only in terms of insulation, but also against damage from sudden storms – typical at higher elevations during summer.

4.3 Fragile pappus & indehiscent achenes: dispersal brakes

Structures like pappus and carpopodium have a clear functional advantage related to dispersal strategies. The modified calyx or pappus of the Asteraceae is demonstrably an adaptation to anemochory. Anemochory is generally considered typical of pioneer vegetation (VAN DER PIJL 1969, PRACH & PYŠEK 1999), particularly in alpine environments. Improvement in dispersal technique and protection against seed predation is further attributed to the barbs on the pappus setae (STUESSEY & GARVER, 1996). Indehiscent pappus and plumose pappus setae are considered adaptations promoting long-distance dispersal. The existence of a carpopodium is said to promote dispersal by providing a specialised ring of cells allowing the clean dehiscence of the propagule. According to MUKHERJEE & NORDENSTAM (2004), the complete absence of a carpopodium is plesiomorphic, while a poorly differentiated carpopodium and a ring-like carpopodium represent stages along the evolutionary development line.

The pappus of *Chaetanthera* subgenus *Chaetanthera* species is united at the base and effectively indehiscent. The barbs are appressed and densely arranged. The pappus of *Chaetanthera* subgenus *Tylloma* species is free at the base and may or may not be indehiscent. The barbs are less densely arranged and can tend toward the ciliate, especially at the pappus base. Often the more high altitude species, such as *C. villosa* or *C. euphrasioides* have dehiscent pappus setae. These species exhibit secondary barochory, i.e. pappus that becomes dehiscent before or during wind-dispersal, depositing the achenes near the parent plant. The carpopodium development seems to be strongly linked to the life cycle strategy. Annuals in *Chaetanthera* have well-developed carpopodia, while the perennials have poorly developed to non-existent carpopodia.

Secondary barochory is typical of the achenes of the *Oriastrum* species, all of which are high altitude inhabitants. Neither annual nor perennial species of *Oriastrum* have well-defined carpopodia, except for the pseudo-carpopodium on some achenes of *O. apiculatum*.

4.4 Xeromorphic adaptations in pollen, pericarp & testa epidermis

Robust pollen grain structure, a myxogenic pericarp and scleroid testa epidermis thickenings confer functional advantages related to xeromorphic adaptation. The adaptive significance of *Chaetanthera* (including *Oriastrum*) pollen types and other features was discussed by TELLERÍA & KATINAS (2004). They suggest that the structure of the pollen grain wall with its thick exine and a well developed columellate internal tectum and the occasional occurrence of parasyncolpate pollen in *Chaetanthera* are significant. These two features are thought to promote mechanical endurance of the pollen grain under hydration stress. Although *Oriastrum* species have a thin exine and a generally less robust pollen grain when compared to *Chaetanthera* species, the two genera both have endocinguli – endoapertures united with an equatorial band of cells, also considered to be an adaptation to modulating hydration events.

Chaetanthera and *Oriastrum* achenes with "Zwillingshaare" have a myxogenic pericarp. This myxogenic property has not been observed on glabrous achenes nor on the achenes with single-celled papillae, nor those with filamentous hairs seen on some *Oriastrum* achenes. According to GUTTERMAN (2002) myxospermy is a common survival strategy of desert plants. In areas of limited water resources, the mucilage provides anchorage for the propagule on an otherwise dry substrate. Additionally, it is an effective barrier against seed predation.

FREIRE & KATINAS (1994) found that filamentous non-twin achene hairs appear as multiple parallel developments in the Nassauviinae, where the typical basic-basic subtype is apomorphic. This latter hair type is typical for *Chaetanthera*. By this classification *Oriastrum* subgenus *Oriastrum* taxa have either a basic-basic or basic-rounded subtype. Species in *Oriastrum* subgenus *Egania* do not have twin hairs but instead are often glabrous or have variations of 1- to 2-celled non-twin hairs. The presence of long filamentous (non-twin) achene hairs was considered to be ancestral in the Mutisieae (HANSEN 1991). Here there seem to be three lines of development: 1) *Chaetanthera* with myxogenic twin hairs, 2) *Oriastrum* subgenus *Oriastrum* with myxogenic twin hairs, and 3) *Oriastrum* subgenus *Egania* with ancestral features and no myxogenic capabilities.

The role of testa epidermis cell thickenings was postulated by GRAU (1980) to be a form of embryo protection. Studies using *Glycine max* (L) Merr. (Fabaceae) revealed that the testa epidermis was extremely important in protecting the seed from injury by rapid and slow hydration, and in maintaining seed viability (DUKE et al. 1986). The prevalent *Mutisia – Gochnatia* type scleroid thickenings seen in *Chaetanthera* represents an apomorphic trait, common throughout the genus. The plesiomorphic thin-walled testa epidermis found in *Oriastrum* subgenus *Egania* cannot provide much protection against hydration fluctuations, while the presence of scleroid ribbed thickenings in testa epidermis of the *Oriastrum* subgenus *Oriastrum* species points to the secondary development of xeromorphic adaptation.

4.5 Competitive ability

It is interesting that closely related *Oriastrum* species often coexist in the same habitats without apparently detrimentally competing for resources (e.g., pollinators, nutrients). For example, *O. acerosum*, *O. polymallum* and *O. pulvinatum* all occur at the same altitude in the high-Andean wetlands of Tambo Puquios (Valle del Elqui) (CEPEDA et al. 2006). *O. apiculatum* and *O. lycopodioides* occur on different sides of the ski run at Portillo by the Laguna del Inca (DAVIES pers. obs.). *O. chilense* (= *C. planiseta*) does outcompete the facultatively autogamous *O. lycopodioides* for pollinators, but not to the detriment of its seed set (ARROYO et al., 2006).

According to ARROYO et al. (1982) *Chaetanthera* species (*C. flabellata*, *C. euphrasioides* and *C. microphylla*) are visited by a mixed collection of Lepidopterans, Dipterans and Hymenopterans. *O. apiculatum* and *O. chilensis* (ARROYO et al. 1982) and *O. lycopodioides* (ARROYO et al. 2007) only attract pollinating Lepidopterans, if at all (*O. pusillum* and *O. dioicum* had no recorded visitors). Microclimate plays a significant role in pollinator acquisition and reproductive success of *C. euphrasioides*, *O. apiculatum* and *O. lycopodioides* (TORRES DIAZ 2007). The impact of climate change, specifically warming, on subnival higher elevation plants and their pollinators, is postulated to have both a short term positive and negative impact as the ideal microclimate temperatures for pollinators will migrate up to optimal and above optimal conditions respectively.

A more immediate threat to success may be felt from invasive non-native species such as *Taraxacum officinale*. CAVIERES et al. (2005) found that although *C. euphrasioides*, *O. pusillum* and *O. chilense* were among 17 species from the high-elevation central Chilean Andes not affected by the presence of the invasive weed, *O. lycopodioides* was one of five native negatively affected species. It is not known whether this is a direct result of pollinator competition.

4.6 Reproductive strategy

Seed bank expression was studied for 14 *Chaetanthera* species and 4 *Oriastrum* species (ARROYO et al, 2006). The *Chaetanthera* species typically had persistent seed banks (where the seeds remain viable beyond the year of production), even the perennials *C. serrata* and *C. chilensis*. Only *C. renifolia* had a transient seed bank. The two annual *Oriastrum* species had small persistent seed banks (*O. pusillum*, *O. chilense*) while the two perennials had transient seed banks. Overall, the study concluded that, independent of phylogeny, the persistent seed bank size was a reflection of environmental selection rather than any trade-off with adult longevity, thus lowland arid habitats promoted the development of a large persistent seed bank.

Knowledge about the reproductive biology, e.g., ability to self-fertilise or pollinator specificity, of a species is important for understanding and predicting its responses to environmental change (after ADAM & WILLIAMS 2001). Studies of self compatibility and self pollination have shown an increase in the frequency of this syndrome over increasing elevations while other studies have found a corresponding increase in outcrossing systems (KLOTZ & KÜHN 2002). Some attribute self pollination to colonizing species (PRACH & PYŠEK 1999), but hybridisation and obligate outcrossing is said to allow genetic drift in small [founder] populations (e.g. CARLQUIST 1974), i.e. could be a more successful strategy. Autogamy is postulated to be an advantage in a short growing season where the pollinators are unpredictable or scarce.

Chaetanthera species seem to be highly self-compatible (autogamous). *C. euphrasioides* (ARROYO et al. 2006), *C. flabellata* and *C. glabrata*, (TORRES-DIAZ et al. 2007) all showed that their autogamy was independent of life cycle and pollinator abundance. This implies that another driving force, e.g. rapid reproductive maturation (ARROYO et al. 2006), lies behind the lowland acquisition of autogamy. Despite this evidence of high self-compatibility, there is also good evidence that breeding barriers are weak amongst some *Chaetanthera* taxa. *C. albiflora* and *C. linearis* hybridize frequently, as do *C. chilensis* and *C. elegans*. Morphological polymorphism is also evident in several species, such as *C. glabrata*.

O. apiculatum and *O. lycopodioides* (high altitude perennials) are highly self-compatible, although they have a mixed reproductive system (TORRES-DIAZ et al. 2007). High genetic diversity and heterozygote deficiency were taken to indicate self-compatibility in *O. chilense* (= *C. pusillum*) (TILL-BOTTRAUD et al. 2004). There are several gynodioecious *Oriastrum* taxa (*O. abbreviatum*, *O. polymallum*, *O. dioicum*, *O. pulvinatum*, *O. revolutum*, and *O. stuebelii*), although female capitula are often hard to find and whole plants are seldom entirely female. RENNER & WON (2001) state that a broad definition of gynodioecy [applicable to an Asteraceous capitula] grades into paradioecy when "the interbreeding of reciprocally sex-biased monoecious plants, at least if pure females are rare". In *Oriastrum* one could conclude that high elevation dioecy had evolved from monoecy, as for the Andean Siparunaceae (RENNER & WON 2001).

V The biogeography of *Chaetanthera* & *Oriastrum*

1 The distribution of *Chaetanthera*

1.1 Species distributions

Chaetanthera is distributed from about 10°S in the Peruvian Altiplano (one species) along the Andean cordillera, and throughout Chile to 41°S, where the Valdivian rainforest starts. Although there is some amplitude in longitudinal distribution of the taxa, a quirk of South American geology means that the longitude is not very useful when considering species distributions of *Chaetanthera*. Table 2 gives the altitude and latitude ranges and the countries of origin of all *Chaetanthera* species. Fig. 23 summarises the latitudinal ranges of all *Chaetanthera* taxa and their lowland or upland distribution. At a glance, there are three groups of species: Altiplano endemics, Chilean endemics, and species distributed throughout Chile and into parts of Argentina.

1.1.1 Species endemic to the Altiplano

The northernmost species is an high elevation taxon, *C. peruviana*. Eendemic to Peru, it is found on the western flanks of the Peruvian Andes, around the lower altitude limit of the Puna vegetation. *C. perpusilla*, found in Chile and Bolivia (type material only), is found on the southernmost end of the Altiplano.

1.1.2 Species endemic to Chile

Of the 30 species described in *Chaetanthera*, 20 are endemic to Chile. The Chilean endemics tend to colonize habitats with open sunny terrain, e.g. scrubland, road verges, or dry slopes, and variable substrates (sand, rubble, scree or soils). The Chilean taxa are most commonly found at lower elevations (below 2500 m.a.s.l.) below the Piso Subandino. The species found at higher elevations are restricted to the Subandean zone. There are 4 localised mid to high elevation endemics: *C. renifolia* and *C. flabellata* in the Andes around Santiago, and *C. kalinae* and *C. splendens* in the Coquimbo Andes. Lower elevation localised endemics include *C. glandulosa* from *Austrocedrus* woodlands south of the Rio Elqui, and the central valley taxon *C. ciliata*.

The Norte Grande, specifically the Lomas formations, supports three species: *C. glabrata*, *C. taltalensis* and *C. schroederi*. More than two thirds of *Chaetanthera* species are found in the Norte Chico and the Zona Central, establishing this taxon as a specialist in semi-arid and Mediterranean habitats. They are seldom found in forests or woods or darker, damper places. Even those species, such as *C. frayjorgensis*, which can be found in the Valdivian forest remnants of Fray Jorge, are not associated with the 'bosco', but are opportunistic invaders of the forest glades and margins.

Six species are found in the cooler, wetter Zona Sur. The three annual taxa (*C. microphylla, C. moenchioides* and *C. ciliata*) are widely distributed ephemeral competitors, stretching into their southern limit. All three are polymorphic in habit. *C. ciliata* can be considered a larger headed, more ciliate form of the related, more northern *C. multicaulis*. A pink colour morph of *C. ciliata* has also been reported (HERSHKOWITZ, 2002, pers. comm.). *C. microphylla*, with its broader bracts and red florets, is a more southerly descendant of the *C. linearis – C. albiflora* taxa. The widely distributed *C. moenchioides* also has a colour morph (white v. yellow) and the habit at the two extreme points of its range has led, in the past, to taxonomic distinction (*C. moenchioides* var. *sulphurea* F. Meigen; *C. australis* Cabrera).

Of the three perennial taxa, *C.* x *serrata* is typically a coastal and central valley inhabitant, characterized by stolons. *C. elegans*, related to *C.* x *serrata*, has deep woody branched roots, and inhabits the Lower Andean foothills in soils. *C. villosa*, also with deep branching roots, occurs in the volcanic screes and rubble of the Andes.

1.1.3 Species from Chile & Argentina

Eight species occur both in Chile and Argentina. *C. philippii, C. spathulifolia, C. euphrasioides* and *C. chilensis* are all sporadically recorded from the High Andes of the Argentinean provinces of La Rioja, San Juan and Mendoza. *C. microphylla, C. villosa,* and *C. elegans* can also be found in the Andean foothills in the provinces of Neuquén, and more seldom in the western areas of Río Negro. *C. moenchioides* is recorded at its northern and southern limits from Mendoza and Neuquén respectively. However, in most cases, *Chaetanthera* can be considered to be more of a tourist in Argentina than a native.

1.2 Species hotspots

Diversity hotspots allow one to consider the environmental conditions and niche expectations that bring closely and distantly related species together geographically. Radiation of species considers the variation and dispersal of closely related taxa. While there is a large radiation of closely related taxa focussed in the Norte Chico, Santiago is the centre of species radiations within *Chaetanthera*. In the Santiago area several lineages can be identified as radiating to the south and the north, particularly in the annual species. The lineages are clearly defined by suites of shared morphological traits and genetic similarity, excluding the possibility of confounding homoplasy and evolutionary parallelism.Although *Chaetanthera* is centred in Chile and does not form any affiliations with other phytogeographic associations, *Chaetanthera* has hotspots of species diversity. The centre of diversity for *Chaetanthera* lies between 29°S and 34°S. On average, more than 50% of all *Chaetanthera* taxa occur in this area at any given latitude. The diversity peaks around Norte Chico around La Serena and Combarbalá, and the northern end of the Valle Central, including the Santaginean basin.

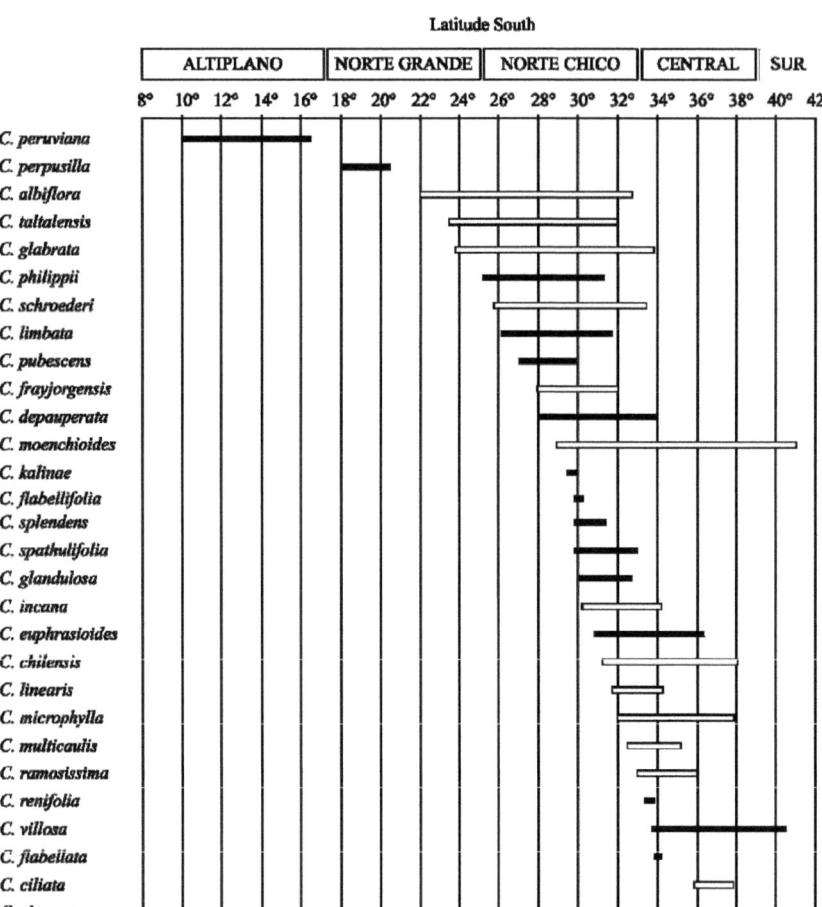

Figure 23: Summary of distributions of *Chaetanthera* species according to latitude range. White bar = mean elevation < 1500 m.a.s.l., black bar = mean elevation > 1800 m.a.s.l.

1.2.1 The diversity hotspot "Coquimbo" (29° - 30°S)

The flora of Coquimbo is incredibly diverse with over 1,700 species represented. Nearly 1,500 of these are native to Chile. The "Libro Riojo" (SQUEO ET AL., 2001) describes an area "situated as it is on the crossover zone between the Mediterranean region and the semi-arid/arid desert region".

Taxon	Altitude (min.)	Altitude (average)	Altitude (max)	Latitude (N)	Latitude (S)	Country
C. perpusilla	2200	2975	3300	18.00	20.50	Bolivia Chile
C. peruvuiana	2700	3150	3800	10.00	16.50	Peru
C. albiflora	0	455	1000	22.00	32.75	Chile
C. taltalensis	20	310	700	23.50	32.00	Chile
C. glabrata	0	553	2200	23.83	33.78	Chile
C. schroederi	265	1616	3200	25.77	33.43	Chile
C. limbata	800	1324	2200	26.13	31.72	Chile
C. pubescens	600	1874	3500	27.45	30.17	Chile
C. depauperata	700	1863	2900	28.00	34.00	Chile
C. frayjorgensis	0	437	2200	28.00	32.00	Chile
C. kalinae	2300	2700	3100	29.50	30.00	Chile
C. flabellifolia	2800	3060	3900	29.78	30.20	Chile
C. splendens	2300	2781	3250	29.83	31.40	Chile
C. glandulosa	1000	1973	3000	30.13	32.75	Chile
C. incana	0	285	1000	30.25	34.17	Chile
C. linearis	150	1289	2300	31.75	34.25	Chile
C. multicaulis	50	600	1500	32.50	35.17	Chile
C. ramosisima	120	875	1700	33.00	36.00	Chile
C. renifolia	2000	3158	4000	33.33	33.83	Chile
C. flabellata	1600	2121	2600	33.83	34.17	Chile
C. ciliata	60	154	1200	35.83	37.83	Chile
C. x serrata	0	200	800	36.50	38.00	Chile
C. philippii	2500	3098	3950	25.17	31.33	Argentina (N) Chile
C. spathulifolia	2800	3232	3720	29.83	33.00	Argentina (N) Chile
C. euphrasioides	1470	2370	3400	30.83	36.33	Argentina (N) Chile
C. chilensis	0	1030	2900	31.25	38.00	Argentina (N) Chile
C. moenchioides	10	953	2300	29.00	41.00	Argentina (N)(S) Chile
C. microphylla	10	699	1520	32.00	37.83	Argentina (S) Chile
C. villosa	1000	1972	3000	33.67	40.50	Argentina (S) Chile
C. elegans	0	930	2000	36.00	40.00	Argentina (S) Chile

Table 2: Altitudinal ranges (minimum, maximum and average recorded altitude m.a.s.l.), latitude ranges (N: northernmost recorded latitude; S: southernmost recorded latitude) and country of origin of *Chaetanthera* taxa. Argentina (north) indicates records from La Rioja, San Juan and Mendoza). Argentina (south) indicates records from Neuquén and Rio Negro).

This encompasses a broad spectrum of biogeographic elements, numerous genera that have experimented with evolutionary radiations and at the same time relicts from the southern forests that once covered Chile. Two thirds of the native taxa recorded from Coquimbo are found at their northern limit.

This hotspot is characterized by unusual and interesting *Chaetanthera*, representing both specialisation and radiation at various elevations. The coastal sands and steppes in the Región de Coquimbo support annual species: *C. glabrata*, *C. schroederi*, *C. frayjorgensis*, *C. taltalensis*, *C. albiflora*, and *C. incana*. These ephemeral taxa escape the seasonal aridity by having an annual life cycle. Further adaptations to the arid conditions include having a densely glandular or lanate indumentum, and subsucculent leaves. Both features allow for water storage and salt tolerance. Several of these taxa also demonstrated the ability for rapid germination (1 – 3 days). The Andean foothills further inland support *C. depauperata*, *C. limbata*, *C. pubescens* and *C. glandulosa*. The densely glandular indumentum of *C. glandulosa* and its reduced, linear sclerophyllous leaves (reduced further in *C. glandulosa* var. *microphylla*) are all indicators of adaptation to seasonal aridity. Woodiness is a repeated element of the perennial species in *Chaetanthera* but is more commonly expressed as rhizomes, or woody stems buried in an unstable or loose substrate. The perennial species that favour woody rhizomes are often found at much higher altitudes (2400 – 4000 m). Examples from Coquimbo include *C. philippii* and *C. spathulifolia*. The Cordillera de Doña Ana and adjacent massifs also support the annuals *C. kalinae* and the closely related *C. splendens* and *C. flabellifolia*.

1.2.2 The diversity hotspot "Santiago" (33° - 34°S)

Santiago, in this instance defined as the area between 33° and 34°S, is frequently seen as a hotspot of diversity and radiations within Chilean genera (*Haplopappus*, *Calceolaria*, *Leucheria*, and *Perezia*). The distance from the coast to the highest peaks in this area is a mere 180 km, but the land rises to over 5000 m.a.s.l. As a consequence the vegetation and climate zones are short (vertically), allowing a high concentration of species to exist over a short geographical distance. The geomorphology of this location also uniquely provides a bridge between the coastal cordillera and the Andes (EHRHART 2005).

The herbaceous annuals of *Chaetanthera* subgenus *Chaetanthera* form the core of the diversity hotspot around Santiago (33°S), comprising 12 of the 14 species found there. The position this location plays in the overall distribution of the species recorded from Santiago is varied. For some taxa (e.g. *C. linearis*, *C. depauperata*, *C. schroederi* and *C. glabrata*) the Cuenca de Santiago represents the southernmost limit of their distribution. For apparent generalists such as *C. moenchioides* and *C. euphrasioides*, the Cuenca is a bridge linking the northerly Andean foothills and the lower, more humid elevations of the Central valley. *C. microphylla, C. ramosissima* and *C. multicaulis* are at their northern limit around Santiago. The two perennials, *C. chilensis* and *C. villosa*, are also mainly distributed southwards from this area. *C. renifolia* and *C. flabellata* are locally endemic. Although *C. incana* is distributed somewhat northwards of 33°S, it is restricted to localities along the Valparaíso coast, and not in the landscape around the Metropolitan conurbation.

1.3 Trichome diversity: evidence of species radiation

Distributed along the coast of Chile, in the foothills of the Andes and to Santiago, there is a complex of six species. *C. glabrata*, *C. limbata*, *C. pubescens*, *C. schroederi*, *C. kalinae* and *C. frayjorgensis* are united by having indistinctly petiolate leaves with ovate to orbicular or cordate laminas with limbate margins. The petiole margins have minute sessile to stalked glandular trichomes, not to be confused with the glandular trichomes seen on the leaf and bract surfaces. The species share similar habit with creeping to ascending spreading flowering branches originated from a short (to 5 cm) erect stem. The leaves cluster in a rosulate fashion below the capitula. The outer involucral bracts are foliaceous, with reflexed laminas. The ray florets are yellow, and showily exerted. The number of ray florets can be very variable both within and between species, and the phenotypic plasticity of the leaves of *C. glabrata*, for example (see Chapter V), can be directly attributed to vegetative output dependant on the seasonal rainfall.

However, these species can be clearly defined according to their lamina indumentum. *C. glabrata* is typically glabrous. *C. schroederi* is conspicuously lanate in the leaf axils; with filamentous hairs each comprising several stacked elongated cells. *C. frayjorgensis* is distinctive because of its densely black-brown glandular indumentum, especially on the laminas and involucral bract surfaces. The capitula buds are apiculate and not spherical as in *C. glabrata*. *C. kalinae* is lanate on all surfaces. The leaves are cordate to nearly reniform. *C. limbata* and *C. pubescens* both have lanate hairs and glandular trichomes on their leaves. Additionally, *C. pubescens* has a dense, nearly farinose indumentum of simple clavate hairs on the leaves. Fig. 24 illustrates the distribution of these species in Chile as identified by their characteristic leaf shape and indumentum.

Several types of distribution can be identified within this group. *C. glabrata* is widely dispersed along the coastal belt of northern Chile, and disjunctly around Santiago. *C. kalinae* has a small isolated distribution in the high Andes of Coquimbo. *C. frayjorgensis* is centred on the coastal areas of La Serena and Fray Jorge, although it also penetrates inland. *C. frayjorgensis* overlaps *C. glabrata* in the southern part of *C. glabrata*'s coastal distribution. *C. schroederi* has a punctuated distribution along the mid-elevations of the semi-arid Chilean Andes, further inland and at consistently higher altitudes than *C. glabrata*. It is sympatric with *C. glabrata* around Santiago. *C. limbata* occurs in several patchy populations where both *C. glabrata* and *C. frayjorgensis* occur, but also in localities further north than the latter is found; refuting any hypothesis that *C. limbata* might be a spontaneous hybrid. *C. pubescens* is found at higher altitude localities more typical of *C. kalinae*. *C. limbata* is morphologically closest to *C. pubescens*, although they are allopatric in their distribution. Where *C. limbata* and *C. pubescens* occur together, the latter is typically at higher latitudes. *C. pubescens* has smaller distributional amplitude than *C. limbata*.

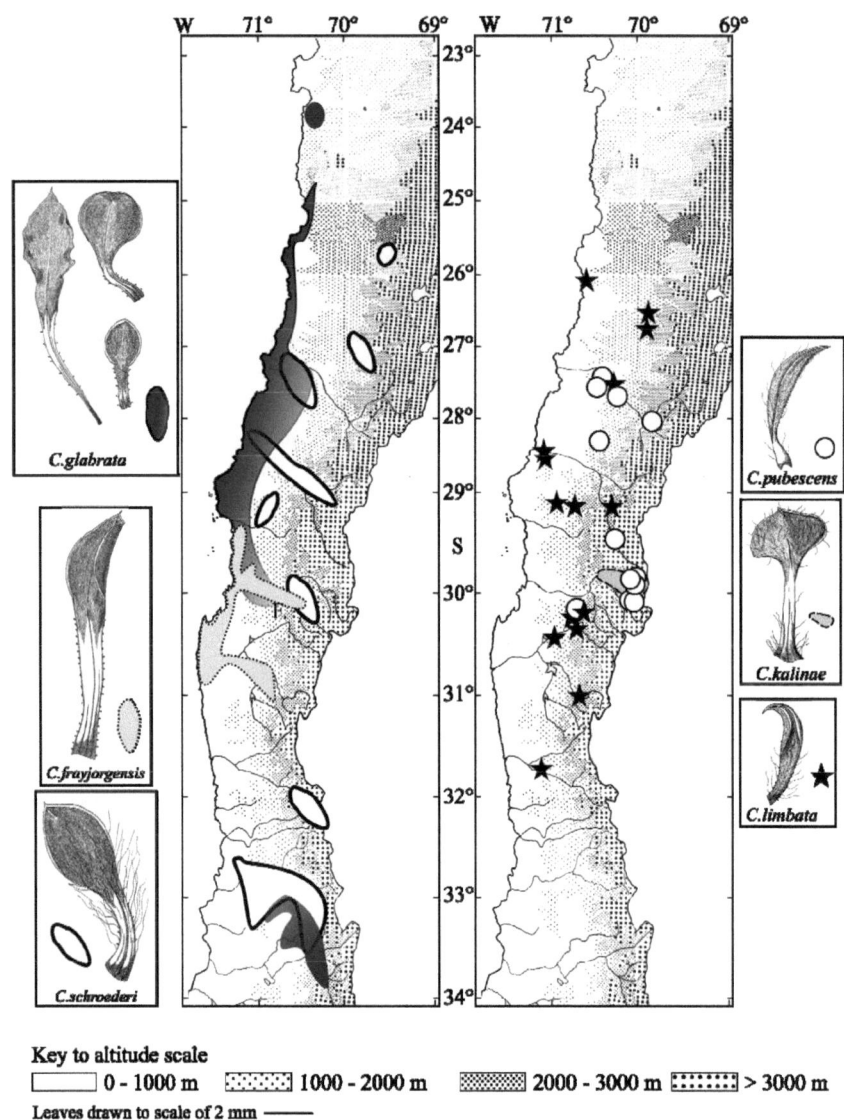

Figure 24: Distribution and leaf indumentum diversity of six *Tylloma* species. *C. glabrata* (no hairs); *C. schroederi* (lanate villous especially in leaf axils and on petioles); *C. frayjorgensis* (lamina usually glabrous, involucral bracts always densely glandular); *C. limbata* (lanate hairs and scattered glandular hairs); *C. kalinae* (lanate on lamina); *C. pubescens* (lanate, glandular and clavate hairs giving mealy appearance). Scale bar = 2 mm

It seems that each species occupies a subtly different niche, implying some form of adaptive radiation. It could be postulated that the area around Baños del Toro represents the centre (focus) of radiation because it has proportionally the highest number of related taxa found together. Assuming that the evolution of each hair type only happens once in this group, then the upland species *C. limbata* (with simple lanate and glandular hair types), *C. pubescens* (with simple lanate, clavate, glandular hair types) and *C. kalinae* (simple lanate hairs) represent one radiation. *C. schroederi* (with compound lanate hairs) is spread further from the radiation focus but the distribution implies a relict event whereby this species became isolated at particular elevations. In this scheme, *C. frayjorgensis* would be a reduced form of *C. limbata* with only glandular hairs, while *C. glabrata* would be the hypothetically most recent evolution with complete loss of all lamina trichome types.

The unusual chromosome complement recorded for *C. glabrata* and *C. frayjorgensis* (2n = 4x-6 = 38) indicates some kind of founder event for this complex. The taxa are extremely well adapted to the semi-arid environment, with their rapid germination ability, myxogenic pericarp/achene hairs, robust testa epidermis, hydrologically stable pollen structures, highly plastic vegetative and reproductive phenotypes capable of rapid response to unpredictable precipitation – a commodity in the geologically modern arid and semi-arid landscape that the plants inhabit.

In conclusion, I interpret this complex as an example of a very recent radiation event, based on a founder event that occurred at higher elevations, with rapid expansion at lower elevations as a result of environmental adaptation to semi-arid conditions.

1.4 Lamina morphology & genetic affinity: evidence of species radiations

Species in *Chaetanthera* subgenus *Chaetanthera* can often be placed suites of morphologically similar taxa. The DNA phylogeny identified two groups of closely related taxa that can be morphologically defined as the "linear – leaved" group and the "dentate-ciliate – leaved" group. The leaves define the phylogenetic grouping and the capitula morphology defines the species. The capitulum illustrations are used here because they show both the leaf and capitula features.

1.4.1 Linear-leaved group

The species in the "linear leaved group" are defined by the possession of linear entire leaves. It consists of five morphologically similar species (*C. albiflora, C. linearis, C. depauperata C. microphylla* and *C. perpusilla*), all with differing floret colour. Their distributions radiate out from Santiago: *C. depauperata* (white rays) spreads northwards from the Cuenca de Santiago along the cold temperate zone in the Andean foothills below 3000 m, while *C. microphylla* (orange-red, seldom yellow rays) is distributed around the Cuenca de Santiago and southwards, below 1500 m (Fig. 25). *C. linearis* (yellow rays) spreads northwards across the Santaginean bridge along the coast, overlapping with the more northerly coastal species *C. albiflora* (Fig. 26) (white rays with blue-green dorsal stripe) between 32° – 33°S (Petorca). *C. perpusilla* (white rays) is found in the far north of the Chilean Atacama desert.

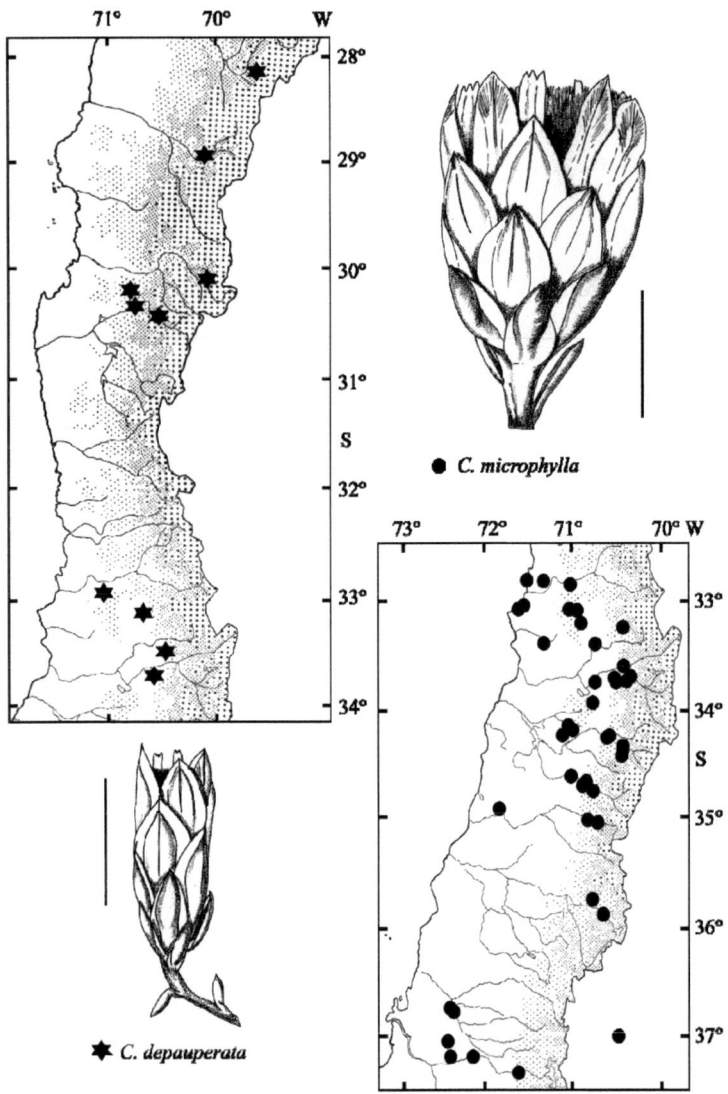

Figure 25: Capitula morphology and distribution of *C. microphylla* and *C. depauperata*. Scale bar = 5 mm.

V The biogeography of *Chaetanthera* & *Oriastrum*

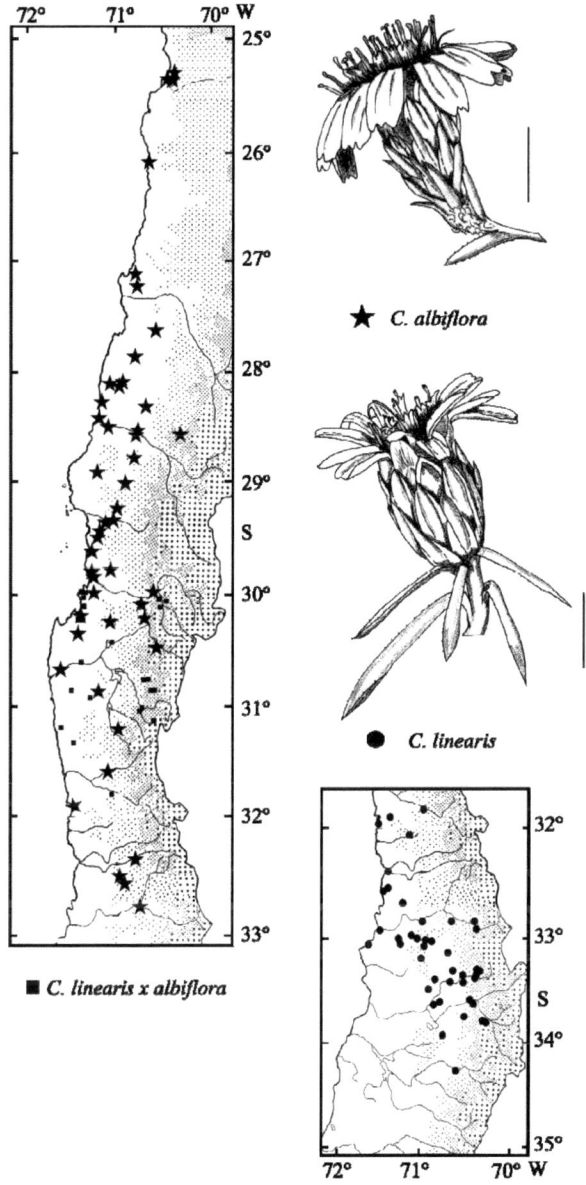

Figure 26: Capitula morphology and distribution of *C. albiflora* and *C. linearis*. Scale bar = 5 mm.

1.4.2 Dentate-ciliate leaved group.

The "dentate-ciliate leaved group" of annual species comprises *C. ramosissima, C. moenchioides, C. incana, C. ciliata, C. multicaulis, C. taltalensis* and *C. peruviana*. They are widely dispersed along the Pacific seaboard of South America The geographic amplitude stretches from the northernmost bastion of *Chaetanthera* in Peru (*C. peruviana*) to its southernmost limit around 42°S in Valdivia (Chile) and Río Negro (Argentina). It is phylogenetically younger than the linear-leaved group, but has more varied evolutions.

The perennial species include the unusual local endemic *C. glandulosa* and a complex of taxa (*C. chilensis, C. elegans* and *C. x serrata*) of which only *C. chilensis* is found around Santiago. The genetic nrDNA places *C. glandulosa* in this lineage, although with its linear, mucronate leaves it seems closer to the "linear-leaved" group. The lanceolate to ovate, yellow-green tipped sclereid bracts are reminiscent of the inner involucral bracts of species in the dentate-ciliate leaved group. Morphologically, *C. incana*, with its somewhat larger capitula than the other annuals, and its spathulate leaves looks very similar to the perennials.

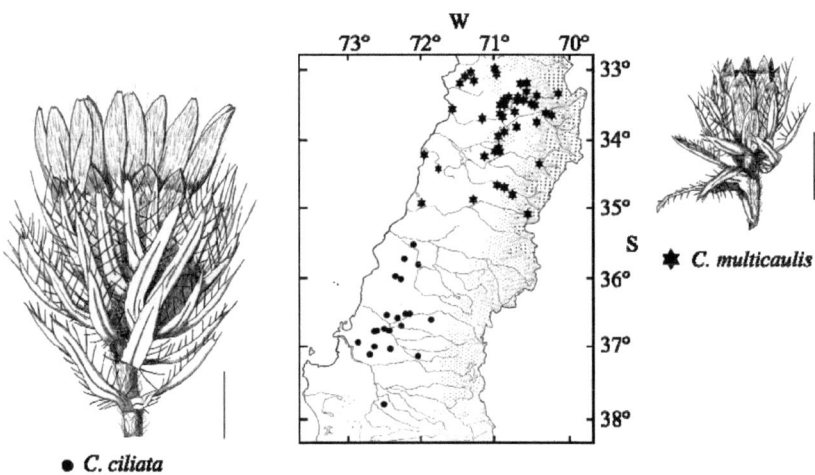

Figure 27: Capitula morphology and distribution of *C. ciliata* and *C. multicaulis*. Scale bar = 5 mm.

C. ciliata and *C. multicaulis* form a pair of closely related species. *C. multicaulis* is found in the northern drier matorral around Santiago, while *C. ciliata* is more or less confined to the southern, wetter matorral towards Concepción (see Fig. 27). *C. multicaulis* is much smaller, has fewer series of involucral bracts, which are also more leafy in appearance, the leaves are dentate to spinose but not ciliate, and plants flower principally between October and November (early spring). *C. ciliata* has much larger capitula, more series of awned, sclerophyllous involucral bracts and the stem leaves tend to be more or less ciliate. The plants flower principally between December and January (summer). These two taxa have a compound glandular trichome in common with the geographically distant *C. peruviana*.

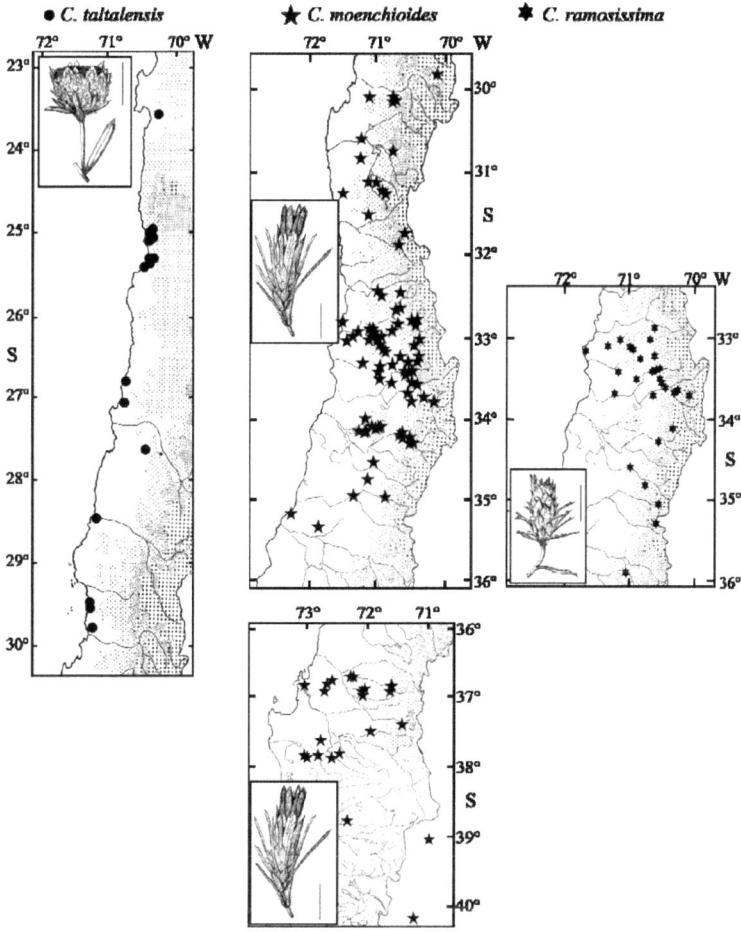

Figure 28: Capitula morphology and distribution of *C. taltalensis*, *C. ramosissima* and *C. moenchioides*. Scale bar = 5 mm.

C. ramosissima, *C. taltalensis* and *C. moenchioides* form a triad of morphologically close species and are illustrated in Fig. 28. *C. moenchioides* appears to be a highly successful generalist species of the mediterranean type habitat. White-flowered *C. moenchioides* is found disjunctly, from Concepción (Chile) to San Carlos de Bariloche in Argentina. *C. ramosissima* is distributed within the range of yellow-flowered *C. moenchioides* around Santiago. *C. taltalensis* stretches along the arid coastline from Coquimbo northwards, taking advantage of the fog banks.

2 The distribution of *Oriastrum*

2.1 Species distributions

Oriastrum is distributed from about 11°S in the Peruvian Altiplano in Lima and Junín, in the Altiplano to the south of Lake Titicaca in Peru, Bolivia, Chile and Argentina (summer rains), and in the high elevation Altoandino habitats along the Chilean and Argentinean Andes to 35°S (winter precipitation, snow melt).

Taxon	Altitude (min.)	Altitude (average)	Altitude (max)	Latitude (N)	Latitude (S)	Country
O. abbreviatum	3400	3982	4400	23.92	27.67	Argentina
O. famatinae	3340	3650	4000	28.50	29.00	Argentina
O. tonatalensis	3000	3635	4000	31.25	34.75	Argentina
O. tarapacensis	4100	4314	4600	18.00	20.00	Chile
O. werdermannii	800	840	880	29.92	29.97	Chile
O. apiculatum	2000	2963	3600	32.25	34.00	Chile
O. pusillum	2300	2858	3400	33.25	33.90	Chile
O. cochlearifolium	4850	4960	5100	11.00	12.50	Peru
O. pulvinatum	3000	4100	5100	23.00	33.00	Argentina Chile
O. gnaphalioides	2000	3309	4300	24.67	33.33	Argentina Chile
O. acerosum	2400	3498	4300	28.67	31.50	Argentina Chile
O. achenohirsutum	3760	3989	4080	29.00	30.00	Argentina Chile
O. lycopodioides	2000	3022	3900	29.83	33.83	Argentina Chile
O. dioicum	2500	3360	4320	30.00	34.00	Argentina Chile
O. chilense	2150	2832	3800	33.00	34.00	Argentina Chile
O. polymallum	1900	4307	5500	21.00	30.50	Argentina Bolivia Chile
O. revolutum	3680	4409	4800	21.25	25.50	Argentina Bolivia Chile
O. stuebelii	3800	4588	5200	16.00	23.00	Argentina Bolivia Chile Peru

Table 3: Altitudinal ranges (minimum, maximum and average recorded altitude m.a.s.l.), latitude ranges (N: northernmost recorded latitude; S: southernmost recorded latitude) and country of origin of *Oriastrum* taxa.

Species in *Oriastrum* (with one exception) are confined to these high elevation habitats found above 2000 m.a.s.l. and most typically between 3000 – 5000 m.a.s.l. The longitude range of many species is only a matter of 1 – 2 degrees, most commonly between 69°W - 71°W, whereas the latitude amplitude is diverse, and a good indicator of dispersion. Geographical endemism is often an underpinning criterion for assessing species importance in global biodiversity terms. However, in the case of *Oriastrum*, a genus whose natural habitat straddles a mountain range with several geopolitical allegiances, geographical endemism is not very meaningful, as shown below (Table 3).

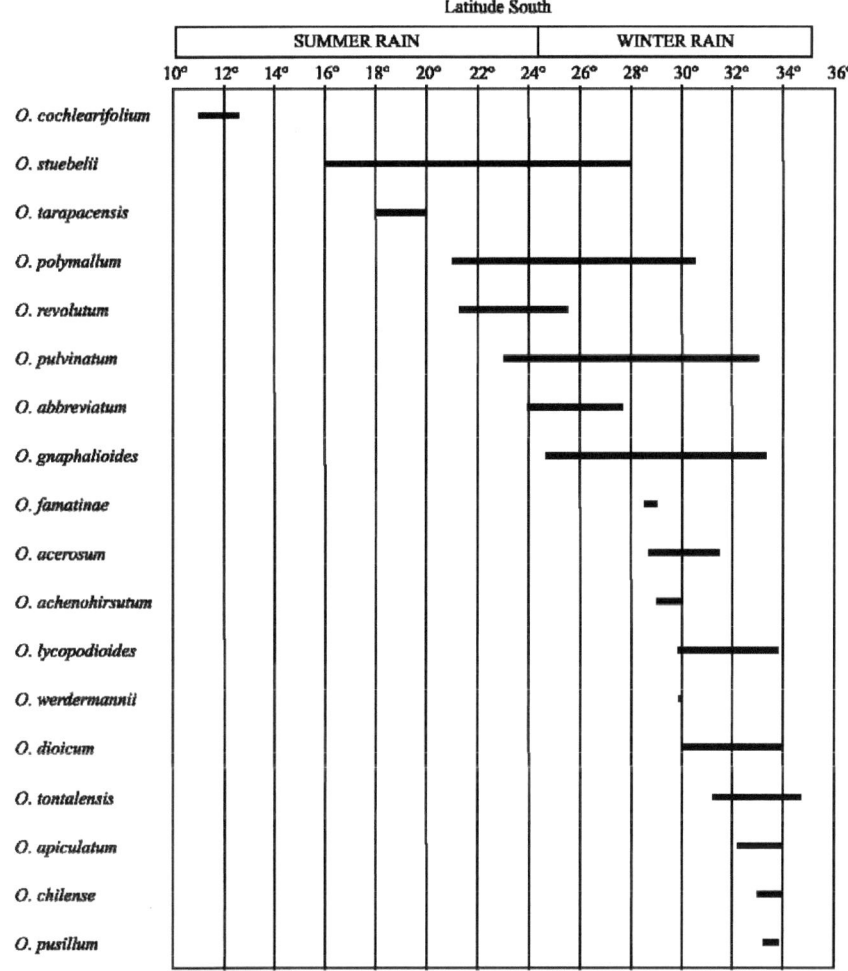

Figure 29: Summary of distributions of *Oriastrum* species according to latitude range.

2.1.1 Species endemic to either Argentina or Chile or Peru

There are several species that are so far only recorded from either Argentina or Chile or Peru. However, this does not rule out the possibility of distributions beyond the political geographic boundaries. *O. tontalensis, O. famatinae* and *O. abbreviatum* are restricted to Argentina. The latter two are typically found in the Cordilleras to the East of the main Andean chain. *O. apiculatum, O. pusillum, O. tarapacensis* and *O. werdermannii* are considered endemic to Chile. Of these four, only *O. werdermannii* is localised well within the Chilean geopolitical boundary, at lower elevations of the Andean Pre-cordillera, and therefore would be unlikely to be collected from other countries. *O. cochlearifolium* is endemic to Peru, and quite disjunct from the other Peruvian-Bolivian species *O. stuebelii*.

2.1.2 Species from Argentina & Chile

O. acerosum, O. achenohirsutum, O. chilense, O. dioicum, O. gnaphalioides, O. lycopodioides and *O. pulvinatum* are found in both Argentina and Chile. Their distributions cross the high passes in the Central Andes. *O. dioicum* and *O. pulvinatum* are the only two that are more commonly recorded from the eastern side of the Andes, i.e. Argentina.

2.1.3 Species from Argentina, Chile, Bolivia & Peru

O. polymallum O. revolutum and *O. stuebelii* are the most widespread species, being recorded from Argentina, Chile, Bolivia and Peru, although they are not necessarily the most numerously collected.

2.2 Species hotspots

Throughout the distribution of the genus between 28°S and 34°S it is possible to locate 5-7 of a possible 18 species at any point of latitude (see Fig. 29). A number of species can be found in and around the mountains of the Valle del Elqui and on the Argentinean (Prov. San Juan) flank of the Andean Cordillera, which lie between ca. 29°30'S and 30°30'S. West of the Cordillera there is one low elevation Chilean endemic (*O. werdermannii,* ca. 800 m). *O. dioicum* and *O. lycopodioides* are at their northern limit, at around 3500 m elevation. *O. achenohirsutum* and *O. polymallum* are at their southern limit, around 4000 m elevation. *O. acerosum, O. gnaphalioides,* and *O. pulvinatum* are all distributed to the north and south of the Cordillera de Doña Ana. Although the elevation ranges overlap, there is a tendency for the different species to be dispersed at differing altitudes. *O. gnaphalioides* is found between 2000 – 4000 m, although this is the only latitude where *O. gnaphalioides* descends below 3000 m altitude. *O. acerosum* is generally found between 3000 – 4000 m and *O. pulvinatum* between 3500 – 4500 m.

When assessing the possible diversity hotspots, or geographical concentrations of taxa, it is important to consider the environmental conditions and niche expectations that bring the taxa together geographically. The mountains in the North of Coquimbo and San Juan are hotspots for floristic diversity. The edaphic and ecological factors important in driving the floristic diversity in

V The biogeography of *Chaetanthera* & *Oriastrum*

the mountains around the Valle de Elqui are discussed by SQUEO et al. (2006). In general, the ability to tolerate cold is thought to determine habitation of the different vegetation belts, while edaphic microclimate factors give the vegetation a high spatial heterogeneity. Biotic factors such as plant – pollinator relationships are also considered to be highly relevant to distribution.

Given the linear, nature of the geographical range of this genus, it is hardly surprising that hotspots are not easy to describe. On the one hand, the extreme cold, dehydration, insolation, and short growing season severely limit the potential of the higher elevation Andean environment as a habitable habitat. On the other hand, the concatenation of related groups of species along the Andean zones of the Altiplano, the Altoandino and the Andino indicate that despite the severity of the environment adaptive fitness means that vicariant dispersal throughout the habitat is possible.

2.3 Species radiations

Radiation of species considers the dispersal of closely related taxa. Taxa in *Oriastrum* subgenus *Oriastrum* are concentrated around the Andes of Santiago in Chile. Three groupings can be identified within *Oriastrum* subgenus *Egania*: the Altiplano group (*O. cochlearifolium, O. stuebelii* and *O. abbreviatum*), the Altoandino group (*O. tarapacensis, O. revolutum, O. famatinae, O. tontalensis, O. polymallum* and *O pulvinatum*), and the Andino group (*O. dioicum, O. acerosum, O. achenohirsutum* and *O. apiculatum*). A grey-scale satellite map is used as the basic background for showing the distributions. Dark areas are graded as low elevation and white as the permanent summer snowline in the high elevation Andes. In this way the high elevation zones are highlighted. These are some of the few maps that show the detail across the geopolitical boundaries.

2.3.1 Santiago

The species in *Oriastrum* subgenus *Oriastrum* show a localised radiation around the eastern Andes of Santiago (*O. chilense* and *O. pusillum*) (see Fig. 30). These two morphologically very similar species are also sympatric with *O. lycopodioides* – a quite widely dispersed species that shows polymorphism in its size and the extent of its indumentum. *O. gnaphalioides* is also a highly polymorphic species. In this case the anecdotal evidence (KALIN, pers. comm.) seems to support a phenotypic plasticity in overall size (leaves, florets, capitula) possibly correlated with water supply. The greater the elevation, the more water there is available and the larger the plants tend to be. Thus, the somewhat larger but lower elevation *O. werdermanni* can be considered a different entity because its size contradicts the environmental hydrological gradient.

2.3.2 Altiplano group

O. cochlearifolium is the only species that occurs somewhat disjunct from the rest. It is found between 11°S and 12°30'S on the eastern flank of the Cordillera Occidental in Peru. It is considered to be most closely related to *O. stuebelii*, a white-flowered, highly polymorphic, trailing dwarf herb found from the Altiplano to the Piso Andino Inferior in Bolivia and Argentina. *O. stuebelii* is found on shrubby or grassy (*Lepidophyllum*) slopes in loose sandy soil or stony/rocky ground ("cantizal"). It is one of the few *Oriastrum* species that seems to be able to take advantage of the different precipitation regimes across its distribution (summer rains in the north, and winter rains in the south). *O. abbreviatum* occurs along the far eastern edge of the Argentinean Cordillera. The distribution of these three species is shown in Fig. 31.

2.3.3 Altoandino group

O. tarapacensis, O. revolutum, O. famatinae, O. tontalensis, O. polymallum and *O pulvinatum* occur sympatrically between 21° and 33°S, and are restricted to the Piso Andino Superior and Inferior. The comparative distributions are given in Fig. 32. They are all dwarf cushions with dense whorls of leaves around the stems, densely pubescent and are mostly gynodioecious. *O. tarapacensis* and *O. tontalensis* are two newly described taxa occurring at the north-eastern and southern extremes of the group's distribution respectively. *O. tarapacensis* lies on the boundary between the desert belt and the High Andean formation of the north. It is restricted to the elevations north of Iquique on the Liparite plateau on which the northern Chilean volcanic formations are found and is clearly related to *O. revolutum,* which inhabits the tundra vegetation surrounding the high altitude lakes on the plateau east of the Salar de Atacama. *O. tontalensis* occurs in damp high "alpine" meadows at the southern end of the Argentinean distribution of *O. pulvinatum*. All the taxa in this group are found around the permanent snow line in rocks, screes and moraine. The dominant vegetation types on these gravelly gelifluxional slopes are dry and open tussock-grasses with low bushes and and cushion-like subshrubs (*Stipa* spp. and *Fabiana* spp.) (NAVARRO 1993; TEILLIER 1999). *O. polymallum* is characteristic of the association *Nototricho auricomae-Chaetantheretum sphaeroidalis* NAVARRO (1993).

2.3.4 Andino group

The "Andino" group has four species (Fig. 33). Three of them occur almost exclusively on the western flank of the Central Andes (*O. acerosum, O. achenohirsutum* and *O. apiculatum)*, while *O. dioicum* is more or less restricted to the eastern flank. *O. dioicum,* the only gynodioecious species in this group, bridges the distribution gap between the more southerly *O. apiculatum* and the more northerly *O. acerosum. O. apiculatum* and *O. acerosum* are morphologically very close at this point, but are separated by a clear disjunction over the Cordillera de Ansilta (ca 32°S). *O. apiculatum* is relatively restricted in its distribution, occurring between 32°30'S and 34°S, south of the Ventana de Horcones massif around the Laguna del Inca and continues along the west flank of the Andes to the Cordillera de San José. It grows in steep scree and moraine, on dry slopes above the snow line, but free of snow in summer. The distribution over the elevation range within the Piso Andino (sur) can be seen in adjacent valleys (Río Aconcagua – Río Molina – Río Colorado – Río Maipo). *O. acerosum* is polymorphic, particularly at the northern end of its range. Variation is seen both in size and the achene hairs. *O. achenohirsutum* is recorded from both Chile and Argentina, but is localised in the Cajón de la Brea (Argentina) and the Cuencas de Laguna Grande and Laguna Chica (Chile). Clearly related to *O. acerosum* and *O. apiculatum,* it has distinctly spathulate leaves, and is not very polymorphic.

V The biogeography of *Chaetanthera* & *Oriastrum*

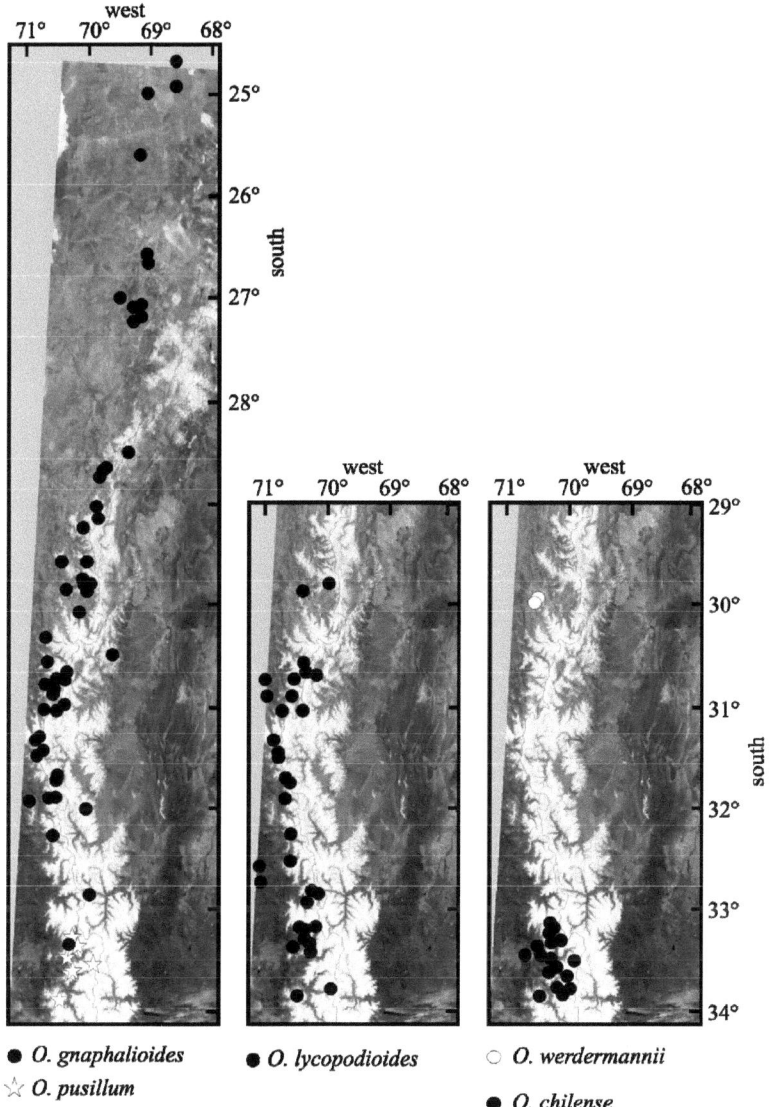

- ● *O. gnaphalioides*
- ☆ *O. pusillum*
- ● *O. lycopodioides*
- ○ *O. werdermannii*
- ● *O. chilense*

Figure 30: Distributions of the species in the "Santiago" group (*O. chilense, O. gnaphalioides, O. lycopodioides, O. pusillum, O. werdermanni*). Credit: Satellite image map derived from NASA GSFC (True colour image, MODIS, 05.02.2001) (http://visibleearth.nasa.gov/). The grey mask indicates the boundary of the satellite image.

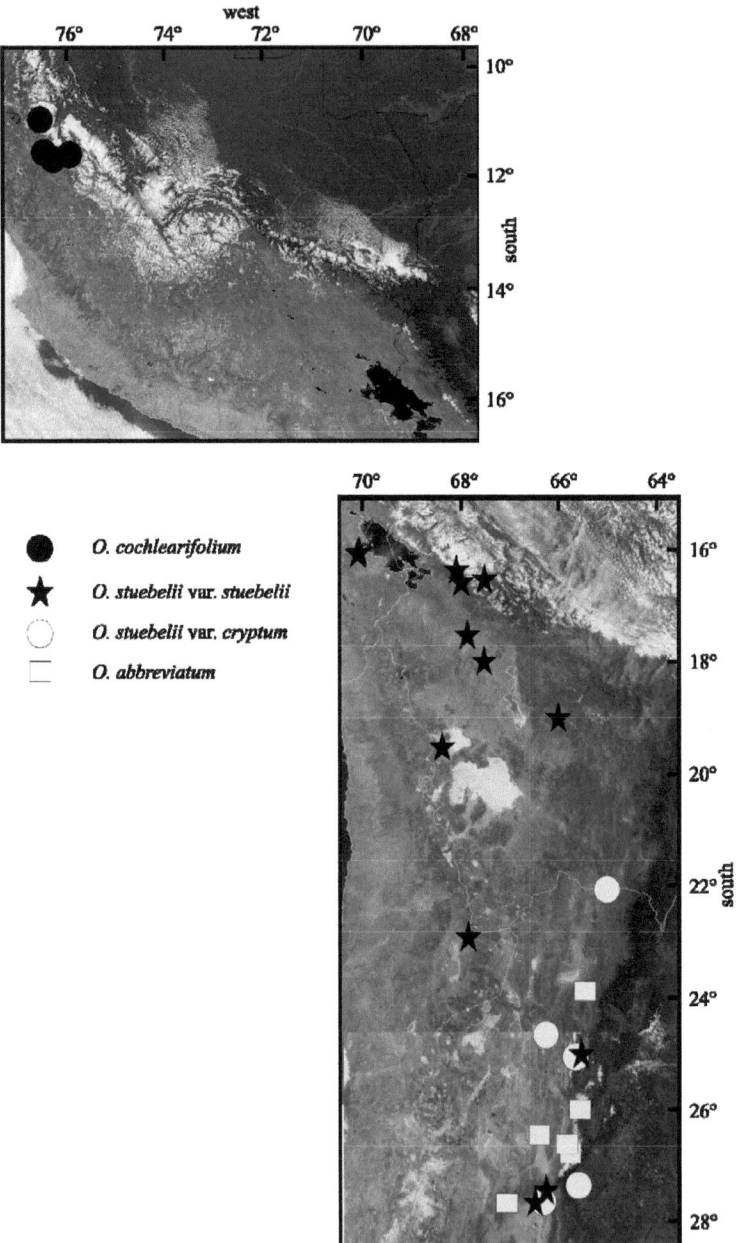

Figure 31: Distributions of the three species in the "Altiplano" group (*O. cochlearifolium*, *O. stuebelii* & *O. abbreviatum*). Credit: Composite satellite image map derived from NASA GSFC (True colour image, MODIS, 05.02.2001 & 30.10.2001) (http://visibleearth.nasa.gov/).

V The biogeography of *Chaetanthera* & *Oriastrum*

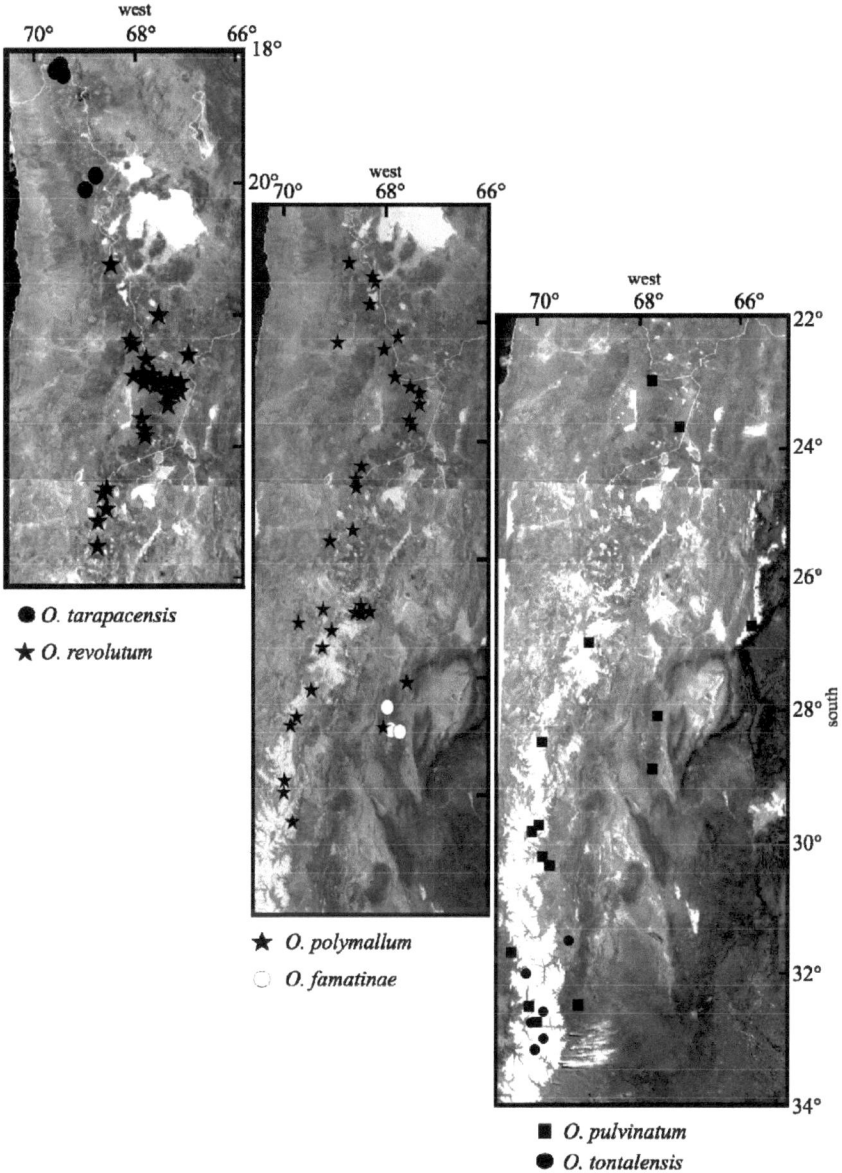

Figure 32: Distributions of the six species in the "Altoandino" group (*O. tarapacensis, O. revolutum, O. famatinae, O. tontalensis, O. polymallum & O pulvniatum*). Credit: Composite satellite image map derived from NASA GSFC (True colour image, MODIS, 05.02.2001 & 30.10.2001) (http://visibleearth.nasa.gov/).

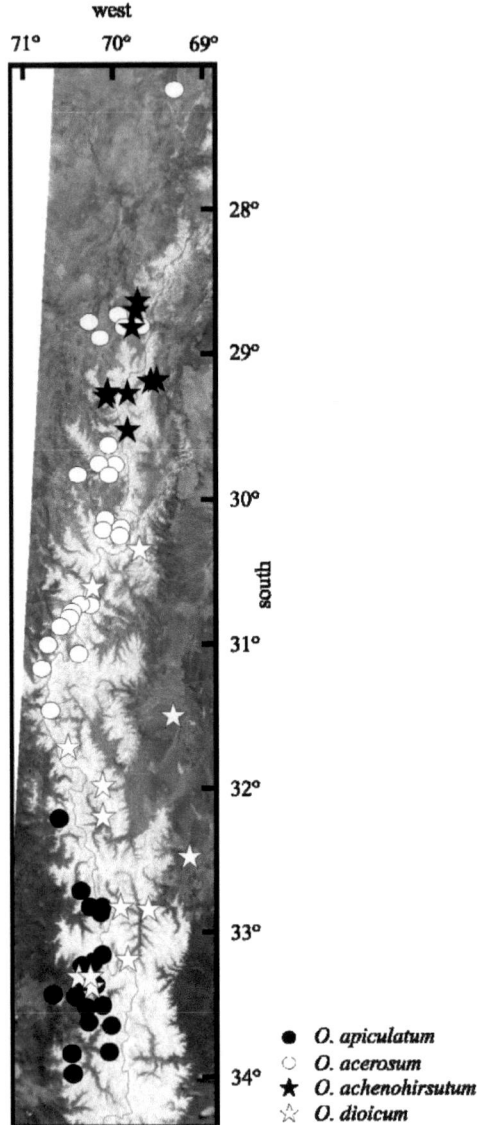

Figure 33: Distributions of the four species in the "Andino" group (*O. acerosum*, *O. achenohirsutum*, *O. apiculatum* and *O. dioicum*). Credit: Composite satellite image map derived from NASA GSFC (True colour image, MODIS, 05.02.2001 & 30.10.2001) (http://visibleearth.nasa.gov/).

VI "The variety of all things forms a pleasure"

EURIPIDES *Orestes* 234

1 Introduction

Systematics and biogeography can be brought together to investigate patterns of isolation and speciation (STUESSEY et al. 2003), and in particular to investigate patterns of variation that are not supported by the current taxonomy. This chapter highlights three case studies concerning species of *Chaetanthera*. Some suites of species in the genus *Chaetanthera* are phenotypically very plastic and consequently species delimitation is blurred. Often, a significant number of collections could only be described as intermediate between two or more species.

Two important factors were found to play a significant role in explaining the observed variation between and within taxa. Firstly, the climatic phenomenon of the El Niño is a key issue for the Chilean flora, especially in the more northerly latitudes of the country along the Pacific coastal strip. The response of plants to this quasi-periodic phenomenon has been studied (SQUEO et al. 2006, LÓPEZ et al. 2006), although there seem to be few documented examples where phenotypic plasticity has been directly attributed to El Niño events.

The second factor is the ability of *Chaetanthera* species to hybridise. Interspecific hybridisation is thought to be common among plants, especially angiosperms and ferns (MCDADE, 1992). ESTABROOK et al. (1996) provide a précis of the techniques used to identify and present hybrids. The simplest and earliest ordination technique was pioneered by ANDERSON (1949) who used pictorialized diagrams and hybrid indices to analyse hybridisation and introgression events. A more complex approach using multivariate analysis (discriminant functions) is also a useful technique, relying on inferring hybridity from morphological intermediacy. Despite the popularity of the canonical approach (e.g., *Armeria*; TAULEIGNE-GOMES & LEFEBVRE 2004; *Gentianella*, GREIMLER & JANG 2007) ESTABROOK et al. (1996) find that this method does not distinguish between ancestral intermediates, where the descendants have mixed character states, while hybrid intermediacy is evidenced by character states intermediate between those of the parents.

There two processes are thought to maintain hybrid populations or swarms. The first, the tension zone model (after BARTON & HEWITT, 1989), states that the hybrid zone is maintained via a balance between selection against hybrids and their continual production as a result of gene exchanges between the parents, even though the hybrids themselves are genetically less fit than the parent species. The second, expanded upon by ARNOLD et al. (1999), allows for the greater fitness of the hybrids in comparison to the parents; that is, natural selection favours the hybrid genotype.

Hybrids have been documented for several Chilean genera (*Calceolaria* L, EHRHART 2000, 2005; *Baccharis* L, HELLWIG 1990; *Haplopappus* Cass. KLINGENBERG, 2007). Hybridsation has never been discussed as an active process in the genus *Chaetanthera*.

2 *C. glabrata*: polymorphism & the El Niño effect

2.1 Introduction

"El tamano y forma de las hojas son muy variables... A la vista de material abundante es imposibile delimitar estas formas."

CABRERA (1937, p.146)

This comment refers to polymorphic material of the taxon *C. glabrata* – *C. limbata* complex interpreted by R.A. PHILIPPI to be six discrete taxa (1894) and more conservatively by REICHE (1904) to be varieties. So far revisions of *Chaetanthera* material have focussed on describing the variation and fitting it into discrete taxonomic entities. Here, the possible reasons driving the observed variation are questioned.

2.2 Variation & distribution

C. glabrata is very polymorphic in its leaf and involucral bract shape. The leaves and foliaceous involucral bracts from the north and south extremes of the distribution tend to have larger, rotund (rarely even cordate) laminae (here referred to as type A), while those in the centre of the distribution (around La Serena) tend to have smaller rotund to ovate (type B) or oblanceolate laminae occasionally with undulate margins (type C). Examples of capitula with lamina types A and B are shown in Fig. 34a. Note that the leaves of the cultivated material depicted here are not as distinctly limbate as in the field specimens. Fig. 34b illustrates the three lamina types.

C. glabrata is distributed from Paposo near Antofagasta (23°S) to Santiago (34°S) along the coast and somewhat inland to 2000 m.a.s.l. It shows a curious disjunction between north-east Santiago (33°S) and La Serena (29°S). In the northern part of its range the plants are found at low elevations between sea level and 750 (rarely over 1000) m.a.s.l. In the southern part of its range the plants are found between 750 and 2000 m.a.s.l. It is cited by DILLON ET AL. (http://www.ChlorisChilensis.cl) as a typical plant of the Chilean Lomas formations.

Germination of *C. glabrata* is rapid (3 – 10 days) and highly successful (>90% germination even after 3 – 4 years in storage). Cultivated material from Carrizal Bajo (*Ehrhart* et al. *2002/142*) and Loma del Viento (*Kalin Arroyo* et al. *25163*) maintained its leaf shape for at least one generation. However, new leaves produced towards the end of the growing season were smaller. Similar reduction was also observed in ray number and capitula size. The polymorphisms could not be wholly explained away by variation in greenhouse regime or loss of vigour.

VI "The variety of all things forms a pleasure"

Figure 34a: Examples of leaf variation in *C. glabrata*. **A.** Leaf morph type 'A' from Loma del Viento (2190 m.a.s.l., 33°22'S 70°20'W, *Kalin Arroyo 25163*). **B.** Leaf morph type 'B' from Carrizal Bajo (50 m, 28°26'S 71°11'W, *Ehrhart 142*). **C.** Leaf morph typh 'C'. La Serena©J. Grau. Images A & B ©A. Davies, cultivated material, Munich 2003.

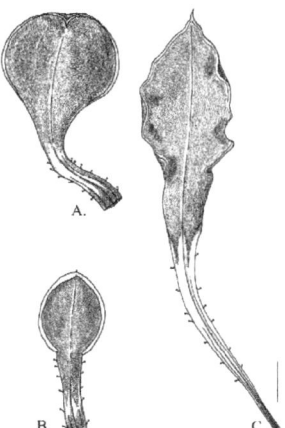

Figure 34b: Examples of leaf variation in *C. glabrata*. **A.** "Type A" *Rosas 1022*; **B.** "Type B" *Geisse 89*; **C.** "Type C" *Looser s.n.* Scale bar = 2 mm.

The latitudinal and altitudinal distribution of all specimens with leaf morph types A, B and C was plotted (see Fig. 35). The original can be found in Appendix I, Table 15.

If polymorphism were wholly due to variable water relations then one would expect that where elevation played a role, the larger leaves would be found higher up, where the snow-melt from the Andes supplies water. Along the coast in the arid and semi-arid regions, where the fog banks often influence local vegetation dynamics, one might expect larger leaves on the coast, and the smaller leaves inland. In the Mediterranean zone the climate is more warm temperate with winter precipitation. Additionally, Chile is also periodically dramatically affected by intermittent "rainy" years caused by the El Niño Southern Oscillation (ENSO). In this case, one could expect to find large-leaved collections only in El Niño years, especially in the dry north.

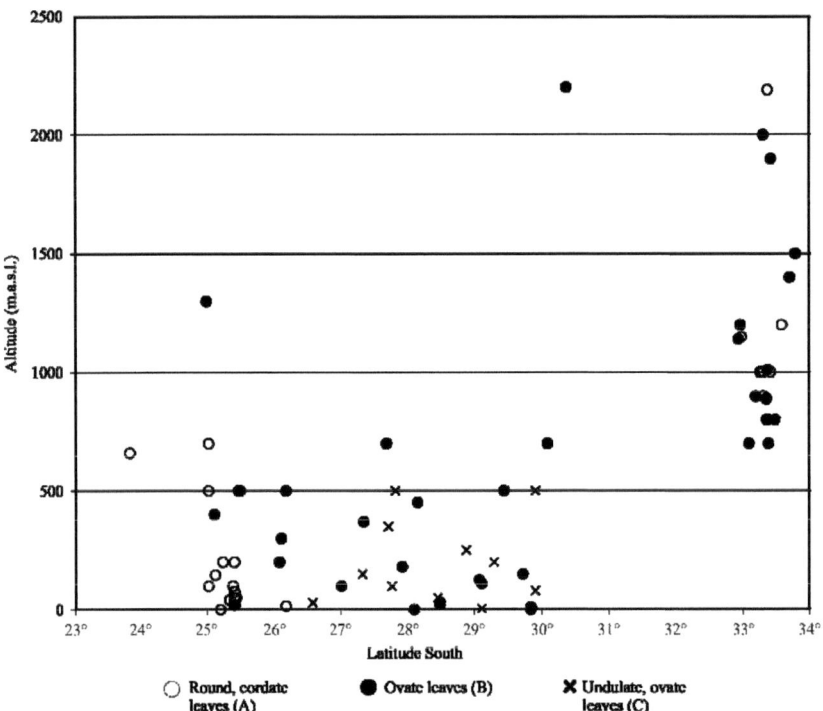

Figure 35: Altitude-Latitude distribution of *C. glabrata* leaf morphs.

As the north and south of the distribution range are separated by over 10 degrees of latitude, it may be possible that seasonal differences could account for the polymorphism. Hypothetically, this could be expressed as larger-leaved collections occurring at the beginning of the season (spring) and smaller-leaved collections towards the end (drier summer conditions), as seen in the greenhouse. This could be found in one locality, e.g. Santiago. This could also hypothetically be seen as a clinal change in leaf form along the coast as the change in season is staggered from north to south; i.e. summer starts later in the south, and so larger leaves earlier in the north and later in the south. Finally, it may also be possible that seasonal progression is not responsible for the geographical distribution of leaf types seen here.

2.3 El Niño Southern Oscillation (ENSO)

Extensive regions around the world are influenced by the ENSO, amongst these the northern coastal areas of Chile. El Niño events occur irregularly but now typically once every three to six years (ALLAN et al. 1996). During an El Niño episode, rainfall dramatically increases in certain areas of the world, whereas severe droughts occur in other regions. The best-studied effects are those in marine environments, where this climatic phenomenon is correlated with dramatic changes in the abundance and distribution of many organisms, and the collapse of fisheries (JORDAN 1991). By contrast, the effects on terrestrial ecosystems have been relatively poorly explored. Only the spectacular greening and flowering of deserts (DILLON & RUNDEL 1990) and the crash of agricultural crops (TAYLOR & TULLOCH 1985) in the core region of El Niño had been commonly noted. In recent years, however, results from several systematic long-term studies have become available, revealing how ENSO events can have pronounced effects on plant and animal communities and their dynamics in many terrestrial ecosystems ranging from arid and semiarid ecosystems to tropical and boreal forests (HOLMGREN et al. 2001).

2.4 Results: observed phenotypic plasticity

The collection month (taken as analogous to flowering time) was plotted against lamina category in 73 collections. The collections were then divided into north and south. 'North' was defined as the area between Antofagasta and La Serena (23° – 30°S). 'South' was defined as the area around Santiago (33°S) (Fig. 36).

As expected, there is a seasonal shift as the northern plants appear first (September), compared to the southern collections, which are first recorded in October. In the northern part of its range the low elevation "C" plants generally flower first (September to October). The "A" and "B" plants generally flower over the same period, between October and November, over a greater elevation range. In the southern part of its range the "A" plants tend to be found at the beginning of the season and the "B" plants towards to the latter end of the season. This offset seasonal effect was also observed in the greenhouse.

Despite the expected effects of north-south seasonal shift, and seasonal variation in leaf type in a fixed locality, there is no explanation why A-type laminas should be found at such geographically separated localities. Therefore, the driving force behind the phenotypic amplitude must be happening on a different time scale.

Along the Pacific coastline of Chile, the quantifiable Southern Oscillation Index (SOI) is a direct measure of the El Niño effect where the more negative the index the more likely an El Niño event will occur. The number of collections and the corresponding mean SOI per annum were plotted for each leaf type in both the north and south categories (Fig. 37; data in Appendix I, Table 16). The graphs are summarized in Table 4.

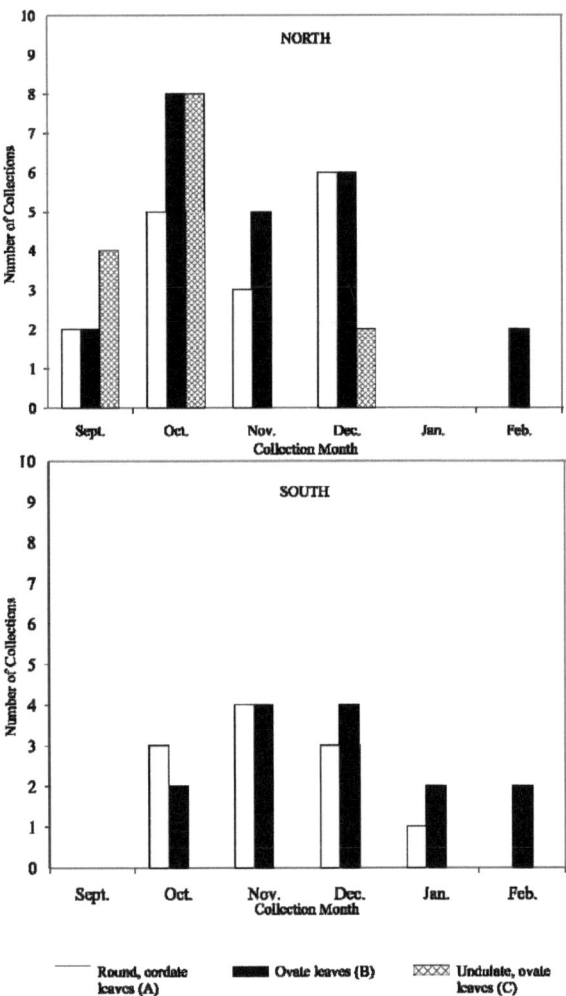

Figure 36: Phenology of leaf morphs. Collection months (analogous to the flowering time) of the lamina types in the "north" (upper) and "south" (lower).

VI "The variety of all things forms a pleasure"

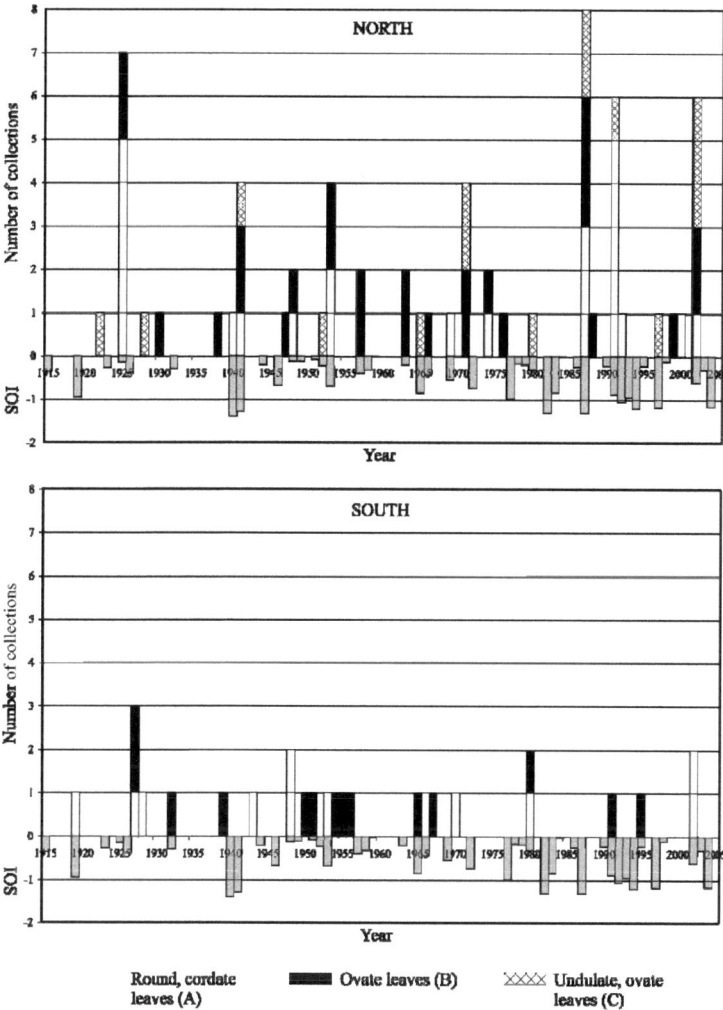

Figure 37: Mean annual SOI and number of collections of different leaf morphs yearly from 1915 – 2005 in the north and south.

		A	B	C	Total (A+B+C)
NORTH	ENSO	(21) 44%	(17) 35%	(10) 20%	48
	NOT ENSO	(3) 18%	(9) 56%	(4) 25%	16
SOUTH	ENSO	(8) 57%	(6) 43%	-	14
	NOT ENSO	(4) 33%	(8) 67%	-	12

Table 4: Numbers and percentages of collections made of each leaf type in the north and south in ENSO and Non ENSO years.

Twice as many collections have A leaves in El Niño years. This is true both in the North and South (44%:18% and 57%:33%). Conversely, there are only two-thirds as many B leaves in El Niño years, both in the north and south (35%:56% and 43%:67%). C type leaves appear to be unaffected by the El Niño (20%:25%). The large round leaves in the north are almost exclusively found between Antofagasta (23°S) and Copiapó (27°S), an area classified as desert. The smaller round leaves and ovate leaves are found between Copiapó (27°S) and La Serena (30°S), which is classified as sub-desert. There is a strong bias towards collections in the north made in El Niño years (a botanical collecting effect). However, as the same El Niño effect is seen in the south of the distribution where the collecting bias can be ruled out, the change in leaf shape according to El Niño climate conditions is probably a real phenomenon.

2.5 Conclusion

The phenotype of *C. glabrata* reacts visibly in response to El Niño events. The most sensitive area is in the far north of the distribution, near Antofagasta. In this desert region where the ephemeral rains fall, the phenotype is expressed in its most extreme form (large round fleshy leaves). This is replicated in the more southerly, temperate, wetter (relatively) Mediterranean region at higher elevations. In the middle of the distribution the species tolerates more punitive water conditions, directly reflected in leaf size and shape. The low elevation, coastal form (ovate undulate leaves) is not so affected, possibly because it grows in the environmentally and hydrologically buffered "neblinas".

The phenotype is so plastic that it can alter over a growing season in response to a spring-summer change – viz. larger leaves earlier in the season and smaller leaves later in the season, and the reduction in leaf size, capitula and ray number seen in the greenhouse[4].

---

[4] The work of Bull-Hereñu & M.T.K. Arroyo about *Chaetanthera moenchioides* was published (Plant Syst. Evol. (2009) 278:159 – 167) after this thesis was submitted for examination. The population differentiation and reaction norms of *C. moenchioides* agreed with predictions of genetic assimilation given that the plastic response of the species under water stress mimics phenotypic differentiation thta has evolved along the environmental gradient.

3 *C. linearis* & *C. albiflora*: active hybridisation

3.1 Introduction

One of the first published records that touched on the possibility of a character complex in *Chaetanthera* refers to variation in the annual species *Chaetanthera linearis*, observed by Rudolfo Philippi. When Philippi (1894) first described *C. linearis* var. *albiflora*, he wrote:

> "*Es singular que se encuentre igualmente en las mismas provincias una variedad de esta especie con ligulas blancas encima, pero que casi siempre tienen en la cara inferior, en el medio, una ancha faja de un purpúreo casi negro. Por lo demas na hai diferencia entre esta variedad i la forma normal de flores amarillas*".[5]

Although this refers to taxonomic entities now considered to be two species, the observed floret variation became the first point of reference for closer examination of the plants.

3.2 Variation & distribution

C. albiflora and *C. linearis* are two superficially similar herbs found along the coastal and lower to mid-elevation Andean areas of northern Chile to Santiago. *C. albiflora* is glaucous to grey-green with slightly succulent leaves and a rounded leaf cross-section. It is lightly pubescent and the ray florets are white with a darker, purple-black dorsal stripe. It has a diploid chromosome complement of 2n = 24. *C. linearis* is not succulent, and has a virid, yellow-green appearance. It is glabrous and has yellow ray florets. It has a diploid chromosome complement of 2n = 22.

The latitude/altitude coordinates of all collections are plotted in Fig. 38 (data in Appendix I, Table 17). *C. albiflora* and *C. linearis* are partly spatially (different altitudes, different latitudes) separated. *C. albiflora* is distributed in the north of Chile along the Pacific coastline between 22°S and 33°S. It is distributed from 31° - 34° from the coastal hills to the lower and mid elevations of the Andes. They are sympatric, in their pure form, between 31°45'S and 32°45'S. They are also temporally separated (phenologically); the more northerly, lower elevation *C. albiflora* flowers earlier (September to November) than the more southerly, higher elevation *C. linearis*, which flowers between October and March.

A number of collections show intermediate characteristics: white flowers but glabrous and / or no dorsal stripe, or yellow flowers and filamentous indumentum, or glabrous plants with unknown flower colour. Figure 38 illustrates examples field variation between the true types and the intermediate forms.

---

[5] **English translation.** It is unusual that in the same provinces one finds a variety of this species [*C. linearis*] with ligules that are white on the upper side and have a central darker purple, nearly black, stripe on the lower side. Otherwise there are no differences between this variety [*C. linearis* var. *albiflora*] and the normal yellow-flowered form.

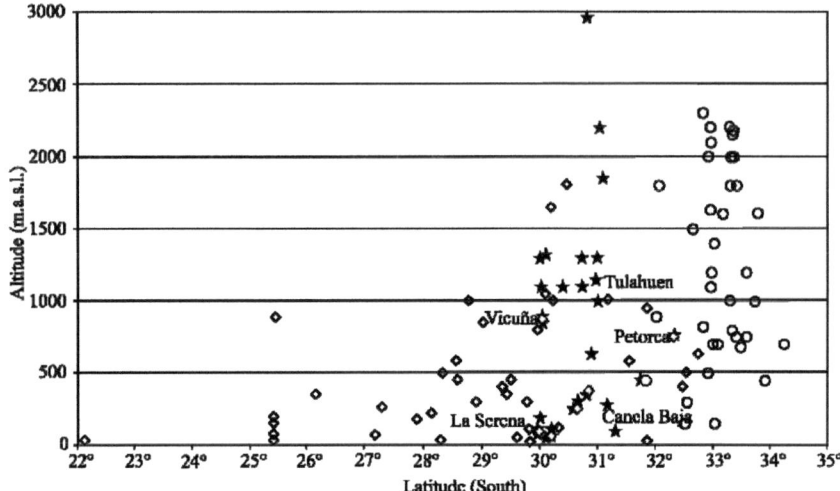

Figure 38: Altitudinal and latitudinal distribution of *C. linearis* (circles), *C. albiflora* (diamonds) and intermediate individuals (stars).

True type *C. albiflora* (Fig. 39, A – B) are pubescent, glaucous plants with white ray florets with a green-grey dorsal stripe [cultivated from *Ehrhart 6.86*: La Serena 29°47'S, 71°17'W]. The intermediate characteristics were also observed in the greenhouse (July 2006) when plants grown from seed from Portezuelo de Chincolco (32° 10'S, 70°48'W, *Grau* s.n.) gave rise to lightly pubescent, grey-green individuals with white ray florets with no dorsal stripe (Fig. 39, C – D), and glabrous, virid green individuals with white-yellow ray florets (Fig. 39, E – F). The same collection also gave rise to true type *C. linearis* (Fig. 39, G – H).

3.3 Results: hybrid zones

The sympatric areas between *C. albiflora* and *C. linearis* are characterised by mixed populations of *C. linearis*, *C. albiflora* and intermediate individuals. These are split into three groups or hybrid zones: inland hills near Vicuña, Río Elqui to Tulahuen, Río Limarí (29°54' – 31°00'S; 70°30' – 70°42'W); coastal areas from Guanqueros (La Serena) to Río Limarí and Canela Baja (30°00' – 31°00'S; 71°00' – 72°00'W); inland hills between Salamanca and Petorca (32°00' – 32°30'S; 70°30' – 71°00'W). These are illustrated in Fig. 40.

VI "The variety of all things forms a pleasure"

Figure 39: Capitula Variation in *C. albiflora*, *C. linearis* and intermediates. **A, B.** *C. albiflora* - white rays. **C, D.** an intermediate form - creamy rays. **E, F**: an intermediate form. **G, H.** *C. linearis* – yellow rays. Photos: A, B (A. Davies, 07.2003, München); C – H (J.Grau, 07.2006, München).

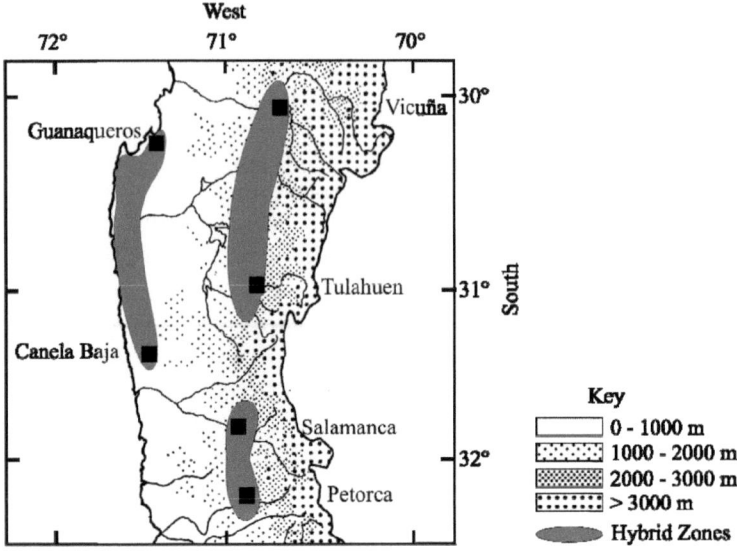

Figure 40: Hybrid zones of *C. albiflora x linearis.*

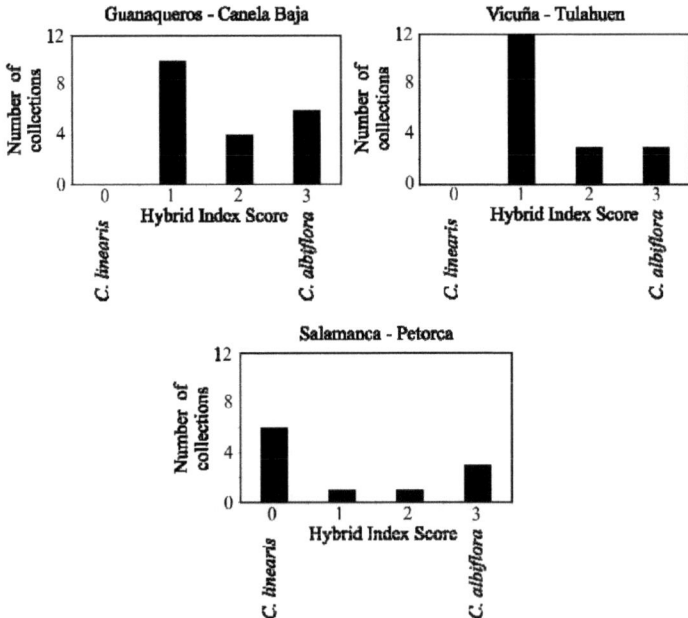

Figure 41: Hybrid index scores for three sympatric localities of *C. albiflora* and *C. linearis.*

Gene flow between these two species can be more closely analysed using the simple application of hybrid indices (ANDERSON 1949), whereby individuals are scored according to the presence of a character state possessed by either parent. In this case the material was scored for flower colour, pubescence and ray floret dorsal stripe. *C. albiflora* parental types scored 3 (white flowers = 1, dorsal stripe present = 1, indumentum absent = 1) while *C. linearis* parental types scored 0 (yellow =0, no dorsal stripe = 0, glabrous = 0). Intermediates were scored with a hybrid index of 1 or 2. The hybrid index scores are illustrated in Fig. 41 (data in Appendix I, Table 18).

3.4 Conclusions

C. linearis forms a dominant component of southern, higher elevation populations, reaching the northern limit of its influence in the hills around Vicuña, although it is not recorded in its pure form further north than Salamanca (Río Choapa). *C. albiflora* forms a more dominant component of the northern coastal populations, reaching its southern limit of its influence in the hills around Petorca. Vicuña and Petorca are separated by over 300 km (as the crow flies). Over this distance each individual population does not seem distinct from the next. This indicates that there is a large scale shallow cline between the two parental species.

Gene flow from *C. linearis* seems to be occuring northwards into the *C. albiflora* distribution. This introgression is suspected to be brought about by hybridisation. The sample of collections observed shows that the intermediate forms, or hybrids, are more numerous than the parental types. The hybrid form can be interpreted as being selectively fitter than both parent species. This is particularly evident in higher elevation habitats north of Salamanca, where true type *C. linearis* does not occur.

Many characters, e.g. leaf colour and cross section, are lost on herbarium material. This is potentially a very good field indicator as *C. albiflora* leaves are a glaucous, grey-green colour, and somewhat fleshy or succulent. *C. linearis* has more virid, yellow-green leaves that are not fleshy. No doubt live population analysis would reveal more characters to support the herbarium study reported here. The cytological nature of the hybrids was not determined. It would be interesting to establish how the different diploid sets (2n = 22, 2n = 24) are reconciled in the hybrids. The putative hybrids are not noticeably morphologically more robust than the parent species. They do not seem, superficially, to be a product of allo-polyploidy (heterosis), as this often brings increased vigour and size (HESLOP-HARRISON, 1953; 93).

4 Perennial, scapose *Chaetanthera*: polymorphism or hybridisation

4.1 Introduction

Within *Chaetanthera* subgenus *Chaetanthera* there are two perennial life forms. The first is a dwarf subshrub, with glandular, sclerophyllous leaves highly adapted to the xeric habitat of the lower montane zones of Coquimbo. The second is a caespitose, scapose herb with glabrescent to sericeous or villous leaves, found in the mesic habitats of the montane, valley and coastal zones from Santiago to the northern perimeter of the Valdivian rainforest. The variation within the latter group of perennials has generated a lot of taxonomic confusion: prior to this revision there were 5 species with 8 varieties, although these did not adequately resolve the variation seen in the field or herbarium.

The material was analysed in three stages. First, the traditional characters were considered. Second, a Principle Components Analysis (PCA) was applied to a subset of the material using twenty morphological characters and a priori identifications to test the taxa groupings. Third, a Hybrid Analysis program (HYWIN, ESTABROOK 1996) was applied to the same data set in an effort to explain the type of variation seen.

4.2 Variation & distribution

The perennial, monoecious caespitose taxa are united by having loose, basal pseudo-rosettes of linear lanceolate to spathulate leaves with denticulate to spinose serrate margins. The large showy capitula are entirely yellow, subtended by multiple series of sub-lignified coriaceous involucral bracts, the outermost with reflexed, foliaceous apices and the innermost of which have acute to obtuse darkly maculate apices. The plants can vary from glabrescent with tufts of villous hairs in the axils to densely sericeous - villous silvery pubescent on all parts. Fig. 12 shows two of the most commonly seen forms.

Figure 42: Perennial scapose herbs. **A.** Cordillera de Santiago 2006©M.Bonifacino. **B.** Laguna del Maule 2000©J.Grau.

Variation can be seen in the rhizomatous and/ or stoloniferous habit. The length and width of the leaves, their dentition and indumentum is also polymorphic. The pedicels can be short or long. The capitula vary in width and height as do the outer and inner series of involucral bracts. The carpopodium can be poorly-developed or well-defined. There may or may not be a pink-red dorsal stripe in the ray florets.

The plants are found widely distributed along the Andean Cordillera from the Cuenca de Santiago (32°30'S) and Valparaíso southwards along the Andes up to 3000 m.a.s.l., and the Cordillera de la Costa, and along the central valley, especially south of Talca to Temuco. The Termas and Lagunas of the south Andean volcanoes between 36°S and 40°S are also colonised as are the Cordillera de la Costa and the western parts of the central valley from Concepción to Temuco and Valdivia.

They do not appear to be competitors in anthropogenic habitats, but are found in open, exposed sunny habitats on a variety of substrates, from rocky scree to sands and deep soils. They are always associated with the native Chilean woodlands such as the "El Roble" (*Nothofagus obliqua, Cryptocarya alba*) forest remnants, the matorral (*Acacia caven* steppe) or the *Araucaria* woodlands to the south.

4.3 Analysing variation: are the traditional characters truly diagnostic?

At the most simplistic level these perennials have been traditionally separated based on three main characters. These are vegetative habit (rhizomes or stolons), leaf indumentum (glabrous or pubescent) and leaf margin serrations (dentate or spinose serrate).

These three characters were assessed for over 250 collections. The vegetative habit varied from clearly rhizomatous plants, to those whose rhizome was compacted into a small dense mat and those plants with woody roots, compacted stems and stolons. The type of leaf serrations varied from shortly dentate serrate on the upper leaf margins to spinose serrate along more than half the length of the leaf margin. The indumentum of the plants was arbitrarily quantified as densely pubescent (leaves, axils, stems), pubescent (leaves and axils or stems) or glabrous (or sparsely hirtellous leaves). The locality of the plants was also recorded.

4.3.1 Results

The specimens could be divided into groups sharing suites of qualitative traits (e.g. plants with rhizomatous roots, shortly serrate leaf apices and generally densely pubescent), and less common suites (e.g. plants with rhizomatous roots, spinose serrate leaf margins and generally densely pubescent). Three localities/zones were identified where the plant collections were particularly variable. The suites of character traits and their localities are given in Table 5. Example collections are also cited.

The collections from Laguna del Maule generally have rhizomatous vegetative parts and spinose serrate leaf margins (occasionally the leaf margins are shortly serrate at the apices only). However, the collections vary widely in their indumentum, from entirely glabrous plants, to those with long silky hairs in the leaf axils, to plants that are entirely densely pubescent. The collections from Chillán generally have spinose serrate leaf margins and are mostly glabrous or have a sparse indumentum on the stems or in the leaf axils. Greatest variation is seen in the vegetative habit, which can be either clearly stoloniferous or rhizomatous. The collections from the Cordillera de Nahuelbuta are extremely variable in the density of indumentum and the leaf margin serrations, although they always have stoloniferous vegetative parts.

No single suite of characters was restricted to a particular habitat or geographical location, although broad trends could be seen. Stoloniferous plants are found south of the Rio Ñuble, but rhizomatous plants can found throughout the distribution range.

Vegetative habit	Leaf margins	Densely pubescent (leaves, axils, stems)	Pubescent (leaves and axils or stems)	Glabrous (or sparsely hirtellous leaves)
Roots	Spinose serrate	1. Maule, Antuco	2. Maule, Chillán, Malleco, Pulmarí	3. Maule, Talca, Chillán, Los Angeles
	Shortly dentate	4. Quillota, Santiago, Catillo, Termas Talca, Malleco	5. Rinconada de Alcones, San Rafael-Litú, Agua de la Vida, Paso Cruz	
Stolons	Spinose serrate	6. Concepción, Cord. de Nahuelbuta, Valdivia to 1100 m.a.s.l.	7. Hualqui, Cord. de Nahuelbuta, Mininco, Llaima	8. Chillán, Llaima, Villarica
	Shortly dentate	9. Concepción, Angol (Cord. de Nahuelbuta), Valdivia to 500 m.a.s.l.	10. Valdivia	

Table 5: Geographical distribution of character variation. The localities of suites of character traits including vegetative habit, leaf margin serration and indumentum variation. Example collections. 1. *Poeppig 208, Grau 2923p.p., Rosas 1917.* 2. *Ehrhart & Grau 95-919, Comber 372, Rosas 1875.* 3. *Gardner, Knees & DeVore 4590, Phliippi s.n.* 4. *Ricardi 2831, Goodspeed 23322.* 5. *Kuntze s.n., Ehrhart & Grau 95-577, Matthei & Quezada 786.* 6. *Grau 3185, Höllermayer 160b.* 7. *Werdermann 1256, Grau 2998 p.p.* 8. *Philippi s.n., Böhnert s.n., Grau s.n.* 9. *Pennell 12836, Grau 2998 p.p., Grau 2991.* 10. *Gay 379, Lechler 488.*

Plants with spinose serrate leaf margins are found south of the Rio Maule, and then rarely in the central valley. Plants with dentate leaf margins can found throughout the distribution range. Two clines were defined for the variation in indumentum. One cline accounts for indumentum variation along an East-West gradient at latitudes south of the Rio Ñuble (36°30'S), mostly on a line from Concepcion to Laguna de la Laja. The coastal Cordillera at these latitudes (including Cordillera de Nahuelbuta) and the central valley support stoloniferous specimens that are medium to densely pubescent with dentate or spinose leaf serrations. Nearer the Andes the specimens tend to be glabrescent, with spinose leaf serrations, but may be stoloniferous or rhizomatous. The other cline accounts for variation in indumentum changes along a North – South cline in the Chilean Andes. The rhizomatous plants of the Andes are densely pubescent with dentate leaf serrations in the north of their distribution (e.g. Santiago). They change around Laguna del Maule and Chillán to having spinose leaf serrations, and become increasingly glabrescent towards Malleco and Antuco.

4.3.2 Summary

The vegetative habit seems to be determined by the local microclimate and habitat that the plants occupy. The prescence of stolons has been considered in the past as a good species indicator, but the current analysis supports what Reiche wrote in 1905:

> *"A mas de los ejemplares con el rizoma rastrero existen otros cuyo rizoma corto y grueso emite varias rosetas largamente sedoso-peludas en la base i dispuestas en céspedes densos. En vista de existir formas intermediarias entre ámbos tipos i teniendo presente, que tales diferencias de órganos vejetativos pueden ser consecuencias de la localidad ± húmeda, de terreno suelto o pedregoso etc., no me parece conveniente establecer especies o variedades sobre el carácter aludido".*

REICHE (1905, p. 338)

The apparently clinal change of the indumentum with latitude, longitude and altitude reduces its value as a discrete taxonomic character.

The type of leaf serrations is very distinctive, as already noted by PHILIPPI (1856). This character seems to be closely associated with indumentum in that shortly dentate leaves are never glabrous. This is the only one of the three traditional characters that could be considered taxonomically useful.

Transition zones are often areas where plants express more variation. The adjacent volcanic entities of Laguna del Maule and the foothills to the Termas de Chillán mark the transition zone where the montaña starts and where the Andes significantly drop in elevation from the airy 5000 m plus summits to the more modest 3000 m peaks. The Cordillera de Nahuelbuta is a coastal massif islanded from the Andes but supporting relict Andean vegetation (i.e., *Araucaria* woodland).

4.4 Analysing variation: principle components analysis

4.4.1 Methods

Multivariate statistics, specifically Factor Analysis, can often help identify characters useful for separating individuals whose relationships are otherwise unclear. Additionally, the "Factors" comprise suites of correlated variables and these also have applications in key construction. Standardising the data ensures that large numerical values (such as obtuse angles) do not have more weighting on a variance axis than measurements of $1 - 2$ mm. This also allows inclusion of binary data (0, 1) encoding presence / absence information of qualitative variables. Eighty-seven herbarium specimens were selected from the entire geographical range, including twelve collection duplicates (e.g. *Grau 2923, Grau 2495, Grau 2998*), and several type collections (e.g. *Chaetanthera nana* Phil. 34.23°S 70.71°W; *Chaetanthera andina* Phil. 39.90°S 72.63°W). A Principle Components Analysis (PCA) was computed on the standardised quantitative and qualitative data.

Figure 43: Variation of Quantitative characters used in the Multivariate Analysis dataset. "A", "B" and "C". Box is 50%, outer line 75%.

The results were used produce Component plots with the a priori groupings superimposed upon the data points. The Eigenvalues were analysed to consider which variables contributed the most variance to each component axis.

Twenty qualitative and quantitative morphological characters were studied. The variation in the quantitative characters is presented graphically in Fig. 43. The characters with their ranges, means and 1 standard deviation are given in Table 6. The collections were assigned a priori taxonomic groupings "A", "B" and "C" with the latter group including all those specimens that seemed intermediate. The aim of the study was to use a more detailed analysis to test whether there are three discrete entities in this complex. The complete data set is given in Appendix 1 (Table 19).

4.4.2 Results

The quantitative data shows that the ranges overlap for all characters and all three groups, i.e. there is no discrete disjunction in any of the quantitative variables. However, there is a tendency for "A" specimens to have narrower capitula and shorter pedicels than "C" specimens, for example. "B" specimens seem to be a subset of "A" specimens in terms of the quantitative variables. The "C" group shows a huge degree of variability in different characters, caused by having oddities and intermediates, e.g. specimens with longer leaves or particularly heavy capitula.

The results from the PCA (87 collections, duplicates) are summarized in Table 7 and Fig. 44. The first two components accounted for nearly 45% of the total variance. Although usually this percentage could be expected to be higher, this indicates that there is a high level of natural variation in the characters used for this analysis. To test the hypothesis that there are three taxonomic entities among the sampled specimens each component is plotted with the next most variable one and the a priori groupings superimposed on the plot.

CPT	EV	%V	CV
1	5.62	28.10	28.10
2	3.30	16.55	44.65
3	1.91	9.54	54.19
4	1.26	6.03	60.53
5	1.21	6.03	66.57

Table 7: Eigenvalues (EV), percentage of variance (%V) and cumulative variance (CV) of the first five Components (CPT) after PCA of 20 morphological characters for 87 collections.

The characters with the greatest Eigenvector scores (EVS) are those which contribute the most to the variation in a component. This can indicate which characters have played a more significant role in the separation of taxonomic units, i.e., the specimens. The four characters responsible for the maximum separations along the first component are: the type of leaf margin serrations (dentate v. spinose) (EVS = 0.83) and their extent (partially v. entirely) (EVS = 0.82), mean leaf width (EVS = 0.75), and mean capitula width (EVS = 0.78). The characters causing the most separation along the second component are the pubescence of the stems (EVS = -0.79) and leaves (EVS = -0.76), and the mean pedicel length (EVS = -0.67). The habit and bract morphology do not play a significant role in the separation of the points in this analysis.

From the variation in the quantitative characters (Fig. 43) we can see that Mean Leaf Width divides the material into two groups (<3mm or >3mm), and Mean Capitula Width could also be used to divide the material into 2 groups (< 15mm or > 16 mm), despite the large overlap in mean measurements for both characters. The importance of the indumentum characters needs to be considered in the light of the clinal evidence presented earlier. Pedicel length is affected by microclimate in the northern part of the distribution. The scatter of points (= specimens) with the a priori labels shown in Figure 44 clearly shows that, despite being arranged in two major clusters and a third minor cluster, the points are not comparable to the a priori groupings in any way. Again, as indicated by the preliminary analysis, this refutes the use of the habit or indumentum as a taxonomic discriminator between these plants.

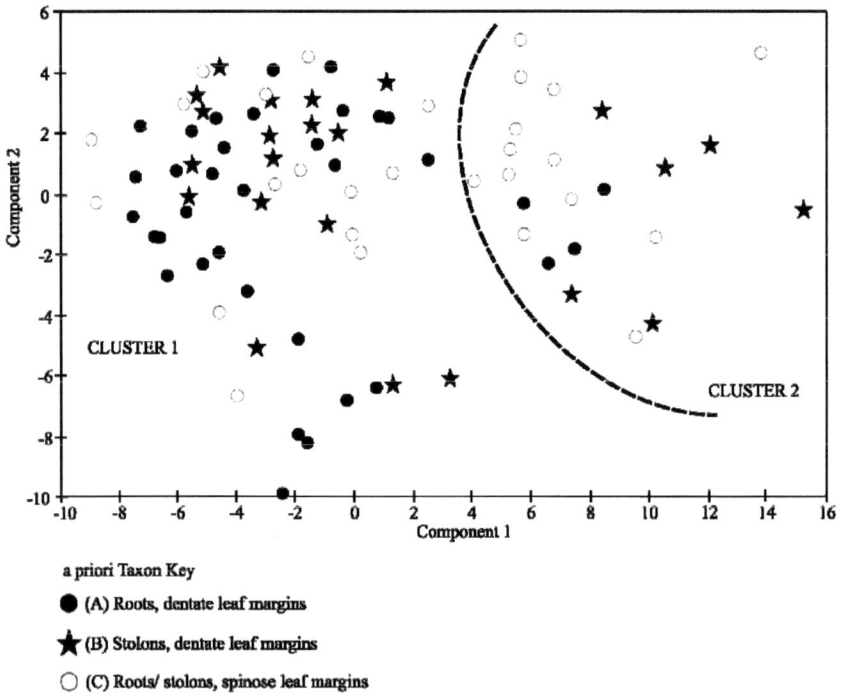

Figure 44: Principle Components Plot. Plot generated by a PCA, labelled with a priori groupings A, B & C and the two posteriori clusters.

VI "The variety of all things forms a pleasure"

		"A" N = 35	"B" N = 23	"C" N = 29
Habit		Roots	Stolons	Roots/ Stolons
Indumentum	Pubescent Leaves	Yes	Yes	Yes (no = 9)
	Pubescence sparse leaves	No (yes = 1)	No (yes = 4)	No (yes = 6)
	Pubescent Axils	Yes	Yes	Yes (no = 6)
	Pubescent Stems	Yes	Yes	Yes (no = 7)
	Pubescence > 3mm	Sometimes (7)	Sometimes (9)	No
Pedicel	Pedicel L (cm)	9.7 ± 6.2 1.5 – 29.5	12.6 ± 4.7 3.3 – 21.0	13.3 ± 7.1 6.5 – 36.0
Leaves	Leaf L (cm)	2.9 ± 0.9 1.8 – 5.0	3.2 ± 0.8 1.8 – 4.8	4.0 ± 1.2 2.0 – 8.5
	Leaf W. (mm)	2.0 ± 0.6 1.0 – 3.5	2.7 ± 1.0 1.3 – 5.0	3.8 ± 1.2 1.5 – 7.0
	Type of dentition	dentate	dentate	Spinose serrate
	Extent of dentition	apical	apical	Entirely
Capitula	Width (mm)	13.4 ± 2.7 8.0 – 20.0	16.3 ± 3.5 9.0 – 20.0	19.4 ± 4.7 10.0 – 30.0
	Length (mm)	11.8 ± 1.8 8.5 – 15.0	12.2 ± 1.9 10.0 – 15.0	13.8 ± 2.6 10.0 – 20.0
Outer Involucral Bracts (OIB)	Length (mm)	8.8 ± 2.3 4.5 – 16.0	7.3 ± 1.6 4.3 – 11.2	9.0 ± 2.0 5.3 – 13.3
	Width (mm)	1.8 ± 0.4 1.1 – 2.9	1.6 ± 0.3 1.2 – 2.4	2.2 ± 0.6 1.3 – 3.2
	Number of individuals with 1;2;3 foliaceous series	5;19;11	15;8;0	13;14;2
	Apical angle	87 ± 40 25 – 150	68 ± 16 30 – 90	84 ± 22 60 – 120
Inner Involucral Bracts (IIB)	Length (mm)	13.8 ± 2.3 9.0 – 19.0	14.5 ± 2.1 18.7 – 11.0	16.0 ± 1.8 11.2 – 18.1
	Width (mm)	1.4 ± 0.4 0.8 – 2.5	1.3 ± 0.3 0.8 – 1.8	1.6 ± 0.5 0.8 – 3.2
	Apical angle	59 ± 32 20 – 150	42 ± 19 25 – 90	52 ± 21 25 – 90

Table 6: Characters with their ranges, means and variation to 1 standard deviation in the a priori groupings A, B and C.

The multivariate analysis shows three groups. The collections are broadly segregated into two groups on the first component axis by entirely serrate spinose leaf margins, broad leaves and broader capitula on the right (Cluster 2), and apically dentate, narrow leaves and narrow capitula on the left (Cluster 1). The environmentally and clinally changeable characters of indumentum and pedicel variables on the second component axis account for the third minor cluster of specimens having such disparate collection localities (Valparaíso, Santiago, Angol and Hualqui). Geographical distribution of the plants does not provide any clear cut explanations either. Cluster 1 contains material typically found to the north of Río Maule, but not exclusively, and in the southern parts of the Central Valley. Cluster 2 is especially well represented by material from the Chillán Andes, although generally represents montane material from the south of Río Ñuble. There is no pattern correlating to altitude.

4.5 Analysing variation: HYWIN

Polymorphic but intrinsically indistinguishable taxa can be characteristic of micropopulations arising through the random fixation of non-adaptive gene patterns (HESLOP-HARRISON p.60, 1964). The identification of intermediate patterns can be indicative of introgressive hybridisation. This is a process, whereby "gamodenes developed in isolation come together with local or regional blurring of pattern" BRIGGS & WALTERS (p.213, 1984). Alternatively, this pattern of variation could be explained by species undergoing separation, i.e. incompletely isolated from each other.

In order to assess the putative hybrid nature of the plants we need to identify both the hybrids, if there are any, and potential parents. HYWIN, a simple method for screening morphological variation in study sets suspected of containing hybrids was devised by ESTABROOK et al. (1996). Discriminant functions (such as Principle Components Analysis) can identify specimens that fall between parental species. The advantage of HYWIN is that no a priori assumptions need to be made regarding parental species or the sampled distribution character variation. A Hybrid Optimality score is generated where one of any three specimens is considered to be potentially a hybrid of the other two; and a probability hypothesis concerning statistical certainty. In cases where introgression is suspected the Parental Distance (PD) statistic can be weighted so the stringency for parenthood is so high that that specimens embedded in the middle of the putative introgressive swarm are not in turn identified as parents of nearby specimens.

Several preliminary runs using HYWIN were completed. Parameters include IN = hybrid intermediacy (the hybrid is quantitatively intermediate to the two parents), EQ = Parental Equality Parental distance (PD) (the parents should be different enough to belong to different species). Two sets of results are given here. In the first instance the results of an equally weighted data set are presented; that is, all specimens were assigned equal rank ($1.0*IN + 1.0*(1-[EQ]) + 1.00* PD$). In the second instance, a heavily weighted parental distance (PD) was used to reduce the likelihood of intermediates being classed as parents for 1000 highest ranking triples ($P>0.95$). The Rank Criterion was calculated as follows: $0.10*IN + 0.10*(1-[EQ]) + 1.00* PD$. The HYWIN output is in Appendix I, Table 20.

4.5.1 Results of the HYWIN analysis
Equal weighting (Figure 45):

The equal weighting analysis gave five categories of results: two parents, Y and Z, weak parents, weak hybrids and hybrids. On the PCA axes, although the picture is not clear, there seems to be two clusters of parent Z and two of parent Y with a large cluster of weak parents and weak hybrids in the top left quadrant. Geographically, Parent Y is found in the southern Andes and Nahuelbuta. Parent Z is found mainly around Santiago with a second centre around Concepción and inland. A swarm of weak parents and weak hybrids, with only a few true hybrids lays geographically inbetween the two parents. The weak parents and weak hybrids represent material that is statistically uncertain and could indicate possible F1 hybrids: i.e. an introgression scenario.

Weighted for possible Introgression scenario (Figure 46):

Two parent types Y and Z representing 16 of the 87 collections were identified. The remaining collections were classed as hybrids. Three collections are identified as Parent Y (collection numbers 80, 81 and 57) coming from Llaima, Villarica and Chillán. Parent X generally comes from around Valparaíso and Santiago (collection numbers 10, 14, 22, 36, 37, 39, 40, 83 and 84). Two collections (9, 70) come from Angol and Los Angeles in the southern Central valley, and two collections (22, 37) from the low Andes west-southwest of Curico. These two latter collections are positioned quite separately from the remaining parent collections on the PCA plot. As there is no specific locality given for Z39 and Z83 (duplicates of *Cuming 182*) these are not shown on the geographical distribution.

Eight of the ten top-ranked hybrids originate between Laguna del Maule (Río Maule) and Chillán down to Concepción (Río Bio-Bío). All the Nahuelbuta material is also of hybrid origin. The hybrids from Santiago Andes (arbitrarily defined as the Andes south to 35 degrees south) are positioned closely to the left hand cluster of Parent Z, while material from the Chillán Andes fall mostly within the right hand cluster dominated by Y_{80}. The hybrids from Andes around Laguna el Maule, Volcán Llaima and Cordillera Nahuelbuta are scattered between the left and right clusters. All the stoloniferous material was identified as being of hybrid origin. One record from Antuco (*Poeppig s.n.*) and one from Valparaíso (*Markham 336*) were oddly placed in the PCA, and subsequently omitted from the analysis.

Discussion

The preliminary HYWIN analysis proposes a scenario involving a large swarm of hybrids arising from two parents with a statistical likelihood of $p > 0.95$. The material selected for analysis is dominated by material of putative hybrid origin arising from two interbreeding parent types; Y and Z. The PCA spread of points indicates that these two types may be four micro-types: Y_{80} (Chillán type); $Z_{22,\ 37}$ (montane Curico type); Z_{all} (Santiago type); $Y_{57,\ 81}$ (south Andes type). Parent Y is self-incompatible and never generates hybrids amongst itself. Parent Z appears to form weakly defined hybrid/parents – possibly an F1 generation, where the hybrids are all closer to the Z parent than the Y parent. This could be evidence of introgression, or it could indicate a ploidy event. The placement of a number of morphological outliers beyond the parent positions in the PCA might also be indications of heterosis – hybrid vigour. This would require a different IN (intermediacy) weighting in HYWIN. Heterosis can sometimes be identified by ploidy events. However, there is only one relevant published chromosome count for *C. chilensis* (from Quillón) of 2n = 22 (BAEZA & SCHRADER 2005b).

Omitting the lower lying central valley and coastal cordillera populations there is a left-to-right shift on the PCA diagram corresponding loosely to the north – south grading of Andean populations between a northern-central parent and a southern-montane parent.

4.5.2 Summary

The hybrid analysis supports the initial analysis of traditional characters and defines the areas around Laguna de Maule and Chillán as critical to the understanding of this group.
The three traditional characters - habit, indumentum type and leaf serration type – are, by themselves, not discrete enough to separate the material into clear taxonomic entities. Instead, leaf serration type and their extent, together with mean leaf width and capitulum width are better markers for taxonomic segregation. The reinterpretation of the characters used in defining the species has not greatly changed the nomenclatural circumscription of the taxa, allowing for three taxonomic units: *C. chilensis* s.l., *C. elegans* s.l. and *C. x serrata*. The distribution of these taxa, as understood in this thesis, is shown in Figure 47. All collections are included.

C. chilensis is spread from the Cuenca de Santiago southwards along the coastal Cordillera, the Central valley and montane zones. *C. chilensis* is highly polymorphic and shows great phenotypic plasticity in its form (particularly in leaf size and pedicel length) throughout its range. *C. elegans* is a montane micro-species whose genetic effect demonstrably radiates outwards from Chillán to the north, west and south, wherever it meets *C. chilensis*. Material that is probably introgressive, but closest to *C. elegans* in terms of morphological appearance has been allocated to *C. elegans* s.l. *C. x serrata* is centred on Concepción and the surrounding Coastal Cordillera and central valley zone.

The complex pattern of variation is partially resolved by assumptions of hybridisation and introgression. *C. elegans* is distinguishable from a highly polymorphic *C. chilensis*, but the two species are incompletely isolated. Furthermore, there appears to be a stable hybrid – *C. x serrata* – that has a distinctive vegetative dispersal feature (stolons), with reduced seed dispersal ability (poorly developed carpopodia). This hybrid (and possibly F1 hybrids also) occupies drier more lowland niches than either of the parents. The hybrid appears to be genetically fit and does not seem to compete against the parents (seed bank studies, ARROYO et al. 2006). It seems likely that this analysis is confused by the incipient reticulate evolution of micro-species, especially in the southern volcanic montane areas of Chile.

The DNA-based phylogeny presented these perennial scapose taxa as the most novel in the genus *Chaetanthera* (HERSHKOVITZ et al. 2006). The generally unresolved nature of the relevant sampled collections could be further evidence of hybridisation. On closer inspection, the analysed collections in this subclade divided into a cluster of unresolved units and one outlier – *Arroyo & Humana 26000*. This outlying collection is *C. elegans*. If *C. elegans* were a result of an anomalous cytogenetic event this might skew the resolution of the DNA variation in this group.

VI "The variety of all things forms a pleasure"

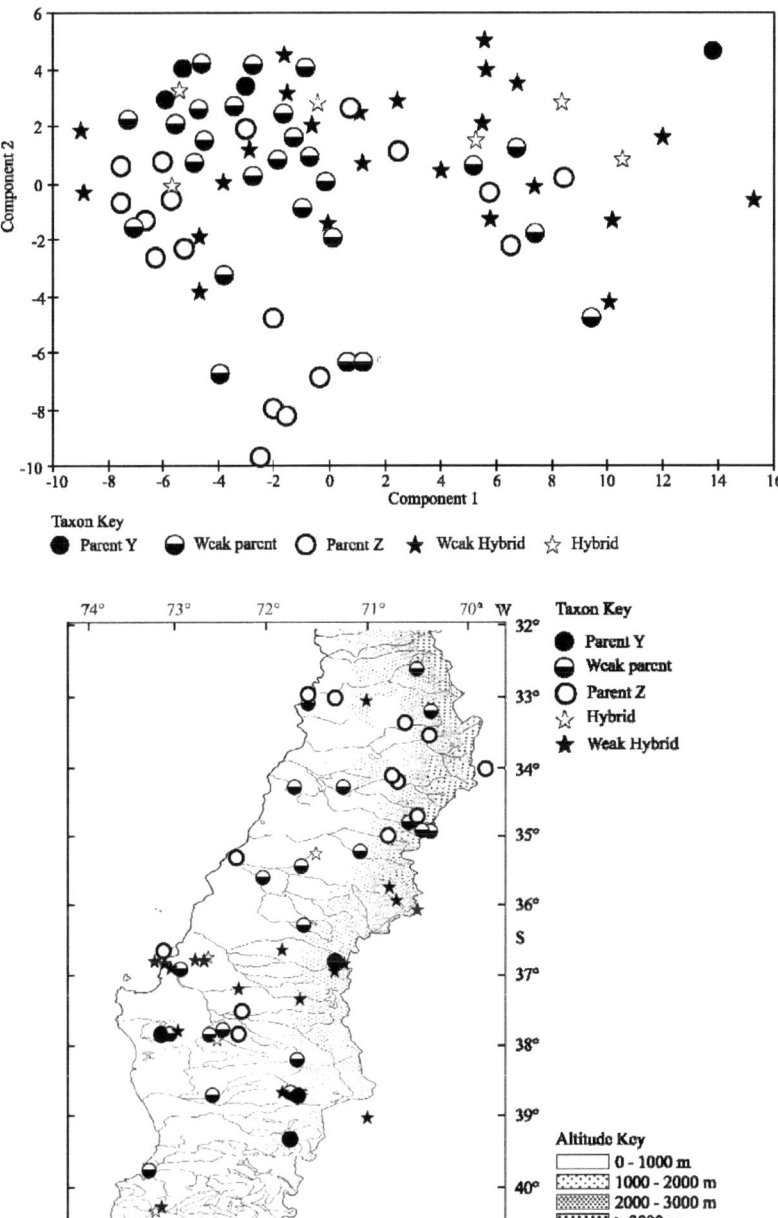

Figure 45: Equal weighting HYWIN Hybrid – Parent Analysis. PCA axes and geographical distribution.

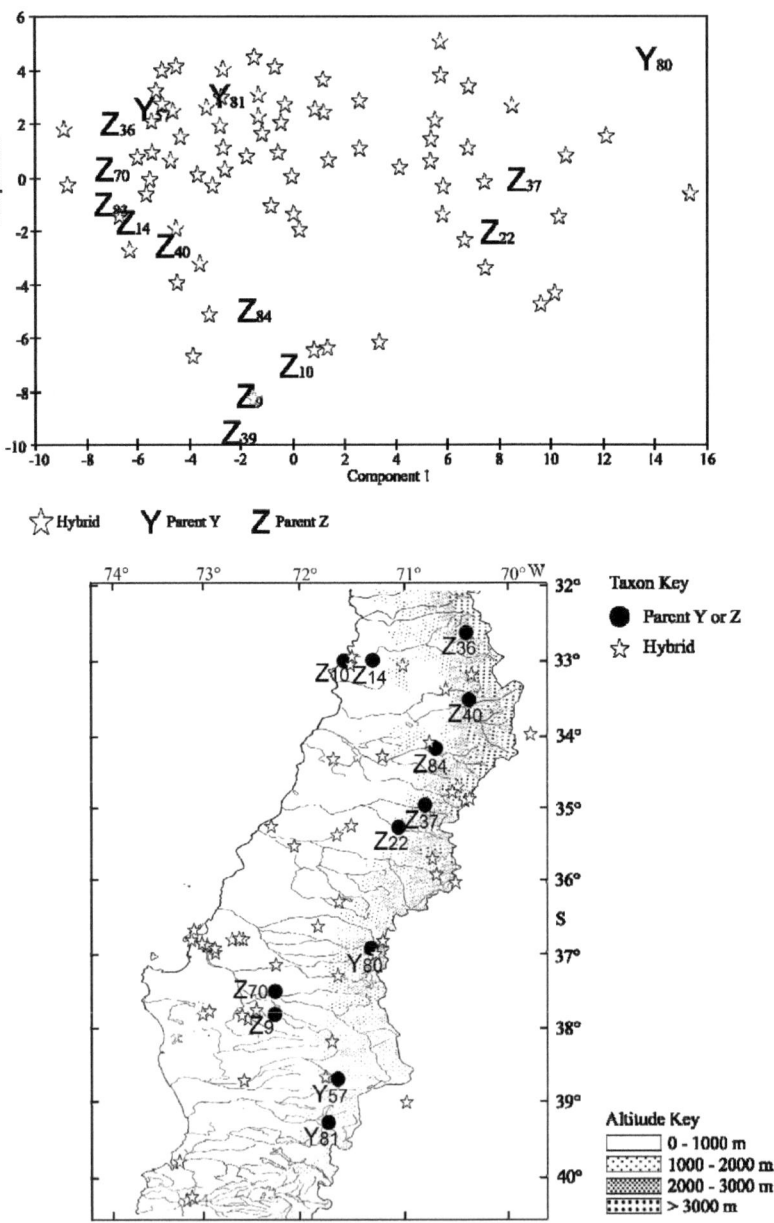

Figure 46: Parental Distance weighting for HYWIN Hybrid – Parent Analysis. PCA axes and geographical distribution. Y and Z are two parent groups. Subscript numbers are the analysis identifier.

4.6 Extinction or botanist's pride?

There is an array of disjunct collections from Valdivia, representing the southernmost locality of the species and the genus. Apart from one collection (*Hoellermayer 160b*) the remaining Valdivian collections are from around the Fundo San Juan (the Philippi residence). They occupy a very concentrated locus in the PCA graphic (not shown). The hybrid analysis (not shown here) places these collections as weak hybrids (probable introgressants) of southern "Chilensis" material and *C. elegans*. No collections of this morph have been made in over 150 years. Collection data implies that a spurious local population was founded by Philippi with hybrid seed from the Concepción area, and subsequently collected by his guests.

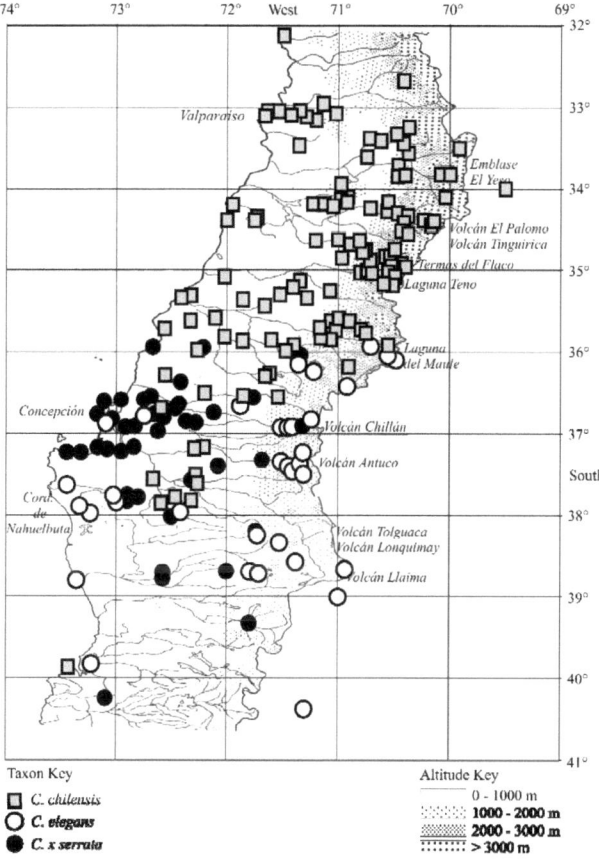

Figure 47: Distribution of *C. chilensis* s.l., *C. elegans* s.l. and the hybrid *C. x serrata*.

VII Discussion & conclusions of systematic analysis

1 Uniting characters - grouping species

Characters that unite species or groups of species are more useful both morphologically and phylogenetically than those characters that merely define a group by exclusion from another. Recent genetic analysis by PANERO & FUNK (2008) has clarified many hitherto paraphyletic generic relationships in the Asteraceae, including those within the Mutisieae. Traditionally, *Chaetanthera* s.l. has been morphologically associated with the genera *Brachyclados* D.Don, *Pachylaena* D.Don ex Hook. & Arn. (KATINAS 2008), and *Trichocline* Ruiz & Pav. within the tribe Mutisieae (e.g., CABRERA 1937, 1977, HANSEN 1991, BREMER 1994). Affinities based only on ndhF sequences (cpDNA) to *Duidea* S.F. Blake, *Plazia* Ruiz & Pav. and *Onoseris* Willd. were tentatively postulated by KIM et al. (2002) and HERSHKOVITZ et al. (2006). This relationship was revoked by PANERO & FUNK (2008) [using 10 cpDNA markers] and LUEBERT et al. (2009), following HIND (2007), who placed *Chaetanthera* s.l. as a part of an unresolved clade of the Mutisioideae subtribe Mutisiinae (= central core of tribe Mutisieae after KATINAS et al. 2008b) sharing bilabiate corollas, caudate anthers with long conspicuous tails, distally papillose style branches and shared pollen features. The presence of foliaceous outer involucral bracts and the alate middle/inner involucral bracts is a morphological feature of both *Chaetanthera* and *Oriastrum*, but not of the next closest genera that include *Adenocaulon* Hook., *Brachyclados*, *Mutisia* L fil., *Pachylaena* and *Trichocline*. These almost entirely South American genera show a variety of habit forms including sub-shrubs, climbers, and acaulous rosettes, with or without scapose inflorescences. The identification of further supra-generic characters was not in the remit of this study. CABRERA (1937) used habit/ life cycle, leaf shape, coloration of bract apices, ray floret exertion and achene pericarp features to distinguish seven subgenera within *Chaetanthera* s.l. This revision has shown that his subgeneric construct was highly artificial. For example, the annual/perennial switch is not phylogenetically useful in this group, and neither is ray floret exertion (see Chapter III, Section 2.6).

Although *Chaetanthera* and *Oriastrum* uniquely share two aforementioned involucral bract features and are genetically seggregated from other Mutisiinae, there are several novel characters that indicate that they form two independent groups of species that deserve generic recognition. The genera *Chaetanthera* and *Oriastrum* are each newly defined by their habit (stem architecture), inner involucral bract anatomy, pappus structure, pollen structure, stigmatic hair types and testa epidermis structure. Subgeneric divisions are recognised in both genera and are defined by variation in leaf shape, involucral bract features, pappus characters, testa epidermis structure, and achene hair type.
A synthesis of the systematically important character variation described in Chapter III is given in Table 8. The generic and subgeneric hierarchy is natural and phylogenetically robust.

Chaetanthera species all show some form of rosette or rosulate architecture.This can take the form of either an ephemeral basal rosette in juvenile plants (annuals) or a degree of lax to compact phyllotaxy below the capitulum (perennials), and rosettes formed by the foliaceous outer involucral bracts. The bracts have thin membranous alae and have no sclerenchymatous cells (apart from surrounding the vascular bundles) in the apical regions of the inner involucral bracts. *Chaetanthera* pappus setae have barbs that are usually appressed, closely spaced (frequent) and longly adhered (50 – 60% barb cell attached to seta). The pollen exine is thick (11 – 20 µm), the ectosexine is columellate (sometimes compact) and the nexine is dumb-bell shaped. The stigma lobes are bifid and are truncate to round. The stigmatic hairs are triangular with a smooth surface. All *Chaetanthera* species have 'U'-form or 'O'-form thickenings in the testa epidermis. The annual species have a distinct carpopodium, while the perennial species have a poorly defined carpopodium or lack one altogether.

The species in *Chaetanthera* subgenus *Chaetanthera* have leaves that are linear to narrowly spathulate. The inner involucral bracts are linear to narrowly lanceolate, or rarely rhomboid, the apices are acute to aciculate, nearly always with a mucro, and usually dark green to brown or black. The pappus is united at the base and generally indehiscent. The achenes have large (80 – 120 µm long) lanceolate twin hairs on the achenes and have 'U'-form thickenings in the testa epidermis. The species in *Chaetanthera* subgenus *Tylloma* are united by having indistinctly petiolate to spathulate, often limbate leaves. The inner involucral bracts are ovate and more or less acute with green, pink or pale translucent apices. The pappus is free at the base, and may or may not be dehiscent. The pappus setae may be basally to entirely ciliate or sub-plumose. The achene twin hairs vary from 10 – 100 µm in length and can be spherical or lanceolate. In some species there are no achene hairs present. The testa epidermis has 'U'- or 'O'-form thickenings, where the O-form has villi-like protrusions in the upper half.

Oriastrum taxa form dwarf densely compact to laxly branching cushions. The involucral bracts are initially foliaceous, progressively with robust membranous alae, and then reduced to entirely membranous. They have sclerenchymatous cells in the apical regions of the inner involucral bracts. The barbellate pappus setae have barbs that are usually lax and spreading, widely spaced (infrequent) and shortly adhered (40 – 50% barb attached to setae). The pollen exine is thin (5 – 10 µm), the ectosexine is always compact and the nexine is elliptic. The stigmas are bilobate with short, rounded (obtuse) lobes, and the stigmatic hairs are oblong with a textured surface.

The species in *Oriastrum* subgenus *Oriastrum* have leaves that are more or less spathulate and indistinctly petiolate. They are defined by having small (ca 30 µm long) spherical to conical twin hairs on the achenes, and internally ribbed testa epidermis cells. The species in *Oriastrum* subgenus *Egania* are defined by the presence of perennating basal stem buds. The leaves are usually linear, occasionally distally dilated to spathulate. There is little or no thickening of the testa epidermis. Several species have evolved the gynodioecious habit. These same species can also have uni-cellular papillae on their achenes. Several species have long (450 µm) filamentous achene hairs

Genus		Chaetanthera	
Subgenus		*Chaetanthera*	*Tylloma*
Character			
Breeding System		Monoecious	Monoecious
Habit	Form	Stem Rosettes (reduced in annuals)	Stem Rosettes (reduced in annuals)
	Cold-tolerance strategy	Woody stems and/or roots	Woody stems and/or roots
Bracts	Morphology	Alae thin membranous Apical region maculate, but contiguous with entire bract Central mesophyllous zone	Alae thin membranous Apical region maculate, but contiguous with entire bract Central mesophyllous zone
	Anatomy	Apices with clusters of sclerenchymatous cells around vascular bundles	Apices with clusters of sclerenchymatous cells around vascular bundles
Pappus		Setae indehiscent Setae = (2) 5 (7) mm long Basal cilia never present Barbs/100µm = 9 – 12 Barbs appressed	Setae in/dehiscent Setae = (4) 8 (14) mm long Basal cilia often present Barbs/100µm = 5 – 6 Barbs appressed
Stigma lobes		Lobes long (ca 0.5 mm), truncate Hairs triangular, apices acute, surface smooth	Lobes long (ca 0.5 mm), truncate Hairs not observed
Achenes	Carpopodium	Annuals: ± symmetrical ring Perennials: poorly-developed – absent	Annuals: ± symmetrical ring Perennials: poorly-developed – absent
	Hairs	Obovate-elliptic or lanceolate twin hairs 90 – 120 µm long	Obovate-elliptic or lanceolate twin hairs 90 – 120 µm long Spherical twin hairs 8 – 20 µm long
	Testa Epidermis	Parallel, linear epidermal cells Walls continuously thickly U-shaped	Parallel, linear epidermal cells (rarely with sinuous margin) Walls continuously thickly U- O shaped
Pollen	Grain	elliptic (Eq.) and subtriangular (Pol.)	elliptic (Eq.) and subtriangular (Pol.)
	Structure	Exine thick (11 – 20 µm) Nexine dumbbell shaped Ectosexine columellate	Exine thick (11 – 20 µm) Nexine dumbbell shaped Ectosexine columellate/ compact

Table 8: Summary of the character variation important at supra-specific levels within and between *Chaetanthera* and *Oriastrum*.

	Genus	*Oriastrum*	
	Subgenus	*Oriastrum*	*Egania*
	Character		
Breeding System		Monoecious	Monoecious/gynodioecious
Habit	Form	Compact to laxly spreading dwarf cushions	Compact to laxly spreading dwarf cushions
	Cold-tolerance strategy		Perennating stem buds
Bracts	Morphology	Alae robust membranous Apical region distinct	Alae robust membranous Apical region distinct
	Anatomy	Apices packed with sclerenchymatous cells	Apices packed with sclerenchymatous cells
Pappus		Setae dehiscent Setae = (2) 4 (6) mm long Basal cilia always present Barbs/100µm = 3 – 5 Barbs spreading	Setae dehiscent Setae = (3.5) 5 (7) mm long Basal cilia always present Barbs/100µm = 3 – 4 Barbs spreading
Stigma lobes		Lobes short (ca 0.1 – 0.25 mm), obtuse Hairs not observed	Lobes short (0.1 – 0.25 mm), obtuse Hairs elongated, apices rounded, surface echinate
Achenes	Carpopodium	Poorly-developed – absent	Poorly-developed - absent
	Hairs	Spherical twin hairs (20 – 50 µm long) Deltoid twin hairs (10 – 20 µm long)	Filiform hairs (450 µm long) Oblate flattened hairs (25 – 45 µm long)
	Testa Epidermis	Tesselated sinuous (seldom parallel, linear) epidermal cells Walls strengthened with U-shaped ribs.	Parallel, linear epidermal cells Walls thin, irregularly thickened O-shaped X-section
Pollen	Grain	subrectangular (Eq.) and subcircular (Pol.)	subrectangular (Eq.) and subcircular (Pol.)
	Structure	Exine thin (5 – 7 µm); Nexine elliptical Ectosexine compact	Exine thin (5 – 7 µm) Nexine elliptical Ectosexine compact

Table 8: Summary of the character variation important at supra-specific levels within and between *Chaetanthera* and *Oriastrum*.

2   Evolutionary adaptation to environmental stress

By considering the form, function and habitat of the species it is possible to identify trends in character variation that have evolutionary significance. The morphological and anatomical arguments reviewed in Chapter IV are summarised and discussed in the context of the nrDNA based lineages and biogeographic scenarios postulated by HERSHKOVITZ et al. (2006).

Chaetanthera species seem to be primarily cold-adapted and secondarily adapted to aridity. The general trend of progressions in several characters throughout the genus, i.e., life cycle, rosette form and function, pollen and testa structure, pericarp myxogeny, pappus and carpopodium features, indicate that the lowland species are more derived than the high elevation taxa.

The genus is so derived that the rosette form and leaf function has devolved upon the foliaceous imbricate outer involucral bracts. The decoupling of this feature from the aridity of the habitat implies that the original function of the rosette in this genus was primarily as protection against cold and for photon capture. The relict stem rosette seen in the annuals, in most cases only present in the juvenile phase, is greatly reduced and used for the harvesting of precipitation. The development of xerophilic features in pollen and testa structure, and pericarp myxogeny indicates taxa adapted to arid conditions. Secondary barochory and poor carpopodium development is found in several high elevation annual and perennial species. The lowland arid-tolerant annual species have a distinct carpopodium and indehiscent pappus, indicating a predisposition for wide-spread dispersal. These species are highly autogamous (self-compatible) and have a persistent seed bank. They also have the ability to germinate rapidly in response to water; an advantage in water-stressed habitats.

Chaetanthera species show indisputable evidence of a current evolutionary cradle in lowland Chile. The lineage divergence dates of HERSHKOVITZ et al. (2006) strongly support a biogeographic scenario involving adaptations of their *Chaetanthera* s.str. clade to increasing lowland aridity as a result of Andean orographic activity between 16 and 13.5 mya. It is considered unusual to find examples of young modern lineages occupying lower elevation older habitats.

Oriastrum species are adapted to high elevations: that is elevations characterised by having a shorter growing season and lower growing temperatures resulting in lower evapo-transpiration, and less water stress than at low elevations. *Oriastrum* species have the stem and leaf architecture of a cushion, (although this is laxly formed in a number of species) that provides thermal insulation. Most perennials have perennating stem buds that are an over-wintering, cold tolerant strategy. All species have a protective indumentum. The bract apices are anatomically strengthened, providing protection and insulation of reproductive organs. *Oriastrum* species are not particularly adapted to arid conditions, or to conditions of extreme hydration fluctuation. They have thin walled pollen grains, and only secondary strengthening in the testa epidermis of the annuals. The annuals also show trends towards secondary adaptation to drier conditions in having myxogenic twin hairs on their achenes. They show secondary barochory (rapid loss of pappus post-dispersal) and have no functional carpopodia.

The nrDNA evidence of HERSHKOVITZ et al. (2006) shows that the *Oriastrum* clade split from *Chaetanthera* s.str. 13.5 mya, before the modern high elevation Andean habitat existed. HERSHKOVITZ et al. postulate that the *Oriastrum* annuals and perennials escaped the increasingly water-stressed lowland habitats of the late Miocene by migrating upwards with the juvenile central Andes, forming a – counterlogical – relictual node in a geologically novel habitat. In the light of the analysis of adaptive characters considered here this is only true in part. In contrast to the modern *Chaetanthera*, the relictual *Oriastrum* annuals (*Oriastrum* subgenus *Oriastrum*) are only somewhat adapted to aridity, but are also not well adapted to frigid high elevation environments. They do seem to be relicts of an earlier age. However, the rapid, recent speciation seen in an apparently ancestorless, high elevation *Oriastrum* subgenus *Egania*, dated from 5 mya, shows novel successful adaptations to the young high elevation habitat. This contradicts the scenario that *Oriastrum* (as a whole) is stuck in a high elevation museum.

Oriastrum subgenus *Egania* demonstrates several expressions of Carlquist's "insular syndrome" (1974, cfr. 2008). This syndrome, characterized by dioecy in plants, polymorphism within species and low rates of reproduction, derives from the high incidence of these features observed in island floras, and may relate to selection for loss of dispersability after lineages have become established on remote islands. Selection for reduced dispersal potential has been demonstrated in some island plant populations (CODY & OVERTON 1996) as evidence of short-term evolution. Those features that preclude long-distance dispersal such as fragile pappus or length reduction of pappus setae are, in some instances, typical of taxa found in mountain areas (e.g. some genera within the Cardueae subtribe Carduinae - HÄFFNER 2000). Trends in the "insular syndrome" can be identified as follows. Dwarfism is often considered to be an aberrant form. *Oriastrum* cushions are dwarfs, especially compared to many of the other taxa found in the high elevation habitats of the Andes. In several species the viable achenes are comparatively large, and seed set is poor (e.g. *O. apiculatum*, *O. cochleariifolium*). Secondary barochory and lack of carpopodium contributes to a loss of dispersibility, which increases genetic isolation of populations and leads to subsequent speciation. Several perennial species are gynodioecious (e.g. *O. polymallum*), a state that can be interpreted in some instances as an apomorphic feature (after RENNER & WON 2001). Obligate outcrossing (gynodioecy) raises genetic diversity and helps avoid inbreeding depression. Morphological (*O. acerosum*) and genetic (*O. stuebelii*) polymorphism within species and wide ranging dispersal throughout the ecozone (*O. polymallum*) indicates dynamic evolutionary entities. They only seem ecologically isolated because of the lack of adaptations to lowland arid conditions. Understanding in this latter group is hampered by the relative amplitude, altitude and inaccessibility of the preferred habitats.

3 Current dynamic change in *Chaetanthera*

The biogeographic distribution of species can be interesting from a biological or an evolutionary perspective. In the past, priority areas for conservation were often selected on the basis of the number of species in a given area. In recent years this has changed towards prioritising efforts according to entities, such as number of endemics (e.g. CAVIERES et al. 2002) or supra-specific species richness. It follows that the conservation value of a genus will be reflected by its endemic or diverse properties. *Chaetanthera* is principally endemic to Chile (20 of 30 species), with all but one of the remaining ten species also recorded from, but not restricted to, Chile. More interestingly, *Chaetanthera* has two diversity hotspots (see Chapter V, Section 1.2); in Coquimbo and Santiago. They are both located somewhat to the north of the main Chilean biodiversity hotspot (situated between 34°S and 40°S, according to BARTHLOTT et al. 1998). By collating hotspots of species – that is the number of species coexisting in the same area (ecological or geographical) – one can consider the environmental influences driving the sympatric aggregation of related taxa.

Species radiations are evidence of past dynamic change in a genus. There are examples in *Chaetanthera* of historical radiations, spanning outwards from both hotspots. The variation in leaf indumentum of the limbate, indistinctly petiolate-leaved species complex within *Chaetanthera* subgenus *Tylloma*, (Chapter V, Section 1.3) or the shared lamina morphology of several suites of annual species in *Chaetanthera* subgenus *Chaetanthera* (Chapter V, Section 1.4) are two examples of past radiation events. However, *Chaetanthera* also has some excellent examples of current dynamic change, as evidenced by polymorphism within the collected material. From an evolutionary biology perspective, areas where species boundaries break down or are incomplete are as interesting as those areas where there is distributional congruence amongst different endemic taxa.

Polymorphism is a well known phenomenon, in particular heterophylly (*Ranunculus* subgenus *Batrachium*; *Polygonum amphibium* after BRIGGS & WALTERS 1984). Phenotypic variation and plasticity can often be an indicator of a plant's response to stress. In *Chaetanthera glabrata*, the observed leaf polymorphism is demonstrably linked to the climatic phenomenon of El Niño (Chapter VI, Section 2). There is some evidence in the literature regarding the potential effect of the El Niño on phenotypic plasticity of plants. The effect of ENSO-linked growth has been recorded for 2 woody *Prosopis* species (*P. pallida* and *P. chilensis*) from the coastal regions of Peru and Chile respectively (LÓPEZ et al. 2006). Changes in primary productivity in vegetation in Coquimbo (Chile) as a result of ENSO events were observed by SQUEO et al. (2006). They also found that El Niño has the greatest impact at low elevations, while the effect is buffered by cooler air temperatures at mid-elevations. In the instance of El Niño-driven polymorphism, it is important to realise that the hydration stress typically experienced by the plants occupying arid/semi-arid regions is alleviated, not that the El Niño event itself is a cause of stress (after HOLMGREN et al. 2006). Thus, in the case of *C. glabrata* the leaf polymorphism (= increased vegetative productivity) is a result of removal of hydration stress. *C. glabrata* avoids the huge competitive surge in plant species recruitment (HOLMGREN et al. 2001) during El Niño events by having a very rapid germination response (1-3 days after hydration).

Phenotypic mosaics in taxa with porous genomes – i.e., those taxa that are incompletely separated and experience gene flow – are especially challenging for evolutionary taxonomy, raising issues such as character conflict in phylogenetic studies, biased sampling of traits in morphological studies and cryptic cases of ecological speciation. These are all issues raised by the case studies of *C. albiflora* – *C. linearis* and *C. chilensis* – *C. elegans*. A good explanation of analytical issues encountered when studying taxa with porous genomes is given by LEXER et al. (2009).

As outlined in Chapter VI, Section 3, a large scale shallow cline of hybrids stretching over 300 Km between 30° – 33°S exists between two parent species: *Chaetanthera albiflora* and *C. linearis*. The apparent introgression of the upland southern species into the lowland (coastal) northern species, to the exclusion of the southern parent in northern upland areas, indicates that natural selection favours the hybrid genotype in that environment.

Not all polymorphism is so easily explained. Complex patterns of variation between only a few species can be generated by incompletely isolated taxa and subsequent reticulate hybridisation and introgression. The perennial scapose *Chaetanthera* present one such scenario, as analysed in Chapter VI section 4. *C. chilensis*, a highly polymorphic, wide-spread, southern Chilean montane-mid elevation species seems incompletely isolated from the montane micro-species *C. elegans*. *C. elegans* has an active genetic effect radiating out from its localities in southern montane areas. Additionally, these two species have generated a stable distinctive hybrid, *C. x serrata*. *C. x serrata* prefers more lowland, southerly niches than the parents, and is characterised by having stolons and poorly developed carpopodia. It is unclear whether the hybrid itself produces viable seed. It seems likely that, even with a reduced viability in the hybrid offspring, the swarm could be maintained via backcrossing (introgression) with the parents. Although the analyses implied that several montane micro-species could be involved in a reticulate configuration, a reassessment of the material would be necessary first. The use of AFLP's to calculate genetic distances and polymorphism in metapopulations is one technique that can aid understanding of spatial variation within and between populations (e.g. *Sisybrium austriacum*, JACQUEMYN et al., 2006). Analysis of cpDNA variation is increasingly used as a tool to aid interpretation of plant evolutionary biology. Directional introgression, initially indicated by multivariate analysis, between two *Populus* species has been successfully demonstrated using single nucleotide polymorphisms from both nuclear and chloroplast genomes (HAMZEH et al. 2007). Variation in the intergenic cpDNA regions was also studied in *Crataegus* hybrids but with less clear results (ALBAROUKI & PETERSON, 2007). It is possible that these techniques might also be suited to this *Chaetanthera* complex.

Hybrids often occur at ecotones or boundaries between different habitats (HARRISON 1993). Phenotypic variation as a result of active hybridisation between two (or more) species is recorded in the literature concerning native Chilean species, although it has not been considered in terms of ecological boundaries found in Chile. For example, various hybrids within the Chilean *Calceolaria* (Scrophulariaceae) are said to be ephemeral or exist as stable swarms (EHRHART 2005) where the parents coexist sympatrically, but little is recorded about the ecological significance of the sympatric zone. Similarly, the species-rich genus *Haplopappus* forms only a few hybrids (7 recognised hybrid forms) where the species distributions are sympatric (KLINGENBERG 2007). *Chaetanthera* has two instances of species instability over different boundary zones. The significance of hybridisation events over the arid – semi-arid boundary in Coquimbo seems to be largely undocumented, although the high number of species, especially endemics, characterising this region is well known. The hybrid swarm between the annuals *C. linearis* and *C. albiflora* seems stable (not ephemeral) and selection appears to favour the hybrid genotype in ecozones where the parent species are less fit. The north-south climatic/hydrological boundary zone between Maule and Chillán seems to be a locus of change in the caespitose *C. chilensis/ C. elegans* species. The genus

Baccharis L (Astereae) forms 27 hybrids, scattered throughout Chile (HELLWIG 1990), including a hybrid specific to the Chillán area. *Nothofagus obliqua* and *N. glauca* also form natural hybrids in this area (DONOSO & LANDRUM 1979). The dwarf hybrid shrub *Haplopappus glutinosus* x *paucidentatus* also occurs as hybrid swarm reported from the Chillán area.

4 Conclusions

Chaetanthera and *Oriastrum* are two independent but closely related South American Mutisioid genera distinguished from each other by morphological, anatomical and genetic variation. The array of novel characters observed is particularly rich for a study of the Asteraceae. Together with the reassessment of traditional features they have enabled the re-defining of the generic boundary of a primarily Chilean *Chaetanthera*, and resulted in the reinstatement of the south Andean *Oriastrum*.

Analysis of form (character variation), function (adaptive strategy) and habitat (source of environmental stress), together with phylogenetic lineages, shows there are several contrasting modes of evolutionary development in *Chaetanthera* and *Oriastrum*. *Chaetanthera* has reduced adaptations to cold and secondary adaptations to aridity, especially in the lowland species. This corresponds to a biogeographic scenario of migration and recolonisation from cooler wetter higher (but geologically younger) elevations in the western Andes to more arid (geologically older) lowland habitats in Chile. *Oriastrum*, particularly *Oriastrum* subgenus *Egania*, is a phylogenetically young group of dynamic taxa very well adapted to but also islanded in the Piso Andino/ Altoandino/ Andino Superior of the modern Central Andes. *Oriastrum* subgenus *Oriastrum* forms a relict group of mid-elevation annuals that is only weakly adapted to both dry and cold conditions.

Endemism and polymorphism can be valuable indicators of historical and current dynamism in a genus. Species diversity hotspots in the endemic Chilean genus *Chaetanthera* demonstrate historical dynamic change. These radiation events have stabilised into well-defined suites of taxa with shared leaf morphologies. Current dynamic events within and among species are identified by polymorphism. The polymorphism is driven by two different mechanisms: 1) phenotypic flexibility is a response to periodic changes in water stress and 2) the polymorphism is a result of weak species boundaries or porous genomes, resulting in hybrisation events and microspeciation.

Part B:

Taxonomy of *Chaetanthera* & *Oriastrum*

VIII *Chaetanthera* & *Oriastrum*: an historic review

1 Introduction

Chaetanthera has always had an unstable generic concept. Since its description in 1794, various genera have been alternately re-instated or merged under the umbrella of *Chaetanthera* s.l. no less than 6 times. The main points about the nomenclature and publication of *Chaetanthera* are outlined here. Omitting taxa that have been relocated, infra-generic and generic names from seminal publications are shown in the Taxonomic History Outline (Fig. 48). The generic and sub-generic hierarchy is given for eight publications in chronological order. This shows the instability or fluidity of the placement of groups of species belonging to *Chaetanthera* and *Oriastrum*. The publications were selected on the basis of their overview of the genus sensu lato and do not necessarily represent all the most significant publications about the genus, nor does the diagram include all publications that have described new infra-generic taxa.

2 Chronological changes in *Chaetanthera* & *Oriastrum*

Chaetanthera Ruiz & Pav. (Asteraceae: Mutisieae) (non Nutt. 1834, Asteraceae: Astereae) was first described in 1794 (RUIZ & PAVON). At that time, no species within the genus were validated and *Chaetanthera ciliata* was the sole species depicted in the original illustration (1794, Lam. 23) accompanying the generic description (PRUSKI & DAVIES 2004). The later description of the two species *Chaetanthera ciliata* and *Chaetanthera serrata* by RUIZ & PAVON in 1798 is taken as the starting point. In 1830 David DON included three new genera in his paper entitled "Mr. D. Don's Descriptions of new Genera and Species of the Class Compositae" (*Proselia*, *Euthrixia* and *Tylloma*). LESSING's "Synopsis Generum Compositarum", published in 1832, was one of the first *Chaetanthera* publications to use infra-generic subdivisions in *Chaetanthera*, dividing the eleven existing taxa among four subgenera based on leaf and bract shape. By 1838 DE CANDOLLE had reworked the material, describing the uni-specific genera *Carmelita* and *Elachia*, retaining *Chaetanthera* and *Tylloma* as separate generic entities. The novel genus *Oriastrum* was described in 1842 by POEPPIG & ENDLICHER. This genus has been intrinsically linked with *Chaetanthera* since its description. In 1848 J. RÉMY used the rich collections of Claudio Gay and upheld 7 genera, including 25 taxa that are currently recognised. As well as using leaf and bract shape, he sourced novel characters for the circumscription of the infra-generic ranks from habit, florets and achenes, although these were later shown to be inconsistent in his configurations.

The next overview of *Chaetanthera* was not published until 1873, in BENTHAM & HOOKER's "Genera Plantarum". They united all eleven of the hitherto published genera (*Cherina* Cassini, *Proselia* D. Don, *Euthrixia* D. Don, *Elachia* Remy, *Tylloma* D. Don, *Carmelita* C. Gay, *Oriastrum* Poepp. & Endl., *Aldunatea* J.Rémy, *Egania* J.Rémy and *Chondrochilus* Phil.) within *Chaetanthera*, bringing the total number of taxa to 26. However, within 8 years this was counteracted. In 1881, F. PHILIPPI published a "Catalogus Plantarum Vascularium Chilensium adhuc descriptarum" in which he reinstated nine of the genera (omitting *Cherina*, *Proselia* and *Euthrixia*, their taxa, erroneously or otherwise, cited as synonyms) for the now forty-one taxa. Thus, by the late 1800's, opinion of the generic limits surrounding these taxa was divided across the Atlantic. On the one hand, the European "lumpers" had access to and actively exchanged both literature and material, especially that based on the types, and on the other hand, the Chilean-based botanists were actively "splitting", based largely on the rapidly expanding exsiccatae and collections of the Philippi family.

K. REICHE reached something of a compromise in his Flora de Chile (1905). He outlined *Chaetanthera* with 18 taxa (divided unequally between two subgenera), *Tylloma* with 5, *Carmelita* with 2 and *Oriastrum* with 11 taxa (arranged in 3 sections). This collection of small but clearly related genera remained like this for another 30 years before the Argentinian botanist A. CABRERA tackled it. He was the first to combine both type material and literature from both sides of the Atlantic, resolving many synonyms, and clarifying species boundaries. His solution to the multitude of inter- and infra-generic taxa was to unite them all under *Chaetanthera*, based on the prescence of foliaceous outer involucral bracts. He grouped thirty-seven species among seven subgenera (*Proselia*, "Eu"*chaetanthera*, *Tylloma*, *Carmelita*, *Egania*, *Oriastrum* and *Glandulosa*) based on life cycle, habit and inner involucral bract form, as well as the more "traditional" characters drawn from achene indumentum, leaf and bract shape.

In the following 70 years five new *Chaetanthera* species were described – *C. boliviensis* Kost. (1945); *C. chiquianensis* Ferreyra (1953); *C. leptocephala* Cabrera (1954); *C. aymarae* Martic. & Quezada (1974); *C. kalinae* A.M.R.Davies (2006). The novel variety *C. pulvinata* var. *acheno-hirsuta* Tombesi (2000) was raised to species level *C. achenohirsuta* (Tombesi) Arroyo, Davies & Till-Bottraud (2004). *C. andina* var. *pratensis* (Phil.) Reiche was moved to *C. elegans* var. *pratensis* (Phil.) Cabrera (1971) and *Luciliopsis perpusilla* Wedd. was moved to *C. perpusilla* (Wedd.) Anderb. & Freire (1990).

Chaetanthera (including *Oriastrum*) has been numbered among the top 5 largest Asteraceae genera in Chile with *Senecio*, *Haplopappus*, *Leucheria* and *Baccharis* (MARTICORENA & QUEZADA 1985; DAVIES & FACHER 2001). The present work defines *Chaetanthera* with 29 species, 1 variety and 2 hybrid taxa and *Oriastrum* with 18 species and 1 variety. This places both genera amongst the twelve most species-rich Asteraceae genera in Chile, along with genera such as *Nassauvia*, *Perezia* and *Mutisia*.

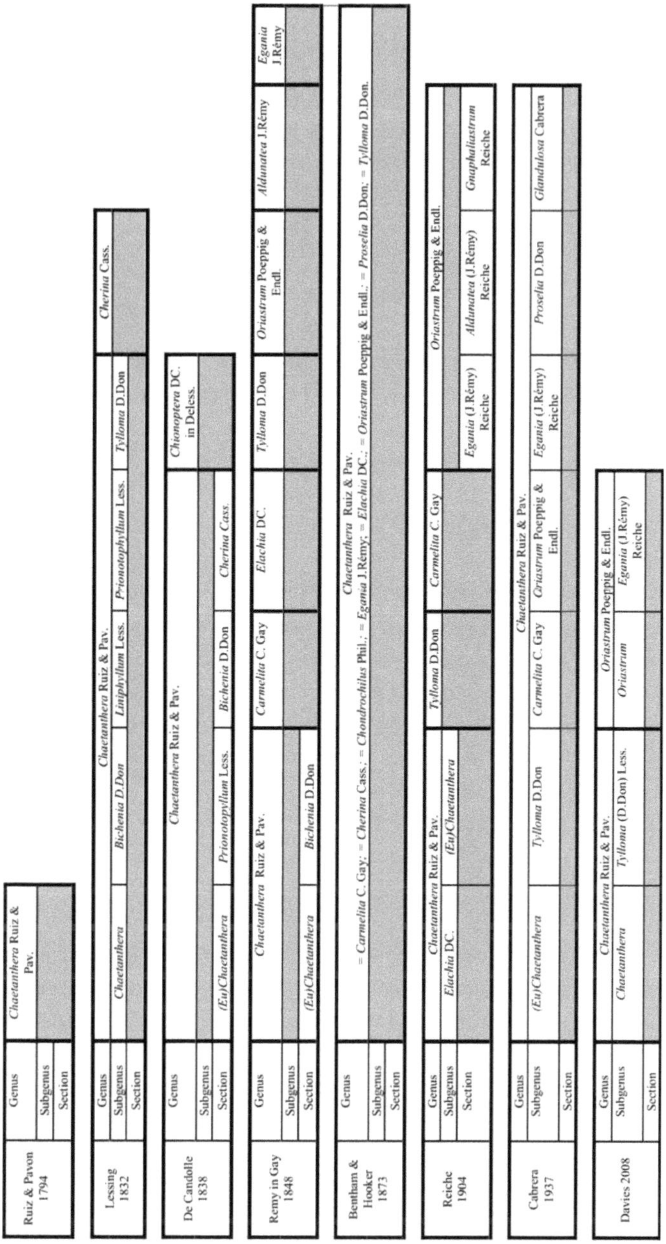

VIII *Chaetanthera* & *Oriastrum*: an historic review

Figure 48: Taxonomic history outline. The figure illustrates the fluidity of the supra-specific ranks of section, subgenus and genus as applied to the taxa in this revision.

3 Nomenclatural notes for *Chaetanthera* & *Oriastrum*

The species descriptions have a nomenclatural note attached to them where appropriate. However, there are two major issues that affect the nomenclature at a generic level. The first issue concerns the generitypes of both *Chaetanthera* and *Oriastrum*. The most significant authors (excluding the present work) in terms of *Chaetanthera* and *Oriastrum* nomenclature have been D. Don (1799 – 1841), C.F. Lessing (1809 – 1862), J. Rémy (1826 – 1893), R.A. Philippi (1808 -1904) and A.L Cabrera (1908 – 1996). The second issue concerns the impact that the authors D. Don and R. Philippi had on the nomenclature of these genera.

3.1 *Chaetanthera ciliata*: lectotype of *Chaetanthera* Ruiz & Pav.

PRUSKI & DAVIES (2004) discussed the lectotypification of *Chaetanthera* in detail. The two original species of *Chaetanthera* (*C. ciliata* and *C. serrata*) described by Ruiz and Pavon have been variously treated as belonging to different genera (e.g., DON 1830), to different infra-generic taxa of *Chaetanthera* (e.g., CABRERA 1937), or to the same infra-generic taxa of *Chaetanthera* (e.g., LESSING 1832; CANDOLLE 1838). DON (1830) removed *C. serrata* to a uni-specific *Proselia* D.Don, leaving *C. ciliata* Ruiz & Pav. and *Chaetanthera spinulosa* Cass. (named in 1826) as the only two names at that time within *Chaetanthera*. That the two original species belong to two different groups was echoed by CABRERA, who treated *C. serrata* within *Chaetanthera* subgenus *Proselia* (Don) Cabrera and *C. ciliata* within *Chaetanthera* subgenus *Chaetanthera*. Neither taxonomic treatments by Don or Cabrera technically effected nomenclatural typification of the name *Chaetanthera*. Similarly BENTHAM & HOOKER (1873: 496) noted that *C. serrata* was the sole species attributed to *Proselia*, but neither did they effect lectotypification of *Chaetanthera*. Two years after Don described *Proselia* LESSING (1832) reduced Don's genus into the synonymy of *Chaetanthera*. Specifically, Lessing treated both *C. ciliata* and *C. serrata* as members of *Chaetanthera* subgenus *Chaetanthera*. Candolle (1838: 30-31) did the same, albeit at the rank of section, within *Chaetanthera* section *Chaetanthera*.

Ruiz and Pavon published 12 new Compositae genera (48 species) in the Systema vegetabilium Florae Peruvianae et Chilensis (1798). The generic characters were illustrated in their 1794 Florae Peruvianae et Chilensis Prodromus, but the series was not completed, and the species plates were never published (CABRERA 1960; STAFFLEU 1979). *C. ciliata* is definitively the subject of the illustration accompanying the original publication of the name *Chaetanthera* in the 1794 Florae Peruvianae et Chilensis Prodromus of Ruiz and Pavon, although no specific epithet was attached to the illustration. CASSINI (1817: 53) wrote "la chaétanthère ciliée devra toujours, selon nous, être considéreé comme le vrai type du genre". It could be debated whether this entry constitutes an effective lectotypification, although Cassini clearly intends to signify this name as the preferred "true" type. Additionally, by applying the unofficial but useful "doctrine of residues", it could be inferred that as a result of Don's description of *Proselia*, and *C. ciliata* being left as the only original Ruiz and Pavon name in *Chaetanthera*, *C. ciliata* was rendered the type by default. However, in 1978, CABRERA did indeed cite *Chaetanthera* as typified by *C. ciliata*, in his treatment of the Argentinean "Flora de Jujuy". Its omission in the generitype listings of FARR et al. (1979) and FARR & ZIJLISTRA (1996) will hopefully be rectified in future editions of the Index Nominum Genericorum.

3.2 *Oriastrum pusillum*: generitype of *Oriastrum* Poepp. & Endl.

FARR et al. (1979) correctly quoted *Oriastrum pusillum* as the generitype of *Oriastrum*. The reinstatement of *Oriastrum* as a genus is quite straightforward and well supported by morphological, anatomical and genetic studies. There is, however, a confusion of epithets defining two taxa belonging to this genus. For the sake of simple argument these are designated the red entity and the white entity (referring to the colouration of the inner involucral bract apices).

The white entity, *Oriastrum pusillum* Poepp. & Endl. is the type species of *Oriastrum*, and was comprehensively described and illustrated in 1842. The red entity, first described as *Tylloma pusillum* D.Don ex Taylor & Phillips (1832), on its transferral to *Oriastrum* would result in the illegitimate combination *Oriastrum pusillum*. However, a substitute name for this red entity already exists – *Oriastrum chilense* (J. Rémy) Wedd. (1855), based on *Aldunatea chilensis* J. Rémy in Gay (1849).

3.3 David Don: nomenclatural types & nomenclatural obscurity

The work of David Don is significant for the genus *Chaetanthera* Ruiz & Pav. Don's most prolific botanical work was done while he was employed by A.B. Lambert between 1820 and 1836. However, there have been difficulties associated with the treatment of Don's names, specifically the citation and distribution of his nomenclatural types, and nomenclatural obscurity. There are two relevant publications. The first, in the Transactions of the Linnaean Society London (vol. 16, 1830), describes 3 new genera: *Proselia* D.Don (generitype: *Proselia serrata* (Ruiz & Pav.) D.Don), *Tylloma* D.Don (generitype: *Tylloma limbatum* D.Don) and *Euthrixia* D.Don (generitype: *Euthrixia salsoloides* D.Don). The second was published in 1832 (Philosophical Magazine, vol. 11), in which 10 relevant names were published: *Chaetanthera argentea*, *Chaetanthera eryngioides*, *Chaetanthera flabellata*, *Chaetanthera prostrata*, *Chaetanthera ramosissima*, *Chaetanthera scariosa*, *Chaetanthera tenuifolia*, *Chaetanthera villosa*, *Euthrixia affinis* and *Tylloma pusillum*.

3.3.1 Don's nomenclatural types

MILLER (1970) is the authoritative publication about Lambert's herbarium, Don's role in its dispersal before Lambert's death in 1842, and the present whereabouts of the herbarium. This was invaluable in determining possible locations of type material (viz. BM, E, G, K and OXF). The type material Don used was sourced primarily from material collected by J. Gillies, although H. Pavon, H. Cuming and M. Caldcleugh also contributed. The type collections made by Gillies were found to be signed, either with a handwritten label or on the reverse by Gillies himself. Duplicates are unusual. The Caldcleugh collection (*Tylloma limbatum* D.Don) was traced to G. Descriptions of *C. argentea* D.Don ex Taylor & Phillips and *C. flabellata* D.Don ex Taylor & Phillips were not attributed to any collector, although potential Cuming material was found in several herbaria, and subsequently has been lectotypified. Pavon material was found in one instance (*Proselia serrata*) but not for *Euthrixia*.

3.3.2 Nomenclatural obscurity of Don names

The ramifications of the 1832 publication for *Chaetanthera* names were first brought to light by MABBERLEY (1981). Don gave a presentation to the Linnaean Society on March 20 1832. Don supplied an abstract of his (unpublished) Descriptive Catalogue in the form of a list of rather brief species descriptions, but the manuscript was prepared and published by the editors TAYLOR & PHILLIPS for inclusion in the Philosophical Magazine. The same abstract was reproduced the following year (1833) by GUILLEMIN in the Bulletin Bibliographique. The names are correctly attributed to D.Don ex Taylor & Phillips.

Parallel to Don's work in England, C.F. Lessing in Berlin was studying Chilean Compositae material collected by Poeppig. This inevitably led to nomenclatural clashes. The date of effective publication of the Philosophical Magazine was April-May 1832. The date of effective publication of Lessing's Synopsis Generum Compositarum was July-August 1832. Lessing's name, *Chaetanthera tenella*, has been used unequivocally by the botanical community since its publication, while Don's nomenclaturally correct name, *Chaetanthera ramosissima* lapsed into obscurity and remained there ever since. A petition for rejection would not be appropriate because the legitimate, valid and effective criteria for publication are satisfied. There is no adequately designated type in the protologue, but designating historical material would reinforce the correct usage of the name. In this case there is well-dispersed material collected by Cuming (*Cuming 656*) and annotated by Hooker. MABBERLEY (1981) already commented on the obscurity of this name, but it remains to find its way into the *Chaetanthera* literature.

There is a second instance where a pertinent Don name has never become properly incorporated into the botanical literature. The details surrounding the untangling of *Euthrixia salsoloides* and *Chaetanthera linearis* were so complex that the rejection of the name *Euthrixia salsoloides* D.Don was formally proposed by DAVIES (2005). The official rejection of the *Euthrixia* generitype *Euthrixia salsoloides* avoided an otherwise necessary but disadvantageous nomenclatural change and allowed the continued and thus correct use of the later name *Chaetanthera linearis* Poepp. ex Less. that has been in common usage since 1832. The attendant loss of the obscure junior synonym *Euthrixia affinis* causes no nomenclatural problems.

3.4 Rudolfo Philippi: lectotypes & nomina nuda

"R.A. Philippi was one of the most important scientists in the history of Chile"
TAYLOR & MUÑOZ SCHICK, 1994

R.A. Philippi was also one of the most important scientists involved in the collection and description of *Chaetanthera* and *Oriastrum* material. Of the 3720 plant names published by R.A. Philippi, 64 pertain to the modern concept of *Chaetanthera* and *Oriastrum*. MUÑOZ PIZARRO (1960) listed type or potential type (typoid) material from the collections in the Museo National de Historia

Natural in Santiago (SGO). The basionyms, along with the type locality taken from the protologue, modern 4-Fig. grid references (approximate) of the localities, and the SGO numbers of appropriate sheets are given. However, a number of the sheets were erroneously cited, or typographic errors had crept in. The publication quotes the numbers in numerical order. According to TAYLOR & MUÑOZ SCHICK (1994), the intention of MUÑOZ PIZARRO (1960) was not to select lectotypes, but to organize the material to facilitate work by specialists.

Those collections where more than one isotype exists, and the holotype is not designated on the sheet or in the protologue and those instances where Muñoz Pizarro listed syntypes have been lectotypified. Thirteen *Chaetanthera* names are lectotypified here: *Carmelita spathulata* Phil., *Chaetanthera andina* Phil., *Chaetanthera argentea* Phil., *Chaetanthera elata* Phil., *Chaetanthera foliosa* Phil., *Chaetanthera humilis* Phil., *Chaetanthera involucrata* Phil., *Chaetanthera obtusata* Phil., *Chondrochilus grandiflorus* Phil., *Chondrochilus involucratus* Phil., *Tylloma ciliatum* Phil., *Tylloma glabratum* var. *microphyllum* Phil., and *Tylloma rotundifolium* Phil. Six *Oriastrum* names are lectotypified here: *Chondrochilus parvifolius* Phil., *Egania revoluta* Phil., *Oriastrum polymallum* Phil., *Tylloma albiflorum* Phil., *Tylloma gnaphalioides* Phil., and *Tylloma minutum* Phil.

A typical nomenclatural problem of Philippi names is the inadvertent creation of nomina nuda: e.g *Tiltilia pungens* Phil. ex sched., *Minythodes umbellatum* Phil. ex sched. There were only a few examples of this amongst the *Chaetanthera* and *Oriastrum* collections.

IX *Chaetanthera* Ruíz & Pav.

1 Diagnosis

Chaetanthera Ruíz & Pav. in Ruíz, H. & Pavón, J. A. Fl. Peruv. Prodr. 106. t. 23; 1794. – **Generitypus**: *Chaetanthera ciliata* Ruíz & Pav. In Syst. Veg. Fl. Peruv. Chil. 190 – 191, 1798 (see PRUSKI & DAVIES, 2004) **Lectotype** designated by Cassini 1817: 53; Cabrera 1971: 311; Cabrera 1978: 627.

Etymology From the Greek "chaite", bristle, long hair and "anthera", anther, alluding to the papillose tails of the anthers.

Plants herbaceous or sub-shrub, monoecious annual and perennial species. **Habit** lax, decumbent to ascending or erect form constructed of a short erect stem and longer, loosely spreading, branched, stems either supporting several sessile or shortly pedunculate capitula or as moncephalous scapes. Some species have branched rhizomes supported condensed stems with whorls (pseudo-rosettes) of leaves with sessile capitula or ± elongated, monocephalous scapes. **Leaves** below the first node are opposite or loosely whorled. They are generally larger than the upper leaves, and frequently die back during the flowering season. Following germination, the primary leaves have the same form as these basal leaves. **Stem** leaves are arranged on the stems originating from the first node. These are alternate or whorled, and can be distant or densely arranged. Leaf shape varies from linear to indistinctly petiolate and spathulate; leaf margins are dentate to ciliate or entire ± limbate. **Capitula** are radiate and bisexual; shape varies from crateriform or campanulate to elongate (cylindrical) or urcinate. The closed capitula buds can be globose to longly ovoid, i.e. with obtuse or acute apices. **Involucral bracts** are imbricate and can be distinguished into three series: **outer involucral bracts** foliaceous, and shortly alate (to ½ way up), initially with dominant lamina component but the ratio of lamina to membranous margin decreases along the series. The length of the bracts also decreases along the series. The lamina component may be reflexed or appressed. **Middle involucral bracts** have alate margins that are ⅔ to entirely membranous, with a reduced or vestigial apical lamina component. The length of the bracts increases from their initial minimum. **Inner involucral bracts** are entirely membranous and are translucent to pink or red, or have striate maculate (darkly coloured) mucronate apical regions. In some species the size of the capitula and number of bracts are reduced. **Ray florets** single series, 4–30 florets, pistillate, white, pale yellow to orange-yellow, brick red or bright pink. Ligule length generally exceeds involucrum when fully mature; some species have reduced ligules barely exerted from capitulum. **Disk florets** yellow, bisexual, (3)20–120. **Styles** longly lobed, truncate to acute arms; hairs flattened triangular, hair surfaces smooth. **Anthers** sagittate, yellow ± greenish or red, sterile acute apical appendages and ciliate tails. **Pollen** grains equatorially elliptic, large sized (polar diameter x equatorial diameter) 41–80 x 27–56 µm, exine with inconspicuous microspines, exine 9–14(20) µm thick, nexine dumbbell shaped, sexine divided into compact or columellate ectosexine and ramified columellate endosexine. **Achenes** turbinate or flattened cylindric; pericarp pellucid, dense indumentum (seldom glabrous) of ovate-lanceolate, obtuse to acute myxogenic twin hairs, 90–120 µm L; indehiscent ring-like carpopodium poorly to well-developed; testa epidermis with long, oblong parallel cells (rarely with sinuously, tessellated cell) with U–O form strengthening. **Pappus** 1–2(3+) series, setae white, free or united at base, sometimes with basal cilia to 1/3 length of setae, in/dehiscent, barbs appressed, frequent, and medium to longly adhered.

2 Subgeneric division of *Chaetanthera*

2.1 *Chaetanthera* Ruíz & Pav. subgenus *Chaetanthera*

= *Chaetanthera* subgenus *Chaetanthera* Less. in Syn. Gen. Compos. 112, 1832. = *Chaetanthera* subgenus *Euchaetanthera* Hook. & Am. in Companion Bot. Mag. 1: 106, 1835. et Reiche Anales Univ. Chile 115: 332, 1904. = *Chaetanthera* sect. *Euchaetanthera* (Less.) DC. Prodr. (DC.) 7 (1): 256, 1838.
= *Chaetanthera* subgenus *Glandulosa* Cabrera in Revista Mus. La Plata, Secc. Bot. 1: 96, 1937. nom. invalidum.
= *Chaetanthera* subgenus *Prionotophyllum* Less. in Syn. Gen. Compos. 115, 1832. – **Generitypus**: *Chaetanthera incana* Poepp. ex Less. l.c. = *Chaetanthera* sect. *Prionotophyllum* (Less.) DC. in Prodr. (DC.) 7 (1): 256, 1838.
= *Chaetanthera* subgenus *Liniphyllum* Less. in Syn. Gen. Compos. 112, 1832. – Typus *Chaetanthera linearis* Poepp. ex Less. l.c. = *Chaetanthera* sect. *Liniphyllum* (Less.) Endl. in Gen. Pl. pp. 485, 1836 – 1840.
= *Chaetanthera* subgenus *Proselia* (D.Don) D.Don ex Taylor & Phillips in Philos. Mag. Ann. Chem. n.s. 11: 391 – 2, 1832. ≡ *Proselia* D. Don, in Trans. Linn. Soc. London 16: 234, 1830. – **Generitypus**: *Proselia serrata* (Ruiz & Pav.) D. Don, in Trans. Linn. Soc. London 16: 235, 1830. = *Chaetanthera serrata* Ruíz & Pav. in Syst. Veg. Fl. Peruv. Chil. 191, 1798.
= *Cherina* Cass. in Dict. Sci. Nat. (ed. Cuvier, G-F.) 8: 437, 1817 et Bull. Sci. Soc. Philom. Paris pp. 67, 1817. – **Generitypus**: *Cherina microphylla* Cass. in Dict. Sci. Nat. (ed. Cuvier, G-F.) 8: 438, 1817.
= *Euthrixia* D. Don in Trans. Linn. Soc. London 16: 257, 1830. – **Generitypus**: *Euthrixia salsoloides* Don, D. in Trans. Linn. Soc. London 16: 259, 1830.
= *Perdicium* Willd. in Sp. Pl. 3, p. 2118, 1763. – **Generitypus**: *Perdicium chilense* Willd.
= *Minythodes* Phil. ex Benth & Hook. in Gen. Pl. 2: 496, 1873. nom. nudum.

Plants herbaceous, monoecious annual and perennial herbs, one sub-shrub. **Habit** lax, decumbent to ascending or erect form constructed of a short erect stem and longer, loosely spreading, branched, flowering scapes. Some species have branched rhizomes supported condensed stems with whorls (pseudo-rosettes) of leaves with ± elongated, monocephalous scapes. **Leaves** opposite or loosely whorled. Basal leaves of annuals frequently die back during the flowering season. Following germination, the primary leaves have the same form as these basal leaves. **Stem** leaves arranged on the stems originating from the flowering node. These are alternate or whorled along the scape, and can be distant or densely arranged. Leaf shape linear to narrowly linear-spathulate, sessile; leaf margins entire, serrate, dentate, or ciliate, never limbate. **Inner involucral bract** apices striate maculate (darkly coloured green, brown or black); obtuse to acute or rhomboid, seldom fringed, shortly mucronate. **Achenes** brown or seldom black-brown; pericarp pellucid with twin hairs on both disk and ray achenes ovate-lanceolate, obtuse to acute, 90–120 μm L; carpopodium indehiscent, asymmetrical, ring-like, poorly formed (perennials) to well-developed (annuals); testa epidermis of long, oblong parallel cells with U-form strengthening. **Pappus** united at base, indehiscent, setae 2.5–5(10) mm L, (2)3–8 cells wide, cells 5–10 μm wide (*C. elegans; C. peruviana* 15 μm). Setae never ciliate at base with short (50 μm L–*C. ramosissima, C. depauperata, C. glandulosa*) or longer barbs (90–140 μm L) with (5)7–15 barbs/100 μm. **Chromosome numbers** n = 14; 2n = 2x = 22, 24, 26.

Sixteen species, one variety and two hybrids distributed in western South America from Huancavelica (Peru) to Valdivia (Chile) and southwest Argentina (Neuquén).

2.2 *Chaetanthera* subgenus *Tylloma* (D.Don) Less.

Chaetanthera subgenus *Tylloma* (**D.Don**) **Less.** in Syn. Gen. Compos. 115, 1832. et Cabrera in Revista Mus. La Plata, Secc. Bot. 1: 97, 1937. = *Chaetanthera* sect. *Tylloma* (D.Don) Endl. in Gen. Pl. p. 485, 1836-40. ≡ *Tylloma* D. Don, in Trans. Linn. Soc. London 16: 238, 1830. – **Generitypus**: *Tylloma limbatum* D.Don, in Trans. Linn. Soc. London 16: 236, 301, 1830.
= *Chaetanthera* subgenus *Carmelita* (C. Gay in DC.) Cabrera in Revista Mus. La Plata, Secc. Bot. 1: 97, 1937. ≡ *Carmelita* C. Gay, in Prodr. (DC.) 7 (1): 14, 1838. – **Generitypus**: *Carmelita formosa* C. Gay in De Candolle, A. P. Prodr. (DC.) 7 (1): 15, 1838. – Icon. Gay in (1849) in Fl. Chil. [Gay] 3: fig. 37.
= *Chaetanthera* subgenus *Elachia* (DC.) Reiche in Anales Univ. Chile 115: 332, 1904. ≡ *Elachia* DC. in Prodr. (DC.) 7 (1): 256, 1838. – **Generitypus**: *Elachia euphrasioides* DC. l.c. et in Delessert Icon. Select. Pl. 4: fig. 99, 1839.
= *Chondrochilus* Phil. in Linnaea, 28: 711, 1856. – **Generitypus**: *Chondrochilus crenatus* Phil., l.c.
= *Pachylaena* D. Don ex Hook. & Arn. in Companion Bot. Mag. 1: 106, 1835. [pro parte].

Plants herbaceous, monoecious annual and perennial species. **Habit** lax, decumbent to ascending or erect form constructed of a short erect stem and longer, loosely spreading, branched, flowering scapes. Some species have branched rhizomes supporting condensed stems with whorls (pseudo-rosettes) of leaves with sessile or shortly pedunculate capitula. **Leaves** below flowering stems opposite or loosely whorled. **Stem** leaves alternate or whorled along stems, and can be distant or densely arranged. Leaves usually indistinctly petiolate, narrowly to broadly spathulate; lamina narrowly to broadly ovate, cordate, flabellate or reniform; leaf margins ± limbate, entire to dentate.

IX *Chaetanthera* Ruiz & Pav.

Inner involucral bract apices acute to obtuse, green and foliaceous or translucent to pink or red. **Achenes** brown or seldom black-brown; pericarp pellucid, indumentum varies from ovate-lanceolate, obtuse to acute, 90–120 µm L twin hairs on ray and (or only) disk achenes, to scattered, tiny (< 20 µm) spherical twin hairs (often poorly preserved); carpopodium indehiscent, ring-like, poorly formed (perennials) to well-developed (annuals); testa epidermis of long, oblong parallel cells or rarely, sinuously, tessellated cells, with U- or O-form strengthening. **Pappus** setae free at base, in/dehiscent, setae 6–11(13) mm L, 4–8 cells wide (*C. euphrasioides* 1–3; *C. limbata* 10–17), cells 5–10 µm wide (*C. flabellata* 15 µm). Setae can be ciliate at base with shorter barbs (90–140 µm L) in upper ⅔ of setae to entirely ciliate or sub-plumose (> 190 µm L) e.g. *C. philppii*, *C. spathulifolia*, *C. villosa*, *C. renifolia*) with 5–6 (8; *C. limbata*) barbs /100 µm. Barbs of ray floret pappus can be up to 10% shorter and somewhat less dense than disk florets, but not truly dimorphic. **Chromosome numbers** $2n = 2x = 22$; $2n = 4x-6 = 38$.

Fourteen species distributed mainly in Chile, with sporadic collections from Argentina.

3 Key to the species

All *Chaetanthera* taxa recognised in this monograph are keyed out here. As far as possible the characters used should be adaptable to both field and herbarium identifications. However, on herbarium specimens floret colour can be ambiguous, especially the difference between yellow and white, and often underground components are not well collected, if at all.

Abbreviations used in Figure captions and conspectus: IIB = Inner involucral bracts, MIB = Middle involucral bracts, OIB = Outer involucral bracts, PN = Parque Nacional, RN = Reserva Nacional. Where illustrated open circles on distribution maps indicate nomenclatural type localities.

1	Species annual (stems herbaceous, roots filamentous); mature plants with spreading branching stems, sometimes with residual basal leaf rosettes	2
1*	Species perennial (distinctly woody stems or roots); mature plant is a subshrub or with distinct rosettes of leaves	27
2	Rays present	3
2*	Rays apparently absent or not exerted beyond capitulum	**18.** *C. taltalensis*
3	Rays white	4
3*	Rays yellow, yellow-orange, red or bright pink	11
4	Glabrous	5
4*	Indumentum sparse to dense	7
5	Leaves linear; margins entire with inconspicuous triangular hairs on ventral surface	6
5*	Leaves narrowly spathulate; margins dentate	**19.** *C. euphrasioides*

6	Capitula campanulate (sub-radiate), inner involucral bracts broadly elliptic, apiculate	**14.** *C. perpusilla*
6*	Capitula cylindrical, inner involucral bracts lanceolate, acute	**4.** *C. depauperata*

7	Leaves and bracts (at least) sericeous to floccose lanate; leaf margins dentate	8
7*	Plant with filamentous hairs at leaf bases and below capitula; leaf margins entire	**1.** *C. albiflora*

8	Leaf margins flat	9
8*	Leaf margins, especially distally, thickened to slightly involute	10

9	Habit ascending-erect; leaves with fragile actinomorphic two-celled hairs; occurring in Peru.	**15.** *C. peruviana*
9*	Habit decumbent, spreading; leaves hoary; occurring in lowland central Chile	**8.** *C. incana*

10	Plant erect and spreading, branched at ground level; leaves linear or narrowly linear lanceolate, densely silvery pubescent; white flowered form found inland central Chile, south of 36°S.	**12.** *C. moenchioides*
10*	Plant erect branched above ground; leaves truncate, 3 – 5-dentate, sparsely sericeous; found in central Chile between 33° and 35°S.	**16.** *C. ramossisima*

11	Rays yellow, or pink flushed with yellow	12
11*	Rays brick-red	**11.** *C. microphylla*

12	Leaves linear; leaf margins entire	13
12*	Leaves indistinctly petiolate; leaf lamina flabellate to spathulate; leaf margins entire to dentate	15

13	Plant glabrous	14
13*	Plant silvery sericeous	**12.** *C. moenchioides*

14	Plant and leaves glaucous (blue-green); leaves flat in X-section	**9.** *C. linearis*
14*	Plant and leaves virid (yellow-green); leaves roundish in X-section	**10.** *C. linearis x albiflora*

15	Leaves spathulate, leaf margins dentate or ciliate	16
15*	Leaves flabellate	**21.** *C. flabellifolia*

16	Plant entirely glabrous or with tufted villous hairs in leaf axils and/or on petioles only	17
16*	Plant pubescent (stems, leaves, bracts)	24

IX *Chaetanthera* Ruiz & Pav.

17	Outer involucral bracts deeply to shallowly ciliate	18
17*	Outer involucral bracts entire or dentate	19

18	Capitulum disk diameter 8 – 9 mm; outer involucral bract length 9 – 10 mm	**3.** *C. ciliata*
18*	Capitulum disk diameter 3 – 4 mm; outer involucral bract length 5 – 6 mm	**13.** *C. multicaulis*

19	Leaf margins entire	20
19*	Leaf margins dentate	22

20	Leaf lamina and involucral bracts glabrous	21
20*	Leaf lamina and inner involucral bracts covered in black sessile glands	**22.** *C. frayjorgensis*

21	Petioles glabrous	**23.** *C. glabrata*
21*	Petioles with densely villous hairs	**29.** *C. schroederi*

22	Capitulum disk diameter > 2 cm	**31.** *C. splendens*
22*	Capitulum disk diameter <1.5 cm	23

23	Capitula cylindrical, 0.4 – 0.6 cm diameter; leaves nearly truncate; leaf margins 3-many dentate	**19.** *C. euphrasioides*
23*	Capitula campanulate, 0.8 – 1.5 cm diameter; leaves spathulate; leaf margins distinctly dentate	**20.** *C. flabellata*

24	Pubescence sericeous or villous, inner involucral bracts acute, or obtuse, fringed	25
24*	Pubescence lanate, inner involucral bracts obtuse, entire	**24.** *C. kalinae*

25	Leaves succulent, indistinctly petiolate, lamina spathulate to lanceolate; indumentums villous with glandular hairs	26
25*	Leaves not succulent, sessile, spathulate; indumentum silvery-sericeous	**8.** *C. incana*

26	Indumentum with no additional capitate hairs on leaves	**25.** *C. limbata*
26*	Indumentum with additional capitate hairs on leaves	**27.** *C. pubescens*

27	Plant dwarf sub-shrub, stems woody; vegetative parts with densely glandular indumentum	28
27*	Plant with leaf rosettes at surface, often with buried woody stems and roots, vegetative parts never with densely glandular indumentum	29

28	Leaves 10 – 15 mm L x 1.5 – 2 mm W.; ray florets white	
		6. *C. glandulosa* var. *glandulosa*
28*	Leaves 5 – 6 mm L x 1 mm W.; ray florets yellow	
		7. *C. glandulosa* var. *gracilis*

29	Rays white	30
29*	Rays yellow	31

30	Plants emerge as individual rosettes; leaf bases pubescent; leaf lamina reniform, apical margin distinctly notched.	**28.** *C. renifolia*
30*	Plants grow as many-flowered, loose cushions; leaves entirely pubescent; leaf lamina spathulate, margins entire	**26.** *C. philippii*

31	Rosettes (joined by thick branched rhizomes) of more or less sessile capitula; disk diameter usually > 2.5 cm	32
31*	Rosulate plants (joined by stolons or rhizomes) with erect, monocephalous scapes; disk diameter usually < 2 cm	33

32	Rosettes dispersed; leaves linear-spathulate, apices obtuse; indumentum dense, silvery villous	**32.** *C. villosa*
32*	Rosettes clumped together; leaves broadly spathulate, apices truncate; indumentum lanate	**30.** *C. spathulifolia*

33	Plants with rhizomes (long or short woody stems more or less buried in substrate) and /or woody roots; more or less well developed carpopodium	34
33*	Plants with surface stolons; poorly developed to non-existent carpopodium	**17.** *C. x serrata*

34	Leaves pale green to grey-green, indumentum densely sericeous, generally < 3 mm W.; capitulum cylindrical, < 15 mm disk diameter.	**2.** *C. chilensis*
34*	Leaves bright to dark green, indumentum glabrescent to lightly pubescent, never densely sericeous, generally > 3.5 mm W.; capitulum campanulate, > 18 mm disk diameter.	**5.** *C. elegans*

4 Species descriptions

4.1 *Chaetanthera* Ruíz & Pav. subgenus *Chaetanthera*

1. *Chaetanthera albiflora* (Phil.) A.M.R. Davies stat. nov.

≡ *Chaetanthera linearis* var. *albiflora* Phil. Anales Univ. Chile 87: 14. 1894. Typus CHILE "En la provincias de Atacama i Coquimbo" [Coquimbo, 10.1878/ Paiguano, 02.1883] Holotypus SGO 64706! pro parte.

= *Chaetanthera linifolia* var. *albiflora* Phil. Anales Univ. Chile 87: 14. 1894. Typus CHILE "En la provincias de Atacama i Coquimbo"
= *Chaetanthera linearis* var. *taltalensis* I. M. Johnst. Contr. Gray Herb. 85: 132. 1929. Typus CHILE "Chile: dry hillsides just back of Caleta de Hueso Parado near Taltal, 11.12.1925, *Johnston 5637*" Holotypus GH! Isotypus K! S (herb. Regnell) US!
- *Chaetanthera microphylla* var. *albiflora* (Phil.) Cabrera Revista Mus. La Plata, Secc. Bot. 1: 177 – 178. 1937.
- *Chaetanthera albiflora* Phil. Verh. Deutsch. Wiss. Verein Santiago de Chile 2: 107. 1890. nomen invalidum. Typoid material ["*Ch. albiflora*" Ta[u]lahuen, (Ovalle), 1889, *Geisse s.n.*] GH! SGO 64720! [Copiapó "*Ch. albiflora*" *Philippi s.n.*] BM! E! SGO!
- *Chaetanthera leucantha* Phil. ex.sched. (E GH)

Nomenclatural Notes R. Philippi originally published the name *Chaetanthera albiflora* Phil. as part of a collections list in 1890, but without an accompanying description. In 1894 he validy published both *Chaetanthera linearis* var. *albiflora* Phil. and *Chaetanthera linifolia* var. *albiflora* Phil. with short descriptions. Sheet SGO 64706 is the only sheet in Philippi's herbarium with *Chaetanthera linearis* var. *albiflora* Phil. inscripted on it, although unfortunately there are several collections mounted on the sheet. The remaining appropriate SGO material (and that dispersed in other herbaria) merely has *Chaetanthera albiflora* Phil. on the labels. This is cited as "typoid material" under the invlaid name *Chaetanthera albiflora* Phil.

Annual monoecious herb. **Stems** to 20 cm, spreading, lax, brown, glabrescent below capitula. **Leaves** up to 4.6 mm L, 0.5–1 mm W., leaves at nodes larger (to 3 cm x 1 mm) than those on scapes/stems (from 3 mm x 0.5 mm); virid green, slightly succulent, roundish in cross-section; margins with sparse translucent teeth, midrib indistinct; dilated at base ± sparse filamentous white hairs. **Capitula** cylindrical (urcinate), disk diameter (fresh material) 2–4 mm. **Involucral bracts** imbricate, arranged in three series, initially foliaceous then reduced to entirely membranous. **Outer involucral bracts** 3.4–4 mm L; as leaves but with short membranous alae to less than ½ height of bract, alae margins with filamentous white hairs (0.3–0.4 mm long). **Middle involucral bracts** 3.7–4.9 mm L, shape ovate to linear, apex ± emarginated; lamina component reduced to central green, slightly thickened tissue; membranous alate, ratio of lamina to alae decreases from ⅔ to nearly entirely alate, alae margins pubescent. **Inner involucral bracts** entirely membranous, 6.5 mm L, linear, acute to emarginate with mucro, apices striate maculate. **Ray florets** (6) 8–13, pistillate, white with distinctive dark green/brown dorsal stripe; corolla 9.9 mm L; corolla tube 4.3 mm L; outer corolla lip 2.1 mm W.; inner corolla lip 2.5 mm L, filamentous. **Disk florets** bisexual, yellow; corolla 5.6–5.7 mm L; corolla tube 4.5–4.7 mm L **Styles** yellow [ray] ca. 7 mm L, stigma lobes 0.3 mm L [disk] 5.9–6.7 mm L, stigma lobes 0.2 mm L **Anthers** 3.7 mm L (6.1 mm L including filaments). **Achenes** 1.7–2.1 mm L, brown-black, fusiform turbinate; carpopodium anular irregular; pericarp pellucid, densely covered in lanceolate twin hairs, 90–115 x 40 µm. **Pappus** white, 3.7–4.7 mm L, 1–2 rows, indehiscent; setae fused at base, setae 4–8 cells wide, cell width 5–10 µm, no basal cilia; barbellate, barbs 50 µm L, barb base medium adhered (50%), free barb appressed, 12–15 barbs/100 µm). **Chromosome number** 2n = 2x = 24. Davies & Vosyka (new count). *Ehrhart* et al. *2002/040* (MSB).

Distribution and habitat. *C. albiflora* is a coastal species endemic to Chile. It is distributed along the Cordillera de la Costa from Tocopilla and Taltal to the Río Aconcagua, only branching inland from La Serena, following a spur of the coastal mountain range between latitudes 22°00'–32°45'S. It occurs on dunes, sandy roadside banks, verges, and on well drained slopes at all elevations up to 1000 m.a.s.l. (1600–1800 m near Vicuña, Hurtado).

Differential diagnosis. *C. albiflora* and *C. linearis* are very similar, and form morphologically intermediate hybrids, classified under *C. linearis x albiflora*. The differences between these taxa are outlined in Table 9. The linear leaved group, to which these two species belong, also includes *C. microphylla*, *C. depauperata* and *C. perpusilla*. Key characters for identifying these taxa are given Table 12.

Species	*C. albiflora*	*C. linearis*	*C. linearis x albiflora*
Leaf colour	Virid green	Glaucous green	Virid or glaucous
Leaf succulence	Succulent	Not succulent	Succulent or not
Indumentum	filamentous	glabrous	filamentous or glabrous
Floret colour	White + dark dorsal stripe	Yellow	White or yellow ± dorsal stripe.

Table 9: Key characters for identifying *C. albiflora*, *C. linearis* and the hybrid.

IX *Chaetanthera* Ruiz & Pav.

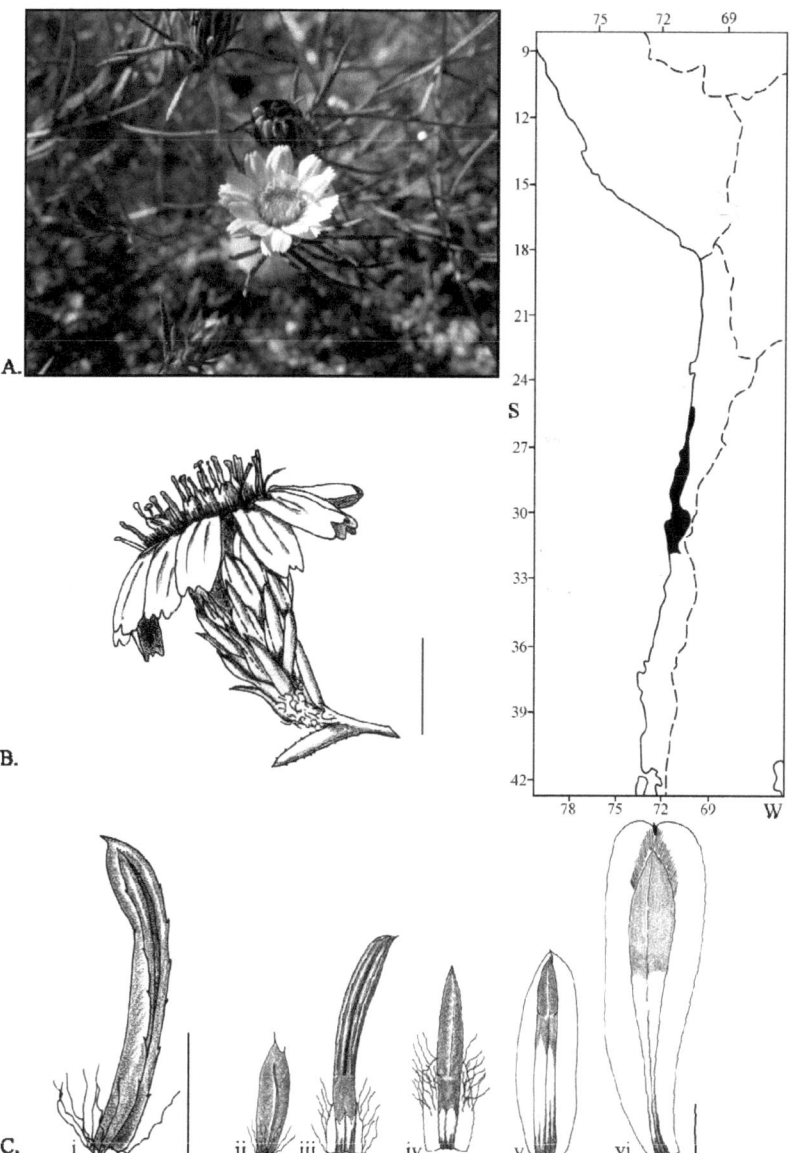

Figure 49: *Chaetanthera albiflora*. **A.** Coquimbo, Las Rojas, 11.2008 ©Mauricio Zuñiga. **B.** Capitulum, *Ehrhart 2002/040*. Scale bar = 5 mm. **C.** Leaf & bract detail, *Ehrhart 2002/040*. **i.** Stem leaf, v.s., scale bar = 2 mm. **ii.** Stem leaf, d.s.; **iii.** OIB, v.s.; iv. MIB, d.s.; v + vi. IIB, d.s. Scale bar (i – vi) = 1 mm.

Material seen. – **Chile. II Región de Antofagasta** Prov. de Tocopilla Tocopilla, 10 m, 10.1932, *Jaffuel 2559* (GH). – Tocopilla, 10 m, 10.1932, *Jaffuel 2565* (CONC GH). – Prov. de Antofagasta Taltal, ca. 10 Km east of Taltal, Quebrada de Taltal, 75 m, 10.1938, *Worth & Morrison 15802* (G GH K). – Taltal, 9-10 Km east of Taltal, Quebrada to North, 150 m, 10.1938, *Worth & Morrison 15842* (GH K). – Quebrada de Taltal, floor of Quebrada, 11.1925, *Johnston 5114* (GH). – Taltal, 20 m, 1938, *Lopez s.n.* (CONC). – Taltal, 200 m, 10.1925, *Werdermann 815* (K NY en parte). – Quebrada El Yeso, aproximadamente 1 Km norte de Taltal, *K. Arroyo et al. 25129* (CONC). – vicinity of Taltal, just back of Caleta de Hueso Parado, 12.1925, *Johnston 5637* (GH K). – Taltal, Quebrada de Infieles about 8 Km south of town, 12.1925, *Johnston 5649* (GH K). – Cerro Perales, este de Taltal, 890 m, 11.1997, *Dillon & Trujillo 8050* (CONC). – **III Región de Atacama** Prov. de Chañaral PN Pan de Azucar, Quebrada de Coquimbo, 350 m, 10.1987, *Teillier 685* (CONC MO NY SGO). – Prov. de Copiapó – Copiapó, *Philippi s.n.* (E). – Vicinity of Copiapó, Quebrada de Chanchoquin, 09.1885, *Gigoux s.n.* (GH). – Camino Caldera-Copiapó, Km 5, 70 m, 10.1971, *Marticorena , Rodriguez & Weldt 1870* (B CONC). – Cerro Bandurrias, 265 m, 1888, *Geisse s.n.* (CONC). – Bandurrias, 265 m, 09.1976, *Zöllner 9282* (CONC MO). – Copiapó, Barros Luco, 10 m, 1914, *Rose & Rose 19326* (NY). – Quebrada de Totoral, Boquerones, 180 m, 11.1941, *Pisano & Bravo 794* (CONC). – Prov. de Huasco Camino Carrizal Bajo, Km 50, Lugar de Leontochir, 11.1991, *Muñoz S., Teillier & Meza 2937* (SGO). – Straße Vallenar-Copiapó Flußbett W der Abzweigung nach Carrizal Bajo, 11.1987, *Rechinger K.H. & W. 63365* (W). – Tal 4 Km E Carrizal Bajo = 75 Km Von Vallenar, 11.1987, *Rechinger K.H. & W. 63366* (M W). – Camino Carrizal Bajo, Km 44, 220, 10, 1965, *Ricardi et al. 1523* (CONC). – Dunas Norte de Huasco, Km 20 del camino entre Huasco y Carrizal Bajo, 10 m, 9.1991, *von Bohlen 1345* (SGO). – ca. 25 Km north of Vallenar, along road to Copiapó, 500 m, 10.1938, *Worth & Morrison 16271* (GH). – Huasco, 11.1930, *Jaffuel 1175* (GH). – Depto. Freirina, Quebrada de Carrizal Bajo, 9.1957, *Cabrera 12669* (K). – Vallenar, Embalse Santa Juana, 585 m, 12.1991, *Saavedra 474* (SGO). – 15 Km N Vallenar längs der Hauptstaße, 11.1987, *Rechinger K.H. & W. 63321* (W). – Vallenar, 450 m, 9.1928, *Barros 2442* (CONC). – Agua Amarga, 1000 m, 10.1952, *Ricardi s.n.* (CONC). – cruce camino de Morado hacia Carrizalillo, 300 m, 9.1977, *Muñoz S., Meza & Barrera 1148* (SGO). – Cachiyuyo, 850, 9, 1957, *Ricardi & Marticorena 4463* (CONC). – Cachiyuyo, 9.1957, *Cabrera 12691* (K). – PA Km 573 Nördl. Incahuasi, am Fuß der Cuesta Pajonales, 10.1980, *Grau 2053* (BM M MSB). – 12 Km al S de Incahuasi, 400 m, 10.1958, *Ricardi & Marticorena 4905* (CONC). – carretera entre El Tofo y Los Choros, 10.1991, *Muñoz S., Teillier & Meza 2666* (MO SGO). – **IV Región de Coquimbo** Prov. de Elqui carretera Panamericana, frente El Tofo, 350 m, 10.1963, *Marticorena & Matthei 211* (CONC). – Mineral La Higuera, N Cuesta Buenos Aires, 450 m, 10.1963, *Marticorena & Matthei 181* (CONC). – Cuesta Buenos Aires, Depto. La Serena, 9.1957, *Cabrera 12594* (MA). – La Serena, Quebrada Honda, 50 m, 11.1961, *Jiles 4570* (CONC). – camino de Agua Grande a La Serena, Km 1, 300 m, 10.1971, *Marticorena et al. 1575* (CONC). – 5 Km Nördlich La Serena, Sanddünen Östlich der Eisenbahnlinie nach El Romeral, 11.2002, *Ehrhart et al. 2002/086* (MSB). – Punta Teatinos - La Serena , 110 m, 10.2002, *K. Arroyo et al. 25048* (CONC). – Punta Teatinos - La Serena , 110 m, 8.2002, *K. Arroyo et al. 25054* (CONC). – Juan Soldado, 10 m, 11.1967, *Ricardi et al. 1812* (CONC). – Rivadavia, 800 m, 11.1923, *Werdermann 1875* (NY). – Rivadavia, 800 m, 9.1980, *Montero 11688* (CONC). – La Serena-Ovalle, por Las Cardas, Km 5, 90 m, 2.1988, *Marticorena et al. 9946* (CONC). – 9 Km desde Vicuña, camino de tierra a Hurtado, 1050 m, 9.2002, *K. Arroyo et al. 25019* (CONC). – Vicuña, 32 Km S of Vicuña on gravel road towards Hurtado (= 6 Km S of Puente del El Pangue; = 4 Km N of culmination of Portozuelo Tres Cruces; = 15 Km N of Hurtado), 1650 m, 10.1997, *Eggli & Leuenberger 3061* (B). – Guanaqueros, PA Km 435 bei der Abzweigung nach Coquimbo, 60 m, 11.1980, *Grau 2542* (MSB M). – Andacollo, 1000 m, 9.1926, *Barros 4570* (CONC). – Coquimbo, *Philippi s.n.* (E GH). – Coquimbo, (La) Serena, 07.1836, *Gay 287* (P). – Coquimbo, *Gay s.n.* (GH). – Prov. de Limarí Weg von Vicuña nach Hurtado, 880 m, 11.2002, *Ehrhart et al. 2002/079* (MSB). – Camarones, 120 m, 10.1961, *Jiles 3897* (CONC). – Quebrada del Toro, Pabellón, 20 Km E Hurtado, Hacienda El Bosque, 1800 m, 11.1993, *Wagenknecht 18489* (F GH). – PN de Fray Jorge, Bosque Fray Jorge, erste Hänge im Landesinneren nach den Steilkurven, 11.2002, *Ehrhart et al. 2002/040* (MSB). – Fray Jorge, lado norte de la desembocadura de La Limary, 250 m, 11.1940, *Muñoz P. & Coronel 1409* (SGO). – Fray Jorge, 09.1935, *Muñoz B-105* (SGO). – Cuesta Punitaqui a 7 Km del Pueblo, 380 m, 09.2002, *K. Arroyo et al. 25022* (CONC). – Cuesta Punitaqui, 6 Km del Pueblo, 380 m, 09.2002, *K. Arroyo et al. 25067* (CONC). – 3 Km al sur de Combarbalá, en camino de Tierra Que Sale del Pueblo, 1010 m, 09.2002, *K. Arroyo et al. 25025* (CONC). – Prov. de Choapa camino entre RN Chinchillas E Illapel, 580 m, 09.2002, *K. Arroyo et al. 25028* (CONC). – camino a Tilama, 890 m, 09.1999, *K. Arroyo et al. 992406* (CONC). – 7 Km al norte de Los Vilos, Ruta 5 Norte, 10 m, 11.2002, *K. Arroyo et al. 25154* (CONC). – 7 Km al norte de Los Vilos, Ruta 5 Norte, 950 m, 09.2002, *K. Arroyo et al. 25012* (CONC). – **V Región de Valparaíso** Prov. de Petorca camino entre Alicahue y La Viña, 760 m, 10.2002, *K. Arroyo et al. 25033* (CONC). – camino entre Cabildo y Guayacan, 400 m, 10.2002, *K. Arroyo et al. 25036* (CONC). – camino entre Cabildo y Guayacan, 500 m, 10.2002, *K. Arroyo et al. 25037* (CONC). – Prov. de San Felipe San Felipe, 630 m, 10.1970, *Zöllner 4416* (CONC).

IX *Chaetanthera* Ruiz & Pav.

2.   *Chaetanthera chilensis* (Willd.) DC s.l. Ann. Mus. Natl. Hist. Nat. 19: 70 fig. 3. 1812.

≡ *Perdicium chilense* Willd. Species Plantarum 3(3): 2118. 1803. Typus CHILE "In Chili" Holotypus Willdenow Herbar (microfiche Cat.# 16092 (MF 1159, Box 26) M!

= *Chaetanthera argentea* D. Don ex Taylor & Phillips Philos. Mag. Ann. Chem. 11: 391. 1832. Typus CHILE "No Type location" "Chilian Andes *Cuming 182*" [fide Hooker & Arnott 1835] Syntypus BM!; E!; GH(photo)!

= *Chaetanthera eryngioides* D. Don ex Taylor & PhillipsPhilos. Mag. Ann. Chem. 11: 391. 1832. Typus ARGENTINA "Las Cuevillas, Andes of Mendoza, Dr. Gillies" Holotypus K! Isotypus "'Chaetanthera eryngioides', *Gillies s.n.*" BM!.

= *Chaetanthera tenuifolia* D. Don ex Taylor & Phillips Philos. Mag. Ann. Chem. 11: 391. 1832. Typus CHILE "*Herb. Gill.*" K!

= *Chaetanthera argentea* Phil. Anales Univ. Chil. (41 – 42):739. 1872. Typus CHILE "Talcaregue" Holotypus SGO (S43716)! Isotypus K! SGO(S64741)!

= *Chaetanthera involucrata* Phil. Anales Univ. Chile 87: 6. 1894. Typus CHILE "Habitat in prov. Santiago, Salto de Conchalí etc." Syntypus SGO (S64712)! (S43742)!

= *Chaetanthera nana* Phil. Anales Univ. Chile 87: 7. 1894. Typus CHILE "In Andibus praedii Cauquenes (prov. O'Higgins) loco dicto 'Agua de la Vida' legimus" Holotypus SGO (S 64734)!

=*Chaetanthera sublignosa* Kuntze Revis. Gen. Pl. 3(2): 140. 1898. Typus CHILE "Chile, Paso Cruz, 2600 m, *Kuntze s.n.*" Holotypus NYBG! Isotypus NYBG!

= *Chaetanthera collina* Phil. ex Reiche Fl. Chile [Reiche] 4: 340. 1905. (Phil. ex sched Herb. Mus. Nac.) Typus CHILE "En colinas secas de las provincias centrales (cerro san Cristobal cerca de la capital; provincia de Colchagua etc.)" **Lectotypus hic loc. designatus** SGO (S71813)! Isolectotypus probabilis "Santiago, 'Chaetanthera collina', *Philippi s.n.*" E!; GH(photo)!; NYBG(photo)!

- *Chaetanthera serrata* var. *nana* (Phil.) Reiche Anales Univ. Chile 115: 319. 1904.
- *Chaetanthera chilensis* var. *argentea* (Phil.) Cabrera Revista Mus. La Plata, Secc. Bot. 1: 166 – 167, fig. 32. 1937.
- *Chaetanthera chilensis* var. *involucrata* (Phil.) Cabrera Revista Mus. La Plata, Secc. Bot. 1: 168, fig. 33. 1937.
- *Chaetanthera chilensis* var. *tenuifolia* (D. Don) Cabrera Revista Mus. La Plata, Secc. Bot. 1: 168 – 170, fig. 34. 1937.
- *Chaetanthera elegans* var. *andina* (Phil.) Cabrera Revista Mus. La Plata, Secc. Bot. 1: 160 – 161, fig. 29. 1937.
- *Chaetanthera serrata* var. *argentea* (Phil.) Reiche Anales Univ. Chile 115: 319. 1904.
- *Chaetanthera tenuifolia* var. *eryngioides* (D.Don) Hook. & Arn. Companion Bot. Mag. 1: 105. 1835.
- *Chaetanthera grandiflora* Steud. Nomencl. Bot., ed. 2 (Steudel) 1: 340. 1840. nomen nudum.

Perennial monoecious caespitose scapose herbs. **Roots** with thick central woody rootstock with short loosely to densely packed woody ± subterranean stems (rhizomes). **Stems** shortly branching, suppporting densely to loosely leafy clumps (pseudo-rosettes) of leaves, often forming a small mat. **Leaves** (10)18–53(70) x (1)2.4(5) mm, linear lanceolate to narrowly spathulate, often plicate, alternate; margins shortly dentate mucronate in upper third; grey green; indumentum densely silvery pubescent (hairs 2–3 mm L) on dorsal and ventral surfaces and in axils (rarely senescent leaves may be glabrous). **Pedicel** (scape) erect to ascending, (2)12.9(30) cm L, sparsely to densely pubescent, leaves as basal leaves, but smaller and infrequently spaced (5–10) up scape. **Capitula** cylindrical to campanulate, (7)8.5(15.0) x (7)8–22.5(25) mm, often subtended by 2 or more rows of reflexed leafy involucral bracts that equal or shortly extend beyond it. **Involucral bracts** imbricate, arranged in three series, initially foliaceous then reduced to entirely membranous. **Outer involucral bracts** 1–3 rows, 4.3–13.3 x 1.1–1.7(2.9) mm; as leaves but with short membranous alae to less than ½ height of bract, alae linear ovate to truncate. **Middle involucral bracts** few, same size as OIB; membranous alate, ratio of lamina to alae decreases from ⅔ to nearly entirely alate. **Inner involucral bracts** 2-3 rows increasing in size, linear lanceolate, 9–18 x 0.8–2.1 mm, entirely membranous to scleroid; apices triangular black-green or brown maculate, broadly acute to narrowly acute, rarely obtuse, mucronate. **Ray florets** 14–24, pistillate, pale to bright yellow with pink-red dorsal stripe, dorsally densely sericeous pubescent; corolla (8)14–19 mm L, corolla tube (3.5)4–5 mm L, outer ligule (1.5)2–3.5 mm W., inner ligule flattened for 2-3 mm then twisted and tightly curled for 2-3 mm. **Disk florets** 20–30, bisexual, yellow; corolla (6.5)7.5–10 mm L; corolla tube 5–7 mm L **Styles** [ray] 10–11 mm L, stigma lobes to 1 mm, slightly spathulate. [disk] 9–10 mm L, stigma lobes 0.5–1 mm, spathulate. **Anthers** [ray] sterile, 8–9 mm L [disk] (7.5)9–10 mm L **Achenes** brown, 1.5–2.5 mm L, angled turbinate, carpopodium almost always well developed, 30–120 μm high (smaller in collections around Santiago, larger in central valley and precordillera); pericarp pellucid, densely covered in lanceolate twin hairs, 90–115 x 40 μm; testa epidermis surface of elongated parallel cells, margins entire, cells with U-formed sclerenchymatous thickenings. **Pappus** white, (5)7–10 mm L, 1–2 rows, indehiscent; setae fused at base (corona 1–2 mm high), setae 4–8 cells wide, cell width 5–10 μm, no basal cilia; barbellate, barbs 160–190 μm L, barb base shortly adhered (<40%), free barb appressed, 1–2 barbs/100 μm). **Chromosome number** $2n = 2x = 22$ (BAEZA & SCHRADER, 2005b). *C. Baeza 4204* (CONC)

Distribution and habitat. *C. chilensis* is typically distributed from Santiago to Malleco (Angol) in the Central Valley, along the Coastal mountain range (Valparaíso; Cajón Las Leñas) and to the volcanic lakes (e.g. Cajón Morales, Cerro Cantillana, Laguna del Flaco, Laguna del Teno), between latitudes (31°16') 32°00'–38°00'S. It occurs in open sunny patches on sandy soil, dry banks and cuttings, hillsides and ravines of the lower Andes at all elevations up to 2900 m.a.s.l.

Species Notes. *C. chilensis* is highly polymorphic and shows great phenotypic plasticity in its form (particularly in leaf size and pedicel length) throughout its range. There are occasional collections from the Andes around Santiago with distinctly obtuse, gongyloid inner involucral bracts with only a coloured central spot on the apices. Formerly applied names (*C. argentea* D. Don ex Taylor & Phillips; *C. collina* Phil. ex Reiche) recognised this morph as a distinct species. Further examples of this morph include e.g. *K. Arroyo* et al. *205915*, *Cuming 182*, *Elliott 292*, *K. Arroyo* et al. *211821*. Some plants in the Central Valley have extremely leafy pedicels (>10 leaves per pedicel), and very long leafy OIB (2 x capitula height) e.g. *Bertero 472, Grau 2375*. The plants found at higher altitudes (above 1500 m.a.sl.) often have shorter pedicels (c. 6–7 cm L), somewhat narrower capitula (c. 12–14 mm W.) and slightly longer outer involucral bracts (c. 9 mm L). These have been recognised as separate species or varieties in the past (e.g. *C. nana* Phil.; *C. argentea* Phil.) but are considered to be phenotypic variants of *C. chilensis* s.l. (e.g. *K. Arroyo* et al. *99439, K. Arroyo* et al. *980838, Werdermann 507*). The tendency to form a cushion at higher altitudes, or rather the ability to grow longer pedicels in sheltered or lower altitude habitats is simply an environmental effect. Individuals with shorter pedicels, slightly more spathulate leaf and bract apices, and larger bracts have been found at higher altitudes south of Santiago (e.g. *Kuntze s.n., Gillies s.n.*) and Mendoza. Formerly distinguished by the name *C. eryngioides*, this name has also been synonymised with *C. chilensis*.

Differential diagnosis. *C. chilensis* belongs to a group of three perennial caespitose scapose taxa. It differs from *C. elegans* in its grey green, lanceolate to narrowly spathulate (generally < 3 mm wide) densely silvery pubescent leaves with dentate margins, its 3 series of involucral bracts and in having pappus setae composed of more, narrower cells, and that are longly infrequently barbellate. The key features are laid out in Table 10. *C. x serrata* specimens exhibit intermediate characteristics between *C. chilensis* and *C. elegans*, but have stoloniferous instead of rhizomatous systems.

Species	*C. chilensis*	*C. elegans*
Leaf shape; width; form	Lanceolate to narrowly spathulate; < 3 mm wide; plicate	Lanceolate spathulate; > 3 mm wide; flat
Leaf colour; indumentum	Grey green; densely silvery pubescent	Bright green; sericeous to glabrescent
Leaf margin	dentate	Spinulose serrate to mucronate dentate
Involucral bract series	3 series	2 series (MIB rare or absent)
Carpopodium	Always well developed	Very variable, usually present but poorly developed
Pappus setae width (number of cells;cell width μm wide)	4 – 8 cells wide; cell width 5 – 10 μm	1 – 3 cells wide; cell width 15 μm
Pappus barb length (μm); barb frequency.	160 – 190 um; 1 – 2(3) barbs/ 100 μm	95 – 140 um; 5 – 6 barbs/ 100 μm

Table 10: Key features for identifying *C. chilensis* and *C. elegans*.

Figure 50: *Chaetanthera chilensis.* **A.** Habit, photographed Santiago, 2006©Mauricio Bonifacino. **B.** Capitulum detail. Photographed VII Región del Maule, Constitución, 1980©Jürke Grau. **C.** Leaf & bract detail, d.s. **i.** Leaf; **ii.** OIB; **iii – iv.** MIB; **v – vii.** IIB. *K. Arroyo 209821* (i, iv); *K. Arroyo 99698* (ii – v); *K. Arroyo 209763* (vii). Scale bar (i – vii) = 5 mm.

IX *Chaetanthera* Ruiz & Pav.

Material seen. – **Argentina. Prov. de Mendoza** Las Cuevillas, *Gillies s.n.* (K). – **Chile. IV Región de Coquimbo Prov. de Choapa** Corral de Julio, 280 m, 11.1971, *Jiles 5862* (CONC). – Pichidangui, Cerro Silla del Gobernador, 500 m, 02.1961, *Schlegel 3738* (CONC). – **V Región de Valparaíso** Prov. de Quillota La Dormida, 3 Km South of Las Viscachas, La Dormida, 1800 m, 1.1939, *Morrison & Wagenknecht 17111* (K). – Cerro Campaña, 1300 m, 11.1952, *Garaventa 6591* (CONC). – Cerro Tres Puntas, 900 m, 10.1930, *Garaventa 2543* (CONC). – Cerro Vizcachas, 1800 m, 09.1964, *Schlegel 4991* (CONC). – Prov. de Valparaíso Valparaíso, 1832, *Cuming 660* (BM, Ex2). – Valparaíso, 12.1851, *Philippi 402* (BM, P, W). – Valparaíso, 12.1829, *Bertero 907 p.p.* (P). – Playa Ancha nr. Valparaíso, 1832, *Bridges 127* (E K). – Valparaíso, 9.1830, *Markham 336* (K). – Las Cardas, 300 m, 12.1920, *Jaffuel s.n.* (CONC). – Marga Marga, 150 m, 1931, *Jaffuel s.n.* (CONC). – Quebradas de El Salto, 110 m, 12.1950, *Garaventa 5094* (CONC). – Recreo, 25 m, 01.1935, *Behn s.n.* (CONC). – Valparaiso, 20 m, 12.1950, *Muñoz s.n.* (CONC). – Valle de Marga Marga, 05.1910, *Jaffuel 1267* (GH). – Marga Marga, 400m, 12.1928, *Jaffuel 632* (CONC GH). – Marga Marga,, 150 m, 12.1930, *Jaffuel 3096* (CONC GH). – Valle Agua Potable, 10 m, 10.1954, *Schlegel 358* (CONC). – Valle de Marga Marga, 01.1911, *Jaffuel 1268* (GH). – Along road to Laguna Verde, 25 km from Valparaíso, 01.1939, *Goodspeed 23322* (GH). – Viña del Mar, Hacienda. Las Siete Hermanas, 280 m, 02.1938, *Behn s.n.* (CONC). – Prov. de Los Andes Río Colorado 12.18?? *Poeppig 110(62) Diar. 594?* (BM). – **Región Metropolitana de Santiago** Andes of Chili, *Cuming 182* (BM Ex2). – *Cuming s.n.* (E). – *Philippi s.n.* (BM). – *Philippi, 273* (BM). – Prov. de Chacabuco Cerro de las Viscachas, Hacienda de Cauquenes, 1873?, *Philippi s.n.* (M). – Prov. de Melipilla Cerro Cantillana, 1900 m, 03.1979, *Villagran & K. 476* (CONC). – RN Roblería de Cobre de Loncha, 1250 m, 11.2000, *K. Arroyo* et al. *205789* (CONC). – RN Roblería de Cobre de Loncha, Portezuelo de Las Mulas, 1000 m, 12.2000, *K. Arroyo* et al. *206427* (CONC). – Prov. de Santiago Santiago, *Ball s.n.* (E). – Santiago, *Philippi s.n.* (BM W). – Cajón Morales, 2200 m, 03.1953, *Ricardi 2432* (CONC). – Los Condes, 1000 m, 12.1927, *Elliott 292* (E). – Cerro Lo Chena, 650 m, 12.1950, *Recabarren 191* (CONC). – Lo Valdes – Las Yeseras, 2450 m, 02.1963, *Ricardi s.n.* (CONC). – Alto del Río, 832 (CONC). – Río Colorado, 1300 m, 1.1902, *Hastings 419* (NY). – Valle del Río Maipo, 2000 m, 03.1933, *Grandjot s.n.* (CONC). – Valle Quebrada de Morales, 2000 m 03.1933, *Grandjot s.n.* (CONC). – Renca, 10.1890, *Philippi s.n.* (BM). – Cuesta de Manantiales, Pap of the Portillo, *Miers s.n.* (BM). – Cerro Renca, 600 m, 11.1960, *Muñoz 176* (CONC). – RN Río Clarillo, 980 m, 11.2002, *K. Arroyo* et al. *25153* (CONC). – RN Río Clarillo, 980 m, 10.2002, *K. Arroyo* et al. *25153ª*(CONC). – RN Río Clarillo, 1000 m, 11.2000, *K. Arroyo* et al. *206145* (CONC). – RN Río Clarillo, 1800 m, 02.1989, *K. Arroyo* et al. *891175* (CONC). – RN Río Clarillo, 2050 m, 02.1989, *K. Arroyo* et al. *891259* (CONC). – RN Río Clarillo, 900, 12, 1999, *K. Arroyo* et al. *994606* (CONC). – RN Río Clarillo, Cajón del Horno, 2250 m, 12.2000, *K. Arroyo* et al. *206713* (CONC). – RN Río Clarillo, Ladera bajo Vega del Cigarro, 2420 m, 01.2000, *K. Arroyo* et al. *20493* (CONC). – RN Río Clarillo, Sector Loma de los Cipreses, 2150 m, 11.1999, *K. Arroyo* et al. *994428* (CONC). – RN Río Clarillo, Sector Los Cristales, 2650 m, 01.2000, *K. Arroyo* et al. *20737* (CONC). – Cerro de Renca, 600 m, 10.1955, *Navas 768* (CONC). – Las Condes, El Arrayán, 885 m, 12.1939, *Junge s.n.* (CONC). – Manzano, 900 m, 11.1930, *Pirion s.n.* (CONC). – Cerro Provincia, 1500 m, 11.1934, *Grandjot 3648* (GH). – Prov. de Cordillera Cajón del Maipo, Quebrada 3 Km E El Diablo, 1950 m, 01.2000, *Teillier s003* (CONC). – El Canelo, 800 m, 12.1926, *Looser 668* GH). – Valle del Maipo, Quebrada del Manzano, 09.1930, *Pirion 1239* (GH). – El Manzano, F.F.C.C. al Volcán, estero El Manzano, 870 m, 12.1927, *Montero 270* (CONC GH). – Paso Cruz, 2000 m, 01.1892, *Kuntze s.n.* (GH). – Paso Cruz 34°, 2600 m, 01.1892, *Kuntze s.n.* (GH). – *Dr. Gillies s.n.* (BM). – **VI Región del Libertador G.B.O'Higgins** Prov. de Cardenal Caro Camino San Fernando a Alcones, Km 30, 250 m, 12.1989, *Niemeyer 89* (CONC). – Pichilemu, Cahuil, 10 m, 01.1929, *Montero 771* (CONC). – Rinconada de Alcones, 175 m, 11.1989, *Matthei & Quezada 786* (CONC). – Tanume, 350 m, 01.1908, *Jaffuel s.n.* (CONC). – Prov. de Cachapoal tepidaria Cauquenes 3000 – 5000 ft, 05.1882, *Ball s.n.* (E). – Rancagua, Monte La Leona, 11.1818, *Bertero 472* (P). – Rancagua, fl. Cachapual. 12.1839, *Bertero 166* (P). – Rancagua, Monte La Leona, 5.1828, *Bertero 166* (BM). – Cajón Las Leñas, 8 Km al Oeste de la Confluencia con El Cachapoal, 1910 m, 1.1998, *K. Arroyo* et al. *99314* (CONC). – Palmar de Cocalán, fl. entre 01.1964, *Schlegel 4953* (CONC). – Termas de Cauquenes, Quebrada Huinganes, 700 m, 11.1952, *Pfister s.n.* (CONC). – Cajón Las Leñas, 2200 m, 01. 1998, *K. Arroyo* et al. *99439* (CONC). – Cajón Las Leñas, Cajón del Millico, 2890 m, 01.1998, *K. Arroyo* et al. *99666.5* (CONC). – Cajón Las Leñas, Laguna Grande, 2040 m, 01.1998, *K. Arroyo* et al. *99698* (CONC). – Coya, Cajón del Río Coya, 2160 m, 01.1999, *K. Arroyo* et al. *980838* (CONC). – PN Cocalán, 470 m, 11.2001, *K. Arroyo* et al. *211816* (CONC). – PN Cocalán, Loma del Espino, 1060 m, 10.2001, *K. Arroyo* et al. *211329* (CONC). – PN Cocalán, Sector Los Peumos Mancuernados, 550 m, 12.2001, *K. Arroyo* et al. *211842* (CONC). – RN Río de los Cipreses, Cerca del Glaciar Cipreses, 1890 m, 01.2000, *K. Arroyo* et al. *201223* (CONC). – RN Río De Los Cipreses, Ladera En Quebrada Las Terneras, 1800 m, 12.2000, *K. Arroyo* et al. *206793*(CONC). – RN Río De Los Cipreses, Quebrada Los Pangues, 1770 m, 01.2001, *K. Arroyo* et al. *210498* (CONC). – RN Río De Los Cipreses, Sector Rincón De Los Guanacos, 2230 m, 1.2001, *K. Arroyo* et al. *210469* (CONC). – RN Río De Los Cipreses, Sector Urreola, 1580 m, 12.2000, *K. Arroyo* et al. *207012* (CONC). – RN Río de los Cipreses, Paso de La Leona, 1120 m, 10.2000, *K. Arroyo* et al. *202856* (CONC). – RN Río de los Cipreses, Sector Agua de La Muerte, 1560 m, 01.2000, *K. Arroyo* et al. *20287* (CONC). – RN Río de los Cipreses, Sector Agua de La Muerte, 1800 m, 01.2000, *K. Arroyo* et al. *20325* (CONC). – RN Río de los Cipreses, Sector Glaciar El Cotón, 2000 m, 02.2000, *K. Arroyo* et al. *201333* (CONC). – Prov. de Colchagua Colchagua, *Philippi s.n.* (W). – *Gillies s.n.* (BM). – Cajón de Herrera, Horno de La Vieja, Termas del Flaco, 2000 m, 1.1942, *Aravena 33348* (GH). – Cajón de Los Helados, 1600 m, 01.1951, *Ricardi s.n.* (CONC). – Camino San Fernando-Vegas del Flaco, 1750 m, 01.1964, *Marticorena & Matthei 734* (CONC). – Cord. Tinguirírica, 2300 m, 1.1929, *Pirion 86* (GH). – Termas del Flaco, 1790

137

m, 1.1988, *Rosas 1956b* (M). – Camino San Fernando-Vegas del Flaco, Km 35, 950 m, 01.1964, *Marticorena & Matthei 763* (CONC). – Centinela, 349 m, 12.1965, *Montero 7379* (CONC). – Huemul, 330 m, 02.1955, *Barrientos 1896* (CONC). – Las Peñas, 755 m, 12.1954, *Barrientos 1892* (CONC). – Nancagua, 500 m, 01.1951, *Ricardi s.n.* (CONC). – Puente Negro, 540 m, 12.1950, *Ricardi s.n.* (CONC). – Rincon de Tinguiririca, 360 m, 12.1950, *Ricardi s.n.* (CONC). – Talcarehue, 1500 m, 12.1950, *Ricardi s.n.* (CONC). – Vegas del Flaco, 1800 m, 01.1968, *Montero 8091* (CONC). – Vegas del Flaco, 1800 m, 02.1955, *Ricardi 3163* (CONC). – Vegas del Flaco, Cerro del Arroyo, 2300 m, 02.1955, *Ricardi 3204* (CONC). – 5 km östlich La Rufina in Richtung auf die Baños del Flaco vor Villa Don Bosco, 11.1980, *Grau 2495* (MSB, M). – **VII Región del Maule** Prov. de Talca Central Los Cipreses, Quebrada El Ciego, 1600 m, 04.2000, *Finot & Lopez 2096* (CONC). – Curillinque, 1000 m, 02.1943, *Behn, H. s.n.* (CONC). – RN Alto de Vilches, camino Laguna, 1800 m, 07.2000, *Finot & Lopez 1730* (CONC). – RN Alto de Vilches, Cerro Pein, 1500 m, 01.2000, *Finot & Lopez 1687* (CONC). – Río Cipreses, Laguna Invernada, 1300 m, 12.1990, *Leuenberger, Arroyo-Leuenberger, Peñailillo & Rotella 4039* (B CONC). – Río Maule, 2000 m, 01.1991, *Ruthsatz 7087* (CONC). – Cerros al oeste de Pencahue, camino a Curepto, 130 m, 10.2002, *K. Arroyo et al. 25043* (CONC). – Constitución, Quivolgo, 30 m, 11.1958, *Barnier 402* (CONC). – Constitución, San Ramon, 30 m, 11.1958, *Barnier 378* (CONC). – Curepto, 10 m, 01.1926, *Barros 463* (CONC). – Entre Tranque del Maule-Bocatoma del Canal, 430 m, 02.1963, *Ricardi et al. 1036* (CONC). – Espinal de Los Llanos, 300 m, 11.1990, *Matthei & Quezada 1171* (CONC). – Panamericana sur km 229, 11.1980, *Grau 2394b* (MSB). – Reserva Los Ruiles de Empedrado, 300 m, 01.1999, *Matthei 901* (CONC). – Talca, 90 m, 11.1925, *Gunckel 603* (CONC). – San Rafael-Litú, 12.1994, *Ehrhart & Grau 95/577* (MSB). – Vilches, 1700 m, *K. Arroyo et al. 209671* (CONC). – Vilches, 1100 m, *K. Arroyo et al. 209763* (CONC). – Vilches, 1900 m, *K. Arroyo et al. 209821* (CONC). – Straße Villa Alegre-Constitución, westl. der Abzweigung nach Nirivilo, 300 m, 11.1980, *Grau 2348* (MSB, M). – Prov. de Cauquenes 5,8 Km al S del desvio a Sauzal, 125 m, 12.1994, *Bliss & Lusk 809* (CONC). – Camino de Parral a Cauquenes, Km 36, 160 m, 01.1964, *Marticorena & Matthei 484* (CONC). – Chanco, 5 m, 12.1961, *Quiros s.n.* (CONC). – Prov. de Linares – Panimavida, 2.1.1883, *Borchers s.n.* (BM). – Quinamavida, 02.1893, *Philippi s.n.* (BM). – RN Bellotos del Melado, 1520 m, 01.2000, *Humaña et al. 20126* (CONC). – RN Bellotos del Melado, 1520 m, 01.2000, *Humaña et al. 20139* (CONC). – Fundo El Castillo, Río Blanco, 975 m, 03.1999, *Ruiz & Lopez 824* (CONC). – RN Bellotos del Melado, 1000 m, 12.1999, *K. Arroyo et al. 996372* (CONC). – RN Bellotos del Melado, 1300 m, 12.1999, *K. Arroyo et al. 996290* (CONC). – RN Bellotos del Melado, 1200 m, 12.1999, *K. Arroyo et al. 996256* (CONC). – RN Bellotos del Melado, 915 m, 12.1999, *K. Arroyo et al. 994820* (CONC). – RN Bellotos del Melado, 1000 m, 12.1999, *K. Arroyo et al. 994809* (CONC). – Vegas del Molino, 03.1999, *Ruiz & Lopez 1410* (CONC). – E & S de Linares, a lo largo Río Ancoa, 750 m, 01.1993, *Taylor & Gereau 10983* (CONC). – Termas de Catillo, 320 m, 01.1961, *Montero 6279* (CONC). – Termas de Catillo, 225 m, 12.1953, *Ricardi 2831* (B, CONC). – Valle Gualquivilo, 1100 m, 01.1961, *Schlegel 3611* (CONC). – Vega Ancoa, 300 m, 02.1949, *Vasquez s.n.* (CONC). – Villaseca, 130 m, 12.1957, *Montero 5396* (CONC). – Villaseca, Cerro Alto de Caliboro, 1200 m, 11.1955, *Aravena 36* (CONC). – Prov. de Curicó Volcan Peteroa, 02.1896, *Philippi s.n.* (BM). – Camino entre Los Queñes y Paso Vergara, 1250 m, 10.2002, *K. Arroyo et al. 25042* (CONC). – Camino de Curico a la Laguna de Teno, 2520 m, 03.1973, *Marticorena et al. 72* (CONC). – Camino Laguna de Teno-Paso Vergara, Km, 2550 m, 03.1967, *Marticorena & Matthei, 936*, (CONC). – Hacienda Monte Grande, 1600 m, 12.1924, *Werdermann 507* (BM E K GH M NY). – Los Queñes, 650 m, 02.1930, *Barros 3221* (CONC). – 5 Km E de Los Queñes, La Jaula, 750 m, 01.1982, *Ugarte 246* (CONC). – Cerro de Los Huemules, Los Queñes, 1500 m, 01.1933, *Grandjot s.n.* (CONC). – Cerros de Los Queñes, 1600 m, 01.1933, *Grandjot s.n.* (CONC). – Curico-P.Vergara, 6 Km interior de Los Queñes, 750 m, 03.1967, *Marticorena & Matthei 778* (CONC). – Itahue, Fundo El Colorado, 230 m, 12.1954, *Garaventa 5156* (CONC). – Los Queñes, 1100 m, 12.1940, *Milner s.n.* (CONC). – Los Queñes, 1400 m, 01.1933, *Grandjot s.n.* (CONC). – **VIII Región del Bio Bió** Prov. de Concepción – Camino Concepción -Chillán, Cuesta Queime, 75 m, 12.1967, *Marticorena & Matthei 1081* (CONC). – Prov. del Bio Bió Mulchén - Los Angeles, camino, 1.1981, *Grau 2736* (M MSB). – Nacimiento, Fundo Tambillo, 120 m, 11.1950, *Pfister s.n.* (CONC). – San Carlos de Puren, Fundo Natalia, 200 m, 12.1938, *Junge s.n.* (CONC). – Pangal del Laja, östl. der Panamericana am Río Laja, 155 m, 01.1988, *Rosas 1888* (M). – Prov. de Ñuble Camino entre el Longitudinal & San Nicolas, 100 m, 12.1963, *Marticorena & Matthei 445* (CONC). – Quirihue, 255 m, 11.1963, *Zunza s.n.* (CONC). – San Fabian, 450 m, 12.1968, *Montero 8202* (CONC). – Coihueco, Las Pataguas, 200 m, 02.1942, *Pfister s.n.* (CONC).

IX *Chaetanthera* Ruiz & Pav.

3. *Chaetanthera ciliata* Ruiz & Pav. Syst. Veg. Fl. Peruv. Chil. 190 – 191, tab. 23. 1798.

Typus CHILE. "Habitat copiose in Regni Chilensis colibus et campis versus Huilquilemu oppidum" Holotypus MA! Isotypus MO! [Photos B! SGO!] Iconotypus !

Nomenclatural Notes A second, more widely dispersed collection of great historical importance is Dombey 450 (F Px2 W). Dombey was the Chief Doctor on the Ruiz and Pavón expeditions through Chile and Peru in the 1790's (Stafleu et al.). A large quantity of the Expedition's material, albeit duplicates, was gradually sold to Lambert by Ruiz over a period of 14 years. This material was later widely dispersed on the sale of Lambert's herbarium (Miller 1970) and can be found in OXF and BM. BM then transferred duplicates to MO and NY in the 1950's. Although in the systematic literature of *Chaetanthera* there is no mention of Dombey's collections until Cabrera studied the European Ruiz and Pavón types in Europe (Cabrera 1960). A short note on the typification issues surrounding *C. ciliata* was published by Pruski & Davies (2004).

Annual monoecious herb. **Roots** filamentous. **Stems** to 15 cm, erect to spreading, dark reddish to green with sericeous white hairs. **Leaves** 9–30 x 1–1.5 mm (n.i. cilia), linear, but outline including cilia ovate, cilia stiff (awn-like), up to 2.5 mm L; upper leaf part partially inrolled or thickened; ventral surface sericeous (compound hairs two-celled; one small 20–30 µm L basal ± inflated cell, one long 60–110 µm L filamentous cell), dorsal surface glabrous, shiny. **Capitula** campanulate, disk diameter 6 mm. **Involucral bracts** imbricate, arranged in three series, initially foliaceous then reduced to entirely membranous. **Outer involucral bracts** as leaves but with short membranous alae to less than ½ height of bract, 9–10 mm, cilia to 2.5 mm L (7–10 pairs + terminal awn). Lower dorsal surface sericeous (where alae are), upper ventral surface glabrous, shiny, changing to distinctive brown-purple towards tips, upper dorsal surface sericeous. **Middle involucral bracts** 9–10 mm L (n.i. cilia), cilia to 3 mm, 2–4 pairs + terminal awn; indumentum as for OIB; membranous alate, ratio of lamina to alae decreases from ⅔ to nearly entirely alate, distal dorsal surface of alae faintly maculate. **Inner involucral bracts** 9–10 mm L, acute, linear lanceolate, entirely membranous alate; apices striate maculate, dorsally sericeous. **Ray florets** 16–18, pistillate, yellow or bright cerise pink, upper dorsal surface sericeous; corolla 13 mm L, corolla tube 3.5 mm L, outer lip 2 mm W., inner lip 2–3 mm L, bifid, filamentous, tightly rolled. **Disk florets** bisexual, yellow; corolla 7–7.5 mm L, corolla tube 6 mm L **Styles** [ray] 7–7.5 mm L, stigma lobes yellow, 0.5–0.7 mm L [disk] 7 mm L, stigma lobes green, 0.5 mm L **Anthers** [ray] sterile, reduced. [disk] 4.2 mm L (7–8 mm incl. filaments). **Achenes** brown, ca. 2 mm L, turbinate; carpopodium anular, irregular; pericarp pellucid, densely covered in lanceolate twin hairs, 80–100 x 45 µm. **Pappus** white, ca. 4–6 mm L, 1 row, indehiscent; setae fused at base, setae 4–8 cells wide, cell width 5–10 µm, no basal cilia; barbellate, barbs 95–140 µm L, barb base shortly adhered (<40%), free barb appressed, 2–3 barbs/100 µm). **Chromsome count** $2n = 2x = 22$ (BAEZA & SCHRADER, 2005b) *C. Baeza 4205* (CONC).

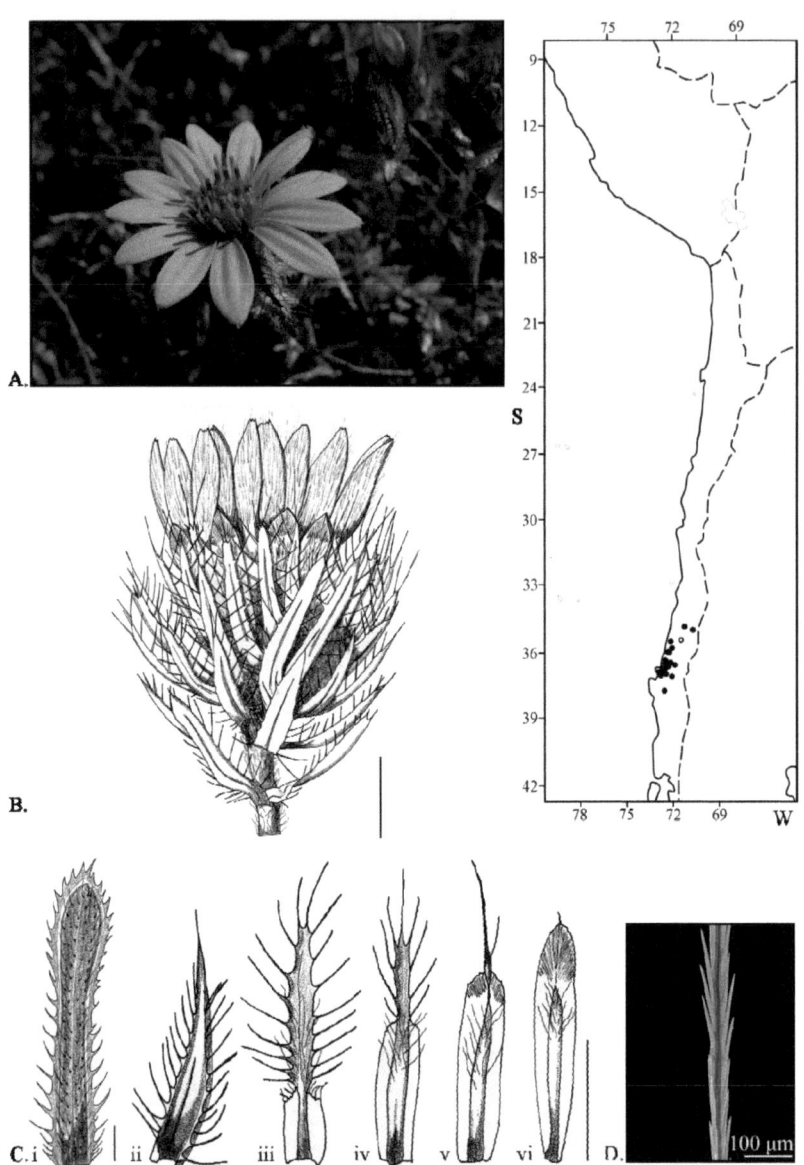

Figure 51: *Chaetanthera ciliata*. **A.** Habit photographed 12.2006©Michail Belov. **B.** Capitulum detail. *Grau 3252*. Scale bar = 5 mm. **C.** Leaf & bract detail, *Grau 3252*. **i.** Basal stem leaf v.s. **ii.** Upper stem leaf v.s.; **iii.** OIB v.s.; **iv – v.** MIB d.s.; **vi.** IIB d.s. Scale bar i = 2 mm; ii – vi = 5 mm. **D.** *Grau 3252*. Pappus setae (central region). Scale bar = 100 µm.

Distribution and habitat. *C. ciliata* is distributed south of the Río Teno in Chile, from Curicó to Mininco (latitudes 35°50–37°50'S). It occurs in the matorral típico, on the black lava sands of the central valley, e.g. Río Itata, Chillán, and the red sands of the Cordillera de la Costa. It is found at elevations between 60–1200 m.a.s.l.

Differential diagnosis. *C. ciliata* is morphologically very close to the more northerly distributed precocious *C. multicaulis*. Both species have distinctly ciliate leaf and bract margins, and share the same compound trichome on the leaf surfaces. However, *C. ciliata* is larger and more robust in every aspect, from leaf size to achene hairs. The pappus setae of *C. ciliata* are narrower, infrequently barbed and have somewhat longer barbs than *C. multicaulis*. Details are in Table 11. A bright cerise form of *C. ciliata* has been recorded by Hershkovitz (pers. comm.) and also in the botanical forum www.chilebosque.cl by Francisco Lira. This is speculated to be a response to grazing pressure as the plants then resemble the less palatable *Erodium* sp. sharing the pasture habitat.

Species	*C. ciliata*	*C. multicaulis*
Leaf L (mm)	up to 30	up to 10
OIB–MIB–IIB L (mm)	9–10	5–6
Pappus width (number of cells; each x μm wide)	4–8; 5–10 μm	10–15; 7–10 μm
Pappus barb length (μm); barb frequency.	95–140; 2–3 barbs/100 μm	ca. 90; 9–11 barbs/100 μm
Achene pericarp: twin hairs	80–100 x 45 μm	70–90 x 35 μm

Table 11: Key characters for identifying *C. ciliata* and *C. multicaulis*.

Material seen. – **Chile. VII Región del Maule** Maule, 02.1892, *Kuntze s.n.* (NY). – Prov. de Curicó Vichuquen, 50 m, 01.1964, *Diaz s.n.* (CONC). – Estación Quinta, 255 m, 02.1954, *Bravo s.n.* (CONC). – Los Quenes (Andes de Curicó), Canyon Río Claro, 1200 m, 01.1942, *Aravena 33391* (G GH). – Comalle, Cerro Chiquíreo, fundo La Vizcaya, 220 m, 11.1963, *Aravena s.n.* (SGOx2). – Prov. de Cauquenes 5,8 Km al Sur del desvio a Sauzal, 125 m, 12.1994, *Bliss & Lusk 795* (CONC). – Cauquenes, 150 m, 11.1975, *Contreras & Oyanedel 593* (SGO). – camino Parral - Cauquenes, 160 m, 01.1964, *Marticorena & Matthei 485* (CONC). – Prov. de Talca Nirivilo, 220 m, 12.1957, *Montero 5405* (CONC). – Prov. Linares Panimavida, 31.12.1882, *Borchers s.n.* (BM). – **VIII Región del Bio Bió** Prov. de Ñuble Environs de Chillán, 1855, *Germain (Philippi) s.n.* (BM G K W). – Cerro Ninhue, 300 m, 12.1998, *Matthei & Bustos 557* (CONC). – camino Cocharcas - San Nicolás, alrededores de San Pedro, 110 m, 01.1983, *Matthei & Bustos 3* (B CONC). – Yungay, Salto del Río Cholguan, 200 m, 01.1950, *Ricardi s.n.* (CONC). – camino El longitudinal - San Nicolas, 100 m, 12.1963, *Marticorena & Matthei 438* (CONC). – Bulnes, 80 m, 12.1968, *Montaldo 4572* (CONC). – Coihueco, Las Pataguas, 200 m, 02.1942, *Pfister s.n.* (CONC). – Panamericana südl. Chillán, 12.1981, *Grau 3252* (Hrb. Grau M). – camino Concepción-Bulnes, 60 m, 12.1945, *Pfister s.n.* (CONC). – camino Concepción-Bulnes, Puente Itata, 60 m, 12.1945, *Pfister s.n.* (CONC). – Puente Larqui, camino entre Chillán y Bulnes, 100 m, 11.2002, *K. Arroyo et al. 25157* (CONC). – Portezuelo, sector matorral típico, 180 m, *Contreras & Oyanedel 630* (SGO). – Prov. de Concepción Concepción, *Neger s.n.* (M). – camino de Rere a Yumbel, 220 m, 01.1959, *Marticorena, Mancinelli & Torres 44* (B CONC). – Subestacion experimental cauquenes, INIA, fundo El Boldo, 177 m, 11.1978, *Ovalle & Avendaño C-015* (SGO). – camino Concepción a Chillán, 65 m, 12.1967, *Marticorena & Matthei 1083* (CONC). – camino de Concepción a Bulnes, 100 m, 12.1967, *Parra & Rodriguez 103* (CONC). – Camino Florida - Yumbel, 200 m, 12.1946, *Pfister s.n.* (CONC). – Hualqui, Pichaco, 200 m, 01.1937, *Junge s.n.* (CONC). – Prov. del Bio Bió Cabrero, 2 km al norweste del Fundo Cabrero, 160 m, 01.1988, *Rosas 1890* (M). – **IX Región de La Araucania** Prov. de Malleco Mininco, 190 m, 12.1952, *Schwabe s.n.* (CONC). **S.l.d.** In aridis chili, *Dombey 450* (P). – *Dombey s.n.* (P W). – 1833, *Gay s.n.* (Kx2 NY P). – In collibus siccis Laguna Tagua, *Gay s.n.* (P). – 1834, *A. de Jussieu 3425* (GH). – Cuchacucha, *Nee s.n.* (MA). – Piunamanita, 1893, *Philippi s.n.* (BM). – *Zea s.n.* (BM). – Hacienda La Muceda, 01.1937, *Looser 6263* (M).

4. *Chaetanthera depauperata* (Hook. & Arn.) A.M.R. Davies stat. nov.

≡ *Chaetanthera microphylla* var. *depauperata* Hook. & Arn. Companion Bot. Mag. 1: 104. 1835. Typus CHILE "Cordilleras, *Cuming 241*" [Chile] Holotypus OXF! Isotypus BM! Ex2! GH! P!

= *Chaetanthera leptocephala* Cabrera Not. Mus., Eva Peron, Bot. 17: 78. 1954. Typus CHILE "Chile, Prov. Coquimbo, Dept. Ovalle, Quebrada del Toro, 20 km al este de Hurtado, Hacienda El Bosque, 1800 m, 30.11.1939, *Wagenknecht 18490*" Holotypus LIL 109456 [Freire & Iharlegui 2000] Isotypus CONC! GH! F! Kx2! LP (photo)! US!
- *Chaetanthera linifolia* Less. var. *depauperata* (Hook. & Arn.) DC. Prodr. (DC.) 7(1): 30. 1838.
- *Chaetanthera diffusa* Poepp. ex sched.

Annual monoecious herb. **Roots** filamentous. **Stems** to 6 cm, lax, ascending, glabrous. **Leaves** (1)2.5–7 x 0.7–1.1 mm, (upper leaves smaller), linear to narrowly ovate, glaucous, slightly succulent, 6–7 prs teeth ± 1–2 apical teeth; midrib visible, ventral surface irregularly sparsely dotted with translucent triangular hairs. **Capitula** sessile, cylindrical, bracts appressed; disk diameter 2 mm. **Involucral bracts** imbricate, arranged in three types, initially foliaceous then reduced to entirely membranous. **Outer involucral bracts** 2.5 mm L, ovate; as leaves but with short membranous alae to less than ½ height of bract. **Middle involucral bracts** 3–4.9 mm L, broadly ovate, lamina part continuous to apex; membranous alate, ratio of lamina to alae decreases from ⅔ to nearly entirely alate. **Inner involucral bracts** 5.9–7.4 mm L, broadly ovate to linear-oblanceolate, entirely membranous; apex ± emarginate, clear central midrib ending in mucro. **Ray florets** ± 4, white, pistillate, glabrous; corolla 4–5.2 mm L, corolla tube 2.5–3.4 mm L, outer corolla lip 0.6 mm W., inner corolla lip 1.5 mm L, deeply bifid. Stigma 4.8 mm L, lobes 0.4 mm L **Disk florets** ± 3, bisexual, yellow; corolla 3.5–4.6 mm L, corolla tube 2.7–3.8 mm L **Styles** [ray] 10–11 mm L, stigma lobes c. 1 mm. [disk] 4.4 mm L, lobes 0.5 mm L. **Anthers** [ray] sterile, 4.4 mm L [disk] fertile, 3 mm L (4.4 mm L including filaments). **Achenes** brown, 2.5 mm L (immature?); fusiform turbinate; carpopodium anular irregular; pericarp pellucid, covered in lanceolate twin hairs, 60–80 x 20 µm. **Pappus** white, (2.5)3.7–4.6 mm L, 1–2 rows, indehiscent; setae shortly fused at base, pappus setae 1–2(3) cells wide, cell width 5–10 µm; no basal cilia; barbellate, barbs 50–100 µm L, barb base shortly to medium adhered (40-50%), free barb appressed, 9–15 barbs/100 µm.

IX *Chaetanthera* Ruiz & Pav.

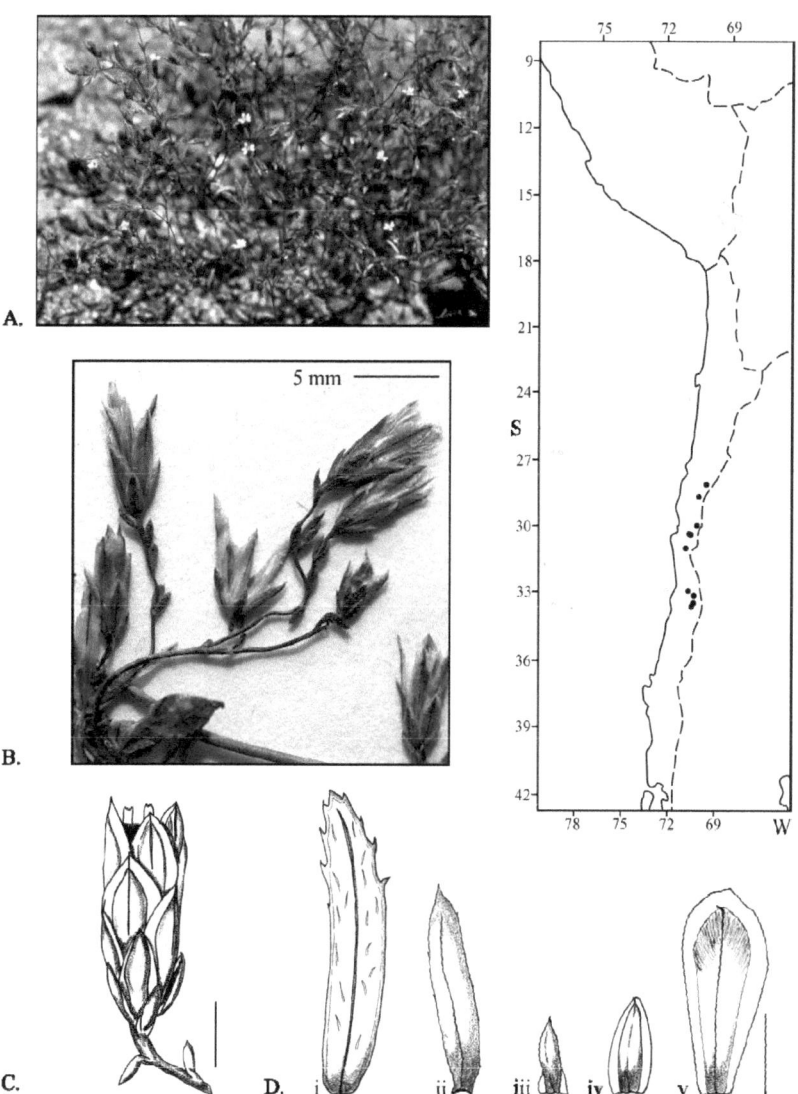

Figure 52: *Chaetanthera depauperata*. **A.** Habit photographed ©María Terese Eyzaguirre. **B.** – **D.** *Poeppig 593*. **B.** Habit detail, photo. **C.** Capitulum detail, scale bar = 2.5 mm. **D.** Leaf & Bract detail, *Poeppig 593*. **i.** Lower stem leaf v.s.; **ii.** Upper flowering stem leaf d.s.; **iii.** OIB d.s.; **iv – v.** IIB d.s. Scale bar = 2mm.

Distribution and habitat. *C. depauperata* forms scattered populations in the lower foothills of the Chilean Andes from Río Ramadilla to Cajón del Maipo (28°00' – 34°00'S). It is found at elevations between 700 – 2900 m.a.s.l. on sandy slopes.

Differential diagnosis. *C. depauperata* belongs to the linear-leaved group that also includes *C. albiflora, C. perpusilla, C. linearis* and *C. microphylla*. Key characters for identifying these taxa are given in Table 12. These species can be identified according to leaf and ray floret colour, leaf indumentum, IIB outline and capitulum shape and the approximate number of imbricate bracts.

Species	*C. albiflora*	*C. depauperata*	*C. linearis*	*C. microphylla*	*C. perpusilla*
Leaf colour	Virid	glaucous	glaucous	glaucous	?
Floret colour	White with dorsal stripe	white	yellow	Red (rarely yellow)	white
Leaf Indumentum	Lanate	Translucent trichomes	Translucent trichomes	Translucent trichomes	Translucent trichomes
IIB	Emarginate or acute	Acute, reflexed mucro on mesophyllous tissue	Emarginate or acute	Acute	Acute with short apical mucro
Capitulum	Cylindrical (urcinate), > 25 bracts	Narrow, cylindrical, few bracts (< 20)	Cylindrical (urcinate), > 25 bracts	Campanulate, > 25 bracts	Open campanulate, few bracts (< 20)

Table 12: Key characters for identifying *C. albiflora, C. depauperata, C. linearis, C. microphylla* and *C. perpusilla*.

Material seen. – Chile. III Región de Atacama Prov. de Copiapó Río Ramadilla, 2870 m, 03.1992, *Arancio 237* (CONC). – Prov. de Huasco Río Laguna Grande, 2400 m, 01.1983, *Marticorena* et al. *83356* (CONC). – **IV Región de Coquimbo** Prov. de Elqui Entre Juntas y Embalse La Laguna, 2300 m, 01.1981, *K. Arroyo 81165* (CONC). – Cerro Tololo, 1800 m, 11.1967, *Jiles 5128* (CONC). – Cordillera de los Llanos de Guanta, 2008 m, 12.1836, *Gay 395* (P). – Prov. de Limarí Cordillera de Ovalle, Serón, 1800 m, 11.1954, *Jiles 3334* (CONC). – Quebrada del Toro, 20 km east of Hurtado, Hacienda El Bosque, 1800 m, 11.1939, *Wagenknecht 18490* (CONC F GH Kx2). – 3 km al sur de Combarbalá, en camino de tierra que sale del pueblo, 1010 m, 09.2002, *K. Arroyo* et al. *25025* (CONC). – **V Región de Valparaíso** Prov. de Quillota Cerro del Roble, 1600 m, 12.1934, *Garaventa 3169* (CONC). – **Región Metropolitana de Santiago** Prov. de Chacabuco Hacienda Chacabuco, 11.1929, *Looser 994* (GH). – Andes de S. Rosa, in montibus aridis ad Río Colorado, 12.1827, *Poeppig s.n. D.593* (W). – Andes de S. Rosa, in montibus aridis ad Río Colorado, 11.1827, *Poeppig s.n. D.594* (W). – Prov. de Santiago Cerro Abanico, 1500 m, 12.1951, *Barrientos s.n.* (CONC). – Cajón del Maipo, Clarillo, 700 m, 02.1951, *Castillo s.n.* (CONC). – **S.l.d.** *Gay s.n.* (GHx2, NY). – *Gay s.n.* (+ illus.) (P). – San Carlos, *Neger s.n.* (M).

5. *Chaetanthera elegans* s.l. Phil. Linnaea, 28: 712. 1856.

Typus CHILE "Prope los Angelos legit Cl. Gay et in herbario sub no. 806. Reliquit" Holotypus SGO (S64729)!

= *Chaetanthera andina* Phil. in Philippi, R. A. Anales Univ. Chile 87: 6. 1894. Typus CHILE "In Andibus provinciae Talca invenit *Fr. Philippi*, Araucaniae (l.d. La cueva c. *Rahmer*) Valdiviae l.d. Pucallu (*Otto Philippi*)" Syntypus *F. Philippi*: GH! NYBG! Syntypus *O.Philippi*: SGO (S43723)! (S64752)!

= *Chaetanthera araucana* Phil. in Philippi, R. A. Anales Univ. Chile 87: 10. 1894. Typus CHILE "Prope Lebu in provincia Arauco legi" Holotypus SGO (S64754)! Isotypus K!

= *Chaetanthera brachylepis* Phil. in Philippi, R. A. Anales Univ. Chile 87: 12. 1894. Typus CHILE "Habitat in montibus Nahuelvuta, loco dicto Vega de Rucapillan. Januario 1877 legi" Holotypus SGO (S64727)! [Photos GH! NYBG!]

= *Chaetanthera comata* Phil. Anales Univ. Chile 87: 8. 1894. Typus CHILE "In monitibus Araucaniae Nahuelvuta dictis l.d. Vages de Chanleo 1877 legi" Holotypus SGO (S64728)!

= *Chaetanthera elata* Phil. Anales Univ. Chile 87: 9. 1894. Typus CHILE "Ad lacum Villarica, v.gr. ad Pucon, frequens esse videtur. (*Otto Philippi*)" Holotypus SGO (S64731)! *Chaetanthera elatior* Phil. Ex sched. (orth.var.) E! GH(photo)! NYBG(photo)!

= *Chaetanthera elegans* var. *pulchra* Cabrera Revista Mus. La Plata, Secc. Bot. 1: 161, fig. 30. 1937. Typus ARGENTINA "Argentina, Neuquén, Chanchahuinganco, 28.1.1935, *Ragonese 310*" Holotypus LP 66857 [Freire & Iharlegui 2000]

= *Chaetanthera foliosa* Phil. in Philippi, R. A. Anales Univ. Chile 87: 11. 1894. Typus CHILE "In montibus Nahuelvutae loco dicto Valle del Palo botado inveni" Holotypus SGO (S64749)! Isotypus SGO (S4371)!

= *Chaetanthera humilis* Phil. Anales Univ. Chil. (41 – 42): 740. 1872. Typus CHILE "Poseo ejemplares de localidades muy diferentes; de la cordillera de Chillán, de la vecinidad de Lota" Syntypus SGO (S 64732)! – mixed collection

= *Chaetanthera pratensis* Phil. Anales Univ. Chile 87: 12. 1894. Typus CHILE "Habitat in pratis nemoralibus Andium chillanensium" Holotypus SGO (S43755)!

= *Chaetanthera valdiviana* Phil. in Philippi, R.A. (1856) Linnaea 28: 712. Typus CHILE "Valdiveae in praedio meo S. Jaun legi" Holotypus SGO #64726! Isotypus probabilis "In provincia Valdiviensi, Jan. 1851, Philippi 488" BM!

= *Chaetanthera volkmannii* Phil. in Philippi, R. A. Anales Univ. Chil. (41 – 42): 740. 1872. Typus CHILE "Traído por *Volckmann*, tal vez de la Araucanía" Holotypus SGO (S43760)!

- *Chaetanthera elegans* var. *pratensis* (Phil.) Cabrera in M. N. Correa Fl. Patagonica 7: 315. 1971.
- *Chaetanthera serrata* var. *humilis* (Phil.) Reiche in Reiche, K. Anales Univ. Chile 115: 319. 1904.

Perennial monoecious caespitose herbs. **Roots** with thick central woody rootstock with short loosely to densely packed woody ± subterranean stems (rhizomes). **Stems** subterranean or on the surface, shallowly branching, creeping, supporting densely to loosely leafy clumps (pseudorosettes) of leaves. **Leaves** (20)28–65(70) x (2)3–6 mm, bright green (not grey green), lanceolate to spathulate, flat, alternate; margins shortly to longly spinulose serrate or mucronate dentate in upper two thirds; indumentum glabrous/ glabrescent to sericeous. **Pedicel** (scape) erect to ascending, (1.0)12.4–14.4(36) cm L, glabrescent, leaves as basal leaves, but smaller and infrequently spaced up scape. **Capitula** cylindrical to campanulate, 10–16 x 10–20(30) mm; often subtended by a few large leafy outer involucral bracts exceeding the height of the capitulum. **Involucral bracts** imbricate, arranged in two types, initially foliaceous then reduced to entirely membranous. **Outer involucral bracts** 1–2 rows, (5.3)9–12.7 x 1.3–2.4(3.2) mm; as leaves but with short membranous alae to less than ½ height of bract, alae linear ovate to truncate. **Middle involucral bracts** absent, no nearly entirely alate bracts. **Inner involucral bracts** 2-3 rows increasing in size, 11.2–15.9–18.1 x 0.8–1.5 (3.2) mm, linear lanceolate; entirely membranous alate; apices acute to broadly acute, black-green or brown maculate, mucronate. **Ray florets** 13–19, pistillate, yellow with no dorsal stripe, dorsally sericeous pubescent; corolla 22 mm L, outer ligule 3 mm W., corolla tube 3.5 mm L, inner ligule 3 + 1.5 mm (upper part curled). **Disk florets** ca. 20, bisexual, yellow; corolla 9–10 mm L, corolla tube 7–8 mm L **Styles** [ray] ca. 10 mm, stigma lobes ca. 1 mm, linear spathulate [disk] not observed **Anthers** [ray] sterile [disk] 10 mm L **Achenes** brown, 2.5–3.5 mm L, angled turbinate; poorly developed carpopodium (seldom absent), variable within capitula; pericarp pellucid, densely covered in lanceolate twin hairs, 75–110 x 45 µm. **Pappus** white, ca. 10 mm L, 1–2 rows, indehiscent; setae fused at base (corona 200 µm high), setae 1-2(3) cells wide, cell width 15 µm, no basal cilia; barbellate, barbs 94–140 µm L, barb base shortly adhered (<40%), free barb appressed, 5–6 barbs/100 µm.

IX *Chaetanthera* Ruiz & Pav.

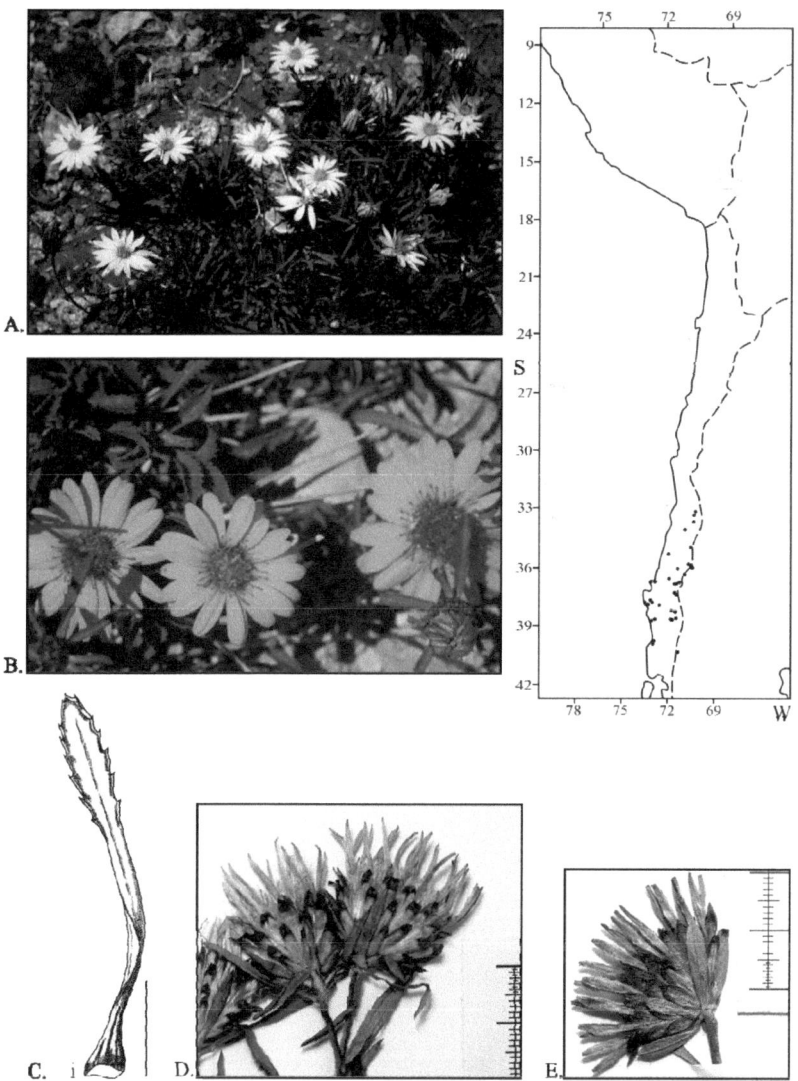

Figure 53: *Chaetanthera elegans*. **A.** Habit Photographed Chillán 1995©Jürke Grau. **B.** Habit Photographed Laguna del Maule, 1981©Jürke Grau. **C.** Leaf & Bract details **i.** Stem leaf. Scale bar = 1 mm. **D.** Capitulum detail *Ehrhart & Grau 95/919*. **E.** Capitulum detail *Grau 2923*.

Distribution and habitat. *C. elegans* is found between latitudes 36°00' – 40°00'S, mostly scattered from volcanic lake to lake (Maule, Laja, Chillán - no lake, Antuco, Lonquimay, Llaima), or on the Cordillera de Nahuelbuta. It occurs at all elevations up to 2000 m.a.s.l. in sunny, dry pastures, above *Austrocedrus* tree-line, and in *Araucaria* woodlands.

Differential diagnosis. Based on the hybrids study of *C. chilensis* and *C. elegans*, most epithets synonymised under *C. elegans* are currently tagged as probable hybrids. The type locality of a Parent in the hybrid analysis, Chillán, is identified with two epithets, *C. humilis* and *C. pratensis*, both of which are described as being glaberrimous and having quite wide (4 – 8 mm) spinulose serrate leaves (PHILIPPI 1872, p. 740; PHILIPPI 1894, p. 12). However, the Typus specimen and the description of *C. elegans* (PHILIPPI 1856, p. 712) best circumscribe the entity considered to be this parent. *C. elegans* is distinguished by having bright green, lanceolate spathulate, > 3 mm wide flat leaves, with a glarbescent to sericeous indumentum. The leaf margins are spinulose serrate to mucronate dentate and the capitula have 2 (not 3) series of involucral bracts. MIB are rare or absent. The pappus setae are only a few cells wide, although the cells are quite broad. The setae are more frequently barbellate with somewhat shorted barbs than *C. chilensis*. Details of these key features distinguishing *C. elegans* and *C. chilensis* can be found in Table 10.

Material seen. – **Argentina**. Neuquén Meli. [Lago Meliquina?] suma, 02.1941, *Bridorelli s.n.* (K). – Pulmari, 3,000 ft., 01.1926, *Comber 372* (E K). – **Chile**. **VII Región del Maule** Prov. de Talca Cuesta Los Condores, Laguna del Maule, 1500 m, 01.1994, *Villagran* et al. *8263* (CONC). – Upper valley of Río Maule, 700 m, 01.1990, *Gardner, Knees & de Vore 4590* (E K). – Zufahrt zur Laguna del Maule, vor der Laguna, 2000 m. 01.1981, *Grau 2923* (M MSB). – Prov. de Linares Camino Reten Achibueno-Las Animas, 03.1999, *Ruiz & Lopez 1086* (CONC). – Laguna Dial, 1520 m, 01.1961, *Schlegel 3656* (CONC). – **VIII Región del Bio Bió** Prov. del Bio Bió Cordillera Polcura, 1000 m, 02.1955, *Ledezma 666* (CONC). – El Abanico, 800 m, 02.1951, *Pfister s.n.* (CONC). – PN Laguna del Laja, lado derecho, base Sierra Velluda, 1000 m, 03.2001, *Baeza & Parra 3522* (CONC). – PN Laguna del Laja, Los Barros, Sector Mallin Florido, 1460 m, 01.2001, *Baeza, Parra & Torres 3125* (CONC). – PN Laguna del Laja, sendero desde Los Zorros al pie de la Sierra Velluda, 1380 m, 01.2001, *Baeza, Parra & Torres 3228* (CONC). – Prov. de Ñuble Cordillera de Chillán, *Philippi s.n.* (W). – Baños de Chillán, 1883, *Borchers s.n.* (BM). – Chillán, XII.1869, *Reed s.n.* (BM). – Camino entre Recinto & Termas de Chillán, 1240 m, 03.1973, *Rodriguez & Torres s.n.* (CONC). – Termas de Chillán, 2000 m, 02.1950, *Barros s.n.* (CONC). – Termas de Chillán, 1600 m, 02.1960, *Pfister s.n.* (CONC). – Chillán, Las Bravas, S. of road to termas, 1750 m, 03.1995, *Ehrhart & Grau 95/919* (MSB). – Cord. Chillán, 01.1904, *Elliott 341* (BM E). – In Chile austr. lecta ad Antuco, 11.18??, *Poeppig 208 Diar. 689* (M W). – **IX Región de La Araucanía** Prov. de Malleco Lonquimay, 1680 m, 02.1940, *Hollermayer s.n.* (CONC). – PN Nahuelbuta, unos 500 m antes de llegar a Piedra de Aguila, 1470 m, 10.2002, *K. Arroyo* et al. *25069* (CONC). – PN de Nahuelbuta, 1250 m, 03.1973, *Rodriguez & Torres s.n.* (CONC). – PN de Nahuelbuta, 1460 m, 02.1967, *Ricardi 5371* (CONC). – PN de Nahuelbuta, 1300 m, 01.1968, *Ricardi* et al. *1947* (CONC). – Rahue, 1680 m, 02.1958, *Montero 5518* (CONC). – Termas Tolhuaca, 1300 m, 01.1935, *Montero 2194* (CONC). – Camino a Laguna Verde entre Tolhuaca & Laguna Malleco 1400 m, 01.1988, *Rosas 1875* (M). – Camino Liucura – Pino Hachado, Km 21, 1500 m, 02.1960, *Ricardi & Marticorena 5073* (CONC). – Prov. de Cautín Volcan Llaima, 1100 m, 02.1927, *Werdermann 1256* (B BM CONC E K M NY). – Lac Villarica "Ch. elatior Ph." *Ball s.n.* (E). – Villarrica, (argent. Abhang), 1897, *Neger s.n.* (M). – *Calvert s.n.* (BM). – **X Región de Los Lagos** Prov. de Valdivia Valdivia, *Hollermayer 160a* (W). – Valdivia, *Hollermayer 160b* (W).

6.   *Chaetanthera glandulosa* J. Rémy var. *glandulosa* in Fl. Chil. [Gay] 3: 311. t. 35. 1849.

Typus CHILE "Se cria en los cerros de la provincia de Coquimbo" [Prov. de Coquimbo, 1838, *Gay s.n.*] **Lectotypus hic loc. designatus** P! Isolectotypus F! G! Other original syntypes include: Prov. de Coquimbo, 1839, *Gay 419* (P!)

= *Tiltilia pungens* Phil ex Reiche Flora de Chile 4: 342. 1905. nomen nudum.

Nomenclature Notes The existence of two possible syntypes from Paris (P), where Rémy worked (TL2), argues for the selection of a lectotype (see Art. 9.2, ICBN) from among the two differently dated Paris collections. Despite a careful comaprison of the illustration in the Altas Botanico (GAY 1849) and the material it was not possible to establish which piece of material Remy might have taken his drawing from (t.35). It was decided that the 1838 sheet was most representative.

Perennial woody dwarf subshrub. **Roots** filamentous (?). **Stems** to 50 cm, erect, densely branched; indumentum dense glandular. **Leaves** 10–15 x 1.5–2 mm, nearly opposite, linear apiculate, slightly dilated at base to connate, margins plicate, midrib ± visible, schlerophyllous, sticky with shortly stalked glands, densely scattered over dorsal and ventral surfaces. **Involucral bracts** imbricate, arranged in three types, initially foliaceous then reduced to entirely membranous. **Outer involucral bracts** 9–11 x 1 mm, linear, apiculate-mucronate; as leaves but with short membranous alae to less than ½ height of bract, 0.5–0.6 mm W., glandular only on margins; lamina as for leaves. **Middle involucral bracts** 7–9 x 2 mm, linear lanceolate, glandular; membranous alate, ratio of lamina to alae decreases from ⅔ to nearly entirely alate. **Inner involucral bracts** 10–11 x 2 mm, linear lanceolate, entirely membranous, green-yellow, with stiff apical mucro. **Capitula** cylindrical, 10 mm L, disk diameter 12 mm. **Ray florets** ca. 9, pistillate, white; Corolla 11.3 mm L, corolla tube 5 mm L, outer ligule 1.7 mm W., inner ligule deeply bifid, 3 mm L Achenes mm L, turbinate, apices apiculate. **Disk florets** ca. 15, bisexual, greenish yellow; corolla 9 mm L, corolla tube 6.8 m L **Styles** [ray] 9 mm L, stigma lobes oblong, 0.5 mm L [disk] 10 m L, stigma lobes oblong, 0.5 mm L **Anthers** [ray] sterile. [disk] fertile, 5.6 mm (9 incl. filaments). **Achenes** brown, 2.5–3.5 mm L, fusiform, turbinate; poorly developed carpopodium; pericarp pellucid, glabrous or sparsely pubescent with lanceolate twin hairs 75–100 μm L; testa epidermis surface of elongated parallel cells, margins entire, cells with U-formed sclerenchymatous thickenings. **Pappus** white, ca. 7–9 mm L, 1–2 rows, indehiscent; setae fused at base, setae 1–2(3) cells wide, cell width 5–10 μm, no basal cilia; barbellate, barbs 50 μm L, barb base shortly adhered (<40%), free barb appressed, 9–11 barbs/100 μm.

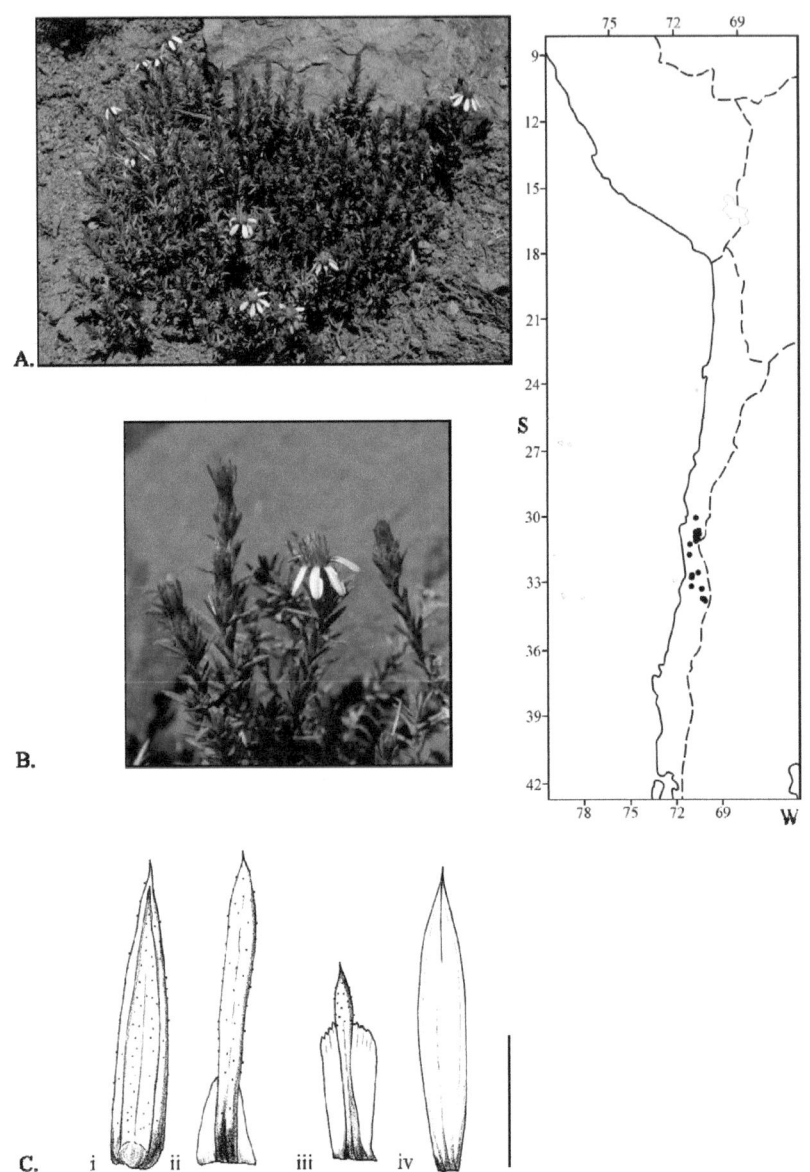

Figure 54: *Chaetanthera glandulosa* var. *glandulosa*. A & B photographed Coquimbo, 01.2009©María Terese Eyzaguirre.**A.** Habit. **B.** Capitulum detail. **C.** Leaf & Bract detail *Rosas 1592*. **i.** Stem leaf v.s.; **ii.** OIB d.s.; **iii.** MIB d.s.; **iv.** IIB d.s. Scale bar = 5 mm.

IX *Chaetanthera* Ruiz & Pav.

Distribution and habitat. *C. glandulosa* is locally endemic in the Chilean pre-cordillera from just south of the Río Elqui to the upper reaches of the Río Maipo between latitudes 30°00 – 33°50'S. It occurs at elevations between 1000 – 3000 m.a.s.l. In the more westerly cordilleran outcrops (e.g. Cerro de Los Caquis, Altos del Roble) it occurs at lower elevations. It occurs in isolated populations in relatively inaccessible locations on rocky soil, in Cipreses [*Austrocedrus chilensis*] woods below the Andean zone.

Differential diagnosis. Optically distinct from most other *Chaetanthera* species because of its subshrub habit, *C. glandulosa* shares the imbricate involucral bracts, achene and pollen characters (ca. 75 x 50 µm, dumbbell shaped nexine, and columellate ectosexine) characteristic of the genus. The U-formed testa epidermal strengthenings with parallel epidermal cells, and pappus characters define its position in the subgenus *Chaetanthera*, It differs from the novel variety by having straight stems, larger leaves, fewer (ca. 9) white ray florets and mucronate leaf and bract apices.

Material seen. – **Chile. IV Región de Coquimbo** Prov. de Elqui 20,2 Km N de Hurtado, 01.1993, *Stuessy & Ruiz 12774* (CONC). – in editillinum andium Los Patos, 01.1837, *Gay 946* (SGO). – in editillinum cordillera de H[G]uanta, 11.1836, *Gay 945* (SGO). – Prov. de Limarí Río Torca, 01.1890, *Geisse s.n.* (BM SGOx2). – H[G]uatalme, cerro, 01.1869, *Volckmann s.n.* (SGO). – Cuesta de Hornos, cerca de la cima SW Combarbalá, 1700 m, 12.1987, *Rosas 1592* (M). – Los Molles, 02.1972, *Zöllner 5573* (CONC). – Cabreria, Morro Blanco, 01.1949, *Jiles 1237* (CONC). – Cord. de Ovalle, Vegas Negras de San Miguel, 01.1959, *Jiles 3645* (CONC). – Vega Negra, 01.1959, *Jiles 3654* (CONC). – Tulahuen, 02.1950, *Collantes s.n.* (CONC). – El Maiten, Río Grande, 01.1954, *Jiles 2444* (CONC). – Río Torca, 02.1961, *Jiles 3776* (CONC). – Portezuelo, Cuesta El Espino, 01.1973, *Marticorena* et al. *633* (CONC). – Prov. de Choapa Limahuida, Salamanca, 01.1950, *Pfister s.n.* (CONC). – **V Región de Valparaíso** Prov. de Los Andes Río Blanco F.C.T.C., 2300 ft, 03.1927, *King 426* (BM). – Laguna del Copin, 02.1926, *Claude Joseph 2746* (CONC). – **Región Metropolitana de Santiago** In andibus Prov. Santiago, *Philippi s.n.* (BM). – Cordillera de Santiago, 1856-57, *Philippi s.n.* (BM G K SGO W). – Cordillera de Santiago, *Philippi s.n.* (W). – Prov. de Chacabuco Altos del Roble, Hacienda de Chicauma, 12.1983, *Villagran 4712* (CONC). – Altos de Tiltil, 04.1895, J.Philippi (SGOx3). – Prov. de Santiago San José de Maipo, Cajón del Río Morales, 10.1989, *Saavedra & Pauchard 91A* (SGO). – San Gabriel, 12.1950, *Gunckel 21866* (CONC). – Valle del Río Maipo, 01.1943, *Grandjot 4711* (CONC). – Cajón del Maipo, campamento Las Gualtatas, 02.1995, *Villagrán, Villa & Hinojosa 8614* (LP). – Cajón del Maipo, 02.1948, *Castillo s.n.* (CONC). – El Volcán, 03.1936, *Espinosa s.n.* (SGO). – Puente Alto, El Volcán, Lo Valdes, 03.1953, *Ricardi 2346* (CONC F). – Valle del Morales, 02.1948, *Castillo 20268* (CONC). – Cerro Buitre, near La Dormida 10 km south of Las Viscachas, 01.1939, *Morrison & Wagenknecht 17104* (K).

7. *Chaetanthera glandulosa* var. *gracilis* A.M.R. Davies var. nov.

Typus CHILE "Chile, V Región de Valparaíso, Prov. de San Felipe, Cerros de Los Caquis, 2000 m, 03.1964, *Zöllner 400*" Holotypus CONC!

A varietate typica differt caulinibus tenuibus flexuosis, foliis caulinis (10–15 x 1.0 –1.5 mm) linearibus in axillis foliorum caulinorum dense fasciculatis (4–6.5 x 0.5–1 mm), bracteis involucri exterioribus et intermediis pilo glanduloso terminale ornatis, interioribus acutis, floribus radii ad 15, aureis. Achenia fusca, turbinata, 2.5mm longa; carpopodium indistinctae; pericarpium pellucidum, sparsum papillosum et pilis didymis ornatum. Pappi setae albidae, 6–7 mm longae, persistentes, uno- ad biseriales, sursum barbellatae setis adpressis. Habitat in Chili.

Perennial woody dwarf subshrub. **Roots** filamentous (?). **Stems** to 30 cm, erect, slender, flexuous, densely branched; indumentum glandular, dense. **Leaves** 10–15 x 1.0–1.5 mm, sclerophyllous, alternate to nearly opposite, nearly all with vigorous compacted axillary growth of smaller shorter leaves (4–6.5 x 0.5–1 mm); linear, slightly dilated to connate at base; apiculate-mucronate, margins plicate, midrib ± visible; upper leaves reduced, angled linear, margins dotted with glands; sticky indumentum of glands with short stalks, densely scattered over both dorsal and ventral surfaces. **Capitula** radiate, sessile, 10 mm L, shortly cylindrical to campanulate, disk diameter 10–15 mm. **Involucral bracts** imbricate, arranged in three types, initially foliaceous then reduced to entirely membranous. **Outer involucral bracts** 4–6 x 0.5 mm, linear, apiculate, apices with a single sessile glandular hair; as leaves but with short membranous alae to less than ½ height of bract, alae narrow, 0.1–0.3 mm W., glandular on margins. **Middle involucral bracts** 6–9 x 1.5–2 mm, linear, angled, lamina as for upper leaves; glandular on margins, dorsally sparsely sericeous; membranous alate, ratio of lamina to alae decreases from ⅔ to nearly entirely alate. **Inner involucral bracts** 10–11 x 2 mm, linear, membranous, green-yellow, longly acute. **Ray florets** ca. 15, pistillate, yellow, dorsally sericeous; corolla ca. 14 mm L, corolla tube 4 mm L, outer ligule 2.5 mm W., inner ligule fused, filamentous, ca. 5 mm L **Disk florets** ca. 20, bisexual, greenish yellow; corolla 7.5–8 mm L, corolla tube 6.8 mm L **Styles** [ray] 9 mm L, stigma lobes oblong, open, 0.5 mm L [disk] 7.6 m L, stigma lobes 0.5 mm. **Anthers** [ray] sterile [disk] 5 mm (7 incl filaments). **Achenes** brown, 2.5 mm L, turbinate; poorly developed carpopodium; pericarp pellucid, sparsely covered with twin hairs. **Pappus** white, 6–7 mm L, 1(2) rows, indehiscent; setae fused at base, no basal cilia; barbellate, barbs appressed.

Distribution and habitat. *C. glandulosa* var. *gracilis* is recorded as isolated on the western spur of cordillera towards Zapallar (Chile) at 33°50'S. at elevations between 1000 – 2000 m.a.s.l. It occurs in rocky soil, in *Austrocedrus chilensis* woods below the Andean zone.

Differential diagnosis. Named for its slender, flexuous stems, *C. glandulosa* var. *gracilis* is only known from one locality. It differs from the typical form by having flexuous stems, smaller outer bracts, more (ca.15) yellow ray florets, and the bracts being tipped with a single large gland.

IX *Chaetanthera* Ruiz & Pav.

Material seen. – **Chile. V Región de Valparaíso** Prov. de San Felipe Cerros de Los Caquis, 2000 m, 03.1964, *Zöllner 1732* (CONC). – Los Caquis, 2000 m, 03.1964. *Zöllner 400* (CONC). – Cerro Caquisito, 1000 m, 04.1966, *Zöllner 1107* (CONC).
7

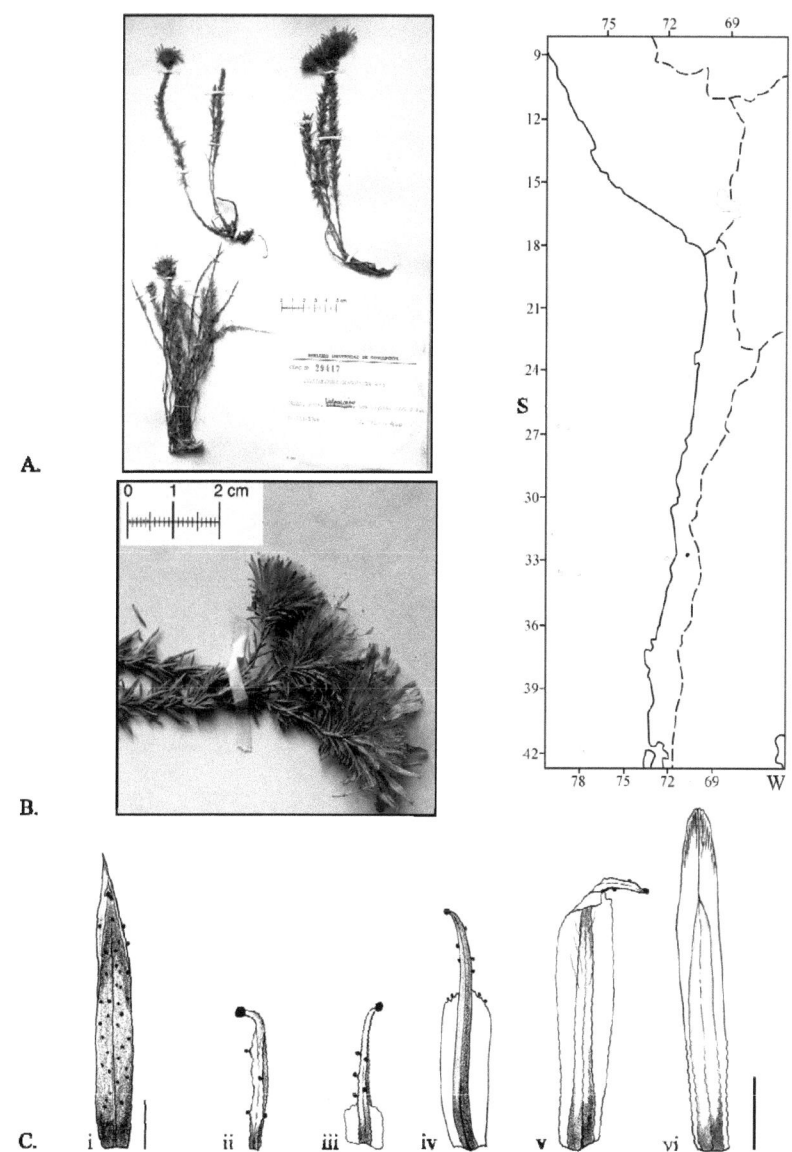

Figure 55: *Chaetanthera glandulosa* var. *gracilis*. Holotypus *Zöllner 400*. **A.** Herbarium sheet showing stem habit. **B.** Capitulum detail. **C.** Leaf & bract detail. **i.** Stem leaf v.s.; **ii.** Upper stem leaf, subtending capitulum v.s.; **iii.** OIB v.s.; **iv - v.** MIB d.s. **vi.** IIB d.s. Scale bar i, ii – vi = 2 mm.

8. *Chaetanthera incana* Poepp. ex Less. Linnaea 5: 284. 1830.

Typus CHILE "Poeppig in montibus arenosis, maritimis pr. Concon Septbr. flor." [Concón, 09.1827, *Poeppig 218 (Diar. 286)*] **Lectotypus hic loc. designatus** W! Isotypus BM! G! GH(photo)! MO! NY(photo)! P! Wx2!

= *Chaetanthera spathulata* Pöpp. ex Less. Linnaea 5: 285. 1830. Typus CHILE "Poeppig in aridissimus collium ad Concon Septbr. flor." [Concón, 09.1827, *Poeppig 219 (Diar. 275)*] **Lectotypus hic loc. designatus** W! Isotypus BM! G! NYx2! P! W!

= *Chaetanthera scariosa* D. Don ex Taylor & Phillips Philos. Mag. Ann. Chem. 11: 391. 1832. Typus ignotus.

= *Chaetanthera obtusata* Phil. Anales Univ. Chile 87: 7. 1894. Typus CHILE "Valparaíso, Quillota, Illapel etc." Syntypus 1 [Valparaíso, Panimavida, *Borchers s.n.*] E! SGO 76383! [Valparaíso, 11.1854, *Germain s.n.*] SGO 71761! [Valparaíso, *Dessauer s.n.*] SGO 64718! SGO 76385! [Valparaíso, 12.1851, *Philippi 393*] BM! G! P! SGO76384! W! Syntypus 2 [Quillota, *Germain s.n.*] SGO 64719! Syntypus 3 [Illapel, 12.1862, *Landbeck s.n.*] SGO 76382! Syntypus BM!

- *Chaetanthera incana* var. *spathulata* (Less.) DC. Prodr. (DC.) 7(1): 31. 1838.
- *Chaetanthera chilensis* Hook. & Arn. non *C. chilensis* (Willd.) DC. Bot. Beechey Voy. 1. 1835. Errata (withdrawn by authors in 1835)
- *Chaetanthera sericea* Lag. ex DC. Prodr. (DC.) 7 (1): 30. 1838. Nomen nudum
- *Chaetanthera ciliaris* Bert. ex Steud. Nomencl. Bot., ed. 2 (Steudel) 1: 340. 1840. Nomen nudum.

Nomenclatural Notes *Chaetanthera scariosa* had no Typus cited although it was probably a collection of H. Cuming. There are two Cuming collections (661 and 662) but unfortunetly the dates on the sheets postdate the publication of the name. The protologue "...involucri appendicibus rotundatis scariosis intergerrimis..." leaves no doubt that this is a synonym of *C. incana*.

Annual monoecious herb. **Roots** filamentous. **Stems** to 10(15) cm, decumbent to ascending, spreading, seldom branched, glabrescent, brown. **Leaves** 13–25 x 4 mm; spathulate, base slightly dilated, midrib visible, hoary, not succulent; margins dentate; indumentum silvery, sericeous (silky). **Capitula** campanulate, bracts loosely appressed; disk diameter 6 mm. **Involucral bracts** imbricate, arranged in three types, initially foliaceous then reduced to entirely membranous. **Outer involucral bracts** 7–7.5 mm L; as leaves but with short membranous alae to less than ½ height of bract; alae oblong. **Middle involucral bracts** 6.2–5.5 mm L; membranous alate, ratio of lamina to alae decreases from ⅔ to nearly entirely alate, alae oblong. **Inner involucral bracts** 5.5–9.2 mm L, entirely membranous, obovate to narrowly obovate-lanceolate; apices distinctly dilated to truncate, striate maculate, tipped with mucro, sparsely pubescent dorsally. **Ray florets** 8–16, pistillate, yellow or white, extending beyond capitulum, dorsal surface of corolla lip sericeous pubescent; corolla 9–10 mm L, corolla tube 3–3.4 mm L, outer lip 2.7–3 mm W., inner lip 3 mm L, curled. **Disk florets** ca. 20, bisexual, yellow; corolla 4.9 mm L, corolla tube 3.5 mm L **Styles** [ray] 6.1 mm L, stigma lobes c. 1 mm. [disk] 5.2 mm L stigma lobes yellow-brown. **Anthers** [disk] 5.4

IX *Chaetanthera* Ruiz & Pav.

Figure 56: *Chaetanthera incana*. **A.** Habit, typical yellow-rayed form photographed Valparaíso, 1981©Jürke Grau. **B.** Capitulum detail of white flowered form photographed ©María Terese Eyzaguirre (Fundación R.A. Philippi). **C.** Leaf & bract detail, d.s. *Gunckel 40598*. **i.** basal stem leaf; **ii.** Upper stem leaf; **iii.** OIB; **iv.** MIB; **v – vi.** IIB. Scale bar i & ii – vi = 5 mm.

mm L **Achenes** brown, 1.8–2.5 mm L, fusiform turbinate; carpopodium anular, irregular; pericarp pellucid, densely covered in narrowly lanceolate twin hairs, 95–140 x 40 μm. **Pappus** white, 4.5–5.5 mm L, 2 rows, indehiscent; setae fused at base, setae 4–8(10) cells wide, cell width 5–10 μm, no basal cilia; barbellate, barbs 94–140 μm L, barb base medium adhered (50%), free barb appressed, 12–15 barbs/100 μm.

Distribution and habitat. *C. incana* is distributed along the Chilean Cordillera de la Costa from Tongoy in Limarí, to Villa Alhué in Melipilla, (latitudes 30°15' – 34°10'S.) growing somewhat inland of Valparaíso. It is found at all elevations up to 1000 m.a.s.l. It occurs in open sunny places, on sandy substrates (coastal sand, roadside dunes), and in matorral/ rolling hills among grasses.

Differential diagnosis. *C. incana* is closely related to the perennial species *C. chilensis*. It is easily distinguished from all other *Chaetanthera* by its spathulate, dentate sericeous leaves and its papery obovate truncate IIB. The white-flowered form was seen after these studies were completed.

Material seen. – **Chile. IV Región de Coquimbo** Prov. de Limarí Tongoy, 10 m, 10.1971, *Gunckel 52562* (CONC). – Prov. de Choapa Quebrada Chiqualoco, 50 m, 10.1966, *Jiles 5001* (M). – El Mollar, 100 m, 10.1966, *Jiles 5001* (CONC). – Huentelauquen, 60 m, 10.1955, *Jiles 2810* (CONC). – near Norte de Los Vilos, 10.1971, *Beckett, Cheese & Watson 4134* (SGO). – Norte de Los Vilos, 30 m, 11.1952, *Jiles 2326* (CONC). – 8 Km al Norte de Los Vilos, 30 m, 10.1963, *Marticorena & Matthei 100* (CONC). – 7 Km al norte de Los Vilos, 15 m, 11.1974, *Marticorena et al. 329* (CONC). – 7 Km al norte de Los Vilos, Ruta 5 Norte, 960 m, 10.2002, *K. Arroyo et al. 25013b* (CONC). – 20 km N of Los Vilos towards Coquimbo, at roadside of Panamericana, 12.1991, *Eggli & Leuenberger 1860* (B). – Cerca del Puente Chivato, Ruta 5 Norte, Los Molles, 900 m, 10.2002, *K. Arroyo et al. 25097; 25011* (CONC). – **V Región de Valparaíso** Prov. de Petorca Los Molles, 10 m, 12.1966, *Schlegel 5787* (CONC). – Los Molles, 5 m, 12.1966, *Schlegel 5792* (CONC). – Los Molles, 20 m, 10.1951, *Gunckel 40598* (CONC). – Los Molles, 20 m, 01.1961, *Gunckel 45414* (CONC). – Panamericana südl. Los Molles, km 190, 11.1980 *Grau 2505* (Hrb. Grau M). – Prov. de Quillota Quillota, *Bertero 908* (P). – Quillota, 1835, *Bertero 909* (BM F G GH Mx2 NY P W). – Cerro de La Cruz, 25 m, 11.1953, *Fuch 10352* (CONC). – Cerro Cruz, 900 m, 10.1927, *Garaventa 1152* (CONC). – Cerro Cruz, 900 m, 11.1928, *Garaventa 1162* (CONC). – Cerro Cruz, 900 m, 01.1932, *Garaventa 6380* (CONC). – Cerro Tres Puntas, 900 m, 10.1930, *Garaventa 2542* (CONC). – on the mountain La Campana, 700 m, 10.1986, *Zöllner 12988* (MO). – PN la Campana, 1000 m, 12.2000, *K. Arroyo et al. 209218* (CONC). – Valle de Ocoa, 300 m, 09.1956, *Sparre & Schlegel 1966* (CONC). – Villa Alemana, Hacienda Moscoso, 140 m, 10.1936, *Behn K s.n.* (CONC SGO). – Prov. de Valparaíso Valparaíso, *Calvert s.n.* (BM). – Common above Mr. MacKays house, Valparaíso, *King s.n.* (BM E). – Valparaíso, 04.1876, *Reed s.n.* (BM). – Valparaíso, Colchetta, *Bridges 130* (BM E GH Kx2). – Valparaíso, Sept., *Matthews 239* (BM E G K). – Valparaíso, *Vieillard 64* (P). – Valparaíso, *King s.n.* (BM). – Prope Valparaíso, 04.1834, *Cuming 661* (BM Ex3 GH Kx2 P W). – Valparaíso, 1832, *Cuming 662* (BM Ex2 GH Kx2 P W). – Valparaíso, *Lechler s.n.* (M). – Concón, *Miers s.n.* (BM). – Concón, *Philippi s.n.* (B). – Fundo El Toqui, nördlich Casablanca an der Straße nach Las Mercedes, am Fuß der Hügel, 11.1984, *Hellwig 3513* (G). – Las Zorras, 50 m, 1902, *Hill 222* (K). – Dünen N Maitencillo, 11.1987, *K.H. & W. Rechinger 63743* (W). – Quintero, Puchuncavi, 25 m, 02.1962, *Gunckel 45603* (CONC). – Quintero, La Ventana, 25 m, 02.1962, *Gunckel s.n.* (CONC). – Los Juanes, 20 m, 02.1953, *Gunckel 50721* (CONC). – Quintero, Loncura, 20 m, 11.1953, *Gunckel 27396* (CONC). – Quintero, Loncura, 25 m, 11.1952, *Barrientos 1905* (CONC). – Quintero, Campiche, 20 m, 02.1962, *Gunckel 35751* (CONC). – Quintero, Playa de las Conchitas, 20 m, 12.1962, *Gunckel 40035* (CONC). – Quintero, 25 m, 11.1942, *Junge s.n.* (CONC). – Quintero, 20 m, 12.1950, *Gunckel 19085* (CONC). – Quintero, 20 m, 11.1952, *Levi 157* (CONC). – Quinteros, 11.1965, *Solbrig, Moore & Walker 3634* (GH). – Quinteros, 03.1914, *Pahlman s.n.* (F). – dunas de Ritoque, 10 m, 11.1952, *Gunckel 23904* (CONC). – El Salto, near Viña del Mar, across from the National Botanical Gardens, 11.1965, *Solbrig, Moore & Walker 3605* (GH). – Viña del Mar, 03.1930, *Behn s.n.* (M). – Near Viña del Mar, 10-15 m, 11.1935, *West 3961* (GH). – Viña del Mar, 11.1918, *Behn s.n.* (F). – Viña del Mar, 11.1930, *Behn s.n.* (MO). – Tranque Vergara, 50 m, 12.1926, *Behn K s.n.* (CONC). – Quebrada El Tranque, 50 m, 11.1922, *Behn s.n.* (CONC). – Castillo de Agua, 100 m, 10.1960, *Schlegel 3018* (CONC). – Cerro Hospital, 20 m, 10.1954, *Schlegel 352* (CONC). – Las Zorras, 50 m, 04.1931, *Jaffuel 1772* (CONC GH). – Quilpué, 10.1975, *Zöllner 8589* (MO). – Quilpué, 80 m, 10.1932, *Behn s.n.* (CONC M). – **Región Metropolitana de Santiago** Prov. de Santiago Cerro Lo Chena, 650 m, 11.1950, *Gunckel 26322* (CONC). – llano La Cuesta, RN Roblería de Cobre de Loncha, 460 m, 08.2002, *K. Arroyo et al. 25002* (CONC).

IX *Chaetanthera* Ruiz & Pav.

9. *Chaetanthera linearis* Poepp. ex Less. nom. conserv. Davies, A.M.R. Taxon 54(3): 838 – 839. 2005; Lessing, C.F. in Syn. Gen. Compos. pp. 112. 1832.

Typus CHILE "Poeppig mss. n. 1. Diar. 414" **Lectotypus hic loc. designatus** [No. 57 *Chaetanthera*, No. 1, Diar. 414, T. *Poeppig*] BM! Isotypus G! Isotypus probabilis [In glareosis ad Río Colorado in Andibus S. Rosae, 11.1827, *Poeppig s.n.* (*D.414*)] W! F!
Dubious collection locality: In collibus graminos ad Talcahuano, Apr. flor., *Poeppig 111(57) (D.414)* BM! G! MO! W! F!

= *Euthrixia salsoloides* D.Don Trans. Linn. Soc. London 16: 259. 1830. Typus CHILE "In Chili Ruiz & Pavon" † Neotypus design. Davies 2005 *l.c.* "Chile. Región Metropolitana de Santiago Santiago, Farellones, switchbacks up to Farellones, on roadside, curve 20, 1800 m, 02.2002, *Davies & Grau 2002/001* MSB".
= *Chaetanthera kunthiana* Less. in Syn. Gen. Compos. pp. 114. 1832. Typus CHILE "In saxosis sterilibus secus flumen Cachapual Rancagua, 10.1828, *Bertero 474*" Holotypus † Isotypus BMx2! F! M! NY! P! W!
- *Chaetanthera microphylla* Kuntze non *C. microphylla* (Cass.) Hooker & Arn. Revis. Gen. Pl. 3 (2): 140. 1898. ["...hat gelbe Blüthen..."]
- *Chaetanthera linearifolia* Poepp. ex Steud. in Steudel, E.G.T. (1840) in Nomencl. Bot., ed. 2 (Steudel) 1: 340. nomen nudum.

Nomenclatural Notes Lessing clearly cites "Poeppig mss. N. 1, Diar. 414". Collections labelled No. 1 have no precise location information but other extant collections of Diar. 414 do. The material purportedly from Talcahuano seems dubious because not only does Talcahuano lie far outside the typical distribution of the species, the late collection month is also suspect. Poeppig lost most of his collections from his northern trips in a catastrophic accident in the Río Aconcagua (BAYER 1998). The Río Colorado – Sta Rosae collections must have been one of the few collections that survived and this places the Río Colorado as that just east of Santiago, not that northeast of Los Andes. Moreover, Lessing worked out of Berlin, and so the holotype would have been destroyed in the 1940's, meaning the selection of a lectotype was necessary. There is no extant Ruiz and/or Pavon material bearing either of the names *Euthrixia* or *salsoloides* in any of the herbaria consulted. The description, particularly in details such as the habit, leaf shape, and flower colour, is explicit enough to place this within the taxon known as *C. linearis*.

Annual monoecious herb. **Roots** filamentous. **Stems** to 20 cm, wiry, long, decumbent to ascending. **Leaves** 6.2–7.4 mm L, glaucous, (blue-green), flat in cross-section, not succulent, glabrous; margins and ventral surface irregularly sparsely dotted with translucent triangular hairs, midrib indistinct. **Capitula** sessile, cylindrical to faintly urcinate; disk diameter 3–5 mm (fresh material). **Involucral bracts** imbricate, arranged in three types, initially foliaceous then reduced to entirely membranous. **Outer involucral bracts** 3.7–4.6 mm L; as leaves but with short membranous alae to less than ½ height of bract. **Middle involucral bracts** 5.5–7.4 mm L lanceolate to narrowly ovate.; lamina reduced eventually to central slightly thickened part; membranous alate, ratio of lamina to alae decreases from ⅔ to nearly entirely alate; apical mucro. **Inner involucral bracts** 6.8–8.9 mm L, lanceolate-truncate, entirely membranous alate; apices emarginate, striate maculate. **Ray florets**

8–10(12), pistillate, yellow, glabrous; corolla 9.1–9.8 mm L, corolla tube 3.4–4.2 mm L, outer corolla lip 1.5 mm W., inner corolla lip 1.9–2.8 mm L, bifid, filamentous. **Disk florets** ca. 20, bisexual, yellow; corolla 5.6–6.8 mm L, corolla tube 4.7–5.6 mm L **Styles** [ray] 6.7–8 mm L, stigma lobes 0.3 mm L [disk] 6.8–8.4 mm L, stigma lobes 0.3–0.6 mm L **Anthers** [ray] sterile, tips dark green or blackish. [disk] fertile, 3.7 mm L (5.5–7.1 mm L incl. filaments). **Achenes** brown, 1.8–2.3 mm L, fusiform, turbinate; carpopodium anular irregular; pericarp pellucid, densely covered in lanceolate twin hairs, 75–100 x 40 µm. **Pappus** white, [ray] 3.4 mm L [disk] 4.4–4.9 mm L; 1–2 rows, indehiscent; setae fused at base, pappus setae 4–8 cells wide, cell width 5–10 µm, no basal cilia; barbellate, barbs 95–140 µm L, barb base shortly adhered (<40%), free barb appressed, 12–15 barbs/100 µm. **Chromosome number** 2n = 2x = 22 Davies & Vosyka new count. *Davies & Grau 2002/001A*

Distribution and habitat. *C. linearis* is distributed in Chile from Los Vilos and Salamanca to Termas de Cauquenes but principally around the "cuenca" de Santiago (coastal hill ranges west and north of Santiago and east of Santiago into the Andes) between latitudes 31°45' – 34°15'S. It is found at elevations between 150 – 2300 m.a.s.l. It occurs in dry sunny areas, in scree, on hillside pastures, roadside banks, anywhere where there is sparse vegetation cover.

Differential diagnosis. Close to *C. albiflora*, these two species form hybrids in their sympatric zone. *C. linearis* is distinguished by having yellow ray florets and glaucous glabrous leaves (not bright green). Key characters for identifying *C. albiflora*, *C. linearis* and the hybrid are given in Table 9, page 119. Table 12, page 133 shows the key features distinguishing *C. linearis* from the other taxa in the linear-leaved group.

Material seen. – **Chile. IV Región de Coquimbo** Prov. de Choapa Río Choapa, Westlich Salamanca, 11.1994, *Ehrhart & Grau 94/295* (MSB). – Depto. Illapel, ca. 33 Km from Illapel and down west slope of Cuesta de Cavilolen, road Illapel to Los Vilos, 450 m, 11.1938, *Worth & Morrison 16470* (G GH K MO). – Los Vilos, 10.1976, *Zöllner 9213* (MO). – Almendrillo, 1800 m, 1.1964, *Marticorena & Matthei 509* (CONC). – **V Región de Valparaíso** Prov. de Petorca Zapallar, Tigre, 300m, 2.1917, *Bohm s.n.* (B). – Zapallar, Tigre, 300 m, 2.1919, *Johow s.n.* (CONC). – Prov. de Los Andes Los Andes, Río Blanco, 2300 m, 3.1927, *King 417* (BM). – Saladillo, Río Blanco, 2000 m, 2.1957, *Silva s.n.* (CONC). – Prov. de Aconcagua Aconcagua, 10.1969, *Swabe s.n.* (B). – Llay Llay, 1.1927, *Edwards s.n.* (BM). – camino La Ligua - Los Molles, 10.1914, *Rose & Rose 19380* (NY). – Prov. de Valparaíso Renaca, ca. 18 Km from Valparaíso, 500 m, 12.1938, *Morrison 16842* (GH). – Valparaíso, 150 m, 10.1910, *Jaffuel 751* (CONC). – Prov. de Quillota Puerta Ocoa near Granizo, 1400 m, 11.1986, *Zöllner 13026* (MO). – Cerro Caquis 15 Km E de Melón, 1500 m, 12.1938, *Morrison 16873* (G GH K). – Las Viscachas, turnoff to Calco - Ramayama Mine, 1200 m, 12.1951, *Hutchinson 79* (F G GH K SGO). – Cerro Roble, 2200 m, 3.1963, *Zöllner 322* (B). – PN La Campana, 1100 m, 12.2000, *K. Arroyo et al. 209102* (CONC). – Cerro El Roble, PN La Campana, 1630 m, 11.1999, *K. Arroyo et al. 994030* (CONC). – Limache, Iman Spitze, 2100 m, 3.1963, *Zöllner 576* (B). – **Región Metropolitana de Santiago** Prov. de Santiago camino de Rungue a Caleu, 700 m, 11.1960, *Schlegel 3120* (CONC F). – Calen, poco más arriba de La Capilla, 3.1975, *Muñoz S. 806* (SGO). – Farellones curve 20, 1800 m, 11.2001, *Davies s.n.* (M). – Farellones, 2000 m, 1.1947, *Wall s.n.* (GH MO NYx2). – Straße nach Farellones, Umgebung der Skipisten, 2.1985, *Hellwig 3156* (G). – Farellones, curve 20, 1800 m, 02.2002, *Davies & Grau 2002/001* (M). – Farellones, curve 30, 2200 m, 02.2002, *Davies & Grau 2002/001a* (M). – Farellones Refugio, 2000 m, 01.1947, *Looser 5214* (CONC M). – Farellones, 2000 m, 3.1956, *Schlegel 1048* (CONC). – Santuario de La Naturaleza Yerba Loca, bajo la curva 30 hacia Farellones, 2160 m, 11.2000, *K. Arroyo et al. 206159* (CONC). – El Arrayán, 800 m, 12.1939, *Junge s.n.* (CONC). – El Arrayán, 10.1959, *Saa s.n.* (SGO). – Loma del Viento, Santuario de La Naturaleza Yerba Loca, 2180 m, 12.2002, *K. Arroyo et al. 25164* (CONC). – Pudahuel, 11.1927, *Looser 666* (GH). – Quebrada de Ramon, 1800 m, 2.1950, *Barros s.n.* (CONC). – Cerro San Cristobal, 750 m, 10.1952, *Gunckel 23475* (CONC). – Quebrada La Plata, 680 m, 12.1960, *Schlegel 3282* (CONC). – Maitenes, 1200 m, 01.1928, *Montero 293* (CONC GH). – Cerro Lo Chena, 750 m, 11.1950, *Gunckel 26331* (CONC). – RN Río Clarillo, 1000 m, 12.2002, *K. Arroyo et al. 25165* (CONC). – Queltehues, 1600 m, 12.1927, *Montero 342*

IX *Chaetanthera* Ruiz & Pav.

(CONC GH). – entrada norte Tunel de Paine, 450 m, 11.1970, *Marticorena & Weldt 522* (CONC). – Prov. de Chacabuco Hacienda Chacabuco, 11.1929, *Looser 995* (GH). – Altos del Roble, Hacienda de Chicau, 1600 m, 12.1983, *Villagran 4761* (CONC). – Umgebung der Baños de Colina, 1000 m, 11.2002, *Ehrhart et al. 2002/002* (MSB). – Prov. de Cordillera Puente El Toyo, Cajón del Maipo, 11.1994, *Muñoz S., Moreira Teillier & Arrigada 3438* (SGO). – Los Piches, Cajón del Yeso, camino al Embalse del Yeso, 1.1995, *Muñoz S., Moreira, Meza & Arrigada 3708* (SGO). – en la cumbre de La Punta Juana, 2200 m, 3.1963, *Zöllner s.n.* (SGO). – Prov. de Melpilla Küstenkordillere Westl. Maipú bei Rinconada, 11.1980, *Grau 2399* (MSB M). – **VI Región del Libertador G. B. O'Higgins** Prov. de Cachapoal Termas de Cauquenes, 700 m, 11.1965, *Mahu 4086* (CONC). – In saxosis sterilibus secus flumen Cachapual Rancagua, 10.1828, *Bertero 474* (BMx2 F M NY P W).

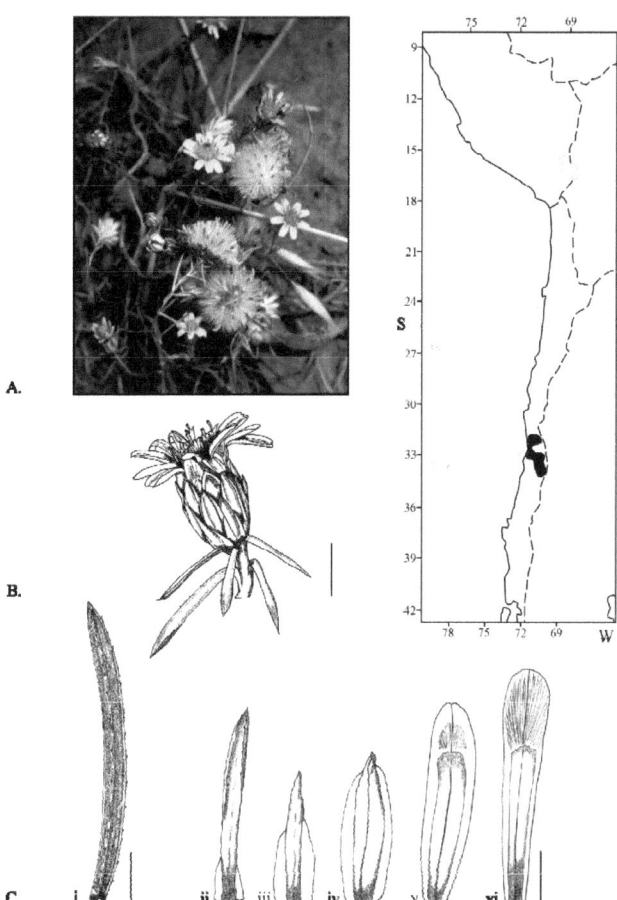

Figure 57: *Chaetanthera linearis*. **A.** Habit, photographed Farellones, 2001©Alison Davies. **B.** Capitulum detail, *Davies 2002/001*. Scale bar = 5 mm. **C.** Leaf & Bract detail, *Davies 2002/001*. **i.** lower stem leaf v.s.; **ii.** OIB v.s.; **iii – iv.** MIB d.s.; **v – vi.** IIB d.s. Scale bar i & ii – vi = 2 mm.

10. *Chaetanthera linearis* x *albiflora*

The following collections show intermediate characteristics: White flowers but glabrous and / or no dorsal stripe, or yellow flowers and filamentous indumentum, or glabrous plants with unknown flower colour (marked by *).

Material seen. – **Chile. IV Región de Coquimbo** Prov. de Elqui Coquimbo, 11.1930, *Jaffuel 1205* (GH). – Coquimbo, 11.1930, *Jaffuel 1206* (GH). – Coquimbo, 10.1931, *Jaffuel 2689* (GH). – Coquimbo, 100 m, 11.1923, *Werdermann 115* (B BM CONC E G GH M NY). – 6 Km S de La Serena, Fundo Peñuelas, 75 m, 1.1948, *Wagenknecht 272* * (CONC). – La Serena, La Vicuña, Paihuano, 1300 m, 10.1927, *Elliott 39* (E). – 15 Km al sur de La Serena, 190 m, 1.1981, *K. Arroyo 81231* * (CONC). – Cordillera de Paihuano, 1100 m, 12.1942, *Gajardo s.n.* * (CONC). – Vicuña (Elqui valley), 10.1927, *Elliott 97* (E K). – Lagunillas, 50 m, 10.1971, *Jiles 5837* * (CONC). – Entre Cochiguaz - Montegrande, 1310 m, 12.1987, *Rosas 1492* (M). – Pisco Elqui, 1290 m, 2.1958, *Pinto 18* * (CONC). – N de Guanaqueros, camping Las Mostazas, 60 m, 10.1984, *Muñoz 1891* * (SGO). – frente a Guanaqueros, 60 m, 11.1980, *Rodriguez & Marticorena 1625* * (CONC). – Guanaqueros (Playa), entre Panamericana y Playa, 100 m, 10.1974, *Garaventa 5553* * (CONC). – Prov. de Limarí Alto(s) de Talinay, 11.1978, *Zöllner 10431* (MO). – Weg von Vicuña nach Hurtado, 850 m, 11.2002, *Ehrhart* et al. *2002/083* (MSB). – Weg von Vicuña nach Hurtado, 880 m, 11.2002, *Ehrhart* et al. *2002/078* * (MSB). – Corral Quemado, 1100 m, 10.1956, *Jiles 3098* (CONC). – Hacienda Tamaya, 250 m, 10.1956, *Jiles 3048* * (CONC). – Fray Jorge, 300 m, 11.1925, *Werdermann 890* (B BM CONC E F G GH K M NY SGO). – PN Fray Jorge, 300 m, 11.1974, *Marticorena* et al. *432** (CONC). – Fray Jorge, 10.1947, *Sparre 2958* * (SGO). – Río Palomo, 1300 m, 10.1957, *Jiles 3259* (CONC). – Hacienda Valdivia, 1100 m, 10.1957, *Jiles 3253* (CONC). – Cordillera San Miguel, 3000 m, 1.1959, *Jiles 3598* * (CONC M). – Zorrilla, 350 m, 2.1948, *Jiles 1536* * (CONC). – Cuesta de Punitaqui, 630 m, 10.1963, *Marticorena & Matthei 365* * (CONC). – Cerro Tulahuén, 1150 m, 12.1961, *Jiles 4064* (CONC M). – Tulahuén, 1000 m, 12.1961, *Jiles 4063* (CONC). – Tulahuén, 1300 m, 10.1948, *Jiles 1032* (CONC). – Cerro Loica, 2200 m, 12.1965, *Jiles 4722* * (CONC). – Cerro Loica, 2200 m, 12.1965, *Jiles 4723** (CONC). – El Maitén, 1850 m, 1.1954, *Jiles 2447** (CONC). – Prov. de Choapa Corral de Julio, 280 m, 11.1971, *Jiles 5861** (CONC). – El Mollar, 100 m, 10.1966, *Jiles 5002* * (CONC). – Chuchiñi, 450 m, 10.1965, *Montero 7274* * (CONC). – **V Región de Valparaíso** Prov. de Petorca camino entre Alicahue y La Viña, 760 m, 10.2002, *K. Arroyo* et al. *25034* (CONC). – Prov. de Valparaíso – Llay llay on sandy plains, 1844, *Bridges 128* (BM E GH K). – Valparaíso, 04 – 06.1856, *Harvey s.n.* (GH). – In collibus graminosis circum Concón, 10.1827, *D.383 Poeppig s.n.* (W). – Prov. de Quillota Quillota, 1829, *Bertero s.n.* (P). – Prov. de Aconcagua Valparaíso, *Bridges s.n.* (BM). – *Guillemin s.n.* (P). – **Región Metropolitana de Santiago** Prov. de Santiago *Philippi s.n.* (B). – Santiago, 11.1856, *Philippi 562* (B G P W). – In pascuis aridis Santiago, *Philippi*? (BM). – Prov. de Cordillera Río Colorado, Decmbr., *D.594? Poeppig 110 (62)* (M). – Chile borealis, Andes de Sa. Rosa, in sterilibus ad Río Colorado, 12.1827, *Poeppig D.594* (G). – **VI Región del Libertador G. B. O'Higgins** Prov. de Cachapual S. Fernando, *Philippi s.n.* (BM). – Colchagua, S. Fernando, *Heike? Philippi s.n.* (BM). **S.l.d.** *Cuming s.n.* (BM).

11. *Chaetanthera microphylla* (Cass.) Hook. & Arn. Companion Bot. Mag. 1: 104. 1835.

≡ *Cherina microphylla* Cass. Dict. Sci. Nat. (ed. Cuvier, G.-F.) 8: 438. 1817. Typus CHILE "Herb. M. de Jussieu, Chili" (†?). **Neotypus hic designatus** "Chile, In saxosis sterilibus montis La Leona et secu? Cachapual, 1829, *Bertero 473* (M)" Isotypus BM! E! G! GH! P! SGO! W!

= *Onoseris linifolia* Bertero Mercurio Chileno 16: 737. 1829. Typus CHILE "Otra nace entre las piedras á lo largo de Cachapual" [Chile, In saxosis sterilibus montis La Leona et secu? Cachapual, 1829, *Bertero 473 ± 1192*] Syntypus BM! E! G! GH! M! P! SGO 64705! TO? W! (collections always mixed)
= *Chaethanthera delicatula* Phil. Anales Univ. Chile 87: 13. 1894. Typus CHILE "Prope Angol novembri 1887 legi" Typus ignotus.
= *Chaetanthera pentapetala* Phil. Anales Univ. Chile 87: 13. 1894. Typus CHILE "In Araucania frequens, novembri florens" Typus ignotus.
= *Euthrixia affinis* D. Don Philos. Mag. Ann. Chem. 11: 391. 1832. [H. Cuming ?]
- *Onoseris linifolia* Bertero ex Colla Mem. Acad. Soc. Torino 38: 42. tab. 25, fig. 2. 1833.
- *Chaetanthera linifolia* (Bertero) Less. Syn. Gen. Compos. pp. 112. 1832.
- *Chaetanthera salsoloides* Kuntze [non *Euthrixia salsoloides* D. Don] Revis. Gen. Pl. 3 (2): 140. 1898. "as *salsolodes* orth.var."

Nomenclatural Notes There is no appropriate extant Type material in Paris or Geneva. This taxon is not cited in the elenchus of Jussieu. However, the original description cites '... et la coronne brun-rouge...' which unambiguously identifies this with material commonly known as *Chaetanthera microphylla*. There is only one species with brick-red ray florets in the genus *Chaetanthera*. This argues strongly for the selection of a neotype. In this case, the most widely distributed, appropriate material is also the Typus of *Onoseris linifolia* Bertero (see below). This creates two homotypical names, in this case justified by the wide availability of the collection.

Annual monoecious herb. **Roots** filamentous. **Stems** to 20 cm, lax to ascending, branched. **Leaves** 11–28 x 1 mm, (upper leaves smaller), alternate, glabrous, glaucous; midrib visible; leaf cross-section flat; slightly succulent; margins flat to somewhat inrolled or thickened, with translucent mucros along margins. **Capitula** open, campanulate, bracts appressed. **Involucral bracts** imbricate, arranged in two series (no **outer involucral bracts** with short membranous alae to less than ½ height of bract), initially foliaceous then reduced to entirely membranous. **Middle involucral bracts** (3.5)4.9–8 mm L; membranous alate, ratio of lamina to alae decreases from ⅔ to nearly entirely alate; central tissue mesophyllous, slightly thickened; apices mucronate. **Inner involucral bracts** ca. 9 mm L, lanceolate, entirely membranous; apices striate maculate. **Ray florets** ca. 10, pistillate, brick-red (rarely yellow); corolla 7.2 mm L, corolla tube 3.7 mm L, outer lip 1.2 mm W., 3-dentate, teeth triangular, 0.4 mm high; inner lip 2.4 mm L, filamentous, bifid, fused. **Disk florets** 10–15, bisexual, yellow; corolla 6.2 mm L, corolla tube 5.6 mm L **Styles** [ray] 6.2 mm L, stigma lobes 0.3 mm L [disk] 6.5 mm L, stigma lobes 0.5 mm L **Anthers** [disk] 2.8 mm L (6.5 mm L incl. filaments). **Achenes** black-brown, 2.4 mm L, fusiform turbinate; carpopodium anular irregular, small; pericarp pellucid, covered in lanceolate twin hairs, 70–90 x 35 µm. **Pappus** white, 5.7–6.3 mm L, 1–2 rows, indehiscent; setae fused at base, pappus setae width 4–8 cells, cell width 5–10 µm, no basal cilia; barbellate, barbs ca. 100 µm L, barb base medium adhered (50%), free barb appressed, 10–15 barbs/100 µm. **Chromosome number** 2n = 2x = 24. (BAEZA & SCHRADER, 2005a) *C. Baeza 4177*(CONC).

Distribution and habitat. *C. microphylla* is found from the Chilean Cordillera de la Costa west of Santiago to Concepción; in the Andes foothills from Cauquenes to Antuco, and into Argentina (Neuquén) with the following latitude range: 32°00' – 37°50'S. It occurs at elevations between 10 – 1520 m.a.s.l. It can be found on river banks that are regularly inundated.

Differential diagnosis. *C. microphylla* is close to the yellow-flowered, more northerly distributed *C. linearis* and the higher elevation, precocious, white-flowered *C. depauperata*. It is distinguished by its glaucous, entirely glabrous leaves, brick red rays (seldom yellow) and broadly ovate lanceolate IIB. See Table 12, page 133.

Material seen. – **Argentina. Neuquén.** Andacollo, Mina Guaraco, 12.1952, *Cabrera 11120* (GH). – **Chile. V Región de Valparaíso** Prov. de Quillota Quillota, 11.1829, *Bertero 1191* (G). – Limache, Cerro Cruz, 300 m, 09.1930, *Garaventa 2157* (CONC). – Cerro Cruz, 300 m, 11.1950, *Garaventa 5898* (CONC). – Prov. San Felipe de Aconcagua Llay-Llay, 380 m, 11.1963, *Weisser 429* (CONC). – Prov. de Valparaíso Valparaíso *Cuming 659* (BM Ex2 Kx2). – Valparaíso, 1832, *Cuming 655* (BMx2 Ex2). – Valparaíso, *Bridges 129* (BM K W). – Valparaíso, *Viaillard 63* (P). – Quintero, 10 m, 11.1952, *Borquez s.n.* (CONC). – Cerros del Tranque, 100 m, 11.1922, *Behn s.n.* (CONC). – Viña del Mar, 11.1922, *Behn s.n.* (F M). – Cuesta Zapata, lado oeste, 1520 m, 11.1999, *K. Arroyo et al. 995374* (CONC). – **Región Metropolitana de Santiago** Prov. de Chacabuco Batuco, 480 m, 09.1951, *Gunckel s.n.* (CONC). – Prov. de Santiago Philippi s.n. (W#113689) (B F). – 1862, *Philippi s.n.* (G). – ad rup? Santiago 11.1829, *Gay 578* (P). – Cuesta La Dormida, lado este, 1000 m, 11.2002, *K. Arroyo et al. 25155* (CONC). – Tiltil, 1000 m, 09.1953, *Moreno s.n.* (CONC). – Cerros de Renca, 700 m, 11.1951, *Gunckel 21919* (CONC). – Las Condes, 800 m, 11.1951, *Gunckel 22499* (CONC). – Melocotón, 1200 m, 12.1949, *Muñoz s.n.* (CONC). – RN Río Clarillo, 1040 m, 11.1999, *K. Arroyo et al. 994086* (CONC). – RN Río Clarillo, 960 m, 09.2002, *K. Arroyo et al. 25007* (CONC). – Prov. de Maipo entrada Norte al Tunel de Paine, 450 m, 11.1970, *Marticorena & Weldt 524* (CONC). – Prov. de Cordillera Cajón del Maipo, El Toyo, 1450 m, 01.2000, *Teillier 4641* (CONC). – El Manzano, San José de Maipo, 11.1961, *Novas 2626* (SGO). – Cajon del Maipo, frente desembocadura El Manzano, 3 Km al W, 11.1994, *Muñoz, Moreira, Teillier & Arrigada 3410* (SGO). – Prov. de Melipilla RN Roblería de Cobre de Loncha, Sector La Isla, 450 m, 10.2002, *K. Arroyo et al. 25046* (CONC). – RN Roblería de Cobre de Loncha, Sector La Isla, 1430 m, 11.2000, *K. Arroyo et al. 206278* (CONC). – RN Roblería de Cobre de Loncha, Sector Los Maquis, 480 m, 11.2000, *K. Arroyo et al. 206122* (CONC). – **VI Región del Libertador G. B. O'Higgins** Prov. de Cachapoal Las Peñas, 835 m, 01.1958, *Barrientos 1660* (CONC). – PN Cocalán, Sector Agua de La Gotera, 1200 m, 12.2001, *K. Arroyo et al 211853* (CONC). – Termas de Cauquenes, 700 m, 11.1965, *Mahu 1569* (CONC). – Camino de Rancagua a Caletones, Km 17, 1100 m, 11.1970, *Marticorena & Weldt 652* (CONC G). – RN Río de los Cipreses, Orilla del Río Cipreses, 1260 m, 12.2000, *K. Arroyo et al. 206999* (CONC). – RN Río de los Cipreses, Sector Mal Paso, 1360 m, 01.2000, *K. Arroyo et al. 20473* (CONC). – Prov. de Colchagua Lado norte del Cerro Centinela a 4 Km al sur de San Fernando, 450 m, 10.1973, *Stebbins 8728 pro parte* (SGO). – Centinela, 349 m, 12.1965, *Montero 7369* (CONC). – Rincón de Tinguiririca, 360 m, 12.1950, *Ricardi s.n.* (CONC). – Puente Negro, 540 m, 12.1950, *Ricardi s.n.* (CONC). – Talcarehue, 1500 m, 12.1950, *Ricardi s.n.* (CONC). – **VII Región del Maule** Prov. de Curico Cuchachucha et cordillera del Planchon, (Exped. Malaespina), *Nee s.n.* (MA). – Curico a villas del Zucuquilla, esca del camino a Zapallar, 12.1961, *Aravena 2* (SGO). – Los Queñes, 650 m, 02.1930, *Barros 3212* (CONC). – camino a Paso a Vergara de Los Queñes, 720 m, 09.2002, *K. Arroyo et al. 25009* (CONC). – Puente Colorado, en terreno arenos que el Río Guiquillo inunda, 02.1964, *Aravena s.n.* (SGO). – Prov. de Talca Maule, 2.1892, *Kuntze s.n.* (NY). – Valle del Río Maule, 1000 m, 02.1994, *Villagran s.n.* (CONC). – Talca, camino a Laguna del Maule Río Cipreses, Quebrada al este, 1210 m, 1.1988, *Rosas 1924* (M NY). – **VIII Región del Biobío** Fundo El Boldo, Subestacion Experimental Cauquenes, INIA, 177 m, 12.1978, *Ovalle C-022* (SGO). – Prov. de Concepción Quilacoya, 11.1896, *Neger s.n.* (M). – Prov. de Ñuble Road from Chillán to Concepción, west of the Río Itata, about 5 km east of Quillón, 150 m, 12.1951, *Hutchison 206* (F G GH K SGO). – El Roble, 60 m, 12.1967, *Marticorena & Matthei 1095* (CONC). – Prov. del Biobío entre Cabrero y Tomeco, 105 m, 11.1976, *Marticorena & Rodriguez 8434* (CONC). – Pangal del Laja, oeste de Yungay, 180 m, 01.1959, *Marticorena et al. 94* (CONC). – camino entre Yumbel y Salto del Laja, 120 m, 01.1959, *Marticorena et al. 79* (CONC). – camino Estacion

IX *Chaetanthera* Ruiz & Pav.

Yumbel y Salto del Laja, 120 m, 12.1946, *Pfister s.n.* (CONC). – camino Estacion Yumbel y Salto del Laja, 120 m, 12.1946, *Pfister 7130* (CONC). – Antuco, 650 m, 12.1946, *Montero 4659* (CONC). **S.l.d.** Cordilleras of Chili, *Cuming 206* (BM Ex3 K). – Chile centr., *Ball s.n.* (E). – *Cuming s.n.* (BM). – *Bridges 124* (W).

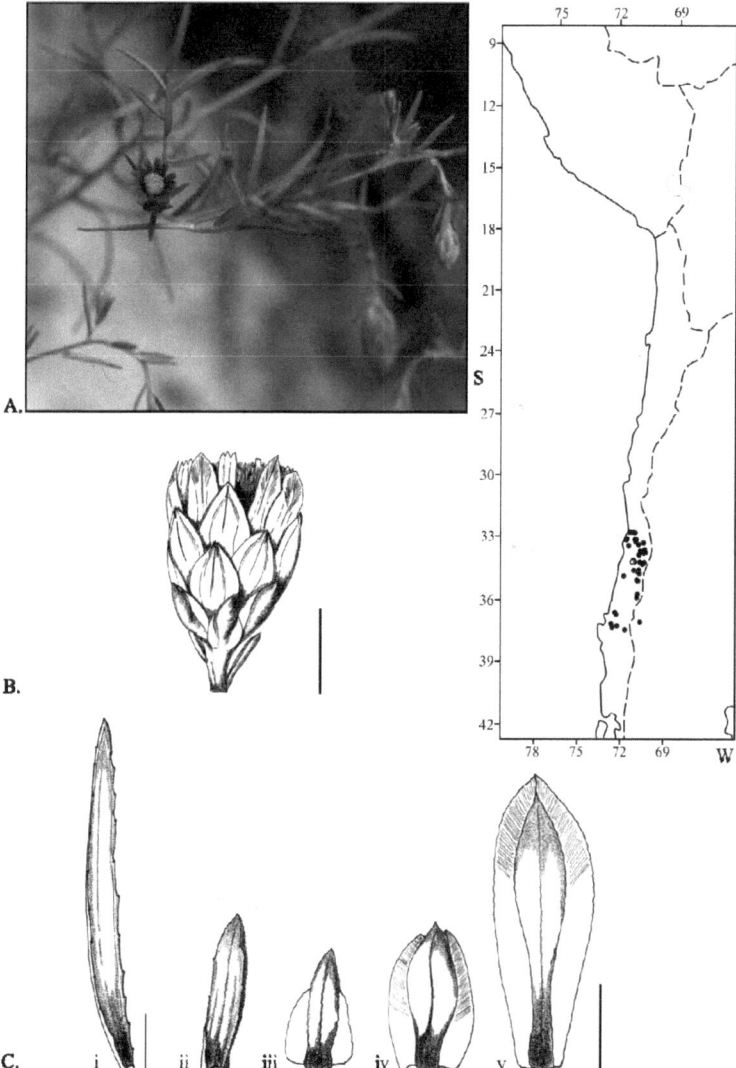

Figure 58: *Chaetanthera microphylla*. **A.** Habit, photographed ©María Terese Eyzaguirre. **B.** Capitulum detail, *K. Arroyo* et al. *206422*. scale bar = 2mm.**C.** Leaf & Bract detail, *K. Arroyo* et al. *206422*. **i.** lower stem leaf v.s.; **ii.** Upper stem leaf v.s.; **iii – iv.** MIB d.s.; **v.** IIB d.s. Scale bar i, ii – v = 2 mm.

12. *Chaetanthera moenchioides* Less. Syn. Gen. Compos. pp. 113. 1832.

Typus CHILE "*Poeppig* in Chile" [In pascuis saxosis lapidosis agri Chilensis ad Antuco, *Poeppig 689*] **Lectotypus hic loc. designatus** G! Isotypus BM! G! GH (photo)! M! NY! P! Wx2!

= *Chaetanthera moenchioides* var. *pauciflora* F. Meigen Bot. Jahrb. Syst. 17: 284. 1893. Typus CHILE "Cerro Gubler, 21.10.1891, *Meigen 460*" (†)

= *Chaetanthera moenchioides* var. *sulphurea* F. Meigen Bot. Jahrb. Syst. 17: 284. 1893. Typus CHILE "Lo Cañas, 900 m, 8.11.1891, *Meigen 461*" [Southeast of Santiago along Canal San Carlos between Peñalolen and Las Vertientes] (†)

= *Chaetanthera australis* Cabrera Notas Mus. La Plata, Bot. 1: 64, fig. 3ª-G. 1935. Typus ARGENTINA "Río Negro: Región del Nahuel Hapi, San Carlos de Bariloche, Arroyo Ñirico, en el cauce seco, arenoso pedregoso del Arroyo, 18.02.1934, *Burkart 6599*" Holotypus BA 12390 Isotypus LP 67042!

- *Chaetanthera filiformis* sin autor, ex sched. (BM Gx2 M NY P Wx2). Nomen nudum.
- *Chaetanthera albicans* Bertero ex sched. Nomen nudum.

Nomenclatural Notes *Chaetanthera moenchioides* var. *pauciflora* and *Chaetanthera moenchioides* var. *sulphurea*, both described by F. Meigen, were both collected in the environs of Santiago. Meigen material for other Chilean species is held in the Museo Nacional de Historia Natural in Santiago (SGO) (fide Ehrhart & Klingenberg pers. comm.). However, there is no Meigen material of *Chaetanthera* in SGO.

Annual monoecious herb. **Roots** filamentous. **Stems** spreading to ascending, branched, pubescent. **Leaves** 8–14 x 1–1.3 mm, bases slightly dilated, narrowly linear-lanceolate, not succulent; margins inrolled/folded ±2–4 pairs short teeth/mucros towards apices, apically mucronate; indumentum densely silvery sericeous both ventrally and dorsally, hairs 1–2 mm L; leaves alternate to opposite, prolific budding even from leaves below flowering capitula, upper leaves subtending capitulum equal or exceed capitulum in length. **Capitula** sessile or shortly pedunculate, faintly urcinate; disk diameter 2.5 mm. **Involucral bracts** imbricate, arranged in three types, initially foliaceous then reduced to entirely membranous. **Outer involucral bracts** 10–8 mm, as leaves but with short membranous alae to less than ½ height of bract; lamina erect; ventral surface of alae glabrous, otherwise bracts entirely covered in dense, silvery indumentum as leaves. **Middle involucral bracts** 8–9 mm L, membranous alate, ratio of lamina to alae decreases from ⅔ to nearly entirely alate; lamina component gradually reduced to mucro. **Inner involucral bracts** 10.5–11 mm, entirely membranous, ovate lanceolate, dorsally sericeous; apices acute to apiculate, dark-green. **Ray florets** (2)3–10, pistillate, yellow or often white in south of distribution; corolla 9 mm L, corolla tube 5 mm L, outer lip dorsally densely white sericeous, inner lip fused, 2.5 mm L **Disk florets** ca. 6, bisexual, yellow; corolla 6.5 mm L, corolla tube 5.8 mm L **Styles** [ray] 6 mm L, stigma lobes ca. 1 mm, [disk] 6–6.5 mm L, lobes 0.5 mm L obtuse-truncate apices with triangular hairs (ca 15 µm L). **Anthers** [ray] 6.5 mm L, sterile [disk] 2.5 mm L (6.3 mm L incl. filaments). **Achenes** brown, 1.5–2.6 mm L, turbinate; well developed carpopodium; pericarp pellucid, densely covered in lanceolate twin hairs, 75–110 x 40 µm; testa epidermis surface of elongated parallel cells, margins entire, cells with U-formed sclerenchymatous thickenings. **Pappus** white, ca. (5)6.5–8 mm L, 1–(2)rows, indehiscent; setae shortly fused at base, pappus setae 4–8 cells wide, cell width 5–10 µm, no basal cilia; barbellate, barbs 80–100 um L, barb base shortly-medium adhered (40-50%), free barb appressed, 7–8 barbs/100 µm. **Chromsome count** 2n = 2x = 26. Davies & Vosyka (new count) *Grau s.n.* (M)

Distribution and habitat. *C. moenchioides* is distributed in Chile in the Cordillera de la Costa from La Serena south to hills around Santiago, foothills of the Andes in valleys from Rancagua, to Temuco, in the Cordillera de la Costa from Constitución to Concepción and Nahuelbuta, and in the southern Andes to Argentina (Junín de Los Andes, San Carlos de Bariloche). It covers a wide amplitude of latitudes: 29°00' – 35°30'; 36°30' – 39°00' (41°00')S. It is found at elevations between 10 – 2300 m.a.s.l. It occurs in sandy (to dunes) or gravelly (to screes) ground or open grassy vegetation in well drained, dry, exposed localities.

Differential diagnosis. *C. moenchioides* is very polymorphic in its leaf size and number and colour of ray florets. It is close to *C. ramosissima* and *C. taltalensis*. It is distinguished from the lanate-floccose, slightly succulent-leaved *C. taltalensis* by its densely silvery indumentum. It is distinguished from the erectly branching *C. ramosissima* with reflexed leaves/bracts by its spreading, ascending branches and ascending leaves/ bracts. Anecdotal evidence suggests that *C. moenchioides* is the only species in this triad with rays conspicuously, if shortly exerted from the capitulum. Key characters for identifying *C. moenchioides* *C. ramosissima* and *C. taltalensis* are given in Table 13.

Species	*C. moenchioides*	*C. ramosissima*	*C. taltalensis*
Branching habit	Spreading, ascending branched	Erect, branched	Spreading, decumbent, profusely branched
Leaves/ bracts	Ascending, appressed	Reflexed	Ascending, appressed
Indumentum	Silvery sericeous	Sparsely lanate	Lanate-floccose
Capituulum	Urcinate	Urcinate	Campanulate
Ray floret	Exerted, yellow, dorsally sericeous	Inconspicuous, (yellow?)	Inconspicuous, (white with dark tips?)
Distribution (°S)	29°00' – 35°30'; 36°30' – 39°00'	33°00' – 36°00'	23°00' – 30°00'

Table 13: Key characters for identifying *C. moenchioides, C. ramosissima* and *C. taltalensis*

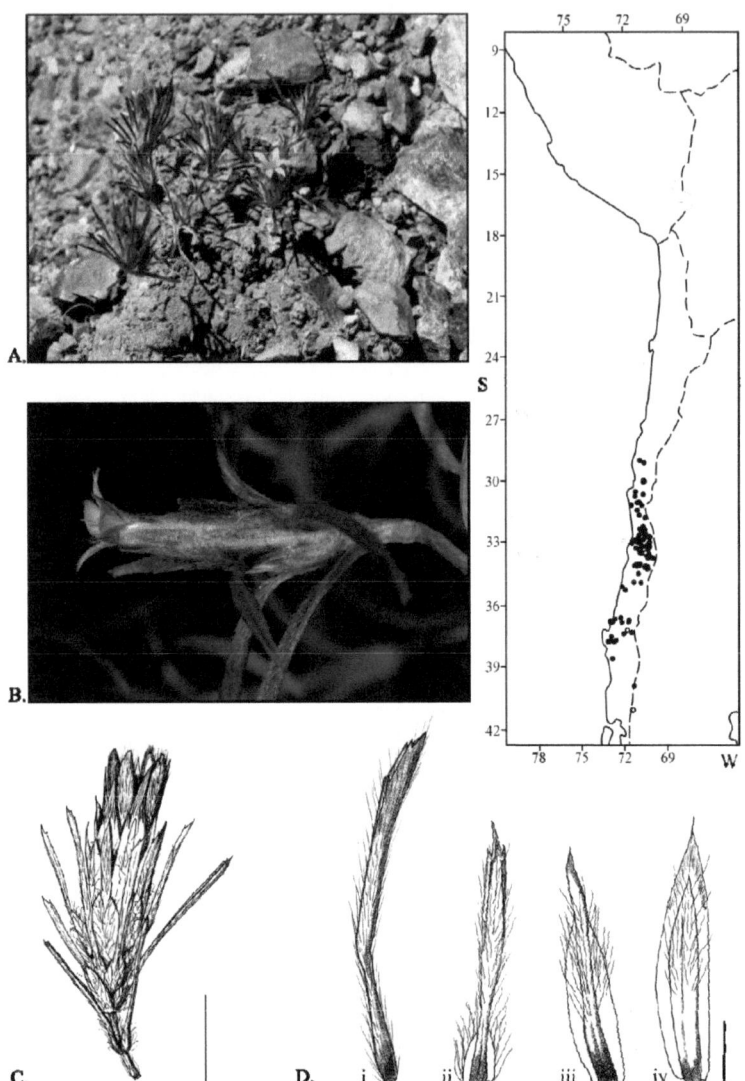

Figure 59: *Chaetanthera moenchioides*. **A.** Habit photographed Santiago, 2001©Alison Davies. **B.** Capitulum detail. *K. Arroyo* et al. *25177* (cultivated) from San Carlos de Bariloche, photographed Munich 2004©Alison Davies. **C.** Capitulum detail, drawn from *Davies s.n.* Scale bar = 5 mm. **D.** Leaf & bract detail. *Davies s.n.* **i.** stem leaf d.s.; **ii.** OIB v.s.; **iii.** MIB d.s.; **iv.** IIB d.s. Scale bar = 2 mm.

IX *Chaetanthera* Ruiz & Pav.

Material seen. – **Argentina. Neuquén** Cerro Collun, 3000 ft, 12.1926, *Comber 874* (E K). – **Río Negro** Cauce del Arroyo Ñireco dentro de la ciudad de San Carlos de Bariloche, 860 m, 02.2003, *K. Arroyo et al. 25177* (CONC). – **Chile. IV Región de Coquimbo** Prov. de Elqui Cuesta Pajonales, 1000 m, 11.1961, *Jiles 398* (CONC). – La Silla, Quebrada Agua de la Mona, 10.1991, *Grenon 22744* (G). – 9 km desde Vicuña, camino de tierra a Hurtado, 1050 m, 09.2002, *K. Arroyo et al. 25018* (CONC). – 19 km al sur de Vicuña, camino a Hurtado, 1730 m, 09.2002, *K. Arroyo et al. 25068* (CONC). – Quebrada Las Trancas, 25 km south of Vicuña, 1350 m, 10.1940, *Wagenknecht 18600* (G GH). – Prov. de Limarí Ovalle, Talahuen, 1889/90, *Geisse s.n.* (GH). – Questa spino, desvio Pola hasta Tunel, 1200-1350 m, 10.1945, *Blese 1860* (SGO). – Cabreria, Morro Blanco, 1600 m, 10.1949, *Jiles 1600* (CONC). – Cuesta Punitaqui, aproximadamente 6 km del pueblo, 370 m, 09.2002, *K. Arroyo et al. 25058^a* (CONC). – Cuesta de Punitaqui, 300 m, 10.1963, *Marticorena & Matthei 381* (CONC). – camino entre Combarbalá y Cogoti, 1160 m, 09.2002, *K. Arroyo et al. 25026* (CONC). – camino entre Cuesta de Punitaqui y Combarbalá, 940 m, 09.2002, *K. Arroyo et al. 25023* (CONC). – Cordillera Combarbalá, El Peñon-Ramadilla, 1400 m, 10.1971, *Jiles 5705* (CONC). – Cordillera Combarbalá, Ramadilla-Las Lajas, 2800 m, 10.1971, *Jiles 5664* (CONC). – Cordillera Combarbalá, Ramadilla-El Morro, 2800 m, 10.1971, *Jiles 5283* (CONC). – Prov. de Choapa Cajón de Hualtata (Choapa), 10.1894, *Philippi s.n.* (BM). – Salamanca, El Canelo, 240 m, 10.1974, *Contreras & Caviedes 302* (SGO). – El Mollar, 650 m, 10.1965, *Jiles 4633* (CONC M). – camino entre RN Chinchillas y Illapel, 580 m, 09.2002, *K. Arroyo et al. 25027* (CONC). – Mina Los Pelambres, camino a Chacay, 2190 m, 02.1999, *K. Arroyo et al. 991559* (CONC). – Camino de tierra al lado este del valle de Choapa, Cuncumen, 1200 m, 10.2002, *K. Arroyo et al. 25029* (CONC). – **V Región de Valparaíso** Prov. de Petorca camino entre Alicahue y Putaendo, 1000 m, 10.2002 *K. Arroyo et al. 25035* (CONC). – camino a Guayacán, 400 m, 09.1999, *K. Arroyo et al. 992215* (CONC). – camino entre Guayacan y Putaendo, 500 m, 10.2002, *K. Arroyo et al. 25038* (CONC). – Prov. de Los Andes camino entre El Tartaro y Reguardo Los Patos, 1230 m, 10.2002, *K. Arroyo et al. 25039* (CONC). – 38 km W of Portillo (i.e. 47 km E of Los Andes), ca. 1.5 km upstream from road-bridge Puente J. Rubina crossing the Río Aconcagua just below Guardia Vieja and following the disused tracks of the Transandean Railway, 1610 – 1650 m, 11.1991, *Eggli & Leuenberger 1709* (B). – Los Andes, 1100 m, 09.1952, *Mancilla s.n.* (CONC). – Río Colorado, Cerro Negro, 2000 m, 11.1974, *Zöllner 7929* (CONC). – Los Andes a Portillo, Km 27, 1400 m, 11.1970, *Marticorena & Weldt 566* (CONC). – Estero Pocuro, 750 m, 09.1951, *Salazar s.n.* (CONC). – 4 km S of turnoff from main road following dirt road over Cuesta de Chacabuco, on first S exposed steep slopes before reaching the culmination of the pass, 1340 m, 11.1991, *Eggli & Leuenberger 1727* (B). – Prov. de Valparaíso Viña del Mar, *Bridges 124* (BM W). – Prope Valparaíso, 1831, *Cuming 656* (Ex2 GH NY W). – Marga-Marga, 40 km de Valparaíso, 250 m, 11.1929, *Jaffuel 231* (CONC GH). – Valle de Marga Marga, 10.1936, *Jaffuel 3687* (GH). – Prov. San Felipe de Aconcagua Cerros de Lo Vargas, 980 m, 09.1933, *Torres s.n.* (CONC). – Jahuel, 800 m, 11.1963, *Valenzuela 378* (CONC). – Catemu, 1250 m, 10.1976, *Zöllner 9107* (MO). – Prov. de Quillota In lapidosis apricis collium Quillota, 10.1829, *Bertero 911* (BM E G GH NY P W). – Limache, Küstenkordillere auf dem Weg zum Robleberg, 1500 m, 12.1963, *Zöllner 441* (B). – Cerro del Roble, 1500 m, 11.1963, *Zöllner 450* (B). – Cerro El Roble, PN La Campana, 1640 m, 11.1999, *K. Arroyo et al. 993977* (CONC). – Estero de Maitenes, 120 m, 10.1931, *Garaventa 2696* (CONC). – Cerro El Roble, Caleu, 1790 m, 11.1994, *K. Arroyo et al. 994068* (CONC). – PN La Campana, Sector La Cascada, 810 m, 10.2000, *K. Arroyo et al. 205025* (CONC). – PN La Campana, Sector Ocoa, camino a la cascada, 750 m, 10.2000, *K. Arroyo et al. 202916* (CONC). – **Región Metropolitana de Santiago** Prov. de Chacabuco Cuesta La Dormida, 10.1990, *von Bohlen 831* (SGO). – Cuesta La Dormida, 1400 m, 11.1958, *Schlegel 1725* (CONC). – Caleu, 1100 m, 10.1960, *Valenzuela s.n.* – Aproximadamente 2 km al sur del pueblo de Caleu, camino a Rungue, 1120 m, 10.2002, *K. Arroyo et al. 25031* (CONC). – Rungue, 650 m, 10.1963, *Saa s.n.* (CONC). – Tiltil, 700 m, 12.1953, *Moreno s.n.* (CONC). – Tiltil, camino a La Capilla, 840 m, 11.1941, *Behn s.n.* (CONC). – camino entre Polpaico y Tiltil, 530 m, 10.2002, *K. Arroyo et al. 25030* (CONC). – Termas de Colina, 10.1943, *Guzmán s.n.* (SGO). – Baños de Colina, 700 m, 12.1950, *Barros s.n.* (CONC). – Batuco, 480 m, 10.1954, *Levi 3125* (CONC). – Laguna Batuco, 480 m, 10.1961, *Schlegel 3967* (CONC). – Prov. de Santiago San Cristobal, 10.1882, *Bertero s.n.* (B). – in collibus Santiago, 09.1829, *Gay 337* (P). – Prov. de Santiago, 1862, *Philippi s.n.* (G). – Prov. de Santiago, *Philippi s.n.* (B W). – Santiago, *Philippi 556* (B BM G P W). – prope la Quinta, 10.1828, *Bertero 163* (BM E G MO NY P Wx2). – Environs de Santiago, 1855, *Philippi s.n.* (BM Gx2 K Wx3). – Cord. Río San Francisco, 3200 m, 1856-57, *Philippi s.n.* (G). – In Andissimus Cord. S. Rosa, 12.1829, *Poeppig s.n.* (G). – Santiago, 1856-57, *Philippi s.n.* (BM K). – Santiago, *Philippi s.n.* (E). – Santiago, *Philippi?* (BM). – Villa Paulina, 2000 m, 10.1951, *Uslar 45* (CONC). – Farellones, 2300 m, 12.1962, *Sierra s.n.* (SGO). – Farellones, curve 6-7, 11.2001, *Davies s.n.* (M). – Arrayán, 1000 m, 11.1951, *Gunckel 22114* (CONC). – El Arrayán, 885 m, 10.1954, *Levi 2251* (CONC). – El Arrayán, 10.1959, *Saa s.n.* (SGO). – Curacaví, 200 m, 11.1953, *Gunckel 37997* (CONC). – Cerro de Renca bei Santiago, 800 m, 10.1937, *C. & G. Grandjot 3049* (GH, M x3). – Cerros de Renca, path m, 11.1951, *Gunckel 22113* (CONC). – Cerro de Renca, 800 m, 10.1953, *Gunckel 26668* (CONC). – Peñaflor, 10.1956, *Gunckel 29786* (CONC). – Cerro del Ramón, 1200 m, 10.1951, *Lopez s.n.* (CONC). – Cerro Abanico, 1500 m, 11.1951, *Gunckel 22081* (CONC). – Monte/Cerro Abanico, 1550 m, 10.1932, *Grandjot s.n.* (CONC 1006 CONC 21092). – Peñalolen, 1300 m, 11.1952, *Bravo s.n.* (CONC). – Quebrada La Plata, 860 m, 11.1957, *Schlegel 1450* (CONC). – Cordillera de Macul, 20 km al este de Santiago, 2300 m, 07.11.1925, *Pirion 492* (GH). – Quebrada de Macul, 1200 m, 11.1952, *Gunckel 25087* (CONC). – Cuesta de Barriga, between Marruecos y Los Cerrillos, 700 – 1000 m, 11.1948, *Killip & Pisano 39655* (K, NY). – Cerro de Lo Chena, 700 m, 10.1950, *Gunckel 18677* (CONC). – RN Río Clarillo, 850-950 m, 02.1989, *K. Arroyo et al. 891087* (CONC). – RN Río Clarillo, 1050 m, 11.1999, *K. Arroyo et al. 993666* (CONC). – RN Río Clarillo, 1100 m, 11.1999, *K. Arroyo et al. 993701*

(CONC). – RN Río Clarillo, Estacionamiento, 900 m, 02.2000, *K. Arroyo* et al. *20810* (CONC). – San Gabriel, 1500 m, 02.1950, *Barros s.n.* (CONC). – RN Río Clarillo, Cajón de Los Cipreses, 1800 m, 11.1999, *K. Arroyo* et al. *994480* (CONC). – RN Río Clarillo, Sector El Potrillo, 1600 m, 11.1999, *K. Arroyo* et al. *994305* (CONC). – Refugio Lo Valdes, 3350 7005, 1950 m, 12.1940, *Schwabe 187* (CONC). – RN Río Clarillo, Sector Loma de Los Cipreses, 2150 m, 11.1999, *K. Arroyo* et al. *994446* (CONC). – Prov. de Melipilla Melipilla, *Philippi s.n.* (W). – RN Roblería de Cobre de Loncha, Cerro Llivi-Llivi, 1820 m, 12.2000, *K. Arroyo* et al. *206520* (CONC). – RN Roblería de Cobre de Loncha, Cuesta del Llivi-Llivi, 1490 m, 12.2000, *K. Arroyo* et al. *206468.5* (CONC). – RN Roblería de Cobre de Loncha, Sectos Los Maquis, 1180 m, 11.2000, *K. Arroyo* et al. *205803* (CONC). – RN Roblería de Cobre de Loncha, Sector Los Maquis, 1130 m, 11.2000, *K. Arroyo* et al. *203422.5; 203423; 203464* (CONC). – RN Robleria de Cobre de Loncha, abajo del Sector Los Maquis, 480 m, 11.2000, *K. Arroyo* et al. *206123* (CONC). – RN Roblería de Cobre de Loncha, Sector La Isla, 480 m, 08.2002, *K. Arroyo* et al. *25003* (CONC). – RN Roblería de Cobre de Loncha, Sector Los Huairavos, 1420 m, 11.2000, *K. Arroyo* et al. *206224* (CONC). – Santiago, 09.1918, *Claude-Joseph 511* (GH). – Prov. de Cordillera San José de Maipo, 09.1961, *Novas 2581* (SGO). – Las Vertientes, Cajón del Maipo, 09.1962, *Novas 2502* (SGO). – Clarillotal bei Puente Alto, 750 m, 10.1933, *Grandjot s.n.* (MO). – **VI Región del Libertador G.B. O'Higgins** Prov. de Cachapoal Cocalán, 400 m, 09.1968, *Muñoz s.n.* (CONC). – PN Cocalán, Morro Mal Paso, 1200 m, 10.2001, *K. Arroyo* et al. *211380* (CONC). – PN Cocalán, 660 m, 11.2001, *K. Arroyo* et al. *211734* (CONC). – PN Cocalán, 500 m, 11.2001, *K. Arroyo* et al. *211614* (CONC). – Termas de Cauquenes, 1100 m, 11.1952, *Pfister s.n.* (CONC). – Termas de Cauquenes, 11.1965, *Behn s.n.* (SGO). – Rancagua a Caletones, Km 17, 1100 m, 11.1970, *Marticorena & Weldt 651* (CONC). – RN Río de los Cipreses, Camino entre Ranchillo y Administración, 1150 m, 10.1999, *K. Arroyo* et al. *993401* (CONC). – RN Río de los Cipreses, Cascada sendero Los Peumos, 1120 m, 10.2000, *K. Arroyo* et al. *202847* (CONC). – RN Río de los Cipreses, Quebrada Las Terneras, 1970 m, 12.2000, *K. Arroyo* et al. *206869* (CONC). – Prov. de Colchagua lado norte de Cerro Centinela a 4 Km al sur de San Fernando, 450 m, 10.1973, *Stebbins 8728* (SGO). – **VII Región del Maule** Prov. de Curicó Curicó, 200 m, 01.1925, *Barros 632* (CONC). – Aproximadamente 4 km antes de Los Queñes, orillas Río Teno, 570 m, 10.2002, *K. Arroyo* et al. *25040* (CONC). – Aproximadamente 4 km antes de Los Queñes, orillas Río Teno, 530 m, 09.2002, *K. Arroyo* et al. *25008* (CONC). – Los Queñes a Curicó, 700 m, 11.1981, *Marticorena & Rodriguez 1724* (CONC). – Prov. de Talca Terra Pehuenchorum, 12.1854, *Lechler 2934* (G K P W). – Putú, 11.1980, *Grau 2360* (MSB). – Cerros al oeste de Pencahue, camino a Curepto, 130 m, 10.2002, *K. Arroyo* et al. *25044* (CONC). – Prov. de Cauquenes Baños de Cauquenes, 1875, *Philippi s.n* (M). – Subestación Experimental Cauquenes, INIA, Fundo El Boldo, 177 m, 11.1978, *Avendaño & Ovalle O-035* (SGO). – Subestación Experimental Cauquenes, INIA, Fundo El Boldo, 177 m, 10.1981, *Avendaño & Norambuena C-049* (SGO). – **VIII Región del Biobío** Prov. de Ñuble San Carlos, 1897, *Neger s.n.* (M W). – camino entre Quiriquina y Chillán, al sur de puente Quilmo, 180 m, 10.2002, *K. Arroyo* et al. *25073* (CONC). – Straße zu den Termas de Chillán, Recinto, 10.1990, *Grau s.n.* (MSB M). – Atacalco, 650 m, 11.1944, *Pfister 1006* (CONC 21097; CONC 4353). – Camino de Los Lleuques a Atacalco, 650 m, 11.2002 *K. Arroyo* et al. *25159* (CONC). – Palpal, Camino de Yungay a Bulnes, 200 m, 11.1946, *Pfister s.n.* (CONC). – Prov. de Concepción Concepción, 11.1896, *Neger s.n.* (Mx4). – Puente Queime, 60 m, 11.1972, *Troncoso & Valenzuela s.n.* (CONC). – Camino entre Florida a Roa, 340 m, 10.2002, *K. Arroyo* et al. *25074* (CONC). – Concepción, 10 m, 11.1922, *Barros s.n.* (CONC). – camino Viejo de Florida a Penco, 250 m, 11.1957, *Ricardi s.n.* (CONC). – Puente Santa Isabel, entre Pueblo Seco y Pemuco, 250 m, 10.2002, *K. Arroyo* et al. *25072* (CONC). – Prov. del Biobío Pascua arida circum Antuco 12.1828, *Poeppig 689* (G W) – Lecta ad Antuco, 12.1828, *D.689 Poeppig 208* (BM M NY P W). – Antuco, *Poeppig 1* (G). – Antuco, entrada PN Laguna de la Laja, 1000 m, 01.1988, *Rosas 1821* (M). – 20 – 37 km von Antuco nach Los Angeles, 400 m, 11.1978, *K.H. & W. Rechinger 63826* (W). – **IX Región de La Araucanía** Prov. de Malleco Ad Angol opp., 11.1896, *Düsen s.n.* (G W). – Comunidad Maitenrehue, Sectore Los Lingues, 570 m, 12.1997, *Baeza & Kottirsch 800* (CONC). – Mininco, 190m, 11.1953, *Kunkel 936* (CONC). – Salida de Mininco hacia Ruta 5 Sur, 170 m, 10.2002, *K. Arroyo* et al. *25071* (CONC). – Subida Angol-Vegas Blancas, entrada A[ngol], 800 m, 11.1985, *Rodriguez* et al. *2136* (CONC). – Cordillera de Nahuelbuta, trockene gebüsche am Strassenrand bei Vega de Aguas Blancas, 700 m, 11.1985, *Hellwig 6833* (G). – Angol, entrando a Los Alpes, 25 m, 12.1957, *Montero 5429* (CONC). – Prov. de Cautín Lado del Río Cautin, camino entre Temuco y Nueva Imperial, aproximadamente 7 km de Temuco, 11.2002, *K. Arroyo* et al. *25122* (CONC). – **X Región de Los Lagos.** Prov. de Valdivia Hacienda de San Juan, 12.1857, *Ochsensis s.n.* (P).. – Material in Mini-Herbario: *K. Arroyo* et al. *92579; 205803; 207029, León 101089.* – **S.l.d.** Cordilleras of Chili, *Cuming 291* (E). – Cordilleras of Chili, 1831, *Cuming 321* (BM E Kx2). – Chile australis, *Neger s.n.* (M). – *Gay s.n.* (K). – *Poeppig s.n.* (P). – *Cuming 240* (BM). – *Philippi 726* (B). – *Gay s.n.* (GH).

13. *Chaetanthera multicaulis* DC. Prodr. (DC.) 7 (1): 31. 1838.

Typus CHILE "In pascuis saxosis collium ad torrentea circa Tagua-Tagua Chilensim legit cl. Bertero" [In pascuis sasxosis ? torrente Tagua Tagua, 09-10.1828, *Bertero 164*] Holotypus G! Isotypus SGO 76556! [*s.n.*]

Nomenclature Notes The sheet in G shows both the number 164 and the locality data on the same handwritten label. However, the sheet in SGO, although with the same handwritten locality information, has no number. The remaining sheets with the number 164 printed on them have somewhat different locality data: [In saxosis aridis montis la Leona Rancagua Chili, 10.1828, *Bertero 164*] BM E F Gx2 GH (photo) M NYx2 P W.

Annual monoecious herb. **Roots** filamentous. **Stems** to 6 cm, ascending, dark reddish-brown to green, sparsely covered in sericeous white hairs. **Leaves** 8 x 0.8 mm, sparsely sericeous on dorsal surface; indumentum of compound 2-celled trichomes: basal inflated cell and elongated filamentous cell. **Capitula** sessile, campanulate, disk diameter 3–4 mm. **Involucral bracts** imbricate, arranged in three types, initially foliaceous then reduced to entirely membranous. **Outer involucral bracts** 5–6 x 1 mm, cilia 1 mm L; as leaves but with short membranous alae to less than ½ height of bract. **Middle involucral bracts** 5–6 mm L; membranous alate, ratio of lamina to alae decreases from ⅔ to nearly entirely alate. **Inner involucral bracts** 6–7 mm L, acute, linear lanceolate, entirely membranous alate; apices striate maculate, dorsally sericeous. **Ray florets** ca. 10, pistillate, yellow; upper dorsal surface sericeous; corolla 10 mm L, corolla tube 3 mm L, outer lip 2.6 mm W., inner lip 2 mm L, bifid, filamentous. **Disk florets** 10–20, bisexual, yellow; corolla 6 mm L, corolla tube 5 mm L **Styles** [ray] 6 mm L, stigma lobes 0.5 mm L [disk] 5.5 mm L, stigma lobes 0.5 mm L **Anthers** [ray] sterile anther tips dark green or blackish. [disk] fertile, 6 mm L **Achenes** brown, 0.8–1.5 mm L, angled turbinate; carpopodium anular, irregular; pericarp pellucid, densely covered in lanceolate twin hairs, 70–90 x 3 µm. **Pappus** white, ca. 4–5 mm L, 1(2) rows, indehiscent; setae very shortly fused at base, pappus setae 10–15 cells wide, each cell width 7–10 µm, no basal cilia; densely barbellate, barbs 90 µm L, barb base shortly adhered (<40%), free barb appressed, 9–11 barbs/100 µm.

Distribution and habitat. *C. multicaulis* is endemic to Chile, distributed from just north of La Ligua, along the Cordillera de la Costa, across the central valley to the foothills of the Andes around Santiago and south to Curicó between latitudes 32°30' – 35°10'S. It occurs between elevations of 50 – 1500 m.a.s.l. It is found in the open *Acacia caven* matorral vegetation of the central valley.

Differential diagnosis. With its long ciliate lamina margins, *C. multicaulis* appears to be a precocious, more depauperate form of *C. ciliata*, which typically grows to at least twice the size. *C. multicaulis* has broader pappus setae that are short frequently barbellate. This species also has somewhat smaller twin hairs on the pericarp. Key characters for distinguishing *C. ciliata* and *C. multicaulis* can be found in Table 11, page 130.

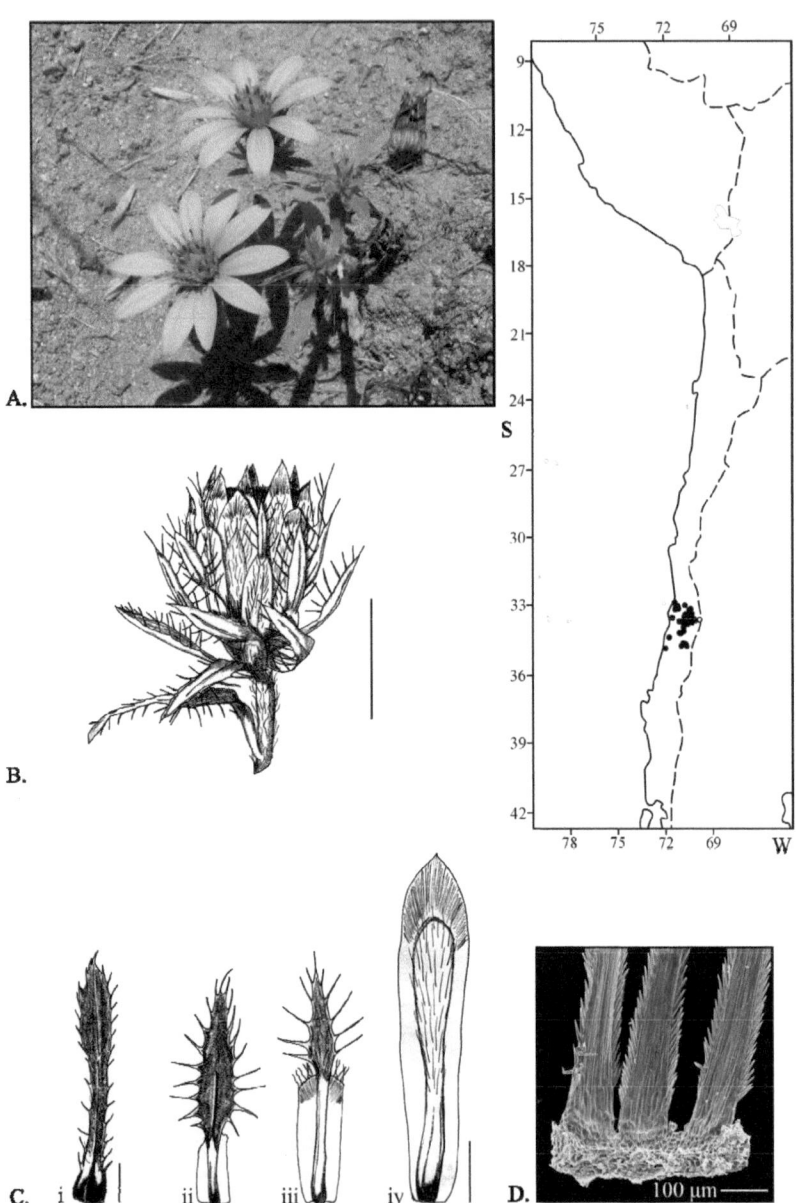

Figure 60: *Chaetanthera multicaulis*. **A.** Habit, photographed Santiago©María Terese Eyzaguirre. B. & C. *K. Arroyo 993434* **B.** Capitulum detail, scale bar = 5 mm. **C.** Leaf & Bract detail, d.s. **i.** Stem leaf; **ii.** OIB; **iii.** MIB; **iv.** IIB. Scale bar i, ii – iv = 2 mm. **D.** *Jaffuel 234*. Pappus setae (basal region). Scale bar = 100 μm.

IX *Chaetanthera* Ruiz & Pav.

Material seen. – Chile. V Región de Valparaíso Prov. de Valparaíso RN Peñuelas, 380 m, 11.2000, *K. Arroyo* et al. *205497* (CONC). – Maipú, Rinconada, Küstenkordillere, 11.1980, *Grau 2400* (M). – Marga Marga, 10.1910, *Jaffuel 1263* (GH). – Marga Marga, 10.1936, *Jaffuel 3631* (GH). – Marga Marga, 250 m, *Jaffuel 234* (GH). – Marga Marga, 1931, *Jaffuel & Pirion 3207* (GH). – Marga-Marga, 150 m, 12.1932, *Jaffuel & Pirion 3092* (CONC GH). – RN Peñuelas, 365 m, 11.2000, *K. Arroyo* et al. *203577* (CONC). – Perales, 390 m, 11.1929, *Jaffue l234* (CONC). – Prov. de San Antonio Cartagena, (Cerro ello), 10.1954, *Navas 927* (SGO). – Cartagena, 60 m, 02.1958, *Gunckel 36716* (CONC). – **Región Metropolitana de Santiago** Prov. de Chacabuco Chacabuco, near Collina, 1832, *Bridges 132* (BM E Kx2 W). – Colina, 600 m, 11.1949, *Acevedo s.n.* (SGO). – Prov. de Santiago *Philippi s.n.* (B). – Santiago, *Philippi 557* (BM G K P W). – Santiago, *Philippi s.n.* (W). – Prov. de Santiago, "Chaetanthera ciliata var floribus purpureis", *Philippi s.n.* (W). – In collibus aridis, Santiago, IX.1839, *Gay 608* (P). – Santiago, *Ball s.n.* (E). – Las Condes, 900 m, 11.1951, *Frödin 42* (BM). – Cuesta La Barriga, en lomaje con matorral, 500 – 600 m, 11.1956, *Kausel 4267* (F). – 1-2 km de Caleu, 10.2002, *K. Arroyo* et al. *25124* (CONC). – Cuesta La Dormida, 10.1976, *Montenegro s.n.* (SGO). – El Arrayán, 11.1966, *Mahu 1514* (SGO). – Colina, 900 m, 10.1919, *Behn K s.n.* (CONC). – Baños De Colina, 1000 m, 11.1950, *Castillo s.n.* (CONC). – El Arrayán, 885 m, 11.1960, *Gunckel 69558* (CONC). – El Arrayán, 885 m, 11.1966, *Mahu 1515* (CONC). – Cerro Renca, 600 m, 10.1937, *Junge s.n.* (CONC). – Cerro Renca, 650 m, 10.1961, *Luck s.n.* (CONC). – Cerro Renca, 540 m, 10.1922, *Montero 1365* (CONC). – Cerro Blanco, 640 m, 11.1950, *Gunckel 18865* (CONC). – San Cristóbal, 800 m, 11.1900, *Hastings 151* (NY). – San Cristóbal, 800 m, 11.1950, *Levi 1262* (CONC). – Navia, 500 m, 11.1914, *Baeza s.n.* (CONC). – Camino de Rungue a Caleu, 700 m, 11.1960, *Schlegel 3120* (CONC). – Peñaflor, 400 m, 10.1922, *Montero 1752* (CONC). – Quebrada de Peñalolen, 1000 m, 12.1953, *Bravo 545* (CONC). – Peñaflor, 400 m, 10.1951, *Collantes s.n.* (CONC). – Pudahuel, 11.1961, *Gunckel 37358* (CONC). – Cuesta Lo Prado, 490 m, 10.1999, *K. Arroyo* et al. *993434* (CONC). – Quebrada de Peñalolen, 1000 m, 10.1952, *Bravo 194* (CONC). – Quebrada de Peñalolen, 720 m, 11.1951, *Guncke 25123* (CONC). – parte final Quebrada de Los Maquis, 860 m, 11.1957, *Schlegel 1448* (CONC). – Quebrada de Macul, 850 m, 11.1952, *Gunckel 25006* (CONC). – Cuesta Lo Prado, lado oeste, 550 m, 9.2002, *K. Arroyo* et al. *25062* (CONC). – Cerro Lo Chena, 650 m, 11.1950, *Gunckel 18682* (CONC). – Quebrada de Macul, 850 m, 10.1953, *Gunckel V s.n.* (CONC). – Mallarauco, 700 m, 11.1964, *Gunckel 42563* (CONC). – Cajón del Maipo, 400 m, 10.1960, *Gunckel 69559* (CONC). – Melipilla, 250 m, 10.1951, *Gunckel 25278* (CONC). – RN Río Clarillo, Quebrada El Almendro, 1000 m, 11.1999, *K. Arroyo* et al. *99 4152* (CONC). – Prov. Cordillera Cordilleras, Questo Potrero, 2nd Range, 04.1834, *Cuming 202* (BM Ex2 GH Kx2 P W). – El Manzano, Cajón del Maipo, 800 m, 11.1981, *Landrum 3803* (SGO). – Los Canelos, Cajón del Maipo, 12.1961, *Navas 2720* (SGO). – Río Colorado, frente desembocadura, Cajón del Maipo, 11.1994, *Muñoz S., Moreira Teillier & Arrigada 3419* (SGO). – Peñalolen, 900 m, 11.1962, *Sierra s.n.* (SGO). – Lagunillas lado del camino, 1500 m, 12.1992, *Von Bohlen 1497* (SGO). – Quebrada Las Coles, frente San Juan del Maipo, Cajón del Maipo, 11.1994, *Muñoz S., Moreira Teillier & Arrigada 3424* (SGO). – Prov. de Maipo Laguna de Aculeo, 356 m, 10.1942, *Pisano* et al. *1558* (SGO). – Laguna de Aculeo, 350 m, 10.1974, *Montero 9355* (CONC). – Prov. de Talagante Cuesta de Mallarauco, en la cima, hacia el poniente, 700 m, 10.1979, *Muñoz 1428* (SGO). – Prov. de Melpilla Milipilla, *Philippi s.n.* (W). – Chile borealis Cordillera de Sa. Rosa in glareosis, XII.1828, *Poeppig s.n.* (W). – Melipilla, a 8 M al sur en falda con espinales, 12.1952, *Kausel 3498* (F). – Cerro Cantillana, Casa de Piedra, 1050 m, 10.1994, *Villagran & Maldonado 8366-B* (SGO). – RN Roblería de Cobre de Loncha, 560 m, 11.2000, *K. Arroyo* et al. *207079* (CONC). – RN Roblería de Cobre de Loncha, Llano La Cuesta, 470 m, 10.2002, *K. Arroyo* et al. *25047* (CONC). – Quebrada La Plata, Hacienda Rinconada lo cerda Maipu, 600 m, 11.1957, *Schlegel 1448* (SGO). – RN Roblería de Cobre de Loncha, Sector Las Represitas, 1000 m, 01.2001, *K. Arroyo* et al. *210020* (CONC). – RN Roblería de Cobre de Loncha, Sector Cuesta Lo Mirandinos, 650 m, 10.2001, *K. Arroyo* et al. *211592* (CONC). – RN Roblería de Cobre de Loncha, Sector Los Maquis, 850 m, 11.2000, *K. Arroyo* et al. *206096* (CONC). – RN Roblería de Cobre de Loncha, Llano La Cuesta, 460 m, 08.2002, *K. Arroyo* et al. *25002* (CONC). – Weg von Melipilla nach Villa Alhue, Cuesta Los Guindos, 11.1985, *Hellwig 6427* (G). – **VI Región del Libertador G. B. O'Higgins** Prov. de Cardenal Caro Alcones, 200 m, 11.1974, *Contreras & Oyanedel 460* (SGO). – Prov. de Cachapoal PN Cocalán, 970 m, *K. Arroyo* et al. *220075* (CONC). – Camino entre Cocalán y entrada al PN Palmas de Cachapoal, 170 m, 10.2002, *K. Arroyo* et al. *25045* (CONC). – Palmeria de Cocalán, bajo el bosque de Jubaea, 10.1942, *Muñoz S., Bartlett & Sudzuki 3501* (SGO). – PN Cocalán, 500 m, 11.2001, *K. Arroyo* et al. *211608* (CONC). – Palmeria de Cocalán, bajo el bosque de Jubaea, 10.1942, *Muñoz S., Bartlett & Sudzuki 3511* (SGO). – PN Cocalán, 660 m, 11.2001, *K. Arroyo* et al. *211793* (CONC). – Prov. de Colchagua *Philippi s.n.* (W). – Río Tinguiririca, oil.1951, *Ricardi s.n.* (CONC). – Rincon del Tinguiririca, 360 m, 12.1950, *Ricardi s.n.* (CONC). – Puente Negro, 540 m, 12.1950, *Ricardi s.n.* (CONC). – **S.l.d.** – Bertero s.n. (GH). – *Gillies s.n.* (NY). – Pudahuel, *Reed s.n.* (K). – Turiéta, *Reed s.n.* (K).

14. *Chaetanthera perpusilla* (Wedd.) Anderb. & S. E. Freire Taxon 39 (3): 431. 1990.

≡ *Luciliopsis perpusilla* Wedd. Chlor. And. 1: 160. t.26. 1855. Typus BOLIVIA "Bolivie: pelouses rases et un peu arides de la Lancha, dans la partie superieure du ravin de Chuquiaguillo, aux environs de La Paz (Wedd.)" Holotypus P!
= *Chaetanthera aymarae* Martic. & Quezada Bol. Soc. Biol Concepcion, 48: 107 – 108, fig.1. 1974. Typus CHILE "Camino de Zapahuira a Puter, Quebrada de Socoroma, 3100 m.s.m. (18°15'S – 69°35'W) 05.05.1972, *Ricardi, Weldt & Quezada 166*" Holotypus CONC 40605! Isotypus CONC!

Annual monoecious herb. **Roots** filamentous. **Stems** to 10 cm, filiform, lax, ascending, simple to branched. **Leaves** 7.5 x 1 mm, sessile, linear, slightly thickened, margins somewhat inrolled, ± 2 dentate to apex, with apical mucro; indumentum on ventral surface irregularly sparsely dotted with short translucent triangular hairs. **Capitula** open campanulate, sessile, few (ca. 10) imbricate bracts. **Involucral bracts** imbricate, arranged in three types, initially foliaceous then reduced to entirely membranous. **Outer involucral bracts** 4.3–5 mm L, ovate, as leaves but with short membranous alae to less than ½ height of bract. **Middle involucral bracts** 3.6–4.3 mm L, ovate; membranous alate, ratio of lamina to alae decreases from ⅔ to nearly entirely alate. **Inner involucral bracts** 4.1–4.8 mm L, lanceolate to broadly ovate, entirely membranous alate; apices striate maculate blue/black, shortly mucronate. **Ray florets** 4–5, pistillate, white, glabrous; corolla 3.5–3.6 mm L, corolla tube 2.1–2.2 mm L, outer lip 0.6 mm W., inner lip 0.7 mm L **Disk florets** 11–16, bisexual, yellow; corolla 2.8 mm L, corolla tube 2.5 mm L **Styles** [ray] 3.3 mm L, stigma lobes 0.2 mm L [disk] 2.9 mm L, stigma lobes 0.2 mm L **Anthers** 1.2–1.5 mm L (2.8 mm L incl. filaments). **Achenes** brown, 2.6 mm L, fusiform turbinate; carpopodium small, anular irregular; pericarp pellucid, densely covered in lanceolate twin hairs, 70–95 x 45 µm. **Pappus** white, 2.5–3.2 mm L, 1–2 rows, indehiscent; setae free at base or very shortly fused, pappus 1–2(3) cells wide, cell width 5–10 µm, no basal cilia; barbellate, barbs 50 µm L, barb base shortly to medium adhered (40-50%), free barb appressed, 10–13 barbs/100 µm.

Distribution and habitat. *C perpusilla* forms scattered populations at higher elevations 2200–3300 m.a.s.l. between 18°00'–20°30'S latitude.
Differential diagnosis. *C. perpusilla* is close to *C. depauperata*. *C perpusilla* has an open, campanulate capitulum, and just a few involucral bracts, all of similar size. *C. depauperata* has a narrow, elongated capitulum with several rows of bracts of increasing size. Key features distinguishing this species from other similar taxa are laid out in Table 12, page 133.

Material seen. – **Bolivia** Dept. de La Paz La Lancha, Bolivie Septentrionale. Ravin de Chuquiaguillo, 1851, *Weddell s.n.* (P). – **Chile. I Región de Tarapaca** Prov. de Arica: Pachica im Tal des Río Camarones, Wüste Atacama, 2200m, 04.1927, *Troll 3286* (B M). – Camino de Zapahuira a Putre, Quebrada Aroma, 3300 m, 05.1972, *Ricardi* et al. *139* (CONC). – Camino Zapahuira a Putre, Quebrada Socoroma, 3100m, 05.1972, *Ricardi* et al. *166* (CONCx2). – Quebrada de Zapahuira, Cuesta Chapiquiña, 2850 m, 03.1961, *Ricardi* et al. *74* (CONC). – Cuesta de Chapiquiña, 3250 m, 03.1961, *Ricardi* et al. *126* (CONC). – Prov. de Parinacota Camino Cuesta de Cardones a Putre, 3100 m, 02.1984, *K. Arroyo 520* (CONC). – Cerca de Putre, 3300 m, 03.1984, *K. Arroyo 585* (CONC).

IX *Chaetanthera* Ruiz & Pav.

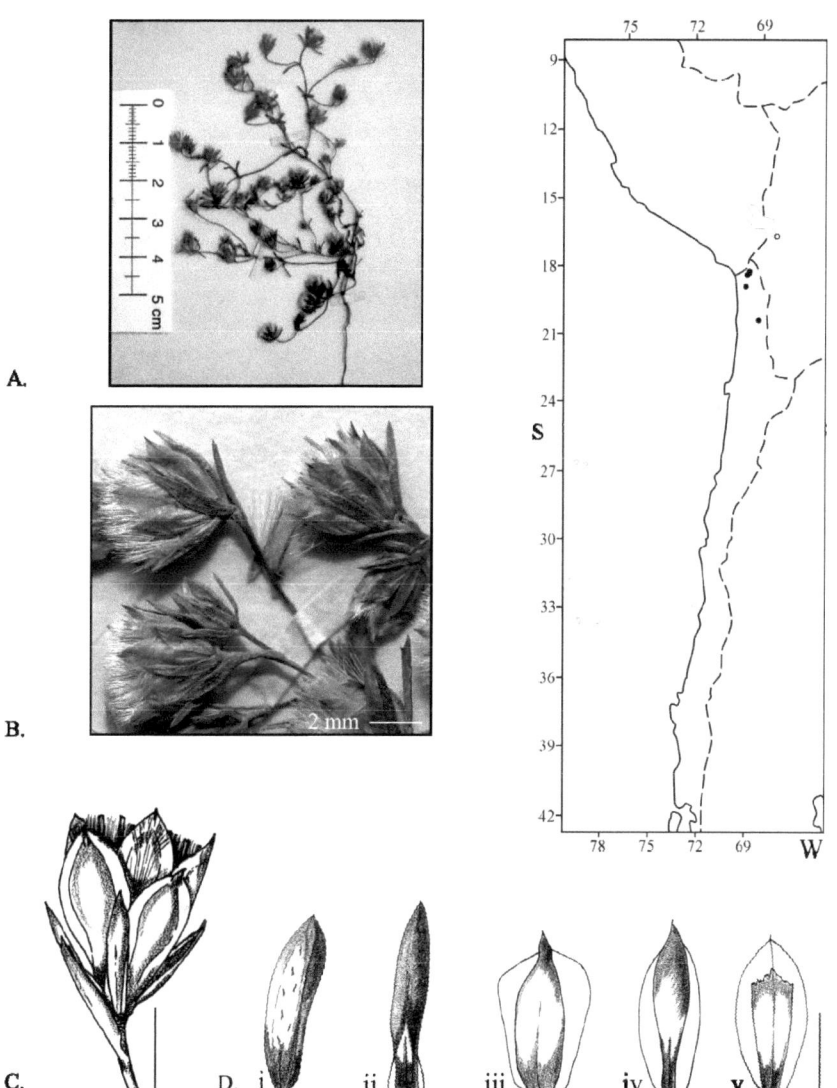

Figure 61: *Chaetanthera perpusilla*. **A.** *Ricardi* et al. *166*. Habit, from herbarium sheet. **B.** *Ricardi* et al. *139* Capitulum detail, photo. **C.** Capitulum detail, sketched. **D.** Leaf & Bract detail. **i.** Stem leaf v.s.; **ii.** OIB d.s.; **iii – iv.** MIB d.s.; **v.** IIB d.s. Scale bar B – D = 2 mm.

15. *Chaetanthera peruviana* A. Gray Proc. Amer. Acad. Arts 5: 144. 1861.

Typus PERU "Andes of Peru above Baños" [Peru, Depto. Lima, Prov. Canta, Baños, *Capt. Wilkes Exped. s.n.*] Holotypus GH! Isotypus F! K!

= *Chaetanthera chiquianensis* Ferreyra Publ. Mus. Hist. Nat. "Javier Prado", Ser. B, Bot. 6: 5 – 7, tab. 3, fig. 1-11. 1953. Typus PERU "Peru, Ancash, Cerro al sur de Chiquián, Provincia Bolognesi, 22.4.1952, 3500 m, *Cerrate 1323*" Holotypus Herb. Cerrate, Lima! Isotypus MO! USM!

Annual monoecious herb. **Roots** filamentous. **Stems** caespitose to erect, branched, (3)10–30 cm, erect, slightly pubescent or pubescent at base but rapidly glabrous. **Leaves** (6)10–22 x (2)3–8 mm, upper leaves smaller (5.5–8 x 2–2.8 mm), opposite then alternate to capitulum; sessile, attenuate, ovate-lanceolate to spathulate, 3–7-dentate, margins somewhat thickened, revolute and pilose, apices mucronate; indumentum sericeous pubescent ventrally and margins, dorsally glabrous; hairs compound two-celled: one small 20–30 µm L basal ± inflated cell, one long 60–110 µm L filamentous cell. **Capitula** open campanulate, solitary, terminal or axillary, shortly pedunculate. **Involucral bracts** imbricate, arranged in three types, initially foliaceous then reduced to entirely membranous. **Outer involucral bracts** 4–7 x 1.5–2.2 mm; as leaves but with short membranous alae to less than ½ height of bract, **Middle involucral bracts** 4–7 mm L; membranous alate, ratio of lamina to alae decreases from ⅔ to nearly entirely alate; thickened or inrolled margins with fewer (3) teeth; indumentum long dorsal filamentous hairs. **Inner involucral bracts** 4.9–8 mm L, entirely membranous alate; ovate–lanceolate; apices acute, shortly mucronate, blue-green maculate, striate. **Ray florets** 6–16, pistillate, creamy to pinkish white, pistillate, dorsally sparsely sericeous above. Corolla 4.5–6 mm L, corolla tube 2.5–4 mm, outer ligule 0.7–1 mm W., glabrous on dorsal surface, inner ligule shortly bifid. **Disk florets** 5–20, bisexual, yellow; corolla 2.5–4.5 mm L corolla tube 2.5–5.4 mm. **Styles** [ray] 3.3 mm L, stigma lobes 0.3 mm L [disk] 3.2 mm L, stigma shortly lobed. **Anthers** [ray] sterile, < 3 mm L [disk] 2–2.5 mm (incl. filaments 3.4 mm L). **Achenes** brown, ca. 2 mm L, turbinate; carpopodium anular irregular; pericarp pellucid, densely covered in lanceolate twin hairs, 80–120 x 40 µm. **Pappus** white, ca. 2.5–4(5) mm L, 1–2 rows, indehiscent; setae fused at base, pappus 3–5 cells wide, cell width 10–15 µm, no basal cilia; barbellate, barbs (50)95–140 µm L, barb base medium adhered (50%), free barb appressed, (9)11–15 barbs/100 µm.

IX *Chaetanthera* Ruiz & Pav.

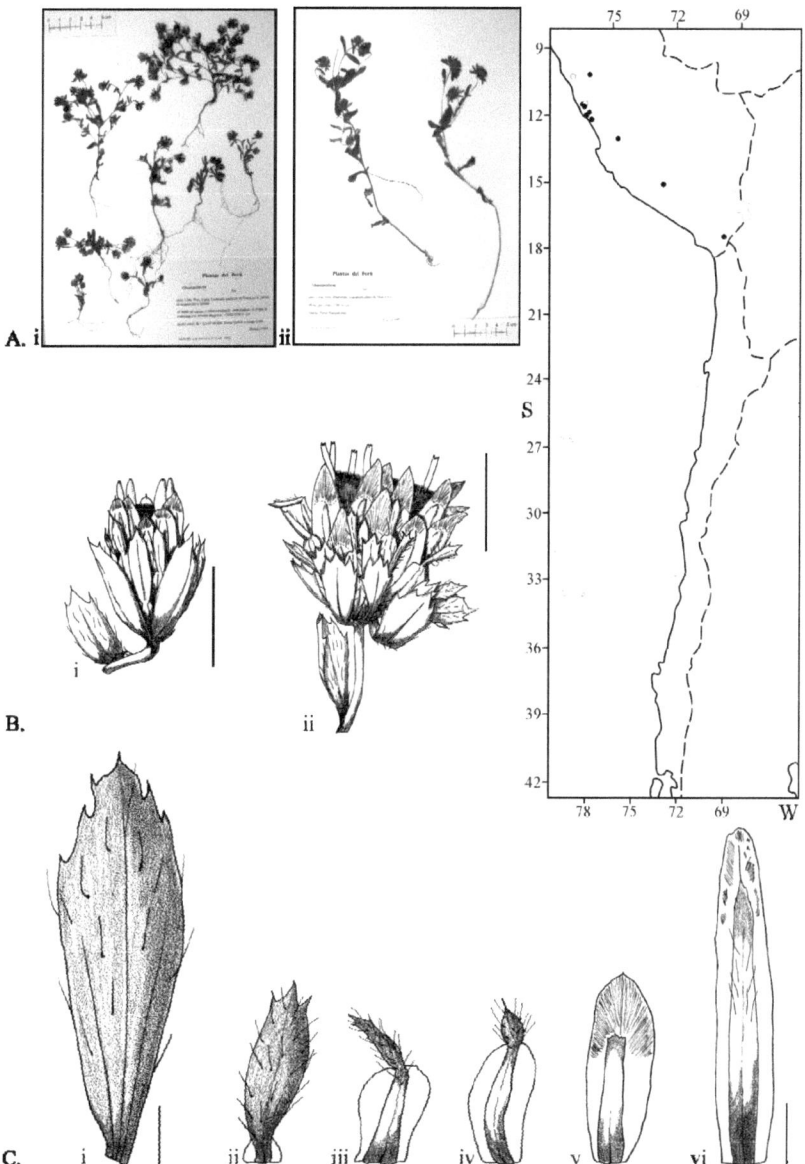

Figure 62: *Chaetanthera peruviana*. **A.** Herbarium sheet, habit. **i.** *Paucar Granda 1433*; **ii.** *Cerrate 1757*. **B.** Capitulum detail, scale bar = 5 mm **i.** *Paucar Granda 1433*; **ii.** *Cerrate 1757*. **C.** Leaf & bract detail d.s. *Paucar Granda 1433* **i.** stem leaf; **ii.** OIB; **iii – iv.** MIB; **v – vi.** IIB. Scale bar i & ii – vi = 2 mm.

Distribution and habitat. *C. peruviana* occurs on the western flanks of the Peruvian Andes, between latitudes 10°00' – 17°30'S. It is found on the lower altitude limit of the Puna vegetation at elevations between 2700 – 3800 (4100) m.a.s.l. It occurs in the "prados" (meadows) in sunny, stony patches between tussocks, in higher matorral, or in dry quebradas.

Differential diagnosis. Formerly considered to be two species, *C. peruviana* and *C. chiquianensis*, no morphological evidence to support this was found during this study. *C. peruviana* is considered here to be a widely distributed polymorphic entity. The key character used to define *C. chiquianensis*, the presence of glandular hairs, was found to be very variable. Furthermore, the glandular hairs themselves are part of a fragile, compound trichome. It is formed of a basal glandular cell and a long filamentous hair. The long hair is very fragile and is easily destroyed. Consequently, the indumentum appears to be only glandular, or even glabrous. This hair type was also observed on the Chilean species *C. ciliata* and *C. multicaulis*.

Material seen. – **Peru. Depto. Castro-Virreyna** Huancavelica, above Huaytara, 2700 – 2800m, 05.1910, *Werberbauer 5419* (F GH). – **Depto. Arequippa** Prov. La Union Encima de Olca, Quebrada Catahuasi, valley of Cotahuasi above Alca, 3700 – 3900m, 03.1914, *Werberbauer 6876* (CONC F GH). – **Depto. Tacna** Prov. Tarata Cerros al SE de la Cordillera del Barroso, 4000 – 4270 m, 03.1998, *La Torre 2197* (USM). – **Depto Lima.** Prov. Canta Lachaqui, arriba de Toma, en el camino de ascenso hacia Quinán. 3800 m.s.m. 05.1995, *Vilcapoma 4253* (PERU). – Lachaqui, quebrada de Toma en el camino de ascenso hacia Quinán, 3600-3700, 05.1995, *Granda Paucar 1433*. – en camino a Chinchán, 3430 m, 04.1995, *Vilcapoma 4242* (PERU). – Arriba del pueblo de San José, camino a Huamantanga, 2450 m, 21.05.1999, *Granda Paucar 2228* (PERU). – **Depto. Lima.** Canta, Huanantaya, sandy soil, *Hooker* (with *548*) (K). – Prov. Huarochirí Huanca a 1 km al SE de Santiago de Tuna, 2700 m, 05.1974, *F. Encarnación 367* (PERU, USM) – Llucanchi, abajo de Huarochiri, 3300 m, 05.1953, *Cerrate 1757* (PERU). – Matucana, 8000 ft, 04-05.1922, *Macbride & Featherstone 197* (F PERU USM). – **Depto Ancash** Prov. Bolognesi Cerro al sur de Chiquián, 3500 m, 04.1952, *E. Cerrate 1323* (MO PERU USM).

16. *Chaetanthera ramosissima* D. Don ex Taylor & Phillips Philos. Mag. Ann. Chem. 11: 391. 1832 (March – April).

Lectotypus hic loc. designatus "Chile, V Región de Valparaíso, Prov. de Valparaíso, Valparaíso, "*C. ramossisima*", *Cuming 656*" K! Isotypus BM! K! OXF s.n.! non Valparaíso, *Cuming 657* (BM Ex2 K P W).

= *Chaetanthera tenella* Less. Syn. Gen. Compos. pp. 114. 1832 (July – August). Typus CHILE "In Chile legit Bertero" [In fruticetis saxosis pascuis sterilibus collium secus flumen Cachapual, Rancagua et Quillota, 08.1828-9, *Bertero 162 ± 910*] **Lectotypus hic loc. designatus** M! Isotypus BM! CONC! F! G! MO! NYBG! P! SGO 76582! W!
= *Chaetanthera kunthiana* Less. Syn. Gen. Compos. pp. 114. 1832 (July – August). Typus CHILE "In Chile Bertero" [Prope la Punta de Cortés, Rancagua, 08.1828, *Bertero 717*] Isotypus P!
= *Chaetanthera berteriana* A. Colla non Less. Mem. Acad. Soc. Torino 38: 21, t. 27, fig. 1. 1833. Typus CHILE "Hab. Chili, Rancagua" [In fruticetis saxosis pascuis sterilibus collium secus flumen Cachapual, Rancagua et Quillota, 08.1828-9, *Bertero 162 ± 910*] Holotypus TO? Isotypus BM! G! GH! P! M! NY! P! SGO (photo)! W!

Nomenclatural notes The protologue of *Chaetanthera ramosissima* identifies it as: "*C. ramosissima*, involucri squamis subulatis tridentatis, radiis involucro ter brevioribus, caule erecto." *C. ramosissima* and *C. moenchioides* have very similar involucral bracts, being 3-dentate and 3 – 5-dentate respectively. The ray florets of *C. ramosissima* are never exerted beyond the involucral bracts while those of *C. moenchioides* are conspicuously exerted. *C. ramosissima* stems are erect, while the stems of *C. moenchioides* are ascending. Although Don specified no type, type material was almost certainly collected by J. Gillies or H. Cuming. There are several potential extant collections of Cuming's: *Cuming 656 & 657* and *Cuming 231 & 240*. These are all mixed collections of *C. ramosissima* and *C. moenchioides*. The citation of *Cuming 856* and *857* by Hooker & Arnott (1835) is an error. Noted in erratum by Cabrera (1937) as a synonym of *C. tenella*, the significance of the name *C. ramosissima* D.Don was belatedly recognised by D. J. Mabberley (1981). The selection of a lectotype for *Chaetanthera tenella* Less. was necessary because Lessing worked out of Berlin, and the material he worked with was destroyed. Several syntypes are cited. The inscription on the labels always includes both numbers (viz *162* and *910*), and also all three Typus localities (Cachapual: 34°20'S 71°15'W; Rancagua: 34°10'S 70°46'W; Quillota: 32°52'S 71°15'W).

Annual monoecious herb. **Roots** filamentous. **Stems** erect, to 25 cm, with many (10–15, sometimes more) simple spreading flowering branches, ca. 8 cm L **Leaves** 22 x 3 mm, linear to narrowly oblanceolate, sparsely dentate at apex (± 6 teeth); midrib visible; indumentum sericeous on dorsal and ventral surfaces; leaves below capitulum 7 x 0.8 mm, margins folded inwards; indumentum on ventral surface only. **Capitula** narrowly urcinate, monocephalous on branches. **Involucral bracts** imbricate, arranged in three types, initially foliaceous then reduced to entirely membranous. **Outer involucral bracts** 4.9–6 mm L; as leaves but with short membranous alae to less than ½ height of bract; lamina linear, 3-dentate, part reflexed, ventrally pubescent; alae ovate, dorsally pubescent, hairs to >1 mm L; apical mucro recurved, revolute. **Middle involucral bracts** 4.6–4.9 mm L; membranous alate, ratio of lamina to alae decreases from ⅔ to nearly entirely alate, alae truncate.

Inner involucral bracts 5.5–6.8 mm L, linear lanceolate, entirely membranous alate; proximally dorsally pubescent; apices acute, longly mucronate, maculate striate. **Ray florets** ca. 10, pistillate, yellow; corolla 5.7 mm L, corolla tube 3.7 mm L, outer corolla lip 0.5 mm W., inner lip 1.9 mm L, bifid, filamentous. **Disk florets** ca. 10? bisexual, yellow; corolla 4.3 mm L, corolla tube 3.6–4 mm L **Styles** [ray] 4.6 mm L, lobes 0.3 mm. [disk] 4.3 mm L **Anthers** 1.9 mm L (4.3 mm incl. filaments). **Achenes** brown, 1.9 mm L, turbinate; carpopodium narrow, anular irregular, variable within capitula; pericarp pellucid, densely covered in lanceolate twin hairs, 70–95 x 45 µm. **Pappus** white, 4.3–4.6 mm L, 1–2 rows, indehiscent, fused at base with corona; pappus width 1–2(3) cells, cell width 5–10 µm, no basal cilia; barbellate, barbs 50 µm L, barb base shortly adhered (<40%), free barb appressed, 7–8 barbs/100 µm.

Distribution and habitat. *C. ramosissima* is endemic to Chile and is found in the low Cordillera to the north, west and south of Santiago and the Cordillera de la Costa of Valparaíso, and along the lower Andean foothills to Talca between latitudes 33°00'–36°00'S. It occurs at elevations between 120–1700 m.a.s.l. It grows in grassy or rocky areas in open matorral and on slopes.

Differential diagnosis. *C. ramosissima* is very similar to *C. moenchioides* and *C. taltalensis*. Key features in identifying the species are laid out in Table 13, page 155.

Material seen. – **Chile.** **V Región de Valparaíso** Prov. de Valparaíso Viña de la Mar, *Bridges 125* (BM E W). – Viña de la Mar, *Bridges 124* (E K). – Marga-Marga, , 400 m, 11.1933, *Behn s.n.* (CONC SGO). – Valle de Marga-Marga, 10.1910, *Jaffuel 1262* (GH). – Laguna Verde, 120 m, 11.1930, *Garaventa 2132* (CONC GH). – Prov. de Quillota PN La Campana, Cerro Guanaco, 870 m, 11.1999, *K. Arroyo et al. 993942* (CONC). – PN La Campana, Sector camino a la mina, 810 m, 10.2000, *K. Arroyo et al. 205039* (CONC). – PN La Campana, Sector Ocoa, camino a la cascada, 750 m, 10.2000, *K. Arroyo et al. 202924* (CONC). – PN La Campana, Sector Cajón Grande, 1670 m, 11.1999, *K. Arroyo et al. 973773* (CONC). – Cuesta La Dormida, 929 m, 11.1999, *K. Arroyo et al. 993730* (CONC). – Cerro Campana, 1500 m, 11.1947, *Bultmann s.n.* (CONC). – Cerro Campana, 1400 m, 11.1936, *Garaventa 6585* (CONC). – Prov. de Los Andes Cuesta Chacabuco, 1150 m, 11.2002, *Ehrhart et al. 2002/009* (MSB). – 10.1964, *Schwabe 046* (B). – **Región Metropolitana de Santiago** Cer. S. Ber., "*Chaetanthera tenuis*" *Philippi s.n.* (K). – Prov. de Chacabuco Caleu, 1100 m, 01.1969, *Castillo 69877* (CONC). – Tiltil, camino a la Capilla, 840 m, 11.1941, *Behn s.n.* (CONC). – Cuesta La Dormida, lado oeste, 910 m, 09.2002, *K. Arroyo 25006* (CONC). – Batuco, 450 m, 10.1954, *Navas 1763* (CONC). – Colina, 900 m, 10.1956, *Gunckel 37899* (CONC). – La Dehesa, 1200 m, 11.1951, *Gunckel 26054* (CONC). – Prov. de Melipilla Südabfall der Cuesta barriga, 550 m, 11.1980, *Grau 2407* (MSB M). – RN Roblería de Cobre de Loncha, Cuesta La Cardita, 980 m, 11.2000, *K. Arroyo et al. 205945* (CONC). – RN Roblería de Cobre de Loncha, Paso de La Canal, 940 m, 12.2000, *K. Arroyo et al. 206474* (CONC). – RN Roblería de Cobre de Loncha, Quebrada del Llivi-Llivi, 980 m, 11.2000, *K. Arroyo et al. 206027.3* (CONC). – RN Roblería de Cobre de Loncha, Sector Los Maquis, 560 m, 11.2000, *K. Arroyo et al. 203538* (CONC). – Prov. de Santiago San Cristóbal, 11.1857, *Philippi s.n.* (SGO). – Salto de Conchalí, 11.1876, *Herb. F. Philippi 1072a* (SGO). – Santiago, *Ball s.n.* (E). – Santiago, *Gay 222* (P). – *Philippi s.n.* (B, W). – In collibus prope Santiago, 11.1855, *Philippi 481* (B BM G K P W). – Andes de Sa. Rosa in lapidosis ad Río Colorado, 12.1827, *Poeppig s.n. D.595* (W). – Río Colorado, 12.1827, *Poeppig 110(62) D.594* (F). – Cerro Manquehue, 1700 m, 12.1976, *Elgueta s.n.* (CONC). – Curacavi, 550 m, 11.1951, *Frödin 93* (BM). – Agua del León, Conchali, 1100 m, 11.1927, *Looser 667* (GH, M). – Quebrada de Peñalolen, 720 m, *Bravo 254* (CONC). – Rinconada de Lo Cerda, 500 m, 11.1960 *Schlegel, 3188* (CONC). – Valle Macul, 850 m, 11.1932, *Grandjot s.n.* (CONC). – Quebrada de Macul, 850 m, 10.1953, *Gunckel s.n.* (CONC). – Las Vertientes, 800 m, 11.1955, *Gunckel 28966* (CONC). – Clarillo, 800 m, 11.1962, *Gunckel 39458* (CONC). – RN Río Clarillo, Quebrada El Almendro, 1000 m, 11.1999, *K. Arroyo et al. 994196* (CONC). – Prov. Cordillera subida a Lagunillas, lado del camino, 1500 m, 12.1992, *von Bohlen 1498* (SGO). – Lagunillas en canchas de esqui arriba del pueblo, 1380 m, 09.2002, *K. Arroyo et al. 25061; 25064* (CONC). – **VI Región del Libertador G. B. O'Higgins** Cuchacucha, Exped. Malaespina, "*Chaetanthera filiformis* Lag (sig)" *Nee s.n.* (MA). – Prov. de Colchagua Puente de Cimbra, Río Tinguirririca, 700 m, 11.1929, *Looser 1093* (GH). – Lado norte del Cerro Centinela a 4 Km al sur de San Fernando, 450 m, 10.1973, *Stebbins 8728 pro parte* (SGO). – Prov de Cachapoal Rancagua, 11.1828, *Bertero s.n.* (CONC). – Colchagua, 2000 - 3000'S.m., 12.1860, *Landbeck s.n.* (SGO). – Termas de Cauquenes, 700 m, 11.1952, *Pfister s.n.* (CONC). – Termas de Cauquenes, 700 m, 11.1965, *Mahu 1566*; *4085* (CONC33303; CONC 33258). – Termas de Cauquenes, 11.1965, *Mahu 1521*

IX *Chaetanthera* Ruiz & Pav.

(SGO). – Cerro Nicunlauta, 360 m, 10.1925, *Montero 1770* (CONC). – PN Cocalán, 470 m, 11.2001, *K. Arroyo* et al. *211750* (CONC). – PN Cocalán, 800 m, 11.2001, *K. Arroyo* et al. *211634* (CONC). – **VII Región del Maule** Prov. de Curicó Cerro Condell, 290 m, 09.1925, *Barros 992* (CONC). – Camino entre Los Queñes y Paso Vergara, 900 m, 10.2002, *K. Arroyo 25123* (CONC). – Prov. de Linares RN Bellotos del Melado, 900 m, 12.1999, *Arroyo* et al. *994903* (CONC). – Prov. de Talca Cordilleras de Maule, 1856, *Philippi s.n.* (P, W). – Cordilleras de Maule, 1855, *Philippi s.n.* "*Minythodes umbellata* Ph." (BM F Gx2 K P). – **S.l.d.** *Cuming 655* (E). – Cordilleras of Chili, *Cuming 240* (BM Ex2 GH K W).

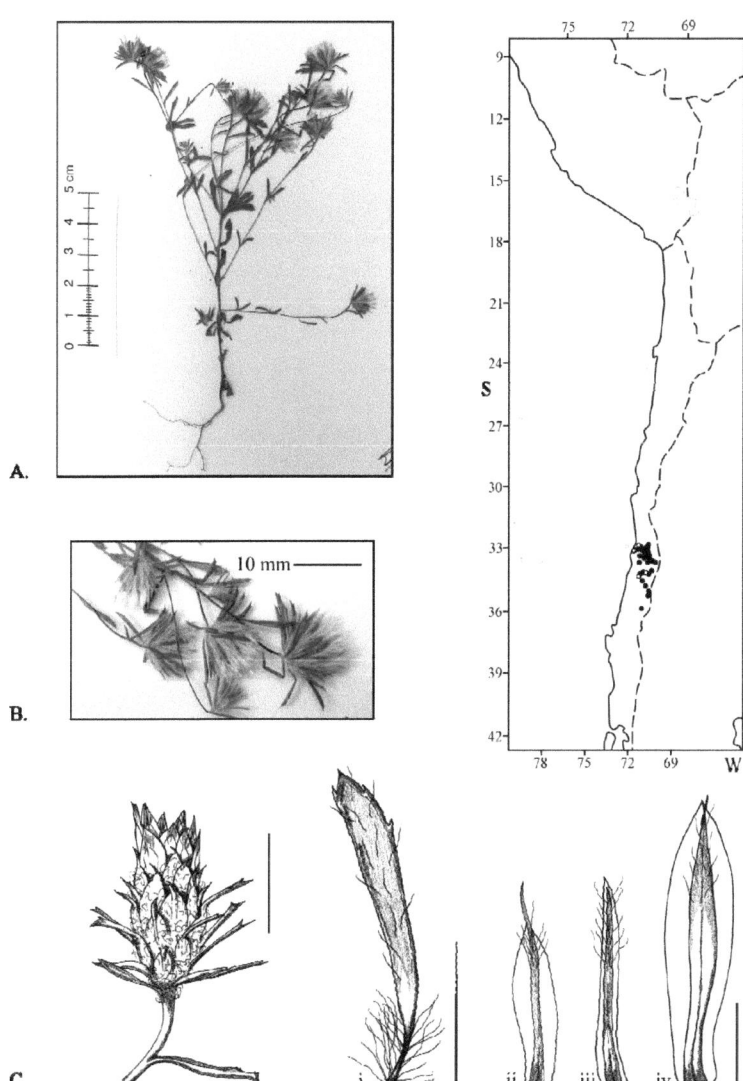

Figure 63: *Chaetanthera ramosissima. K. Arroyo 994196.* **A.** Habit, herbarium material. **B.** Capitulum detail, scale bar = 10 mm. **C.** Capitulum detail, scale bar = 5 mm. **D.** Leaf & Bract detail. **i.** Stem leaf v.s.; **ii.** MIB d.s.; **iii – iv.** IIB d.s. Scale bar i = 5 mm; ii – iv = 2 mm.

17. *Chaetanthera* x *serrata* Ruíz & Pav. Syst. Veg. Fl. Peruv. Chil. 190 – 191. 1798.

Typus CHILE "Habitat in arenosis Conceptionis Chile et in Rere Provinciae inter praecedentem, *Pavon*" Holotypus BM [fide Cabrera 1960]! "Aster sp. nov. ad Concepcion, *Ruiz s.n.*"

= *Proselia serrata* (Ruiz & Pav.) D. Don Trans. Linn. Soc. London 16: 235. 1830.
= *Chaetanthera valdiviana* Phil. Linnaea 28: 712. 1856. Typus CHILE "Valdiveae in praedio meo S. Jaun legi" Holotypus SGO #64726! Isotypus probabilis "In provincia Valdiviensi, Jan. 1851, Philippi 488" BM!

Perennial monoecious caespitose herbs. **Roots** are short woody rootstocks and/or creeping stolons. **Stems** short compact, suppporting densely to loosely leafy clumps (pseudo-rosettes) of leaves. **Leaves** (15)18–43(50) x (1.0)1.3–2.2(5.0) mm, linear lanceolate, acute, often plicate, alternate, shortly to longly serrate at the apices; lightly to densely pubescent (hairs ~2 mm L) on dorsal and ventral surfaces and in axils. **Pedicel** (scape) erect to ascending, (1.0)6–13.5(17.0) cm L; leaves becoming smaller and infrequently spaced. **Capitula** cylindrical to campanulate, 10–12(15) x 8–14–22 mm, sessile. **Involucral bracts** imbricate, arranged in three types, initially foliaceous then reduced to entirely membranous. **Outer involucral bracts** 1–2 rows, (5.0)7.6–11.2 x 1.2–2.1 mm; as leaves but with short membranous alae to less than ½ height of bract, alae linear ovate to truncate. **Middle involucral bracts** rare, only one or two transitional bracts; membranous alate, ratio of lamina to alae decreases from ⅔ to nearly entirely alate. **Inner involucral bracts** linear lanceolate, 11–18.7 x 0.9–1.8 mm, 2-3 rows increasing in size, entirely membranous alate; apices acute to longly acuminate, black-green or brown maculate, mucronate. **Ray florets** 9–12, pistillate, yellow with pink-red dorsal stripe, dorsally sericeous pubescent; corolla 16–20 mm L, corolla tube 3–5 mm L, outer ligule 2.5–3 mm W., inner corolla ligule 3+2 mm (upper part bifid, tightly twisted and curled). **Disk florets** bisexual, yellow; corolla 8.5 mm L, corolla tube 6 mm L, coloured tips **Styles** [ray] 10–11 mm L, stigma lobes ca. 1 mm. [disk] **Anthers** green tipped (not seen at anthesis).**Achenes** brown, 2–4.4 mm L, angled turbinate; poorly developed carpopodium (seldom absent), variable within capitula; pericarp pellucid, densely covered in lanceolate twin hairs 70–95 x 45 µm. **Pappus** white, ca. 9 mm L, 1–2 rows, indehiscent; setae fused at base (corona 1–1.5 mm high), cell width 5–10 µm, no basal cilia; barbellate, free barb appressed, 5–6 barbs/100 µm.

IX *Chaetanthera* Ruiz & Pav.

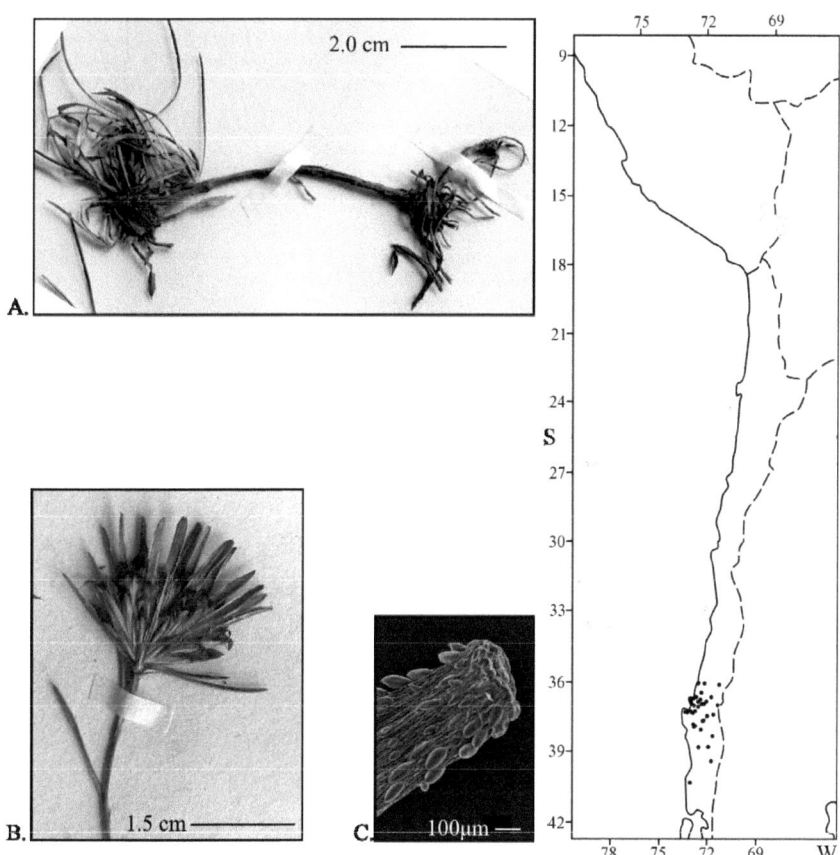

Figure 64: *Chaetanthera* x *serrata. Grau 2998* **A.** Stoloniferous habit detail from herbarium sheet. Scale bar = 2 cm. **B.** Capitulum detail. Scale bar = 1.5 cm. **C.** S.E.M. image of disk carpopodium. Scale bar = 100 μm

Distribution and habitat. It is found in the coastal mountains from Concepción and inland to the central valley on the loose red sandy soils, dry meadows, and degenerate woodlands to Valdivia between latitudes 36°30' – 40°00'S. It occurs at all elevations up to 800 m.s.a.l.

Differential diagnosis. The main characterising feature of this hybrid is the stoloniferous habit and the poorly developed carpopodium. Key is that specimens neither quite fit *C. chilensis*: with narrow (< 3mm wide) densely sericeous dentate grey-green leaves and vegetative parts, and narrowly cylindrical capitula (< 15 mm diam.) nor *C. elegans*: with broader (> 3 mm) lightly pubescent to glabrescent, spinulose, bright green leaves and other vegetative parts and broader, more campanulate capitula.

Material seen. – **Chile. VII Región del Maule** Prov. de Cauquenes 10 Km al Sur de Curanipe, 5 m, 01.1982, *Ugarte 232* (CONC). – Fundo El Boldo, 150 m, 10.1975, *Rodriguez 769* (CONC). – Prov. de Linares Camino Carrizal-Cajon de Pejerreyes, Km 37, 725 m, 02.1987, *Montero 22* (CONC). – **VIII Región del Bio Bió** Prov. de Concepción Cerro Caracol, 200 m, 02.1941, *Behn F s.n.* (CONC). – Hualqui, Pichaco, 200 m, 01.1937, *Junge s.n.* (CONC). – Laraquete, Las Cruces, 25 m, 03.1936, *Junge s.n.* (CONC). – Parque Hualpen, 60 m, 02.1981, *Ugarte s.n.* (CONC). – Cerro Caracol, 200 m, 01.1940, *Pfister s.n.* (CONC). – Fundo Trinitarias, 100 m, 12.1934, *Pfister s.n.* (CONC). – San Pedro, 20 m, 02.1951, *Pfister s.n.* (CONC). – Talcahuano, Parque Hualpen, 60 m, 01.1970, *Carrasco 354* (CONC). – Chile austral, colles graminos apric. circum Talcahuno 04.1828 *Poeppig s.n.* (W). – Aster sp. nov. de la Concepción de Chile, *Pavon s.n.* (BM). – Concepción, *King s.n.* (BM). – Concepción, *Miers s.n.* (BM). – Concepción, *Lesson & Durville s.n.* (P). – Concepción, *Dombey s.n.* (P). – Cerro La Tolvona, 01.1949, *Junge 2728* (GH). – Cerros Laguna Redonda, 20 m, 01.1935, *Junge s.n.* (CONC). – Concepción, 03.1937-38, *Pfister s.n.* (B). – Concepción, *Dombey s.n.* (P). – Straße Concepción-Bulnes, Urwald am Aserrado bei km 36, 02.1981, *Grau 2991* (MSB). – Camino de Florida a Yumbel, 200 m, 01.1944, *Pfister 739* (CONC). – Camino de Florida a Yumbel, 200 m, 01.1944, *Pfister s.n.* (CONC). – Camino de Florida a Yumbel, 200 m, 12.1946, *Pfister s.n.* (CONC). – Hualpén, 02.1940, *Junge 2286* (GH). – Hualpén, 25 m, 02.1940, *Junge s.n.* (CONC). – Camino de Hualqui a Rere, 220 m, 01.1959, *Marticorena et al. s.n.* (CONC). – östlich Hualqui (Richtung Yumbel), 07.11.1981, *Grau 2998* (MSB, M). – Parque Hualpén, 60 m, 01.1970, *Carrasco 377* (CONC). – Parque Hualpén, 60 m, 02.1981, *Ugarte s.n.* (CONC). – Parque Hualpén, 60 m, 03.1980, *Ugarte 128* (CONC). – Parque Hualpén, 60 m, 03.2003, *K. Arroyo et al. 25131* (CONC). – Puente Queime, 75 m, 12.1967, *Parra & Rodriguez 102* (CONC). – 4.4 Km E Junta Camino San Rafael-Coelemu, 480 m, 01.1979, *Solomon J & A 4401* (CONC). – Tome, Lomas de Cocholgue, 70 m, 01.1935, *Junge s.n.* (CONC). – Santa Juana, 60 m, 03.1978, *Oehrens s.n.* (CONC). – Prov. del Bio Bió Fundo Santa Olga, 265 m, 02.1954, *Gautier s.n.* (CONC). – Abanico, El Canelo, 750 m, 01.1982, *Montero 12165* (CONC). – El Abanico, 800 m, 02.1951, *Pfister s.n.* (CONC). – El Abanico, 800 m, 02.1968, *Zöllner 3037* (CONC). – Antuco, 650 m, 12.1946, *Morales s.n.* (CONC). – Camino de Los Angeles a Mulchen, 150 m, 02.1951, *Pfister s.n.* (CONC). – Camino entre Los Angeles y Mulchen, a 2 km al sur del puente Duqueco, 150 m, 10.2002, *K. Arroyo et al. 25070* (CONC). – Prov. de Ñuble – Bureo, 250 m, 02.1926, *Barros s.n.* (CONC). – Cerro Cayumanqui, 200 m, 12.1977, *Oehrens s.n.* (CONC). – Cerro Ninhue, 600 m, 02.2000, *Le Quesne s.n.* (CONC). – Diguillin 650m 01.1934, Pfister *s.n.* (CONC). – General Cruz, 95 m, 12.1913, *Stuardo s.n.* (CONC). – Guarilihue, 150 m, 02.1977, *Quezada 188* (CONC). – Nueva Aldea, Sta Lucia, 25 m, 12.1936, *Behn s.n.* (CONC). – Puente Meco, camino de Chillán a Pemuco, 160 m, 11.2002, *K. Arroyo et al. 25158* (CONC). – Prov. de Arauco Arauco, 5 m, 02.1977, *Werlinger 85* (CONC). – Desembocadura Río s Tubul & Raqui, 15 m, 12.1949, *Ricardi s.n.* (CONC). – Camino Contulmo-Cañete, 50 m, 02.1961, *Ricardi & Matthei 5340* (CONC). – Contulmo, 10 m, 01.1919, *Behn K s.n.* (CONC). – Los Alamos, 170 m, 11.1955, *Montero 4859* (CONC). – Costero de Arauco – Lebú, "Ch. glabra", 02.1885, *Borchers s.n.* (BM). – Lebú, camino 8 km despues del Cruce a Tubul, 90 m, 01.2001, *Lopez, Finot & Torres 2175* (CONC). – **IX Región de La Araucanía** Prov. de Malleco Termas de Tolhueca, 1000 m, 01.1947, *Gunckel 16531* (CONC). – Collipulli, Convento San Francisco, 250 m, 02.1947, *Ricardi s.n.* (CONC). – Lolco, 800 m, 02.1972, *Zöllner 5519* (CONC). – Collipulli, 150 m, 02.1947, *Ricardi s.n.* (CONC). – PN de Nahuelbuta, 800 m, 03.1973, *Rodriguez & Torres s.n.* (CONC). – Prov. de Cautín Puerto Saavedra, 10 m, 01.1951, *Aravena 36* (CONC). – Puerto Saavedra, 10 m, 12.1919, *Hollermayer 160* (CONC). – Temuco, Cerro Ñielol, 180 m, 03.1937, *Montero 3091* (CONC). – Cerro Ñielol, 150 m, 03.1976, *Montero 10132* (CONC). – Cherqueco, 520 m, 03.1933, *Montero 1230* (CONC). – Truf-Truf, 110 m, 01.1958, *Montero 5593* (CONC). – Cerro Ñielol, 180 m, 02.1941, *Gunckel 9787* (CONC). – **X Región de Los Lagos** Prov. de Valdivia Valdivia, *Philippi s.n.* (B, W). – In arvis sterilibus Daglipulli, 01.1835, *Gay 379* (P). – In provincia Valdiviensi, La roble Trumedatha, 01.1851, *Philippi 488* (B BM GH W). – Hacienda San Juan, 07.1830, *Ochsensis s.n.* (GH). – In collibus apricis, San Juan, 02.18?? *Philippi 359* (P). – Prope S. Juan in pr. Valdiviensi, 02.18??, *Philippi 359a* (P#141350, #141300 p.p.). – La Hacienda de Guit. Privina of Valdivia, *Bridges 604* (E). – Valdivia, "Ch. valdiviana Ph." *Ball (Philippi) s.n.* (E). – In collibus apricis, Valdiviae, *Philippi s.n.* (BM). – San Juan Feb.1898?, *Philippi s.n.* (BM).

18. *Chaetanthera taltalensis* (Cabrera) A.M.R. Davies stat. nov.

≡ *Chaetanthera tenella* var. *taltalensis* Cabrera Revista Mus. La Plata, Secc. Bot. 1: 192. 1937. Typus CHILE "Chile, Antofagasta, Taltal, 200 m, X.1925, *Werdermann 815* (en parte)" Holotypus LP 67038 Isotypus Bx2! BM! E! F! G! GH! K (en parte)! MO! NYBG (p.p.)! SI

Nomenclature Notes According to the ICBN 2000, the correct name for any taxon below the rank of genus is the combination of the final epithet of the earliest legitimate name of the taxon in the same rank. (§11.4). The varietal epithet "taltalensis" was first ascribed by Johnston (1929) (*Chaetanthera linearis* var. *taltalensis* I. M. Johnst.). *Chaetanthera tenella* var. *taltalensis* Cabrera (1937), although described later, is legitimate (§53.4, Note 2.: the same final epithet may be used in names of infraspecific taxa within different species). A name does not have priority outside its own rank (§11.2) thus *Chaetanthera taltalensis* may be correctly applied here.

Annual monoecious herb. **Roots** filamentous. **Stems** short (< 10 cm) erect, branching, sparsely leafy, floccose lanate below capitula. **Leaves** 7–12 x (1)1.5–2 mm, basal leaves form rosette that senesces during season, upper leaves smaller; leaves alternate, midrib visible, apical mucro plus 1 pair teeth towards tip on inrolled surface; margins thickened, slightly involute; indumentum on ventral surface floccose lanate. **Capitula** campanulate, shortly pedunculate, disk diameter ca. 5 mm. **Involucral bracts** imbricate, arranged in three types, initially foliaceous then reduced to entirely membranous. **Outer involucral bracts** 3.4–4 mm L; as leaves but with short membranous alae to less than ½ height of bract; indumentum floccose lanate on ventral surface and on dorsal surface at base around membranous alae. **Middle involucral bracts** 3.4–4.3 mm L; membranous alate, ratio of lamina to alae decreases from ⅔ to nearly entirely alate; lanceolate; apices acute, sometimes with reddish-brown patches; indumentum OIB. **Inner involucral bracts** 4.9 mm L, entirely membranous alate; ovate lanceolate; apices redish-brown maculate with recurved mucro; floccose lanate on upper dorsal surface. **Ray florets** ca. 5, pistillate, white with dark (purple-red) tips, not exerted beyond capitulum; corolla 3.0–3.8 mm L, corolla tube 2.2 mm L, outer lip 0.3 mm W., inner lip 0.8 mm L, bifid. **Disk florets** ca. 20, bisexual, pale yellow-cream with dark (purple-red) tips; corolla 2.8–3.3 mm L, corolla tube 2.5–3.0 mm L **Styles** yellow [ray] 3.1 mm L, [disk] 2.8 mm L; stigma lobes 0.2 mm L oblong truncate, dark coloured, with hairs. **Anthers** [disk] 1.4 mm L (2.8 mm L incl. filaments). **Achenes** brown, ca. 1.8 mm L, broadly turbinate; carpopodium anular irregular, narrow; pericarp pellucid, densely covered in lanceolate twin hairs, 80–120 x 50 μm. **Pappus** white, 2.5–3.4 mm L, 2 rows, indehiscent; pappus fused at base, pappus setae width 4–8 cells, cell width 5–10 μm, no basal cilia; barbellate, barbs 95–140 μm L, barb base longly adhered (>60%), free barb appressed, 12–15 barbs/100 μm.

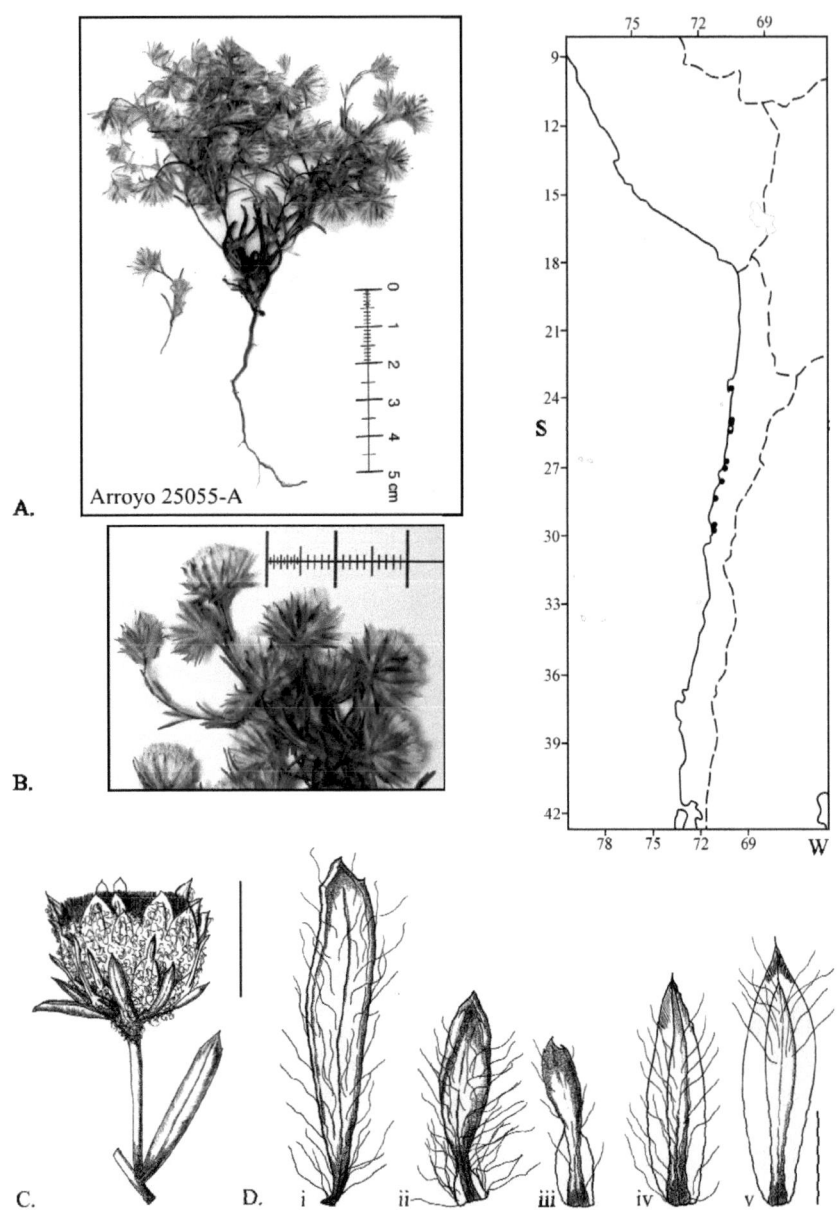

Figure 65: *Chaetanthera taltalensis*. K. Arroyo 25055-A. **A.** Habit, herbarium specimen. **B.** Specimen detail. **C.** Capitulum detail, scale bar = 5 mm. **D.** Leaf & Bract detail. **i.** Lower stem leaf v.s.; **ii.** Upper stem leaf v.s.; **iii.** OIB d.s.; **iv.** MIB d.s.; **v.** IIB d.s. Scale bar = 2 mm.

Distribution and habitat. *C. taltalensis* is distributed along the Cordillera de la Costa, from Antofagasta to La Serena, between latitudes 23°00'–30°00'S. It occurs between elevations up to and around 500 m.a.s.l. Its typical habitat is sandy.

Differential diagnosis. *C. taltalensis* is similar to *C. moenchioides* and *C. ramosissima*, but distinct due to its floccose lanate indumentum and the reduced, inconspicuous ray florets. Key features in identifying the species are laid out in Table 13, page 155. It is geographically disjunct from *C. ramosissima*. Its campanulate capitula are different from the narrow urcinate capitula of *C. moenchioides*.

Material seen. – **Chile. II Región de Antofagasta** Prov. de Antofagasta Antofagasta, 360 m, 10.1930, *Jaffuel 1142* (CONC GH). – Paposo, camino a Mina Julia, 300 m, 10.1991, *Taylor* et al. *10707* (CONC). – Cuesta Paposo, 700 m, 1969, *Jiles 5429* (CONC). – Taltal, Paposo, Quebrada Guenillos, 610 m, 09.1992, *Teillier, Rindel & P. García 2791* (SGO). – Quebrada Matancillas cerca de Paposo, 175 m, 10.2002, *K. Arroyo* et al. *25128* (CONC). – Taltal, Quebrada Peralito, 100 m, 1953, *Ricardi 2479* (CONC). – Taltal, 20 m, 1946, *Vidal s.n.* (CONC). – Taltal, 200 m, 10.1925, *Werdermann 815* (Bx2 BM CONC E G GH). – Taltal, Hueso Parado, 150 m, 1953, *Ricardi 2693* (CONC). – Taltal, Quebrada de Taltal, 09.1992, *Teillier, Rindel & P. García 2625-b* (SGO). – **III Región de Atacama** Prov. de Chañaral PN Pan de Azucar, Quebrada Coquimbo, 350 m, 10.1987, *Teillier 686* (CONC SGO). – Prov. de Copiapó Caldera-Copiapó region, 09-10.1890, *Morong 1302* (E F G GH K NY). – Panamericana Caldera-Chañaral, Km 18, 50 m, 10.1965, *Ricardi* et al. *1289* (CONC). – Panamericana km 908, nördl. Caldera, 10.1980, *Grau 2092* (MSB). – Sierra Atacama a 39 Km al sur de Copiapó, 150-200 m, 09.1941, *Muñoz, P. & Johnson 1945* (SGO). – **IV Región de Coquimbo** Prov. de Elqui Carretera Panamericana, frente al Tofo, 350 m, 09.1957, *Ricardi & Marticorena 4353* (CONC). – Cuesta Buenos Aires, Ruta 5 Norte entre La Serena y Vallenar, 615 m, 09.2002, *K. Arroyo* et al. *25016* (CONC). – Camino del Mineral La Higuera, 500 m, 10.1963, *Marticorena & Matthei 164* (CONC). – Subida sur de Cuesta Buenos Aires, 10.1991, *Munoz, Teillier & Meza 2631* (MO SGO). – Cuesta Buenos Aires, 09.1957, *Cabrera 12594* (MA). – Cuesta Buenos Aires, Ruta 5 Norte entre La Serena y Vallenar, 615, 09.2002, *K. Arroyo* et al. *25056[a]* (CONC). – Caleta Hornos, entre La Serena y Cuesta Buenos Aires, 280 m, 09.2002, *K. Arroyo* et al. *25057* (CONC). – Punta Teatinos cerca La Serena, 92 m, 09.2002, *K. Arroyo* et al. *25055[a]* (CONC).

4.2 Chaetanthera subgenus *Tylloma* (D.Don) Less.

19. *Chaetanthera euphrasioides* (DC.) F. Meigen Bot. Jahrb. Syst. 17: 284. 1893.

≡ *Elachia euphrasioides* DC. Prodr.(DC.) 7 (1): 256. 1838. DC. in Delessert Icon. Select. Pl. 4: t. 99. Typus CHILE "In excelsis Andibus Chilensium legit cl. C. Gay" Iconotypus! Isotypus probabilis [Hautes cordilleres de Talcaregue, 1831, *Gay s.n.* (G – microfiche)]

= *Chaetanthera debilis* Meyen & Walp. Nov. Act. Nat. Cur. 29 (1): 287. 1843. Typus CHILE "Chile, Cordillera de San Fernando" [Cord. S. Fernando, Febr. 1843, *Bustillos s.n.*] **Lectotypus hic loc. designatus** W! Isotypus NY! SGO 76567!

= *Elachia spinulosa* Phil. Linnaea 33: 115. 1864 – 65. Typus CHILE "In Andibus elatioribus prov. Santiago" [Cord. de Santiago, 1862, *Philippi s.n.*] Holotypus SGO (†) Isotoypus F! G! W!

– *Chaetanthera euphrosiodes* (DC.) Kuntze Revis. Gen. Pl. 3 (2): 140. 1898. Orth var.

– *Chaetanthera euphrasioides* (DC.) Reiche. Fl. Chile [Reiche] 4: 343. 1905.

– *Chaetanthera euphrasioides* var. *spinulosa* (Phil.) Reiche Anales Univ. Chile 115: 324. 1904; Fl. Chile [Reiche] 4: 343. 1905.

Nomenclatural Notes Walpers published material from Meyen's collections and notes after Meyen's death in 1840. Bustillos is not mentioned as a collector in "Reise um die Erde", but the location and date would place this material in the right time frame for Walpers to have cited it.

Annual monoecious herb. **Roots** filamentous. **Stems** to 5 cm, glabrous, decumbent to ascending, spreading from central node above a short stem (<1 cm). **Leaves** 5–7 x 2 mm; succulent, indistinctly petiolate, linear-spathulate to truncate; margins polymorphic from notched spathulate to dentate, 6–8 teeth; midrib indistinct; glabrous. **Capitula** cylindrical (open), disk diameter (2.5)4–6 mm. **Involucral bracts** imbricate, arranged in three types, initially foliaceous then reduced to entirely membranous. **Outer involucral bracts** 5–6 mm L; as leaves but with short membranous alae to less than ½ height of bract. **Middle involucral bracts** 5.5–7 mm L, teeth reduced to few or one, membranous alate, ratio of lamina to alae decreases from ⅔ to nearly entirely alate. **Inner involucral bracts** 8.5–9 mm L; entirely membranous alate; ovate-lanceolate, shortly mucronate, apices colourless. **Ray florets** ca. 6, pistillate, white (or pale yellow to yellow) ± pink-red dorsal tips; corolla 5.9 mm L, corolla tube 2.7–2.9 mm L, outer lip 1 mm W., inner lip 1.9–2.2 mm L, fused, rounded at apex. **Disk florets** ca. 8, bisexual, yellow; corolla 4.3–4.5 mm L, corolla tube 2.7–3 mm L **Styles** [ray] 4 mm L [disk] 4.4 mm L, stigma lobes 2.2 mm L **Anthers** 2.2 mm L (4.4 mm L incl. filaments). **Achenes** brown, pericarp pellucid [Rays] 1.9 mm L, sterile, glabrous. [Disks] 2.8–3.2 mm L, fertile, sparsely covered with small globular twin hairs, 35–50 x 40 µm; testa epidermis surface of elongated parallel cells, margins entire, cells with U-formed sclerenchymatous thickenings. **Pappus** white, 4.3–5.5 mm, 2 rows, dehiscent; pappus setae width 1–2(3) cells, cell width 5–10 µm, no basal cilia; barbellate, barbs 95–140 µm L, barb base longly adhered (>60%), free barb spreading, 3–4 barbs/100 µm.

IX *Chaetanthera* Ruiz & Pav.

Figure 66: *Chaetanthera euphrasioides*. **A.** Habit photographed ©Iréne Till-Bottraud. **B.** Leaf & Bract detail, d.s. *Leuenberger & Arroyo 3937*. **i.** Leaf; **ii.** OIB; **iii.** MIB; **iv.** IIB. Scale bar = 2 mm.

Distribution and habitat. *C. euphrasioides* is distributed in the Andes of Argentina and Chile between latitudes 30°50'–36°20'S at elevations between (1500) 2390–3400 m.a.s.l. It grows at lower altitudes in south of distribution. It is typically found on steep scree slopes or in stony soil with *Chuquiraga oppositifolia-Tetraglochin alatum*, *Laretia acaulis-Anarthrophyllum cumingii* and *Anarthrophyllum gayanum-Nassauvia heterophylla* associations (personal observation; HOFFMAN et al. 1997).

Differential diagnosis. Morphologically close to the larger, lower elevation *C. flabellata*, *C. euphrasioides* has very variable lamina shapes, ranging from linear-spathulate dentate to truncate spinulose.

Material seen. – **Argentina. Mendoza** Depto. San Rafael Environs de St. Raphael et de la vallee du Río Atuel, Cajón del Burro, 2500 m, 01/02.1897, *Wilczek 178* (G). – Cord. de Mendoza (Río Salado sup.), Alverjalito, 02.1892, *Kurtz 7128* (NY). – Cord. de Mendoza, Río Salado sup., 02.1892, *Stuckert 7128* (G). – Cord. de Mendoza, Río Salado sup., Los Molles, 01.1893, *Stuckert 7493* (G). – Los Molles, arriba del "Cuchillo", 2400 – 2600 m, 12.1949, *Sleumer 639* (B). – Depto. Malagüe Paso Pehuenche (73 km W de Bardas Blancas), frontera argentina-chilena, 2490 m, 02.1989, *Leuenberger & Arroyo 3937* (B). – **San Juan** Andes de San Juan, Cord. de L'Espinazito, Los Patillos (Herb. Hauman), 02.1897, *Bodenbender s.n.* (G). – **Neuquén** Cordillera del Viento, 01.1935, *Ragonese 157* (LP). – **Chile. IV Región de Coquimbo** 1839, *Gay 952* (G). – Prov. de Limarí Cordillera Ovalle, San Miguel-Río Mostazal, 3400 m, 01.1972, *Jiles 5918* (CONC). – Cordillera de Ovalle, Mantos Grandes, 2600 m, 01.1972, *Jiles 5899* (CONC). – Cordillera de Combarbalá, El Derecho, 3300 m, 01.1966, *Jiles 4793* (CONC). – Prov. de Choapa Dept. Illapel, Quebrada La Vega Escondida, 3 hrs by horse due east of Cuncumén, hacienda at fork of Río Tranquilla and Tencaan creek, SSW from Las Placetas, 2700 m, 11.1938, *Worth & Morrison 16575* (GH). – Cerrro La Yerba Loca, 2 hrs by horse above La Vega Escondida, 2900 m, 12.1938, *Morrison 16945* (G GH K MO). – near Salamanca, Cord. de Qurlén, 2500 m, 01.1984, *Zöllner II 769* (MO). – **V Región de Valparaíso** Prov. de Petorca Junta de Piquenes, Río Sobrante, 3000 – 3100 m, 12.1939, *Morrison 17303* (GH K). – Cerro Chache 5 hrs by horse southeast of Patagua Mine, ca. 18 km east of La Ligua, west slope southeast of Chache, 2200 m, 12.1938, *Morrison 17071* (CONC G GH K). – Prov. San Felipe de Aconcagua ridge of Cerro de Las Viscachas, 2000 – 2200 m, 01.1936, *West 5166* (GH). – Prov. de Valparaíso Colliguay, 470 m, 01.1918, *Jaffuel s.n.* (CONC). – **Región Metropolitana de Santiago** Prov. de Chacabuco Altos del Roble, Hacienda de Chicauma, 1600m, 12.1983, *Villagran 4729* (CONC). – Prov. de Santiago después del Pueblo La Parva, en explanada sobre Escuela de Ski, 01.1979, *Muñoz & Meza 1391* (SGO). – Complejo de Esqui "Valle Nevado", camino a Cerro Franciscano, 3315 m, 02.2003, *K. Arroyo* et al. *25176* (CONC). – about 2 km above Farellones on road to La Parva ski village, ca. 2400 m, 01.1961, *Moore 402* (MO). – La Parva, 2800 m, 01.1993, *Stuessy & Ruiz 12729* (CONC). – cerca La Parva, exp. N-O. 2765 m, 01.1979, *Muñoz & Meza 1297* (SGO). – Road along Valle Nevado, Km 8 from Farellones, 2750 m, 02.2002, *Davies 2002/004* (M). – Camino entre Farellones y Complejo de Esqui "Valle Nevado", Curva 12 camino a Valle Nevado, 2810 m, 01.2003, *K. Arroyo* et al. *25166* (CONC). – Complejo de Esqui "Valle Nevado", laderas cerca del hotel, 3050 m, 02.2003, *K. Arroyo* et al. *25172* (CONC). – Camino entre Farellones y Complejo de Esqui "Valle Nevado", aproximadamente 2 km hacia el este de Casa de Piedra, 2460 m, 01.2003, *K. Arroyo* et al. *25167* (CONC). – El Colorado, after Farellones, 02.2002, *Davies 2002/002* (M). – El Colorado, bajando, 2640 m, 01.1993, *Muñoz & Eggli 3230* (SGO). – San Gabriel, 1500 m, 01.1950, *Gunckel 21413* (CONC). – RN Río Clarillo, Sector Las Flores, 2730 m, 02.2001, *K. Arroyo* et al. *210615* (CONC). – RN Río Clarillo, Cajón del Horno, 2250 m, 12.2000, *K. Arroyo* et al. *206718* (CONC). – RN Río Clarillo, Ladera bajo Vega del Cigarro, 2500-2600 m, 01.2000, *K. Arroyo 20535; 20536* (CONC). – RN Río Clarillo, Sector Cega del Cigarro, 2300 m, 11.1999, *K. Arroyo 994372* (CONC). – RN Río Clarillo, Sector Cega del Cigarro, 2430 m, 12.2000, *K. Arroyo 206625* (CONC). – RN Río Clarillo, Vega de los Manantiales Secos, 2410 m, 01.2000, *K. Arroyo 20623* (CONC). – RN Río Clarillo, Sector Los Cristales, 2650 m, 01.2000, *K. Arroyo 20746* (CONC). – RN Río Clarillo, Sector Los Cristales, 2870 m, 02.2001, *K. Arroyo 210646* (CONC). – Prov. de Cordillera Cordillera, Paso Cruz 34°, 1600 m, 01.1892, *Kuntze s.n.* (NYx2). – Valle del Yeso, *Philippi s.n.* (K W). – Lagunillas, 2250 m, 12.1950, *Barros s.n.* (CONC). – Lagunillas, 12.1971, *Beckett, Cheese & Watson 4521, 4475* (SGO). – Lagunillas, ca. 2 km más arriba del pueblo, 2470 m, 02.2003, *K. Arroyo 25119* (CONC). – Río Yeso, Laguna Piuquenes, 2500 m, 01.1945, *Biese 1008* (SGO). – Cajón del Yeso, Termas El Plomo, 3000 m, 20.01.1995, *Muñoz, Moreira, Meza & Arrigada 3587* (SGO). – Laguna La Encañada, 5 Km antes Laguna Negra, 2350 m, 01.1990, *Teillier et al. 2037* (CONC). – Valle del Yeso, 2500 m, 01.1951, *Ortiz s.n.* (CONC). – PN El Morado, frente a Santiago, 2275 m, 01.1991, *Teillier, Pauchard & P. García 2498* (SGO). – Cajón del Río Maipo, Refugio Cruz de Piedra, 2400 m, 01.2000, *Teillier 4545* (CONC). – Valle del Maipo, 2300 m, 02.1937, *Grandjot s.n.* (MO). – Oberes Maipotal (Puente de Sierra), 2300 m, 02.1937, *C. & G. Grandjot 2758* (GH MO). – Prov. de Melipilla Cerro Cantillana, 2200 m, 12.1995, *Bliss* et al. *2362* (CONC). – RN Roblería de Cobre de Loncha, Altos del Gusano, 2060 m, 01.2001, *K. Arroyo 210150* (CONC). – **VI Región del Libertador G. B. O'Higgins** Prov. de Colchagua RN Río de los Cipreses, Ladera Cerro Morro Blanco, 2070 m, 12.2000, *K. Arroyo 206940* (CONC). – RN Río de los Cipreses, Sector Morro de Piedra, 2470 m, 01.2001, *K. Arroyo 210312* (CONC). – Cajón Las Leñas, Rodales al principio de la laguna, 2170 m, 01.1998, *K. Arroyo* et al. *99686* (CONC). – RN Río de los Cipreses, 2770 m, 03.2002, *K. Arroyo* et al. *220046* (CONC). – RN Río de los Cipreses, Cima Cerro Colorado, 2320 m, 02.2000, *K. Arroyo 201260* (CONC). – RN Río de los Cipreses, Cima Cerro Colorado, 2350 m, 02.2000, *K. Arroyo 201275* (CONC). – Cordilleras above Colchagua, 01.1930, *Pirion 108* (GH). – Tinguiririca, 01.1930, *Pirion 188* (SGO). – Vegas del Flaco, 1900 m, 01.1964, *Marticorena & Matthei 730* (CONC). – Vegas del Flaco, 1800 m, 01.1955, *Ricardi 3178* (CONC). – Baños del Flaco, 1800m, 12.1936, *Milner s.n.* (CONC). – Termas del Flaco, 1700 m, 01.1963, *Montero 6644* (CONC). – Termas del Flaco, 1700 m, 02.1983, *Montero 12531* (CONC). – Termas del Flaco, 2450 m, 12.1994, *Baeza 292*. (CONC). – Prov. de Cachapual Ex regione alpina Andium chilensium prope tepidaria Cauquenes, *Reid s.n.* (E). – **VII Región del Maule** Cordilleras de Maule, 1855, *Philippi s.n.* (BM G K P W). – Prov. de Linares Cajón Troncoso, bajade hacia el Estero Nieblas, 2300 m, 01.1961, *Schlegel 3712* (CONC F). – Prov. de Curicó Camino de Curico a Laguna de Teno, 2520m, 03.1973, *Marticorena* et al. *70* (CONC). – Laguna de Teno, 2500m, 01.3967, *Marticorena & Matthei 905* (CONC). S.l.d. Mts east side of Andes, *Bridges 1158* (K). – Chile, *Gay s.n.* (K NY W). – ad rupe editiorum adella 11.1839, *Gay 514* (P). – In petrosis altissimo andium, *"Elachia euphrasioides"*, *Gay s.n.* (GH). – *Ruiz & Pavon 27* (MO).

20. *Chaetanthera flabellata* D. Don ex Taylor & Phillips Philos. Mag. Ann. Chem. 11: 391. 1832 (March – April).

Typus CHILE [Cordilleras of Chili, 1831, *Cuming 291*] **Lectotypus hic loc. designatus** G! Isotypus BM! Ex2! Kx2 OXF (*s.n.*)! W!
= *Chaetanthera prostrata* D. Don ex Taylor & Phillips Philos. Mag. Ann. Chem. 11: 391. 1832. Typus CHILE "Herb. Gill." [San Pedro Nolasco, *Gillies s.n.*] **Lectotypus hic loc. designatus** BM! Isotypus K! OXF!
– *Chaetanthera flabellata* var. *prostrata* (D.Don) Hook. & Arn. Companion Bot. Mag. 1: 105. 1835.
– *Chaetanthera multicaulis* var. *prostrata* (Hook. & Arn.) DC. Prodr. (DC.) 7 (1): 31. 1838.

Nomenclatural Note After the dispersal of the Lambert herbarium, where David Don worked during the significant period of his *Chaetanthera* nomenclatural publications, H. Cuming's collections were sold to G, thus the designation of Lectotype from G. Type material of both *C. prostrata* and *C. flabellata* belong to the same entity. ["Chaetanthera prostrata" *Gillies s.n.*] and [San Pedro Nolasco, Chile, *Miers s.n.*] are on the same sheet (BM!). However, the labels on the K and OXF sheets clearly have "*Chaetanthera prostrata*, San Pedro Nolasco, *Gillies s.n.*".

Annual monoecious herb. **Roots** filamentous. **Stems** to 12 cm glabrous, decumbent to ascending, spreading from a central node above a short stem (< 2cm). **Leaves** 12–17.5 x 2–5 mm; succulent, indistinctly petiolate, narrowly spathulate, bright blue-green; margins dentate, 9–13 teeth, midrib indistinct; glabrous; leaves at nodes generally larger than those on scapes/ stems; opposite on stem the alternate along flowering branches. **Capitula** campanulate, buds ovate-apiculate; disk diameter 0.8–1.5 cm. **Involucral bracts** imbricate, arranged in three types, initially foliaceous then reduced to entirely membranous. **Outer involucral bracts** 10–6 x 3–4 mm; as leaves but with short membranous alae to less than ½ height of bract. **Middle involucral bracts** 7–10 mm L, teeth reduced to few or one; membranous alate, ratio of lamina to alae decreases from ⅔ to nearly entirely alate. **Inner involucral bracts** 9–10 mm L; entirely membranous alate, ovate-lanceolate, mucronate, apices colourless. **Ray florets** ca. 14, pistillate, yellow (sometimes rather pale or white) with pink-red dorsal tips; corolla 10 mm L, corolla tube 4.3 mm L, outer lip 2.2 mm W., inner lip 4 mm L, fused or parted slightly at apex. **Disk florets** ca. 20, bisexual, yellow; corolla 8 mm L, corolla tube 6.2–6.5 mm L **Styles** yellow [ray] 8.3 mm L, stigma lobes 0.4 mm L [disk] 8 mm L **Anthers** 4.3 mm L (8 mm L incl. filaments). **Achenes** brown, 2–3 mm L, turbinate, ray achenes sterile, disk achenes fertile; no carpopodium; pericarp pellucid, covered with conical to globular twin hairs, 35–45 x 35 µm. **Pappus** 5.9–7.5 mm, white, 2 rows, dehiscent; pappus setae width 4–8 cells, cell width 15 µm, no basal cilia but scabrid spreading barbs; barbellate, barbs 95–140 µm L, barb base medium adhered (50%), free barb appressed, 5–7 barbs/100 µm.

Distribution and habitat. *C. flabellata* is locally endemic to the Andes around Santiago between 33° 55'–34° 15'S. It occurs between 1600–2600 m.a.s.l. It is typically found in open, sparse vegetation, on dry, well-drained sandy or rocky slopes. It grows on the boundary zones of the sclerophyllous matorral (*Colliguaya*) and the subandean matorral, often with *Chuquiraga opposistifolia*. **Differential diagnosis.** *C. flabellata* has distinctly dentate leaves and is generally much bigger than *C. euphrasioides*.

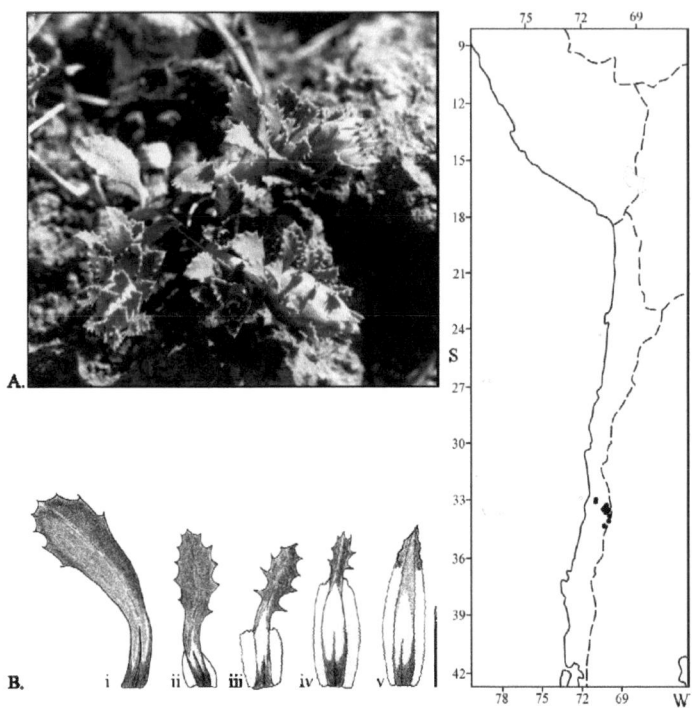

Figure 67: *Chaetanthera flabellata*. **A.** Habit photographed Farellones, 11.2001©Alison Davies. **B.** Leaf & Bract detail d.s. *Grau 2440*. **i.** Stem leaf; **ii.** OIB; **iii – iv.** MIB; **v.** IIB. Scale bar = 5 mm.

Material seen. – Chile. V Región de Valparaíso Prov. de Valparaíso Sierra Bella Vista, Aconcagua, *Bridges 131* (BM E K W). – Prov. de Quillota PN La Campana, Cerro El Roble, 2230 m, 12.1999, *K. Arroyo* et al. *996485* (CONC). – Cerro del Roble, Punta Iman, 2000 m, 12.1934, *Garaventa 3172* (CONC). – Cerro El Roble, 2300 m, 11.1965, *Zöllner 993* (CONC). – Las Vizcachas, ca. 10 km from La Dormida, 1910 m, 12.1938, *Morrison 16763* (CONC G K). – **Región Metropolitana de Santiago** Prov. de Santiago Cordillera, 1861, *Philippi s.n.* (K). – Straße nach Farellones, Kurve 33, 2250 m, 11.1980, *Grau 2440* (BM MSB M). – Farellones, curve 32-33, 2200 m, 11.2001, *Davies s.n.* (M). – Camino entre Santiago y Farellones, curva 34, 2310 m, 12.2002, *K. Arroyo* et al. *25162* (CONC). – Farellones, curva 34, 2200 m, 11.1978, *Villagrán & Meza 419* (SGO). – Farellones, 3 km east of village, 2000 m, 11.1996, *Gardner & Knees 6003* (E). – unterhalb des Ortes Farellones, 2000 m, 12.1981, *Grau 3288* (M). – Loma del Viento, 2200 m, 02.2002, *Davies 2002/003* (M). – al oeste del pueblo Farellones, 2360 m, 03.2003, *K. Arroyo* et al. *25284* (CONC). – 2 kms pasado Farellones, 2150 m, 12.1987, *Rosas 1776* (M). – Cerro de La Provincia, 2000 m, 10.1960, *Schlegel 3078* (CONC). – Cerro Abanico, 2000 m, 12.1950, *Orellana s.n.* (CONC). – Potrero Grande, 1780 m, 01.1948, *Muñoz s.n.* (CONC). – Piuquencillos, Valle Río Colorado, 3500 m, 12.1942, *Pisano* et al. *1641* (CONC). – Lagunillas, 2250 m, 12.1979, *Zöllner 10658* (CONC). – Lagunillas, 2600 m, 12.1966, *Mooney 195* (CONC). – Cajón del Maipo, Estero El Diablo, 1900 m 01.2000, *Teillier 4610* (CONC). – Lo Valdes, 2000 m, 02.1951, *Morales s.n.* (CONC). – **VI Región del Libertador G. B. O'Higgins** Prov. de Cachapoal RN Río de los Cipreses, Quebrada Las Terneras, 1940 m, 12.2000, *K. Arroyo* et al. *206910* (CONC). – **S.l.d.** Andes, Chili, *Bridges 1161* (E K OXF).

21. *Chaetanthera flabellifolia* Cabrera Revista Mus. La Plata, Secc. Bot. 1: 153 – 155, fig. 27. 1937.

≡ *Pachylaena elegans* Phil. Linnaea 33: 113. 1864-65. Typus CHILE "In prov. Coquimbo propre los Baños del Toro, *Volckmann*" Holotypus SGO 64689! Isotypus Kx2!

Annual monoecious herb. **Roots** filamentous. **Stems** to 10 cm, spreading from central node above short erect stem (<2 cm), decumbent-ascending, glabrous, nearly naked of leaves. **Leaves** few, alternate, forming clusters at nodes and rosettes below capitula, glabrous; indistinctly petiolate, 13–15.5 mm L, petiole 9.5–10.5 x 1.6–2 mm W., lamina 4.5 x 7.5 mm.; lamina glabrous, glaucous green, succulent, spathulate/flabellate to rhomboid, margins limbate, loosely to distinctly regularly dentate (teeth 1 mm L, triangular), lightly dotted with minute glands/teeth. **Capitula** sessile, buds obtuse to rounded; disk diameter ca. 1 cm. **Involucral bracts** imbricate, arranged in three types, initially foliaceous then reduced to entirely membranous. **Outer involucral bracts** 9–14 mm L; as leaves but with short membranous alae to less than ½ height of bract; pseudopetiole 4–8 x 2–1.5 mm; lamina as leaves, 4–6 x (3)6–9 mm. **Middle involucral bracts** (7.5)9–15 x 3–6 mm; membranous alate, ratio of lamina to alae decreases from ⅔ to nearly entirely alate; lamina rapidly reduced to apical mucro with green succulent mesophyllous tissue below, broadly ovate-lanceolate, bases truncate to cuneate, apices broadly acute. **Inner involucral bracts** (1 series only) 15.5–17 x 3.5–4.2 mm; completely transparent, entirely membranous alate; linear-lanceolate, bases cuneate, apices broadly acute, green edged with pink-purple. **Ray florets** ca. 25, pistillate, pale to golden yellow; corolla 14.5–16 mm L, corolla tube 4.5–4.8 mm L, outer ligule 3–4 mm W., inner ligule 4.5–5 mm L (not conspicuously bifid). **Disk florets** ca. 30, bisexual, golden yellow; corolla 10–10.5 mm L, corolla tube 8.2–9 mm L **Styles** yellow [ray] 9–10 mm L, stigma lobes 0.5 mm. [disk] 10.8–11.3 mm, stigma lobes 0.3–0.5 L **Anthers** 5.5–6 mm L (10–10.8 mm incl. filaments). **Achenes** brown, turbinate, 2.5–3.5 mm L (immature), carpopodium present; pericarp pellucid, densely covered in lanceolate twin hairs, ca. 170 x 50 µm. **Pappus** [ray] 6–7 mm L; [disk] 9.5–10.5 mm L, white, 2 rows, indehiscent; pappus setae width 4–8 cells, cell width 5–10 µm, no basal cilia, scabrid spreading barbs; barbellate, barb base longly adhered (60%), free barb appressed, 3–4 barbs/100 µm.

Distribution and habitat. *C. flabellifolia* is locally endemic in the Chilean Cordillera Doña Ana, between latitudes 29°47'–30°12'S. It occurs at elevations between 2800–3900 m.a.s.l. in scree. High altitude *C. splendens* (unconfirmed *C. flabellifolia*) is recorded from San Juan, Argentina, (KATINAS, Cat. Pl. Arg., 1996).

Differential diagnosis. *C. flabellifolia* is close to *C. splendens* – a more depauperate, precocious form from lower, more southerly localities. It is distinguished by its spectacular flabellate laminas.

Material seen. – Chile. IV Región de Coquimbo Prov. de Elqui Coquimbo, *Ball (Philippi?)s.n.* (E). – Baños del Toro "Pachylaena elegans Ph." *Philippi s.n.* (K). – Coquimbo, 02.1888, *Philippi s.n.* (K). – Baños del Toro, *Philippi s.n.*

(K). – Km 29, camino a Indio, 3900 m, 01.1988, *Squeo 88046* (CONC). – camino a Indio, Km 29, 3800 m, 02.1988, *Squeo 88090* (CONC). – Cordillera Doña Ana, Baños del Toro, 3450 m, 01.1992, *Arancio 121* (CONC). – Baños del Toro, 3100 m, 01.1981, *K. Arroyo 81212* (CONC). – 2 km al oeste de Baños del Toro, camino entre Guanta y Baños del Toro, 3250 m, 01.2003, *K. Arroyo et al. 25084* (CONC). – Baños del Toro, 3380 m, 01.1948, *Wagenknecht 257* (CONC). – Baños del Toro, 3500 m, 12.1923, *Werdermann 189* (CONC E F G K). – Baños del Toro, 12.1971, *Beckett, Cheese & Watson 4654* (SGO). – faldeos al sur de la entrada de Mina del Indio, 3200 m, 01.2003, *K. Arroyo et al. 25078* (CONC). – 16,5 Km N de Juntas del Toro, 2900 m, 01.1993, *Stuessy & Ruiz 12792* (CONC). – camino al Embalse de La Laguna, 2800 m, 02.1963, *Ricardi et al. 717* (CONC). – cerros al oriente de Embalse La Laguna, 3600 m, 02.1987, *Niemeyer 8703* (CONC). – Paso Aguas Negras, 01.2003, *Aubert, Douzet & Hurstel s.n.* (JAL image). – La Troya, Las Termas Hediondas, 11.2001 *M. Belov* (image www.chileflora.com)

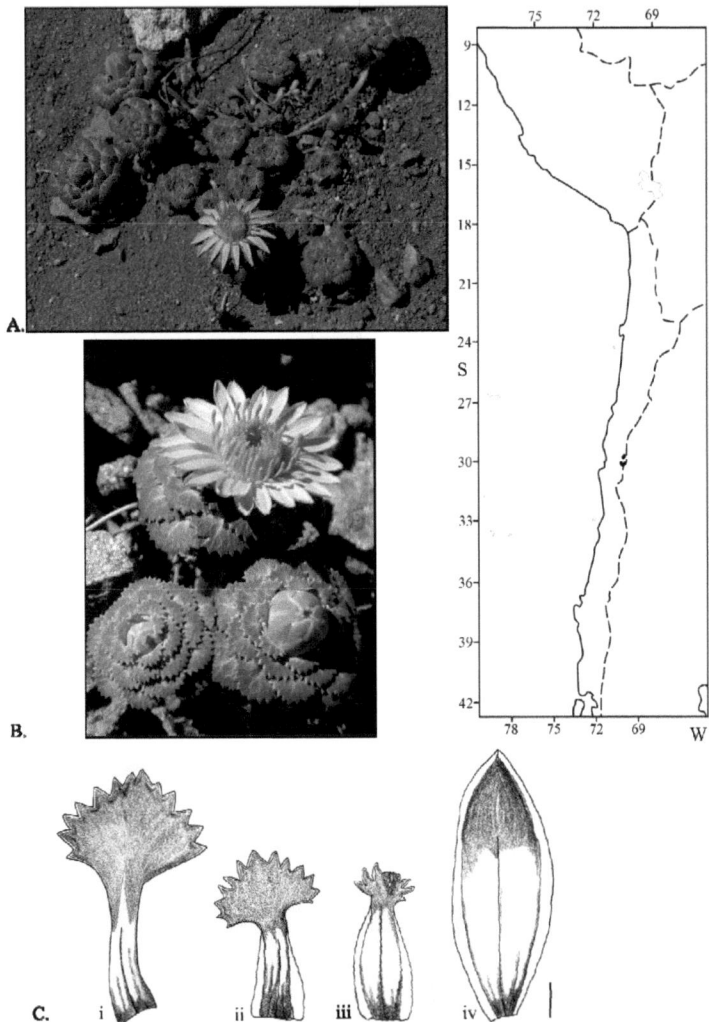

Figure 68: *Chaetanthera flabellifolia*. **A.** Habit, photographed 2001©Michail Belov. **B.** Capitulum detail, photographed 2003©Iréne Till-Bottraud. **C.** Leaf & Bract detail, d.s. *Werdermann 189*. **i.** Stem leaf; **ii.** OIB; **iii.** MIB; **iv.** IIB. Scale bar = 2 mm.

22. *Chaetanthera frayjorgensis* A.M.R. Davies nomen novum

≡ *Tylloma glabratum* var. *microphyllum* Phil. Anales Univ. Chile 85: 842. 1894. Typus CHILE "In litore arenoso ad La Serena et in monte Fray Jorge, januario 1883 invenit *Fr. Philippi*" **Lectotypus hic loc. designatus** SGO 64676! Isotypus SGO 45033! SGO 76575!

= *Tylloma brachylepis* Phil. Anales Univ. Chile 85: 842. 1894. Typus CHILE "Inter specimina T. glabrati jacebat" Holotypus SGO 64680!
- *Tylloma glabratum* var. *brachylepis* (Phil.) Reiche Anales Univ. Chile 115: 328.1904.

Nomenclature Notes The epithet "microphylla" is already in use in the genus *Chaetanthera* (*Chaetanthera microphylla* (Cass.) Hook. & Arn.). Therefore, a new name has been designated for this taxon.

Annual monoecious herb. **Roots** filamentous. **Stems** to 20 cm, ascending, decumbent, glabrous, branched from central node above short erect stem (<2 cm). **Leaves** to 20 mm L, few, alternate, forming clusters at nodes and rosettes below capitula; petioles with shortly stalked or sessile glands scattered on margins and dorsal surfaces; lamina 7 x 4.5–5 mm, succulent, elongate ovate to orbicular, conduplicate, sessile pale or dark glandular hairs on surfaces, limbate ± undulate margins. **Capitula** sessile, urcinate, buds apiculate, disk diameter ca. 0.5 cm. **Involucral bracts** imbricate, arranged in three types, initially foliaceous then reduced to entirely membranous. **Outer involucral bracts** 10–16 mm L; as leaves but with short membranous alae to less than ½ height of bract; lamina 3.5–6 x 3.5–5.2 mm. **Middle involucral bracts** oval to oblanceolate, cuspidate to broadly acute, mucronulate, 8.5–14 mm x 3.5 mm; central tissue mesophyllous; membranous alate, ratio of lamina to alae decreases from ⅔ to nearly entirely alate; dorsal surface densely scattered with sessile glands (pale to dark brown or black). **Inner involucral bracts** oblanceolate, 14.3–14.8 x 2.2–3.3 mm; translucent, entirely membranous alate; dorsal surface ± scattered with sessile glands; apices shortly freely recurved mucronulate, with bright pink or reddish purple colouring. **Ray florets** 19–26, pistillate, bright yellow; corolla 17–18 mm L, corolla tube 4–4.5 mm L, outer ligule 4 mm W., inner ligule 1–1.5 mm L **Disk florets** ca. 30, bisexual, yellow; corolla 9.2–9.6 mm L, corolla tube 7.4–8 mm L **Styles** yellow, [ray] 10.5–11.2 mm L, stigma lobes 0.5 mm L [disk] 9–10.6 mm L **Anthers** 5.7 mm L (7.5 mm incl. filaments). **Achenes** brown, fusiform-turbinate, 3–4.5 mm L; carpopodium anular; pericarp pellucid, densely covered with narrow, oblate-lanceolate twin hairs, 75–100 x 35 µm L **Pappus** white, 6.5–7.5 mm L, 2 rows, free at base, in/dehiscent; pappus setae width 4–8 cells, cell width 5–10 µm, basal cilia; barbellate, barbs 90–140 µm L, barb base medium adhered (50%), free barb appressed, 5–6 barbs/100 µm. **Chromosome number** $2n = 4x-6 = 38$ (Davies & Vosyka, new count) *Lopez s.n.* (CONC).

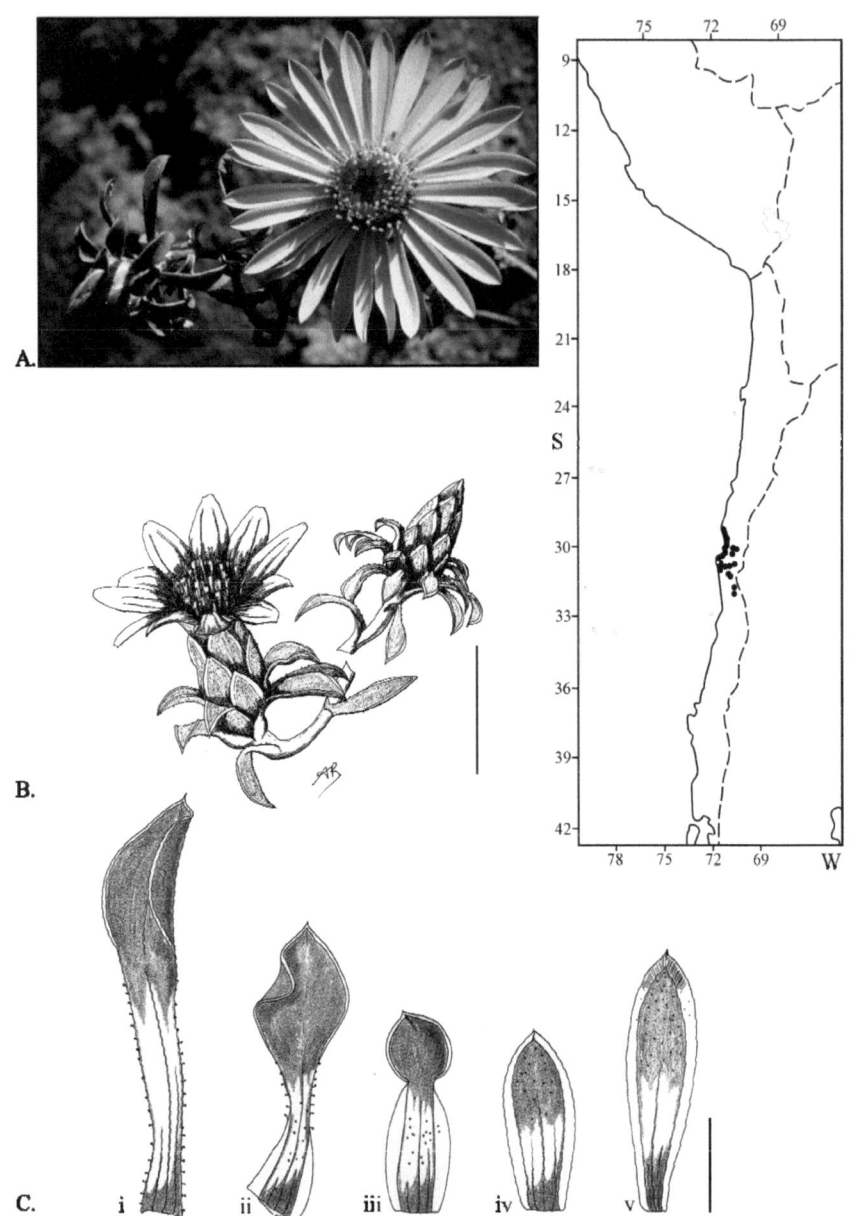

Figure 69: *Chaetanthera frayjorgensis*. **A.** Habit, photographed La Serena©Jürke Grau. **B.** Capitulum detail, drawn from cultivated *Ehrhart 2002/036*, Scale bar = 10 mm. **C.** Leaf & Bract detail, *Ehrhart 2002/036*. **i.** Stem leaf v.s.; **ii.** OIB v.s.; **iii.** MIB d.s.; **iv – v.** IIB d.s. Scale bar = 5 mm.

IX *Chaetanthera* Ruiz & Pav.

Distribution and habitat. *C. frayjorgensis* is endemic to Chile, with a mostly coastal distribution beyween 28° and 32°S latitude. It is typically found at elevations around 500 m.a.s.l., among boulders above coast, on sandy dunes near the sea, and rarely in open matorral vegetation. It is not a component of the "typical" Fray Jorge vegetation.

Differential diagnosis. It is distinguished from other *Tylloma* species by having a dense indumentum of black glandular hairs on all lamina surfaces except stem leaves.

Material seen. – Chile. **III Región de Atacama** Prov. de Huasco Tal 4 Km E Carrizal Bajo, 75 Km von Vallenar, 50 m, 11.1987, Rechinger K.H: & W. 63377 (M W). **IV Región de Coquimbo** Prov. de Elqui Quebrada Los Choros, Sandflächen westlich Choros Bajos, 50 m, 11.2002, *Ehrhart* et al. *2002-107* (MSB). – La Higuera, 455 m, 10.2002, *K. Arroyo* et al. *25152* (CONC). – 5 km nördlich La Serena, Sanddünen östlich der Eisenbahnlinie nach El Romeral, 100 m, 10.1997, *Ehrhart & Grau 97-1110* (MSB). – 5 km nördlich La Serena, Sanddünen östlich der Eisenbahnlinie nach El Romeral, 100 m, 11.2002, *Ehrhart* et al. *2002-085* (MSB). – Punta Teatinos, 15 m, 11.2002, *Ehrhart* et al. *2002-091* (MSB). – Punta Teatinos, cerca de La Serena, 110, 9.2002, *K. Arroyo* et al. *25065* (CONC). – La Serena, Punta Teatinos, 110 m, 12.1987, *Landrum & Matthews 5640* (MO). – Tongoy, 1.1975, *Jiles 6217* (CONC). – Cuesta Churqui, 4 Km S of Vicuña, road to Hurtado, 350 m, 09.1940, *Wagenknecht 18567* (G). – 6 Km S de La Serena, Fundo Peñuelas, 75 m, 01.1948, *Wagenknecht 273* (CONC). – Carretera Panamericana, frente Tofo, 350 m, 10.1971, *Marticorena* et al. *1642* (CONC). – Cerros Punta Teatinos, 5 m, 02.1968, *Ricardi 5461* (CONC). – Carretera Panamericana, entre Cuesta Poroitos y Juan Soldado, 100 m, 10.1991, *Munoz, Teillier & Meza 2610* (SGO). – Elqui, 12.1948, *Collantes s.n.* (CONC). – frente Juan Soldado, 10 m, 01.1967, *Ricardi* et al. *1811* (CONC). – La Serena, 10 m, 10.1965, *Kohler 272* (CONC). – La Serena, 15 m, 11.1957, *Wagenknecht s.n.* (CONC). – Lagunillas, 50 m, 10.1971, *Jiles 5838* (CONC). – Los Choros, 60 m, 11.1961, *Jiles 3964* (CONC). – On road from La Serena to Punta de Teatinos, 20 m, 11.1935, *West 3917* (CONC GH MO). – Sur de La Serena, Km 440, 100 m 09.1965, *Gleisner 91* (CONC). – c. 40 km N La Serena, Quebrada Honda, *Ehrhart & Sonderegger 96962* (MSB). – Coquimbo, between Coquimbo and Serena, 10.1927, *Elliott 80* (E K). – La Serena, Bahia Coquimbo, Punta Teatinos, 15 m, 12.1997, *Gardner & Matthews 73* (E). – Punta Teatinos, 10 km north of La Serena, 11.1940, *Wagenknecht 18106* (G). - salida norte de La Serena, 190 m, 12.1987, *Rosas 1399* (M). – camino zur Playa Temblador y Cruz Grande, 10.1980, *Grau 2024* (BM MSB M). – Tongoy südlich Coquimbo, 11.1987, *K.H. & W. Rechinger 63699* (W). – Prov. de Limarí Carretera PA frente a Tongoy, en suelo arenoso, 10 m, 12.1953, *Kausel 3738* (F). – Ovalle, Llanos de Talinay Alto, lado oriente del Cerro Talinay, 300 m, 11.1942, *Muñoz & Coronel 1312* (SGO). – PN de Fray Jorge, Bosque Fray Jorge, steilhang zur Küste, 200 m, 11.2002, *Ehrhart et al., 2002-036* (MSB). – Fray Jorge, 300 m, 09.1935, *Munoz B-262* (SGO). – Fray Jorge, in *Haplopappus* bushes, 450 m, 5.1941, *Schwabe 228* (SGO). – Guatulame, 10.1914, *Rose & Rose 19355* (NY). – Cuesta Punitaqui, aproximadamente 6 km del pueblo, 380 m, 9.2002, *K. Arroyo* et al. *25021* (CONC). – 3 km al sur de Combarbalá, camino de tierra que sale de pueblo, 1010 m, 9.2002, *K. Arroyo* et al. *25024* (CONC). – 14 Km al Sur de Socos, 300 m, 02.1963, *Ricardi* et al. *782* (CONC). – 7 Km al Norte de Quebrada Los Almendros, 150 m, 10.1971, *Marticorena* et al. *1462* (CONC). – Cerro Sitio, 320 m, 11.1973, *Marticorena* et al. *454* (CONC). – Cordillera de Combarbalá, Ramadilla, 1800 m, 04.1971, *Jiles s.n.* (CONC). – Estancia Camarones, 75 m, 10.1961, *Jiles 3896* (CONC). – Seron, 1400 m, 03.1953, *Jiles 2363* (CONC). – Talinay, 700 m, 03.1950, *Jiles 1692* (CONC). – Zorrilla, 350 m, 03.1948, *Jiles 598* (CONC). – Zorrilla, 400 m, 02.1948, *Jiles 535* (CONC). – Zorrilla, 420 m, 09.1948, *Jiles 819* (CONC). – Fray Jorge, 10.1947, *Sparre 2957* (SGO). – Fray Jorge, 250 m, 10.1961, *Kubiztki 79* (M). – South end of Fray Jorge forest, 500 m, 11.1938, *Worth & Morrison 16420* (GH, K). – Fray Jorge, 300 m, 11.1925, *Werdermann 920* (B E G GH K M NY). – PN de Fray Jorge, camino al Bosque, 240 m, 11.1980, *Grau 2550* (MSB M). – Antes del desvio hacia Ovalle, a lado de servicentro, 11.2001, *Lopez s.n.* (CONC). – Prov. de Choapa Mina Los Pelambres, camino a Chacay, 2190 m, 2.1999, *K. Arroyo* et al. *991566* (CONC). – Almendrillo, 1800 m, 01.1964, *Marticorena & Matthei 507* (CONC). – Monte Redondo, 400 m, 11.1947, *Jiles 471* (CONC). – Quebrada Pajaritos, 280 m, 11.1974, *Marticorena* et al. *375* (CONC).

23. *Chaetanthera glabrata* (DC) F. Meigen Bot. Jahrb. Syst. 18: 456. 1894.

≡ *Tylloma glabratum* DC. Prodr. (DC.) 7 (1): 32. 1838. Typus CHILE "In Chili ad Collina legit *Macrae*" Holotypus G! [Herb. Soc. Hort. Lond., Chili, Collina, 1825, *Macrae s.n.*]

= *Tylloma stolpi* Phil. Anales Univ. Chile 85: 841. 1894. Typus CHILE "In Valle Largo Andium Santiago invenit orn. *Carolus Stolp*" Holotypus SGO 64681!
= *Tylloma rotundifolium* Phil. Anales Univ. Chile 85: 843. 1894. Typus CHILE "Locum ubi repertum ignoro" Syntypus SGO 45025! SGO 64677! SGO 71765!
- *Tylloma glabratum* var. *rotundifolium* (Phil.) Reiche Anales Univ. Chile 115: 328. 1904.
- *Tylloma glabratum* var. *stolpi* (Phil.) Reiche Anales Univ. Chile 115: 328. 1904.
- *Tylloma glabratum* var. *strictum* (Phil.) Reiche Anales Univ. Chile 115: 327. 1904.
- *Tylloma obcordatum* Phil. ex sched. GH (photo)! NYBG (photo)! W!

Nomenclature Notes The Macrae material identified as a Typus of *Tylloma glabratum* DC. in K is not a Typus (the locality on the label is given as Cumbre, Andium Claustrum, Chili, 1825). *Chaetanthera glabrata* as '*glabraia*' in error in Index Kewensis Suppl.1. Icon. in (1849) in Fl. Chil. [Gay] 3: tab. 35.

Annual monoecious herb. **Roots** filamentous. **Stems** to 20 cm, decumbent to ascending, with flowering branches. **Leaves** 12–23 mm L, indistinctly petiolate, always glabrous, petiole margins with pale ± glandular teeth; lamina 4–6.5(10) x 3.5–6.5(7.5) mm, pale to dark green, succulent, distinctly limbate, orbicular to obovate, rarely elongated with undulate margin, apex broadly acuminate to acute; apical mucro often recurved, (lamina often conduplicate in herbarium specimens), opposite to first node where multiple stems spread, then alternate up stems to capitula forming clusters at nodes and rosettes below capitula. **Capitula** size variable according to season and water availability. Mature capitula buds with obtuse apex. **Involucral bracts** imbricate, arranged in three types, initially foliaceous then reduced to entirely membranous. **Outer involucral bracts** few, reflexed from capitulum; 8–14(17) mm L, lamina 2–7.5(9.5) x 2–6(7.5) mm; as leaves but with short membranous alae to less than ½ height of bract. **Middle involucral bracts** 5.5–11(13) mm x 2.1–3(4) mm W.; outline linear-triangular to oblanceolate; central tissue mesophyllous; membranous alate, ratio of lamina to alae decreases from ⅔ to nearly entirely alate; apex cuspidate to broadly acute, shortly freely recurved mucronulate. **Inner involucral bracts** 11–17 mm x 2.2–3.5 mm; entirely membranous alate; oblanceolate, cuneate, apex broadly acute to rounded, ± emarginate, shortly freely recurved mucronulate; mostly translucent, apices coloured pale to darker pink-red. **Ray florets** (10)20–30, pistillate, pale or deep yellow-orange, dorsal surface of ligule dark red-purple with white sericeous hairs (<1 mm L); corolla 14–22 mm L, corolla tube 3.5–5 mm L, outer ligule 2.5–3.5 mm W, inner ligule (1.5) 3–5 mm L conspicuously bifid. **Disk florets** 28–48, bisexual, yellow; corolla 9–10.5 mm L, corolla tube 7.5–8.4 mm L **Styles** yellow, [ray] 9–11 mm L [disk] (8)10–12 mm L; stigma lobes 0.35–0.5 mm. **Anthers** 5–6.5 mm L (9–10.2 mm incl. filaments). **Achenes** brown, fusiform-turbinate, to 4.5 mm L shortly rostrate; carpopodium anular; pericarp pellucid, covered in oblate-lanceolate twin hairs, 95–125 x 50 µm;

testa epidermis with O-formed strengthenings with inclusions, cells linear parallel. **Pappus** ray pappus somewhat shorter than disk pappus, white, (7)8.5–10.5 mm L, 2 rows, free at base, in/dehiscent; pappus setae width 1–2(3) cells, cell width 5–10 µm, basal cilia; barbellate, barbs 50 µm L, barb base medium adhered (50%), free barb appressed, 5–6 barbs/100 µm. **Chromosome number** 2n = 4x-6 = 38. (Davies & Vosyka new count). *Ehrhart* et al. *2002/142*; *K. Arroyo* et al. *25163* (CONC).

Distribution and habitat. *C. glabrata* is found between latitudes 23°30'–33°50'S. In the northern part of its range the species is coastal, restricted to low altitudes in the coastal quebradas. In the southern parts of its range the species spreads inland and reaches higher altitudes, especially inland of La Serena and around Santiago. *C. glabrata* shows an unusually disjunct distribution, growing from Taltal to the Cordillera of Ovalle and then reappearing in the "Cuenca" (after NAVAS BUSTAMENTE, 1973) around Santiago from Cuesta Chacabuco eastwards to Queltehues in the Río Maipo valley. It occurs at elevations between 0–585 (2200) m.a.s.l. on exposed sand, gravel or scree, in dry quebrada courses, or seldom in matorral.

Differential diagnosis. *C. glabrata* is different from the other *Tylloma* species with entire leaf margins because it has no indumentum, although the pseudopetioles are dotted with glandular teeth.

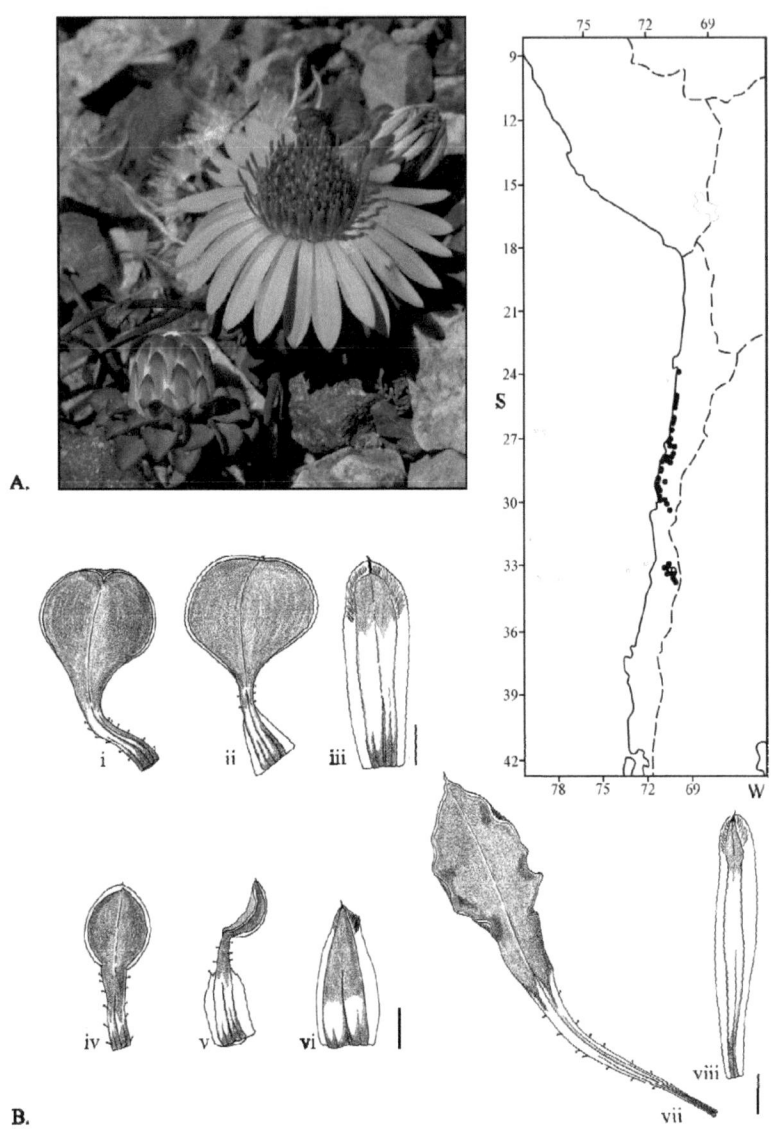

Figure 70: *Chaetanthera glabrata*. **A.** Capitulum detail, photographed Atacama©Michael Dillon. **B.** Leaf & Bract detail. *Rosas 1022* "Type A" **i.** Leaf; **ii.** OIB; **iii.** IIB; *Geisse 89* "Type B" **iv.** Leaf; **v.** MIB; **vi.** IIB; *Looser s.n.* "Type C" **vii.** Leaf; **viii.** IIB. Scale bars i – iii, iv – vi, vii – viii = 2 mm.

IX *Chaetanthera* Ruiz & Pav.

Material seen. – **Chile. II Región de Antofagasta** Prov. de Antofagasta Abzweigung von der Quebrada Taltal nach Cifuncho, Quebrada Los Zanjones, 1 Km nach Abzweigung, 660 m, 12.2002, *Ehrhart* et al. *2002/226* (MSB). – Quebrada Paposo, Mina Abundancia, 1300 m, 1953, *Ricardi 2639* (CONC). – Depto. Taltal, El Rincon, just north of Paposo, along trail to old Parañas Mine, 11.1925, *Johnston 5536* (GH). – Cuesta Paposo, 700 m, 1969, *Jiles 5424* (CONC). – Paposo, Quebrada al sur de caserio toma de Agua Salobre, parada 9.5 km abajo, 100 m, 12.1987, *Rosas 1022* (M NY). – Quebrada Paposo, 500 m, 1991, *Quezada & Ruiz 207* (CONC). – Depto. Taltal, vicinity of Paposo, Quebrada de Guanillo, above Agua de Perales, 12.1925, *Johnston 5598* (GH). – Taltal, Paposo, Quebrada de Guanillos, 16.1992, *Teillier, Rindel & P. García 2771-B* (SGO). – Quebrada Matancilla, 400 m, 1988, *Hoffmann 249* (CONC). – Caleta Bandurrias, N Quebrada Bandurrias, 1991, *Taylor 10763* (CONC). – Taltal, Palo Parado, 200 m, 1953, *Ricardi 2521* (CONC). – 17 km N of Taltal towards Paposo, following the coastal road, 0 – 40 m, 11.1991, *Eggli & Leuenberger 1759* (B). – Quebrada San Ramon, S-E Clta Hueso Parado, 100 m, 09.1941, *Pisano & Bravo 220* (CONC). – Cerca de Taltal, 20 m, 1930, *Jaffuel 978* (CONC). – 7 – 15 km N Taltal, 11.1987, *K.H. & W. Rechinger 63504* (M W). – Taltal, 20 m, 1940, *Grandjot 4400* (CONC). – Taltal, 200 m, 11.1925, *Werdermann 128* (CONC). – Depto. Taltal, Taltal, ca. 10 km east of Taltal, Quebrada de Taltal, 75 m, 10.1938, *Worth & Morrison 15805* (G GH K). – Depto. Taltal, Taltal, 50 m, 10.1925, *Werdermann 826* (B CONC E G GH K M NY). – Taltal, 20 m, 1970, *Zöllner 5374* (CONC). – Taltal, Quebrada Changos, 50 m, 1953, *Ricardi 2576* (CONC). – Depto. Taltal, Quebrada de Taltal, 11.1925, *Johnston 5115* (GH K). – Taltal, Quebrada El Nueve, 500 m, 10.1953, *Ricardi 2729* (B CONC). – Quebrada de Taltal, 1. Parada, 500 m, 12.1987, *Rosas 1076* (M). – Vicinity of Agua Grande ("Cachinal de la Costa" of Philippi), near Antofagasta-Atacama boundary, 12.1925, *Johnston 5793* (GH K). – **III Región de Atacama** Prov. de Chañaral Quebrada La Quiscuda, 300 m, 10.1941, *Pisano & Bravo 554* (CONC). – PN Pan de Azucar, quebrada de Coquimbo, 10.1991, *Muñoz, Teillier & Meza 2838* (SGO). – hasta 12 km del camino acceso costero al PN Pan de Azucar, 10.1991, *Muñoz, Teillier & Meza 2798* (SGO). – Quebrada Coquimbo al sur de las casas de Pan de Azucar, 500 m, 12.1987, *Rosas 1135* (M). – PN Pan de Azucar, 15 m, 10.1991, *Rodriguez 2638* (CONC). – Cerros frente a Puerto Flamenco, 30 m, 10.1991, *Rodriguez 2669* (CONC). – Prov. de Copiapó Caldera, dunas de Ramadas, Cerro Caracoles, 100 m, 10.1999, *Teillier 4725* (CONC). – Vicinity of Caldera, Cantera, 1922, *Gigoux 16* (GH). – El Caseron camino de Copiapó a Caldera, 150 m, 09.1941, *Muñoz & Johnson 1928* (SGO). – 50 Km antes de Copiapó, camino Vallenar, 500 m, 10.1952, *Ricardi 2214* (CONC). – Copiapó, vicinity of Copiapó, sand in quebrada just northwest of depot, 370 m, 11.1925, *Johnston 5035* (GH). – camino Vallenar-Copiapó, 39 Km de Copiapó, 700 m, 02.1963, *Ricardi* et al. *666* (CONC). – entre Copiapó y Vallenar, Km 38, 350 m, 10.1965, *Ricardi* et al. *1499* (CONC). – Panamericana 924, Westhang de Cerro Obispo, 100 m, 10.1980, *Grau 2097* (MSB). – Totoral, 5 Km por el camino hacia la mina, 10.1987, *Muñoz & Meza 2344* (SGO). – Quebrada de Totoral (Boquerones), 180 m, 11.1941, *Pisano & Bravo 782* (CONC). – Prov. de Huasco Carrizal Alto, 450 m, 09.1952, *Ricardi 2271* (CONC). – Quebrada de Carrizal, 10.1987, *Teillier 1007* (CONC SGO). – Carrizal Bajo, 9km nördlich Huasco, 50 m, 12.2002, *Ehrhart* et al. *2002/142* (MSB). – Algarrobal, 450 m, 09.1957, *Ricardi & Marticorena 4415* (CONC). – Huasco, 12.1971, *Beckett, Cheese & Watson 4705* (SGO). – Puerto Guacolda, 30 m, 11.1966, *Schlegel 5714* (CONC). – Huasco, Isla Guacolda, higher parts of island, 10 – 20 m, 10.1938, *Worth & Morrison 16227* (K). – Palincie entre Chañaral de Aceituna y Carrizalillo, 125 m, 12.1987, *Rosas 1326* (M NY). – Depto. Freirina, Carrizalillo, 300 m, 10.1971, *Marticorena, Rodriguez & Weldt 1819* (G). – **IV Región de Coquimbo** Prov. de Elqui Coastal sands near Los Lobitos 50 km west of Pan American highway, north of La Serena, 09.1987, *Hannington 36* (K). – Choros Bajos, 200 m, 10.1971, *Marticorena* et al. *1693* (CONC). – carretera Panamericana, frente al Tofo, 500 m, 10.1963, *Marticorena & Matthei 210* (CONC). – Dunas frente a Juan Soldado, 10 m, 10.1948, *Behn F s.n.* (CONC). – camino Marqueza-Condoriaco, 3 Km antes Talcuna, 500 m, 10.1971, *Marticorena* et al. *1517* (CONC). – La Serena, 80 m, 09.1928, *Barros 3001* (CONC). – Quebrada San Carlos, 700 m, 12.1974, *Edding & Villagran s.n.* (CONC). – Quebrada San Carlos, 700 m, 11.1974, *Torres* et al. *s.n.* (CONC). – Pisco Elqui, Fundo Cochiguaz, 1600 m, 10.1948, *Behn F s.n.* (CONC). – Prov. de Limarí Cordillera de Ovalle, Las Majadas, 2200 m, 12.1976, *Jiles 6415* (CONC). – Fray Jorge, 10.1947, *Sparre 2958* (SGO). – **V Región de Valparaíso** Prov. de Los Andes 1.6 km S of turnoff from main road following dirt road over Cuesta de Chacabuco (11 km S of Rinconada), 1140 m, 11.1991, *Eggli & Leuenberger 1721* (B). – 12 Km from southern entrance and 3 Km from northern entrance to paved road to Los Andes on road around Tunel Chacabuco, 3759 ft, 12.1995, *Gengler & Arriagada 156* (SGO). – Prov. de Quillota Depto. Quillota, Las Viscachas, ca. 10 km from La Dormida, 1600-1900 m, 12.1938, *Morrison 16740* (G GH K). – **Región Metropolitana de Santiago** Prov. de Chacabuco Cuesta de Chacabuco, 1150 m, 11.1970, *Marticorena & Weldt 605* (CONC). – Tiltil, 700 m, 10.1927, *Montero 136* (GH). – Prov. de Santiago Baños de Colina, 1000 m, 10.1919, *Behn K s.n.* (CONC). – Colina, 900 m, 10.1956, *Caceres s.n.* (CONC). – Villa Paulina, 2000 m, 01.1980, *Uslar 121* (CONC). – Las Condes, 2000-3000 ft, 11.1927, *Elloitt 293* (E K). – El Arrayán, 800 m, 11.1966, *Mahu 1519* (CONC). – El Arrayán, 885 m, 12.1939, *Junge s.n.* (CONC). – Straße nach Farellones, östl. Las Condes, 890 m, 11.1980, *Grau 2450* (Hrb. Grau M). – camino del viento, Santuario de la Naturaleza Yerba Loca 2190 m, 12.2002, *K. Arroyo* et al. *25163* (CONC). – Puente Nilhue, 1010 m, 02.1967, *Schlegel 5872* (CONC). – Cerro de Renca, 700 m, 12.1951, *Gunckel 21029* (CONC). – Cerros de Renca, 800 m, 10.1953, *Gunckel 26667* (CONC). – Cerro Provincia, 1900 m, 11.1932, *Grandjot 1007* (CONC). – Cerro Provincia, 1800 m, 11.1938, *C. Grandjot 3825* (GH). – San Ramón, Los Azules, 800 m, 11.1955, *Schlegel 943* (CONC). – Quebrada de Ramón, 1000 m, 10.1948, *Looser 5554* (GH). – El Manzano a Maitenes, 1200 m, 01.1928, *Montero 290* (GH). – Queltehues, 1500 m, 12.1927, *Montero 343* (CONC). – Prov. de Cordillera bei Melocotón, aufwärts von Santiago, 1200-1500 m, 02.1950, *Soyka s.n.* (B).

24. Chaetanthera kalinae A.M.R.Davies Novon 16: 51 – 55. 2006.

Typus CHILE "Chile, Región de Coquimbo, Prov. de Elqui, Straße von Vicuña zum Embalse La Laguna, 30°03'S 70°05'W, 2340 m, 11.2002, *Ehrhart 2002/061*. Holotypus MSB Isotypus CONC K MO.

Annual monoecious herb. **Roots** filamentous. **Stems** to 8 cm, originate from central node, rarely branched, creeping to ascending, decumbent, glabrous to sparsely pubescent with white, filiform hairs. **Leaves** 8–9(13) x 5–6.5 mm, indistinctly petiolate, lamina spathulate, apices obtuse to cordate, mucronate; margins of petioles dotted with glandular teeth; lamina succulent, dark grey-green, densely villous lanate (pubescence ca. 5 mm long), margins limbate, entire; stem leaves few, alternate up scape, with subtle transition to phyllaries surrounding capitula; leaves form densely villous clusters at nodes. At flowering cauline leaves usually die back. **Capitula** crateriform (goblet-shaped), 1.2–1.5 cm wide, 5–7 per plant. **Involucral bracts** imbricate, arranged in three types, initially foliaceous then reduced to entirely membranous. **Outer involucral bracts** 8.5–11.5 x 2–5.5 mm; as leaves but with short membranous alae to less than ½ height of bract; densely white villous on laminar surfaces and on lower ventral membranous margins. **Middle involucral bract** elliptic, obtuse; 9–10.5 x 4–5 mm; membranous alate, ratio of lamina to alae decreases from ⅔ to nearly entirely alate; ventral pubescence short, sericeous, white. **Inner involucral bracts** 13 x 3–4 mm, oblanceolate, hyaline, entirely membranous alate; glabrescent, tinged pale pink at apices. **Ray florets** ca. 20, yellow, single series, pistillate; corolla 14–15 mm L, corolla tube 6.5 mm L; ligule 2.6 mm wide, shortly 3-dentate, ventrally sericeous, inner lip 4 mm L, bifid, acute. **Disk florets** ca. 120, bisexual, yellow; corolla 9.5 mm L, corolla tube 8 mm L, glabrous. **Styles** yellow [ray & disk] 10 mm L, stigma lobes linear, 0.4 mm L **Achenes** brown, turbinate, 2.5–5 mm L; carpopodium anular; pericarp pellucid, densly covered in lanceolate oblong twin hairs, 90–120 x 45 µm; testa epidermis with U-formed strenthenings, cells parallel. **Pappus** white, 8–10 mm L, 2 rows, free at base, dehiscent; pappus setae width 4–8 cells, cell width 5–10 µm, basal cilia; barbellate, barbs 95–140 µm L, barb base medium adhered (50%), free barb appressed, 5–6 barbs/100 µm.

Distribution and habitat. *C. kalinae* is locally endemic between latitudes 29°30'–30°00'S. It occurs in the Cordillera de Doña Ana (Río La Laguna-Río Turbio-Río Elqui) and the Sierra de Tatul del Medio (Río del Carmen-Río Huasco) at elevations between 2300–3100 m.a.s.l. Individuals grow scattered among screes and on dry clay soil.

Differential diagnosis. *C. kalinae* is distinct because of its round tipped buds and oblanceolate obtuse pale pink IIB apices. The yellow rayed, lower altitude *C. limbata* differs from *C. kalinae* in having spathulate-orbicular to elongate-cordate laminas with rounded to attenuate bases. The perennial *C. philippii* has white rays with a dorsal stripe, but has spathulate orbicular reniform laminas with subcordate bases in common with *C. kalinae*.

Material seen. – **Chile. III Región de Atacama** Prov. de Huasco camino entre San Felix y Miña Pascua Lama, sector El Nevado, 2520 m, 01.2003, *K. Arroyo & Till-Bottraud 25090* (CONC). – **IV Región de Coquimbo** Prov. de

Elqui curva 9, sector "Mil Cuevas" camino entre Guanta y Baños del Toro, 2740 m, 01.2003, *K. Arroyo & Till-Bottraud 25076* (CONC). – Rodados Río Seco, 3 km east of Nueva Elqui, 2700-3200 m, 12.1940, *Wagenknecht 18117* (G GH). – Weg von Las Juntas im Elqui-Tal zu den Baños del Toro, 12.1984, *Hellwig 1671* (G). – Straße von Vicuña zum Embalse La Laguna km 112, 2690 m, Flußbett – Schotter, 12.1996, *Ehrhart & Sonderegger 96/1056* (MSB).

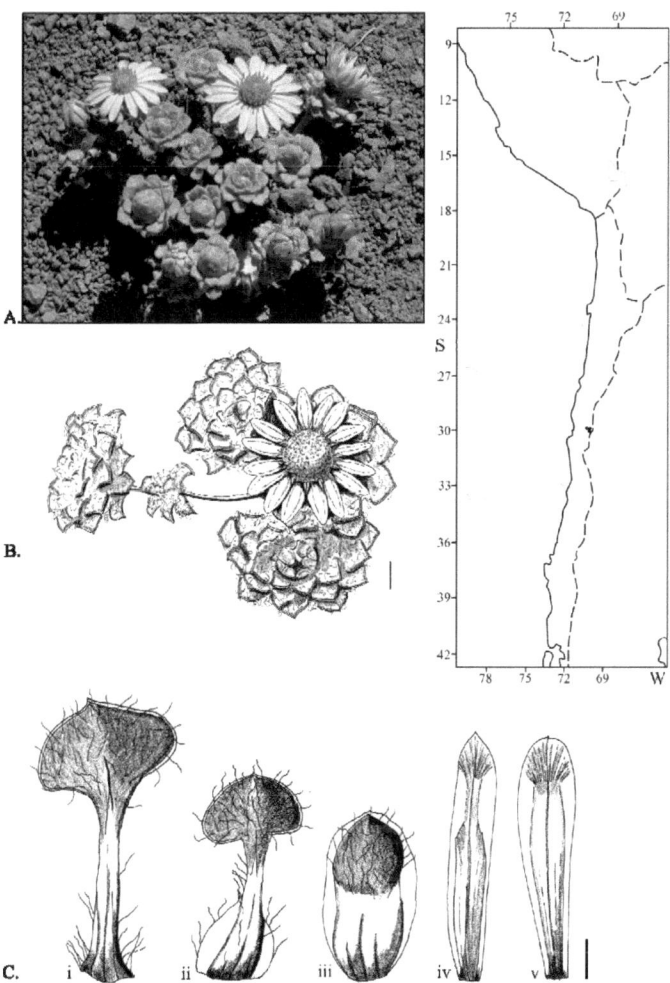

Figure 71: *Chaetanthera kalinae*. **A.** Habit, photographed Paso Agua Negra, 2003©Iréne Till-Bottraud. **B.** Habit, drawn from slide (*Ehrhart 2002*). **C.** Leaf & Bract detail, d.s. *Ehrhart & Sonderegger 961056* **i.** Stem leaf; **ii.** OIB; **iii.** MIB; **iv – v.** IIB. Scale bar = 1 mm.

25. *Chaetanthera limbata* (D. Don) Less. Syn. Gen. Compos. pp. 116. 1832.

≡ *Tylloma limbatum* D. Don Trans. Linn. Soc. London 16: 236, 301. 1830. Typus CHILE "In Chili ad Coquimbo *Caldcleugh*" Holotypus G!

= *Tylloma strictum* Phil. Anales Univ. Chile 85: 840. 1894. Typus CHILE "In deserto Atacama loco Quebrada del Rosario invenit *Franciscus San Ramon*" Holotypus SGO 64678!
= *Tylloma eurylepis* Phil. Anales Univ. Chile 85: 840. 1894. Typus CHILE "Prope Paihuano februario 1883 lectum" Holotypus SGO 45024!
- *Tylloma glabratum* var. *eurylepis* (Phil.) Reiche Anales Univ. Chile 115: 328. 1904.; Fl. Chile [Reiche] 4: 347. 1905.
- *Chaetanthera schmederi* G. F. Grandjot & K. Grandjot in Index Kewensis. Erratum Orth. var.
- *Tylloma obcordatum* Phil. ined. NYBG (photo)!; GH(photo)!
- *Chaetanthera caput-tringae* Less. ex DC in De Candolle, A. P. (1838) Prodr. (DC.) 7 (1): 32. ex sched. /only cited as a synonym. Nomen nudum.

Nomenclatural Notes With the sale of the Lambert Herbarium in which David Don worked, the Cladcleugh material was sold to Genévre and incorporated into the Delessert Herbarium. Although the protologue of Tylloma strictum quotes "ramis erectis, strictis, superius lanatis" the Typus specimen only has sparse lanate hairs on the stems.

Annual monoecious herb. **Roots** filamentous. **Stems** initially short (compacted to 1 cm), glabrous; flowering stems originate from central node, 3–12 cm, rarely branched, creeping to ascending. **Leaves** few, alternate up stems, densely whorled to rosettes below capitula; variable dimensions up to 12 x 2–3 mm; indistinctly petiolate, margins dotted with sessile or shortly stalked glandular hairs (0.2 mm L); leaf axils with dense tufts of simple filamentous hairs (1–2 mm L); lamina narrowly spathulate, apices obtuse to cordate, mucronate, succulent, conduplicate when dried, margins limbate, entire, elongate ovate–orbicular; indumentum sparse, of shortly stalked or sessile glandular hairs and simple filamentous hairs. **Capitula** narrowly obtuse to apiculate when closed; disk diameter 1–2 cm, subtended by dense filamentous hairs. **Involucral bracts** imbricate, arranged in three types, initially foliaceous then reduced to entirely membranous. **Outer involucral bracts** 9 x 3–4 mm; as leaves but with short membranous alae to less than ½ height of bract; lamina component reflexed, ovate–orbicular; indumentum as leaves. **Middle involucral bracts** 7–8 x 1.5–3 mm, ovate, cuspidate–acute, mucronate; membranous alate, ratio of lamina to alae decreases from ⅔ to nearly entirely alate. **Inner involucral bracts** 7–13 x 3–4 mm, hyaline, oblanceolate, entirely membranous alate; glandular hairs on dorsal surface only; apices bright pink. **Ray florets** ca. 15–20(30), pistillate, yellow-orange with pink dorsal stripe, dorsally densely sericeous; corolla 12–16 mm L, corolla tube 5 mm L; ligule 3 mm wide, shortly inconspicuously 3-dentate, ventrally sericeous, inner lip 1–2 mm L, bifid, acute. **Disk florets** 50+, bisexual, yellow; corollas 7–9 mm L, corolla tube 6.5–7.5 mm L, glabrous. **Styles** yellow [ray & disk] 7–9 mm L, stigma lobes, obtuse, 0.4 mm L **Achenes** brown [ray & disk] turbinate, shortly rostrate, 2–4 mm L; carpopodium anular; pericarp pellucid, densely papillose with narrowly oblate-lanceolate twin hairs 50–115 x 35 µm L

Pappus white, 7–9 mm L, 2 rows, in/dehiscent; pappus setae width 10–17 cells, cell width 5–10 µm, basal cilia; barbellate, barbs 95–140 µm L, barb base shortly adhered (<40%), free barb appressed, 7–8 barbs/100 µm.

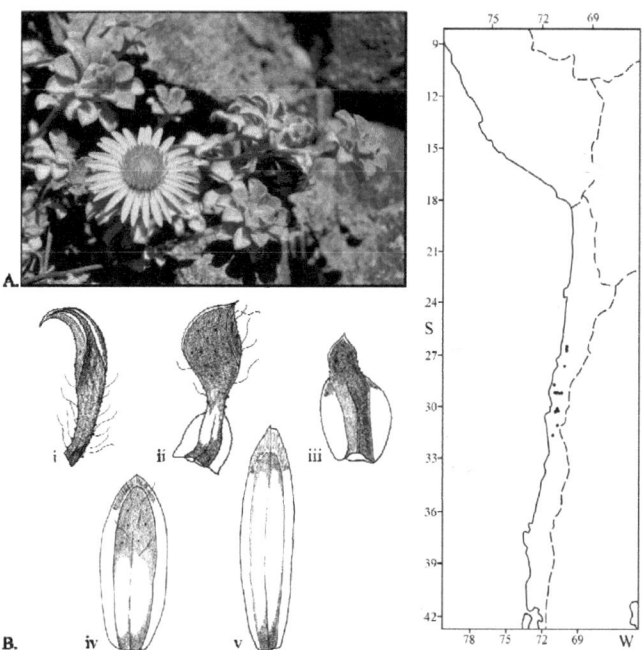

Figure 72: *Chaetanthera limbata*. **A.** Habit, photographed La Silla, 1985©Michel Grenon. **B.** Leaf & Bract detail, *Grenon 22736*. **i.** Stem leaf v.s.; **ii.** OIB v.s.; **iii.** MIB v.s.; **iv – v.** IIB d.s. Scale bar = 5 mm.

Distribution and habitat. *C. limbata* is found between 26°08' – 31°43'S latitude in the Cordillera de la Costa and Precordillera of Atacama & Coquimbo. It occurs at elevations between 800 – 2200 m.a.sl., on gravelly substrates, valley slopes and dried river beds. **Differential diagnosis.** *C. limbata* has a unique combination of glandular and filamentous trichomes on its lamina surfaces, within *Tylloma* species.

Material seen. – **Chile. III Región de Atacama** Prov. de Chañaral Diego de Almagro - Inca de Oro, 1300 m, 10.1997, *Teillier 4701* (CONC). – Inca de Oro, 1500 m, 01.1950, *Pfister s.n.* (CONC). – Prov. de Copiapó Tierra Amarilla - Las Juntas, Quebrada Molle Alto, 3 Km nach Abzweigung zur Minas Tres Marias, 970 m, 10.1997, *Ehrhart & Grau 971359* (MSB). – Prov. de Huasco 33 Km S de San Felix, 1850 m, 10.1997, *Eggli & Leuenberger 2984* (CONC). – Huasco-Carrizal Bajo, 9km nördlich Huasco, Felshänge, 12.2002, *Ehrhart* et al. *2002/142 p.p.* (MSB). – Freirina-Cuesta La Tortora, südlich der Cuesta, 870 m 12.2002, *Ehrhart* et al. *2002/128* (MSB). – **IV Región de Coquimbo** Prov. de Elqui Cerro Tololo, 1800 m, 11.1967, *Jiles 5132* (CONC). – Cuesta de Pajonales, 1000 m, 09.1957, *Ricardi & Marticorena 4385* (CONC). – Vicuña, camino a Río Hurtado, 880, m 11.1967, *Jiles 5045* (CONC). – Observatorio La Silla, Quebrada de Araya, 1700 m, 19.10.1991, *Grenon 22736* (G). – Prov. de Limarí Corral Quemado, 1100 m, 10.1956, *Jiles 3072* (CONC). – Serón, 1250 m, 11.1957, *Jiles 3298* (CONC). – Cerro Loica, 2200 m, 12.1965, *Jiles 4724* (M CONC). – Prov. de Choapa Hacienda Chillesín, 800 m, 01.1932, *Looser 2215* (GH).

26. *Chaetanthera philippii* B. L Rob. Proc. Amer. Acad. Arts 49: 514. 1913.

≡ *Chondrochilus involucratus* Phil. Fl. Atacam. pp. 27, t. 3. 1860. Typus CHILE "In montibus Pingo-Pingo [S1] (23° 40' lat m., 10700 p. s.m.) ad aquam Vaquillas [S2] dictam (25°7' lat. m., c. 9000 p.s.m.) legi" Syntypus 1 GH! NYBG! SGO 64693! W! Syntypus 2 †

= *Chondrochilus lanatus* Phil. Anales Univ. Chile 87: 14. 1894. Typus CHILE "Prope thermas Baños del Toro montis doña Ana invenit *F. Philippi*" [Coquimbo, *Philippi s.n.*] Holotypus SGO 64694! Isotypus E! GH (photo)! K! NY (photo)!
= *Chondrochilus grandiflorus* Phil. Anales Univ. Chile 87: 15. 1894. Typus CHILE "In deserto Atacama crescit, ad fontem "Acerillos" legit orn. *Villanueva* [S1], ad rivulum Río de Piquenes in altitudine 3750 m. orn. *San Roman* [S2], ad S. Andres orn. *G. Fluhmann* [S3]" Syntypus SGO 43682! SGO 64692!
= *Chaetanthera lanata* (Phil.) I. M. Johnst. Physis (Buenos Aires) 9: 325. 1929. ≡ *Tylloma lanatum* Phil. Linnaea 33: 112. 1864-65. Typus CHILE "In cordillera Doña Ana prov. Coquimbo legit orn. *Volckmann*" Holotypus SGO 64690! Isotypus US!
- *Tylloma involucratum* (Phil.) Reiche Anales Univ. Chile 115: 329. 1904; Fl. Chile [Reiche] 4: 348. 1905. (= *Chondrochilus involucratus* Phil., 1860.)

Habit monoecious perennial herb with deep, spreading root system. **Roots** thick, more or less woody. **Stems** often buried in loose talus or gravel, densely clustered to form loose cushion at substrate surface, short (2–6 cm high) lanate, flexible with sparse, whorled leaves. **Leaves** (7.4)11–24 mm L, becoming larger up stem; initially opposite becoming whorled, indistinctly petiolate; lamina (3)4–6 x (4)7–8 mm, grey-green, succulent, spathulate-orbicular, conduplicate, margins purple-red or paler limbate, mucronate; indumentum short, dense, villous. **Capitula** campanulate to broadly cylindrical, disk diameter 17–18 mm, buds broadly obovate. **Involucral bracts** foliaceous imbricate, arranged in three types. **Outer involucral bracts** (9)12–16.5 mm L; as leaves but with short membranous alae to less than ½ height of bract; lamina component 2–4.5(6) x 7–2 mm, spathulate to orbicular; indumentum of densely longly villous (hairs 2.5–4 mm L) both dorsally and ventrally. **Middle involucral bracts** scarce, often like inner bracts; membranous alate, ratio of lamina to alae decreases from ⅔ to nearly entirely alate. **Inner involucral bracts** (10)15–16 x 2.5–4 mm with entirely membranous alae; apices with thicker mesophyllous tissue, acute and shortly mucronate to obtuse and completely hyaline but with brown coloured apices; dorsal apical surface lightly pubescent. **Ray florets** (16)20–30, pistillate, white ± dark pink-purple tips or dorsal surfaces; corolla 16.5–16.9 mm L, corolla tube 6.8–7 mm L, outer ligule 2.1–2.5 mm W., inner ligule bifid, 5–5.3 mm L **Disk florets** >50, red or yellow with reddish tinge; corolla 11–12.5 mm L, corolla tube 10.5–11 mm L **Styles** yellow, [ray] 8–11 mm L; stigmas red, lobes closed, 0.3 mm. [disk] 10–12 mm L; stigmas yellow, shortly bluntly lobed (0.3 mm). **Anthers** 6.5–7 mm (11–12 mm incl. filaments). **Achenes** brown, to 5.5 mm L, both disk and ray achenes fertile; carpopodium poorly formed or absent; pericarp pellucid, glabrous or with scattered twin hairs of two different types: either spherical ca. 20 μm L or rarely oblate–lanceolate 80–100 μm L; testa epidermis surface of

IX *Chaetanthera* Ruiz & Pav.

elongated parallel cells, margins entire, cells with U/O-formed sclerenchymatous thickenings. **Pappus** white, 9–11 mm L, 2 rows, in/dehiscent; pappus setae width 4–8 cells, cell width 5–10 µm, no basal cilia; barbellate, barbs 230 µm L, barb base medium adhered (50%), free barb appressed, 5–6 barbs/100 µm.

Figure 73: *Chaetanthera philippii*. **A.** Habit, photographed Portal del Inca, 1992©Michel Grenon. **B.** Capitulum detail, photographed Vega Juntas, 1992©Michel Grenon **C.** Leaf & Bract detail, d.s. *Ehrhart 2002/194*. **i.** Stem leaf, scale bar = 5 mm; **ii.** OIB; **iii.** MIB; **iv.** IIB. Scale bar = 2 mm.

Distribution and habitat. *C. philippi* is found in the Argentinean and Chilean Andes between latitudes 25°10' – 31°20'S (33°20'S) at elevations between 2500 – 3950 m.a.s.l. It occurs on slopes and benches in amongst gravel, talus or larger rocks.

Differential diagnosis. Unmistakable in the field due to its white-rayed capitula in lanate rosettes of limbate spathulate laminas. Poorly collected/preserved material could be mistaken for the annual species *C. limbata* (yellow rays, laminas lightly covered with glandular and filamentous trichomes) or *C. kalinae* (yellow rays, laminas densely covered in lanate indumentum). The anomalous collection *Morrison 16941* (G K) is recorded as a perennial herb with orange rays with a red dorsal stripe. As such, there are no extant descriptions matching this collection.

Material seen. – **Argentina.** La Ríoja Dept. Gral. Lamadrid Cord. de la Brea, Sepultura, 3400 m, 01.1949, *Krapovickas & Hunziker 5658* (LP). – entre Río Las Cuevas y Portillo El Alto, 3700 m, 01.1949, *Krapovickas & Hunziker 5588* (LP). – **San Juan** Dept. San Juan Baños San Crispin, 3300 m, 01.1926, *Johnston 6125* (GH). – Dept. Calingasta Río Mondaca (o Río blanco o Patillos), Alojo de Mondaca, 2900 m, 02.1990, *Kiesling, Ulibarri & Krapovickas 7537* (MO). – Cuesta de la Ollita (oeste de cerro Castaño), 02.1960, *Fabris & Marchioni 2366* (LP). – Río Manantiales, Los Charqueaderos, 2800 - 3200 m, 02.1990, *Kiesling, Ulibarri & Krapovickas 7526* (LPB). – **Chile. II Región de Antofagasta** Prov. de Antofagasta Quebrada Yerbas Buenas, 3290 m, 02.1997, *Arancio & Squeo 10587* (CONC). – Gran Llano, 2970 m, 12.1996, *Arancio & Squeo 10180* (CONC). – **III Región de Atacama** Prov. de Chañaral entre Chañaral et El Salvador, Km 133, bords de la route, en haute de la Cuesta de Llanta, 2600 m, 12.1993, *Charpin, Grenon & Lazare 23662* (G). – camino Potrerillos a Pedernales, Km 15, 3200 m, 02.1966, *Ricardi* et al. *1585* (CONC). – entre El Salvador et le salar de Pedernales, km 15 de la bifurcation El Salvador/ Potrerillos, sous la station Hidroelectr. Mantandas, 2985 m, 12.1993, *Charpin, Grenon & Lazare 23674* (G). – Potrerillos, 2 Km bajando Cuesta Los Patos, 2600 m, 10.1991, *Munoz, Teillier & Meza 2752* (SGO). – Chañaral, vicinity of Potrerillos, upper part of Los Patos gulch, 2700 m, 10.1925, *Johnston 3661* (GH). – Prov. de Copiapó camino interior Copiapó-Tinogasta, Quebrada Codoceo, 3500 m, 03.1983, *Villagran & K. Arroyo 4663* (CONC). – Copiapó – Laguna Santa Rosa, 2940 m, 12.2002, *Ehrhart 2002/194* (MSB). – Quebrada de Las Vizcachas, 3500 m, 12.1963, *Ricardi* et al. *642* (CONC). – Quebrada de Vizcachas, 3550 m, 11.1956, *Ricardi & Marticorena 3780* (CONC). – camino Quebrada de Las Vizcachas, 43 Km de La Puerta, 3100 m, 02.1963, *Ricardi, Marticorena & Matthei 661* (CONC). – camino a Salar de Maricunga, Km 88, 2800 m, 01.1963, *Ricardi, Marticorena & Matthei 569* (CONC). – Quebrada Chinches, cerca de Burgos, 3020 m, 01.1973, *Marticorena, Matthei & Quezada 523* (CONC). – Quebrada de Paipote, interior Aguada Pastillo, 2500 m, 01.1973, *Marticorena* et al. *505* (CONC). – Quebrada de Paipote, above Pastillo, 2600 m, 11.1925, *Johnston 4855* (GH). – Quebrada el Colorado, 3600 m, 01.1973, *Marticorena, Matthei & Quezada 581* (CONC F). – 10 Km al Sur de Laguna El Negro Francisco, 01.1991, *Torres s.n.* (SGO). – Cordillera Río Turbio, Cerro Cadillal, 3000 m, 01.1926, *Werdermann 934* (CONC E F G GH K Mx2 NY). – Quebrada Peña Negra, 3500 m, 02.1975, *Niemeyer s.n.* (CONC). – Prov. de Huasco Quebrada Cantarito, 3400 m, 02.1981, *K. Arroyo 81560* (CONC). – Río Laguna Grande, 3100 m, 01.1983, *Marticorena* et al. *83419* (CONC). – Quebrada Yerba Buena, 3950 m, 01.1983, *Marticorena* et al. *83615* (CONC). – Weg von Conay zur Laguna Chica, zwischen dem Río Conay und der Paßhöhe, 2300 – 4000 m, 03.1986, *Hellwig 8459* (G). – Vallenar, Quebrada Alfalfa (Quebrada de los Pozos) below pass on west slopes of Cerro Negro, 3800 m, 01.1926, *Johnston 5987* (GH). – Vallenar, vicinity of Laguna Grande, 3150 m, 01.1926, *Johnston 5912* (CONC GH). – camino entre San Felix y Mina Pascua Lama, sector El Nevado, 2520 m, 01.2003, *K. Arroyo* et al. *25089* (CONC). – Quebrada de Paipote s/Vegas Juntas, 3200m, 11.1992, *Grenon 921518* (slide). – **IV Región de Coquimbo** Prov. de Elqui Baños del Toro, 3260 m, 01.1948, *Wagenknecht 259* (CONC). – Baños del Toro, Poseción, 3000 m, 12.1923, *Werdermann 212* (BM CONC E F G K M). – Cordillera Doña Ana, 600 m abajo Baños del Toro, 3200 m, 01.1992, *Arancio 004* (CONC). – Sector "Mil Curvas", camino entre Guanta y Baños del Toro, 2790 m, 01.2003 *K. Arroyo* et al. *25075* (CONC). – Refugio Quebrada Monardes – Afluente Río Figueroa, 3000 m, 02.1968, *Niemeyer s.n.* (CONC). – Prov. de Limarí Loma del Palo, 2500 m, 01.1949, *Jiles 1219* (CONC). – Cordillera de Ovalle, Río Gordito, 2850 m, 29.01.1954, *Jiles 2487* (CONC M). – Río Cenicero, 2800 m, 02.1962, *Jiles 4175* (CONC). – Prov. de Choapa steep screes above La Vega Escondida, 2700 – 2900 m, 12.1938, *Morrison 16941* (G K). – **V Región de Valparaíso** Prov. de Los Andes Portezuelo E of Portal del Inca, 2900-3000m, 11.1992, *Grenon 922323* (image). – **Región Metropolitana de Santiago** Prov. de Santiago Valle Nevado, 2500 – 3500 m, 01.2003, *Aubert, Douzet & Hurstel s.n.* (JAL - www).

IX *Chaetanthera* Ruiz & Pav.

27. *Chaetanthera pubescens* A.M.R. Davies sp. nov.

Typus CHILE "Chile, IV Región de Coquimbo, Prov. de Elqui, Curva 9, sector "Mil Curvas", camino entre Guanta y Baños del Toro, 2740 m, 01.2002, *K. Arroyo* et al. *25076* (CONC).

Planta annua herbacea, ramis florescentibus 1-10 cm altis adscentendibus ornata. Folia remote alterna, indistincte petiolata, lamina anguste spathulata, indistincte carnosa, apice obtusa vel cordata, pilis glandulosis indisticte stipitatis vel pilis lanuginosis vel dens pilis brevibus clavatis ornata. Capitula radiata diametro 1-2 cm. Involucrum pluriseriatum; involcri bracteae exteriores anguste membranacae, reflexae, involucri bracteae intermediate truncatae anguste membranaceae mucronatae, involucri bracteae interiores ovatae vel lanceolatae complete membranaceae. Flores radii feminei, 10-15 raro 20, pallide lutei, ligula indistincte dentata ad 12 mm longa. Flores disci hermaphroditi, ad 30, aurei. Stylus luteus ad 9 mm longus, appendices styli luteae, lineares ad 0,4 mm longae. Achenia turbinata, 3-4 mm longa, fusca; carpopodium annulare; pericarpium pellucidum, dense papillosum et pilis didymis 90-100 µm longis ornatum. Pappi setae albidae, 6-7 mm longae, persistentes, bi- ad triseriales, basaliter ciliatae, sursum barbellatae setis adpressis.

Annual monoecious herb. **Roots** filamentous. **Stems** short, compacted to 1 cm, sparsely covered with filamentous hairs (2–3 mm L). Flowering stems 1–10 cm, originate from central node, rarely branched, creeping to ascending. **Leaves** to 12 x 2–4 mm, remotely alternate, indistinctly petiolate, lamina narrowly spathulate, apices obtuse to cordate, mucronate; leaf axils with dense tufts of simple filamentous hairs, 1–2 mm L; margins of petioles dotted with sessile or shortly stalked glandular hairs, 0.2 mm L; lamina slightly succulent, elongate ovate-orbicular; margins limbate, entire. **Indumentum** of leaf surfaces sparsely covered with shortly stalked or sessile glandular hairs and simple filamentous hairs and dense short (0.1–0.2 mm L) clavate tipped tomentose hairs giving a farinose appearance. **Capitula** sessile, cylindrical, buds ovate apiculate when closed; disk diameter 1–2 cm. **Involucral bracts** imbricate, arranged in three series, initially foliaceous then reduced to entirely membranous. **Outer involucral bracts** 10–12 x 3–4 mm; lamina component reflexed, ovate orbicular; indumentum as leaves; narrowly membranous alate to ½ way up, ratio of lamina to alae decreases along involucral bract progression. **Middle involucral bracts** truncate–broadly acute, mucronate, 7–9 x 1.5–3 mm; narrowly membranous alate, ratio of lamina to alae further decreases from ⅔ to nearly entirely alate. **Inner involucral bracts** broadly ovate-lanceolate, entirely membranous, 7–12 x 2.5–3.5 mm; no filamentous hairs, dorsal glandular hairs and clavate hairs; pink at apices. **Ray florets** 10–15(20), pistillate, pale yellow often with pink dorsal stripe, dorsally densely sericeous; corolla ca. 12 mm L, corolla tube 5 mm L; ligule 2 mm wide, shortly inconspicuously 3-dentate, inner lip 1–2 mm L, bifid, acute. **Disk florets** < 30, bisexual, yellow; corolla 8.5 mm L, corolla tube 7.5 mm L, glabrous. **Styles** yellow [ray & disk] 8.5–9 mm L, stigma lobes yellow, linear, 0.4 mm L **Achenes** brown [ray & disk] turbinate, 3–4 mm L; carpopodium anular; pericarp pellucid, densely papillose with oblate-lanceolate twin hairs 90–100 x 50 µm L

Pappus setae white, 6–7 mm L, indehiscent, 2–3 rows; basal cilia; barbellate, barb base shortly adhered (<40%), free barb appressed.

Distribution and habitat. *C. pubescens* is found in the Pre-cordillera around La Serena and Copiapó between latitudes 27°27' – 30°10'S. It is found at elevations between 600 – 3500 m.a.s.l. in gravel, scree, and stony slopes.

Differential diagnosis. *C. pubescens* is defined by having glandular, long filamentous and short clavate trichomes on its lamina surfaces.

Material seen. – Chile. **III Región de Atacama** Prov. de Copiapó Vallenar – Copiapó, Km 10, Quebrada Cardone, 760 m, 02.1988, *Marticorena* et al. *9920* (CONC). – camino Vallenar – Copiapó, Llano los Lirio, 600 m, 11.1980, *Rodriguez & Marticorena 1618* (CONC). – Tierra Amarilla - Las Juntas, 21 Km nach Abzweigung zur Cerro Blanco. Schutthalden vor Portezuelo Tirado (= P.La Difunto), 1280 m, 10.1997, *Ehrhart & Grau 97/1339* (MSB). – Prov. de Huasco Las Juntas, Vallee du Río Copiapó, 1500 m, 11.1991, *Grenon 22784; 22785* (G). – Quebrada Algarrobal, 28 Kms de la Ruta 5, 950 m, 12.1987, *Rosas 1226* (M). – camino entre San felix y Mina Pascua Lama, Sector El Nevado, 2520 m, 01.2003, *K. Arroyo* et al. *25090* (CONC). – **IV Región de Coquimbo** Prov. de Elqui Baños del Toro, 3100 m, 01.1981, *K. Arroyo 81228* (CONC). – Baños del Toro, 3500 m, 01.1970, *Zöllner 5357* (CONC). – Curva 9, sector "Mil Curvas", camino entre Guanta y Baños del Toro, 2740 m, 01.2002, *K. Arroyo* et al. *25076* (CONC). – entre Juntas y Emblase La Laguna, 2300 m, 01.1981, *K. Arroyo 81162* (CONC). – entre Juntas y Emblase La Laguna, Km 8, 2200 m, 01.1967, *Ricardi* et al. *1729* (CONC). – Prov. de Limarí Weg von Vicuna nach Hurtado, 1420 m, 11.2002, *Ehrhart 2002/071* (MSB).

IX *Chaetanthera* Ruiz & Pav.

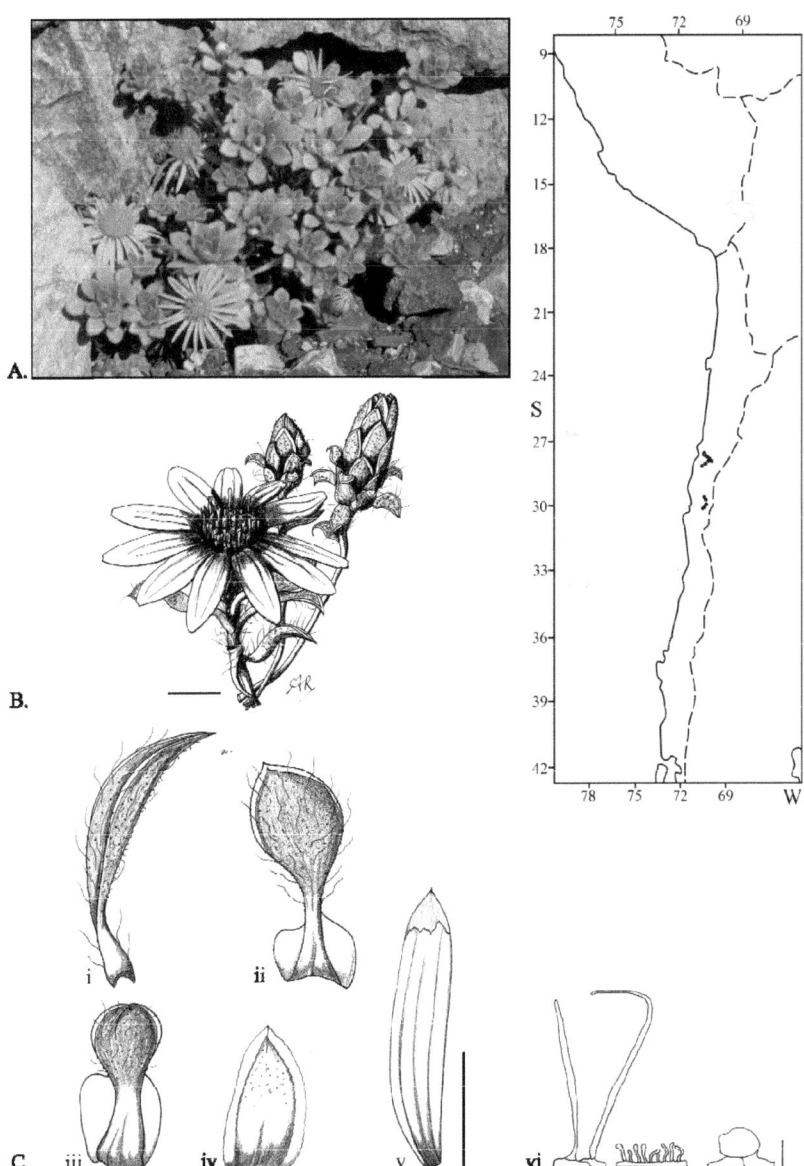

Figure 74: *Chaetanthera pubescens*. **A.** Habit photographed 1991©Michel Grenon (*Grenon 22784*). **B.** Capitulum detail, drawn from photograph taken by Mauricio Bonifacino 2006. Scale bar = 5 mm. **C.** Leaf & bract detail, d.s. *K. Arroyo* et al. *25076*. **i.** Stem leaf; **ii.** OIB; **iii.** MIB; **iv – v.** IIB; **vi.** Hair types: filamentous, tangled clavate & glandular. Scale bar i – v = 5 mm, vi = 0.05 mm.

28. *Chaetanthera renifolia* (J. Rémy) Cabrera Revista Mus. La Plata, Secc. Bot. 1: 149 – 151, fig. 25. 1937.

≡ *Elachia renifolia* J. Rémy in Fl. Chil. [Gay] 3: 315. 1849. Typus CHILE "Se cria en los arenales de las cordilleras de la provincia de Santiago, á la Polvadera, altura de 3,000 m" [*Gay 1000*] Holotypus P! Isotypus SGO (photo)!

= *Chondrochilus crenatus* Phil. Linnaea 28: 711. 1856. Typus CHILE "In Andibus prov. Santiago prope argentifodinam Las Arañas crescit" Holotypus SGO 64687! Isotypus E!
- *Tylloma renifolium* (Remy) Wedd. Chlor. And. 1: 28. t. 8. 1855.
- *Chaetanthera crenata* (Phil.) F. Meigen Bot. Jahrb. Syst. 18: 456. 1894.
- *Chondrochilus crenulatus* Phil. ex.sched. orth. var. *Chondrochilus crenatus* Phil.

Annual or possibly perennial herbs, forming monoecious rosettes. **Roots** unknown. **Stems** short, to 8 cm, mostly naked, usually with one, sometimes 2 flowering rosettes. **Leaves** to 30 mm L, opposite to whorled, forming rosettes below capitula, glabrous, indistinctly petiolate, petioles dilated to base, 20–23 x 3.3–5 mm; lamina succulent, glabrous, venation in living examples distinctly paler than leaf surface, reniform, 8.3–10 x 15–16 mm, crenate, pinkish limbate. Leaves below capitula alternate, ca. 20 mm L, lamina 8 x 12.5 mm. **Capitula** solitary terminal, sessile to shortly pedunculate. **Involucral bracts** imbricate, arranged in three types, initially foliaceous then reduced to entirely membranous. **Outer involucral bracts** 19.7–11.7 mm L, as leaves but with vestigial glabrous membranous alae to less than ½ height of bract; dense long hairs (1.7–2.5 mm L) on margins of pseudopetiole, on smaller bracts also on lower dorsal surfaces. Lamina from 2–10.8 x 1.7–5.8 mm. **Middle involucral bracts** few, 10 mm L, vestigial lamina apices, sparsely hairy below lamina component; membranous alate, ratio of lamina to alae decreases from ⅔ to nearly entirely alate. **Inner involucral bracts** 18–23 x 5–3.3 mm, lanceolate to narrowly lanceolate, glabrous, central tissue mesophyllous, margins entirely membranous alae; apices acute, apical margins tinged pink. **Ray florets** 20–40, pistillate, white ventral surface and entirely purple or reddish with white sericeous hairs dorsally; corolla 18–22 mm L, outer corolla lip 2.5–3.3 mm W., inner corolla lip 2.5–3.4 mm L, acute, apex fused or shortly bifid, corolla tube 5.8–6.7 mm L **Disk florets** 50–70, bisexual, yellow; corolla 11.5–12 mm L, corolla tube 10 mm L **Styles** [ray] 10.5–12.5 mm L, stigma lobes red or reddish brown, 0.3–0.4 mm. [disk] 11 mm L, stigma lobes yellow, 0.4 mm L **Anthers** 7.5 mm (12.5 incl. filaments). **Achenes** brown, 6.5 mm L (mature), rhomboid–turbinate; carpopodium anular but poorly formed; pericarp pellucid, densely covered with twin-hairs 90–120 µm L; testa epidermis with U-O-formed strengthenings, cell margins narrowly crenulate tesselate. **Pappus** setae white, 10.5–11.7 mm L, dehiscent; setae width 4–8 cells, cell width 5–10 µm, basal cilia; barbellate, barbs 230 µm L, barb base medium adhered (50%), free barb appressed, 3–4 barbs/100 µm.

Distribution and habitat. *C. renifolia* is endemic to the mountains east of Santiago (Chile) between latitudes 33°20'–33°50'S. It is found at elevations between 2000–4000 m.a.s.l. above the snow line, amongst rocks. **Differential diagnosis.** Unique due to its reniform crenulate laminas.

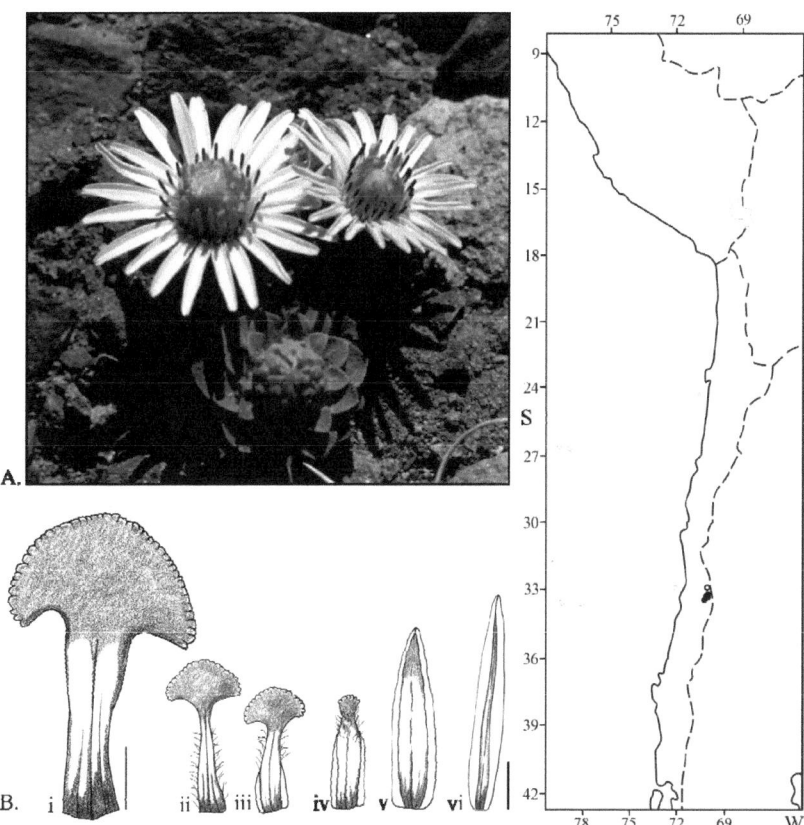

Figure 75: *Chaetanthera renifolia*. **A.** Habit, photographed El Yeso©Gustavo Aldunate. **B.** Leaf & Bract detail d.s. **i.** Lower stem leaf; **ii.** Upper stem leaf; **iii.** OIB; **iv.** MIB; **v – vi.** IIB. Scale bar i, ii – vi = 5 mm.

Material seen. – Chile. Región Metropolitana de Santiago Prov. de Santiago Valle Nevado, última pista, trayecto entre Alto Tres Puntas hasta Piedra Numerada, camino al El Plomo, 3000-3500 m, 01.1993, *Solervicens s.n.* (SGO). – El Plomo, Piedra Numerada, 3000 m, 03.1956, *Schlegel 1080* (CONC). – Cajon de La Yerba Loca, 4000 m, 02.1950, *Morales s.n.* (CONC). – Faldeos al oeste de Cerro La Parva, portezuelo entre canchas y Vega de Las Vacas, 3410 m, 02.2003, *K. Arroyo* et al. *25175* (CONC). – La Parva, Portezuelo de La Parva, 3270 m, 02.1999, *K. Arroyo* et al. *991214* (CONC). – Farellones, 2000 m, 02.1957, *Rassmusen s.n.* (CONC). – Cerro San Ramón, 3000 m, 02.1933, *Grandjot s.n.* (CONC). – Monte San Ramón, 3000 m, 02.1933, *Grandjot 1087* (CONC). – in monte San Ramón, 3000 m, 02.1933, *Grandjot s.n.* (Mx2). – Santiago, *Philippi s.n.* (E G W). – Santiago, 02.1888, *Philippi s.n.* (K). – Cordilleras de Santiago, 1856-1857, *Germain (Philippi?) s.n.* (BM G K W). – *Seibold 2850* (W). – Santiago, 1839, *Gay 1000* (P). – Cordillera de Santiago, 1861, *Philippi s.n.* (G K NY Wx2). – In cordillera de St. Jago, *Philippi 825* (BM). – Prov. de Cordilleras Tal de Los Paramillos, Cepopass am Cordillera de Santiago, Cerro Plomo, 3750 m, 02.1939, *Grandjot 3762* (GH).

29. *Chaetanthera schroederi* G. F. Grandjot & K. Grandjot Verh. Deutsch. Wiss. Verein Santiago de Chile 3: 65. 1936.

≡ *Tylloma ciliatum* Phil. Anales Univ. Chile 85: 841. 1894. Typus CHILE "Specimina duo servo, unum in provincia Colchagua a doctore *Simon* alterum a me ipso prope Vicuña s. Elqui lectum" Syntypus 1 Colchagua, *Simon s.n.* SGO76516! Syntypus 2 Elqui, Coquimbo, 1878, *Philippi s.n.* SGO 65685!

Annual monoecious herb. **Roots** filamentous. **Stems** to 15 cm, flowering stems spread from central node above short erect stem (<2 cm), ascending to decumbent. **Leaves** 12–23 mm L, indistinctly petiolate; opposite to first node where multiple stems spread, then alternate up stems to capitula, forming clusters at nodes and rosettes below capitula; indumentum of long hirsute filamentous hairs on petiole margins rarely with additional pale ± glandular teeth; trichomes composed of several vertically stacked cells; lamina pale to dark grey green, glabrescent, succulent, 4–6.5(10) x 3.5–6.5(7.5) mm, distinctly limbate, orbicular to obovate, rarely elongated with undulate margin, apex broadly acuminate to acute, apical mucro often shortly recurved, (lamina often conduplicate in herbarium specimens). **Capitula** size variable according to season and water availability. Unopened mature capitula buds with obtuse apex. **Involucral bracts** imbricate, arranged in three types, initially foliaceous then reduced to entirely membranous. **Outer involucral bracts** few, reflexed from capitulum; 7–14 mm L, lamina 2–7.5(9.5) x 2–6(7.5) mm; as leaves but with short membranous alae to less than ½ height of bract. **Middle involucral bracts** 5–11 x 2–3 mm; outline linear-triangular to oblanceolate; apex cuspidate to broadly acute, shortly freely recurved mucronulate; membranous alate, ratio of lamina to alae decreases from ⅔ to nearly entirely alate. **Inner involucral bracts** 11–17 x 2–3 mm; entirely membranous alate; oblanceolate, cuneate, apex broadly acute to obtuse, ± emarginate, shortly freely recurved mucronulate; thinly translucent; innermost bract apices coloured pale to darker pink-red. **Ray florets** 20–30, pistillate, yellow-orange, dorsal surface of ligule dark red-purple with white sericeous hairs (<1 mm L); corolla 13–20 mm L, corolla tube 3–5 mm L, outer ligule 2.5–3.5 mm W, inner ligule 3–5 mm L conspicuously bifid. **Disk florets** ca. 35, bisexual, yellow-orange; corolla 9–10 mm L, corolla tube 7.5–8.5 mm L **Styles** yellow [ray] 9–11 mm L [disk] 10–12 mm L **Anthers** 5–6.5 mm L (9–10 mm incl. filaments). **Achenes** brown, 2–4.6 mm L, turbinate; carpopodium present; pericarp pellucid, covered in lanceolate acute twin hairs, 90–100 x 40 µm. **Pappus** white, 8–10 mm L, indehiscent, free at base; pappus setae width 10–17 cells, cell width 5–10 µm; basal cilia; barbellate, barb base medium (50%) adhered, free barb appressed, 5–6 barbs/100 µm.

IX *Chaetanthera* Ruiz & Pav.

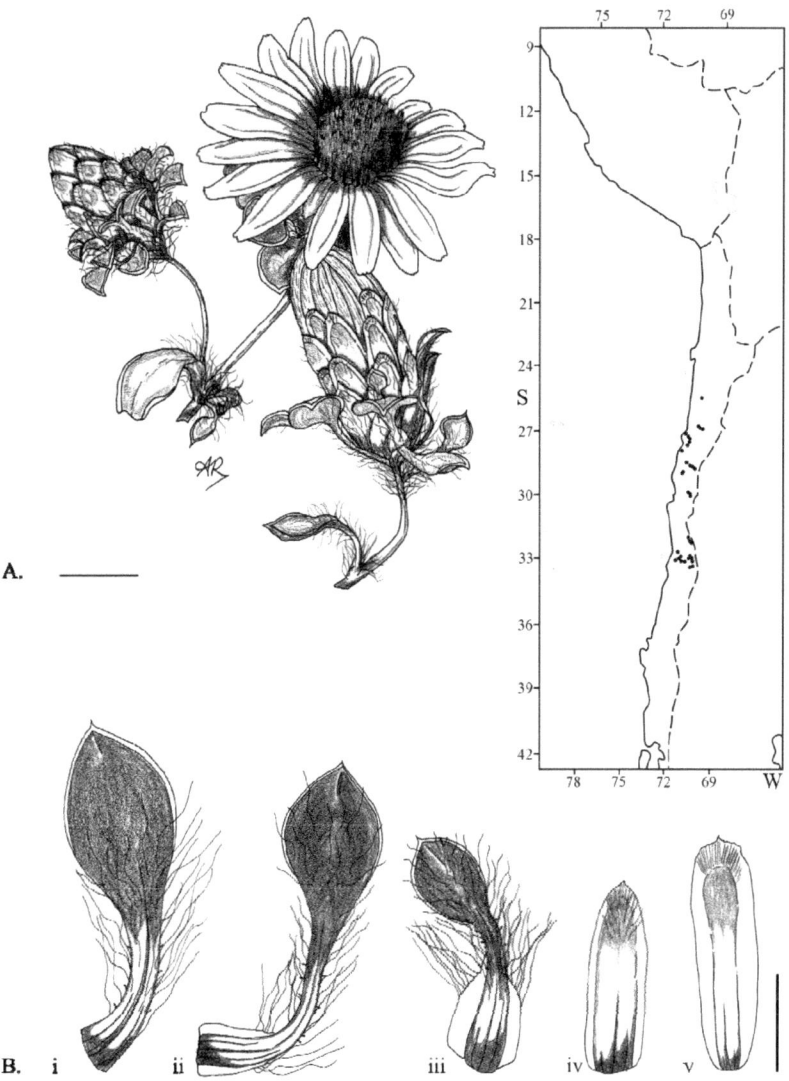

Figure 76: *Chaetanthera schroederi*. *Simon 224*. **A.** Capitulum detail. Scale bar = 10 mm. **B.** Leaf & Bract detail d.s. **i.** Stem leaf. **ii.** OIB; **iii.** MIB; **iv – v.** IIB. Scale bar = 5 mm.

Distribution and habitat. *C. schroederi* has a scattered distribution from inland of Taltal, coastal and inland hills between Vallenar and Huasco, inland of Paihuano and then Cerro del Roble, and around Santiago to Cerro Provincia. It is found at elevations between 265–3200 m.a.s.l. in exposed, sunny places in scree or fine rock slopes, and flats, but rarely in sand.

Differential diagnosis. *C. schroederi* is very close to *C. glabrata* in form and habit, but is distinguished by having long filamentous hairs composed of several stacked cells (up to 5) at leaf bases and along petiole margins. A beautiful image of this species can be found at www.flickr.com, taken by Cecilia Vidal at El Yeso.

Material seen. – **Chile. II Región de Antofagasta** Monte Amarga, 9.1890, *Morong 1215* (E NY). – Atacama desert, 1885-7, *Geisse 89* (NY). – Des. Atac., 1856, *Philippi s.n.* "Tylloma obcordatum" (W). – Prov. de Antofagasta camino a Mina Ciclon, 2310 m, 12.1996, *Arancio & Squeo 10054* (CONC). – **III Región de Atacama** Llano de Varas, 1800 m, 10.1991, *Muñoz, Teillier & Meza 2739* (SGO). – Prov. de Copiapó Cerro Bandurrias, 11.1888, *Geisse s.n.* (CONC). – Vallenar, Embalse Santa Juana, 585 m, 12.1991, *Saavedra 472* (SGO). – Km 100 de Vallenar a Copiapó, 10.1984, *Muñoz S. 1990* (SGO). – camino Salar de Maricunga, 2250 m, 01.1963, *Ricardi* et al. *549* (CONC). – PA km 765, 600 m, 11.1980, *Grau 2539* (MSB). – Prov. de Huasco Sandy plains between Huasco and Copiapó, *Cuming or Bridges 1424* (E P W). – Cachiyuyo, 900 m, 09.1957, *Ricardi & Marticorena 4468* (CONC). – Camino Longitudinal, limíte Atacama-Coquimbo, 1200 m, 01.1950, *Pfister s.n.* (CONC). – Río Laguna Grande, 2400 m, 01.1983, *Marticorena* et al. *83355* (CONC). – Straße Vallenar Copiapó; Flußbett W der Abzweigung nach Carrizal Bajo, 11.1987, *K.H. & W. Rechinger 63337* (W). – Alto de Carmen, 828 m, 10.2002, *K. Arroyo* et al. *25150; 25151* (CONC). – **IV Región de Coquimbo** Prov. de Elqui Environs de La Serena, X.1836, *Gay 388* (P). – Coquimbo, *Gay s.n.* (G Px2). – Coquimbo, *Gay 983* (G). – Coquimbo, *Ball (Philippi) s.n.* (Ex2). – Coquimbo, *Cladcleugh s.n.* (G). – Coquimbo, *Philippi s.n.* (W#22582). – Coquimbo, 1843, *Bridges 1422* (G). – Environs de La Serena, 10.1836, *Gay 388* (P). – Cordillera de Paihuano, 2900 m, 12.1942, *Gajardo s.n.* (CONC). – Paihuano, 1000 m, 10.1937, *Gajardo s.n.* (CONC). – Paihuano, Fundo Cochiguaz, 1600 m, 10.1948, *Behn F s.n.* (CONC). – Pisco Elqui, Fundo Cochiguaz, 1600 m, 10.1948, *Behn F s.n.* (CONC). – Rivadavia, 800 m, 11.1923, *Werdermann 98* (E G GH). – Paihuano, 2000-3000 ft, 10.1927, *Elloitt 38* (E K). – Prov. de Choapa Cordillera de Illapel, Río Totoral, 1400 m, 10.1962, *Jiles 4361* (CONC). – La Vega Escondida, 2700-2900 m, 12.1938, *Morrison 16941* (G K). – **V Región de Valparaíso** Prov. de Petorca Cuesta Alicahue, 10 km south from Chincolco, 1100 m, 11.1970, *Simon 224* (GH). – Prov. de Quillota PN La Campana, Cerro El Roble, 1630 m, 11.1999, *K. Arroyo* et al. *994050; 994024* (CONC). – Cerro del Roble, 1800 m, 12.1934, *Garaventa 3171* (CONC). – Cerro La Campana, 1500 m, 11.1947, *Bultmann s.n.* (CONC). – Cerro Vizcacha, 2000 m, 12.1948, *Garaventa 6419* (CONC). – Cerro Caquis, ca. 15 km east of Melón, 1500 – 1900 m, 12.1938, *Morrison 16835* (G GH K). – Prov. de Los Andes Coquimbito, 850 m, 09.1952, *Mancilla s.n.* (CONC). – Río Blanco, 4000-5000 ft, 10.1927, *Elliott 216* (E K). – **Región Metropolitana de Santiago** Prov. de Melpilla Las Viscachas, ca. 10 km from La Dormida, 1600 – 1900 m, 12.1938, *Morrison 16740* (G GH K). – Prov. de Santiago Santiago, 1856, *Philippi s.n.* (W). – In collibus aridis prov. Santiago, *Philippi 706* (B G P). – Collina, 1825, *Macrae s.n.* (G K). – Santiago, 1861, *Philippi s.n.* (G K). – 5 Km Antes de Perez Caldera, 2400 m, 01.1964, *Marticorena & Matthei 703* (CONC). – Cerro Provincia, 2000 m, 11.1946, *Castillo s.n.* (CONC). – Potrero Grande, 3200 m, 11.1936, *Grandjot s.n.* (CONC). – Potrero Grande, 3200 m, 02.1933, *Grandjot 1088* (CONC). – Cerro Provincia, Kordillere bei Santiago, 1900 m, 11.1932, *C. & G. Grandjot s.n.* (GH M). – Cerro Provincia, 1800 m, 11.1938, *C. Grandjot 3825* (GH). – Prov. de Chacabuco Altos del Roble, Hacienda de Chicauma, 1600 m, 12.1983, *Villagran 4762* (CONC). – Quilpue, auf dem Robleberg, 1500 m, 12.1963, *Zöllner 263* (B). – Prov. de Cordillera Cordillera, Paso Cruz 34°, 1700 m, 01.1892, *Kuntze s.n.* (NY). – Cordilleras de Santiago, 1856-57, *Philippi s.n.* (BM W). – In glareosis montium ad Río Colorado, 11.1827, *Poeppig 109(30) (Diar. 588)* "Chaetanthera caput-tringae" (M NY P W).
S.l.d. Cassinjal, *King #2049?* (E). – La Cuesta de Zapata, Chili, *Dr. Gillies 21* (E K). – Aconcagua and Cuesta Zapata, *Bridges 126* (E K W). – Valparaíso, *Bridges s.n.* (W). – Valparaíso, 04.1834, *Cuming 58* (NY W). – Cordilleras of Chili, *Cuming 314* (Ex2 GH K P). – *Poeppig s.n.* (NY). – Chili, *Dr. Gillies s.n.* (NY). – Chili, *Gay s.n.* (NY W). – *Seibold 3043* (W). – 1839, *Style s.n.* (G).

30. *Chaetanthera spathulifolia* Cabrera Revista Mus. La Plata, Secc. Bot. 1: 141 – 143. fig. 22. 1937.

≡ *Carmelita spathulata* Phil. Anales Univ. Chile 85: 831. 1894. Typus CHILE "Habitat in Andibus Illapelinis loco dicto El Rodeo, aestate 1888 lecto" [El Rodeo, 1888] **Lectotypus hic loc. designatus** SGO 43867! Isotypus SGO 64699! BM!

– *Chaetanthera spathulata* (Phil.) Hauman non Lessing Anales Soc. Ci. Argent. 86: 318. 1918.

Habit perennial, monoecious herb forming clumps of single-flowered rosettes with deep, spreading root system. **Roots** thick, woody, rhizomatous, branching. **Stems** flexible, at substrate surface short (2–4 cm high) with densely whorled leaves; whole rosettes to 6 cm diam. **Leaves** lower stem leaves 4–6 x 2 mm, obtuse, linear, (scale-like), ventral surface with sparse short indumentum; upper stem leaves to 30 mm L, indistinctly petiolate, pale green, succulent, spathulate; petioles to 10 mm; lamina 5–8.5 x 6.5–10 mm, decurrent to petiole, apices truncate, limbate; indumentum dense, lanate, 2.5–3.5 mm, on lamina and petioles, especially margins; leaves initially opposite then whorled. **Capitula** sessile, solitary; disk diameter generally > 2 cm (Argentinian material tends to be smaller). **Involucral bracts** imbricate, arranged in three types, initially foliaceous then reduced to entirely membranous. **Outer involucral bracts** 14–22 mm L, as leaves but with short membranous alae to less than ½ height of bract; barely dilated with alae at bases; lamina 5–8.5 mm W., spathulate, truncate limbate, becoming smaller along series. **Middle involucral bracts** 16–24 mm L; membranous alate, ratio of lamina to alae decreases from ⅔ to nearly entirely alate, alae 0.6–0.9 mm W.; indumentum on distal, dorsal surfaces of alae and both lamina surfaces; innermost series of MIB has distinctly long bracts with terminal obovate limbate lamina, often prominently exerted around capitula. **Inner involucral bracts** 17.5–21 x 2.5–3.5 mm, not entirely membranous; apices sparsely villous, terminated with small, thickened, sometimes narrowly spathulate appendages. **Ray florets** 20–30, pistillate, yellow with red dorsal surfaces, barely exceeding the involucral bracts; corolla 25.4 mm L, corolla tube 8 mm L, outer ligule 3.3 mm W., inner ligule shortly bifid, 0.3 mm L **Disk florets** 50–70, bisexual, orange; corolla 15.5 mm L, corolla tube 12.5 mm L **Styles** yellow [ray] 16.4 mm L; stigma lobes closed. [disk] 16.5 mm L; stigmas shortly bluntly lobed (0.3 mm). **Anthers** 9.5 mm (18.5 mm incl. filaments). **Achenes** brown, 9.2 mm L; no carpopodium; pericarp pellucid, glabrous or with minute spherical twin hairs to 20 μm. **Pappus** white, 2 rows, dehiscent; (9.0)13.5–15 mm L, 4–8 cells wide, cells 5–10 μm W.; no basal cilia; barbellate (5–6 barbs/ 100 μm), barbs 190 μm L, barb base shortly adhered (<40%), free barb appressed.

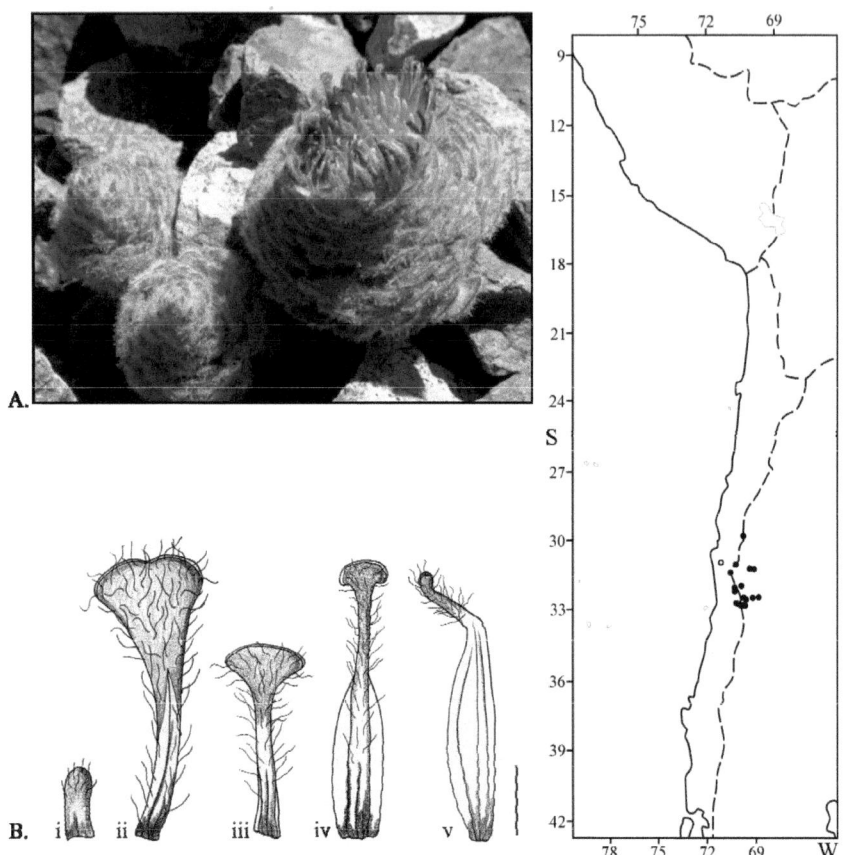

Figure 77: *Chaetanthera spathulifolia*. **A.** Habit – unknown source, www.flickr.com **B.** Leaf & Bract detail d.s. *Malme 2909* **i – ii.** stem leaves; **iii.** OIB; **iv.** MIB; **v.** IIB. Scale bar = 5 mm.

Distribution and habitat. *C. spathulifolia* is distributed in the high Andes on both sides of the Chilean: Argentinean border from the Cordillera de Doña Ana (Chile) to the Cordillera de Espinazito and Aconcagua (Argentina) between 29°50' – 33°00'S latitude. It is found at elevations between 2800–3720 m.a.s.l. in gravel, rocky areas, screes, on slopes or ridges.

Differential diagnosis. Unusual because it has no entirely membranous IIB like all other *Tylloma* species, *C. spathulifolia* is also distinctive because of its truncate spathulate laminas.
Morrison 16943 (G GH K) purportedly has white rays. Together with its sparsely pubescent leaves and bracts, and the obtuse inner bracts this collection is something of an anomaly.

Material seen. – **Argentina. San Juan** Junta l'Uspallata, 3700 m, 01-02.1897, *Wilczek 198* (G). – Cordillera del Espinazito, La Colorada, au dessous de 3000 m, 01.1897, *Bodenbender s.n.* (G). – Dept. Calingasta Río Manatiales, al NW de Calingasta, 3500 – 3700 m, 14.02.1990, *Kiesling, Ulibarri & Krapovickas 7467* (NY). – Calingasta, Río de Palque, 02.1960, *Fabris & Marchioni 2382* (LP M). – **Mendoza** Río Blanco, 3720 m, 01.1980, *Miehe 91/80/22.1* (LPB). – Río Las Cuevas, Cerro Cruz de Cañon, 3250 m, 04.1980, *Miehe 92/80/7.4* (LPB). – Las Cuevas, camino entre Cristo Redentor y Mendoza, 3000 m, 02.2003, *K. Arroyo et al. 25098* (CONC). – Las Cuevas y Cristo Reducto, 3200-4000 m, 02.1940, *Ruiz Leal & Dawson 72* (LP). – Río Vacas, NE of Aconcagua, 3620 m, 03.1980, *Miehe 192/80/31.3* (LPB). – In viciniis montis Aconcagua, Puente del Inca, 02.1903, *Malme 2940* (F G GH). – In viciniis montis Aconcagua, Las Cuevas, 02.1903, *Malme 2909* (K NY). – Los Penitente near Puente del Inca, 3000 m, 02.1931, *King 706* (BM LP). – Puente del Inca, 2700 – 3000 m, 01.1926, *King 51* (BM). – Las Cuevas, 03.1901, *Spegazzini 2546* (CONC).
Chile. IV Región de Coquimbo Prov. de Elqui Baños del Toro, 3260 m, 02.1939, *Wagenknecht s.n.* (CONC). – Prov. de Limarí Cordillera Gordito, 3000 m, 02.1932, *Miranda, M de s.n.* (CONC). – **V Región de Valparaíso** Prov. de Petorca Paso Leiva, 1500 m, *Zöllner 5354* (CONC). – 5 km south of Junta de Piquenes, 3500m, 02.1939, *Morrison 17297* (CONC F G K). – Prov. de Los Andes Aconcagua, 3000 – 3500 m, 1897, *Dessauer s.n.* (M). – Upper valley of Aconcagua, above Inca Hotel, 13000 ft, 1896-7, (Fitzgerald Exped.) *Gosse s.n.* (K). – Río Colorado, Paso Las Minas, 3000 m, 02.1967, *Zöllner 1733* (CONC). – Valle del Aliste, 3000 m, 02.1972, *Zöllner 5502* (CONC). – Portillo, 2870 m, 03.1952, *Ricardi s.n.* (CONC). – Portillo, 3500 m, 29.12.1946, *Sparre 2464* (K). – Portillo, 9400 ft, 02.1935, *Chapin 1099* (NY).
S.l.d. Portezuelo Pehuenche, 02.1882, *Sage s.n.* (BM).

31. *Chaetanthera splendens* (J. Rémy) B. L Rob. Proc. Amer. Acad. Arts 49: 514. 1913.

≡ *Elachia splendens* J. Rémy Fl. Chil. [Gay] 3: 315. 1849. Typus CHILE "Se cria en la provincia de Santiago, collina Chimbarongo etc...." **Lectotypus hic loc. designatus** *Gay 2304* (F!) Icones Weddell (1855) Chlor. And. 1: fig. 8.

– *Tylloma splendens* (J. Rémy) Wedd. Chlor. And. 1: 27, fig. 8. 1855.

Nomenclature Notes There is an inconsistency with the Typus locality. The protologue cites the following: "Se cria en la provincia de Santiago, collina Chimbarongo etc..." This is clearly a listing of several Typus localities, indicating the existence of syntypes. There is one sheet with a fragment that is undeniably *C. splendens*, on which the locality "Santiago" is given (*Gay 2304;* F). However, *C splendens* occurs neither in Santiago nor further south. The only record of a "Chimbarongo" is in the VI Región (34° 40'S 71° 03'W) and it is not even close to the modern or historical distribution of *C. splendens*. One collection exists - "Coquimbo, sur les collines arides et ? des Llanos de Guanta, 2003 m, 11.1836, *Gay 387*" (GH, P) - which matches the *C. splendens* locality (29° 46'S 70° 17'W).

Annual monoecious herb. **Stems** to 12 cm, spreading from central node above short (<2 cm) erect stem; flowering stems ascending-decumbent, glabrous. **Leaves** ca. 15 mm L, glabrous, indistinctly petiolate, petiole 9.5–10.5 x 1.6–2 mm W., margins lightly dotted with minute glands/teeth; lamina 4.5 x 7.5 mm, glabrous, glaucous green, succulent, spathulate/flabellate to rhomboid; margins limbate, loosely dentate to distinctly regularly (teeth 1 mm L, triangular). **Capitula** campanulate to cylindrical, bud tips obtuse to rounded, disk diameter ca. 0.75 cm. **Involucral bracts** imbricate, arranged in three types, initially foliaceous then reduced to entirely membranous. **Outer involucral bracts** 9–14 mm L; pseudopetiole 4–8 x 2–1.5 mm; as leaves but with short membranous alae to less than ½ height of bract; lamina 4–6 x (3)6–9 mm, spathulate-rhomboid, dentate. **Middle involucral bracts** (7.5)9–15 x 3–6 mm, membranous alate, ratio of lamina to alae decreases from ⅔ to nearly entirely alate; lamina rapidly reduced to apical mucro with green mesophyllous tissue below, broadly ovate-lanceolate, bases truncate to cuneate, apices broadly acute. **Inner involucral bracts** (1 series only) entirely membranous alate; completely transparent, linear-lanceolate, bases cuneate, apices green edged with pink-purple, broadly acute, 15.5–17 x 3.5–4.2 mm. **Ray florets** ± 23, pistillate, pale to golden yellow with dark red-pink dorsal apices; corolla 14.5–16 mm L, corolla tube 4.5–4.8 mm L, outer ligule 3–4 mm W., inner ligule 4.5–5 mm L (not conspicuously bifid). **Disk florets** 20–30, bisexual, yellow; corolla 10–10.5 mm L, corolla tube 8.2–9 mm L **Styles** yellow, [ray] 9–10 mm L; stigma lobes 0.5 mm. [disk] 10.8–11.3 mm; stigma lobes 0.3–0.5 mm L **Anthers** 5.5–6 mm L (10–10.8 mm incl. filaments). **Achenes** brown, 2.5–3.5 mm L (immature); carpopodium anular; pericarp pellucid, covered in long, lanceolate acute twin hairs, 170 x 50 μm. **Pappus** white, 2 rows, dehiscent, dimorphic [ray] 6–7 mm L; [disk] 9.5–10.5 mm L

Distribution and habitat. *C. splendens* is endemic to Chile and distributed in the Cordillera Doña Ana, Cordillera de Ovalle, and Cordillera de Combarbalá between 29° 47'–31° 24'S. It is found at elevations between 2000–3250 m.a.s.l.

IX *Chaetanthera* Ruiz & Pav.

Differential diagnosis. *C. flabellifolia* and *C. splendens* are closely related. Their genetic affinity (HERSHKOVITZ et al. 2006) indicates these two taxa probably represent a recent speciation event, rather than a parallel development of leaf form. *C. splendens* is distributed further south, at somewhat lower elevations, and has smaller, more oval leaves than *C. flabellifolia*.

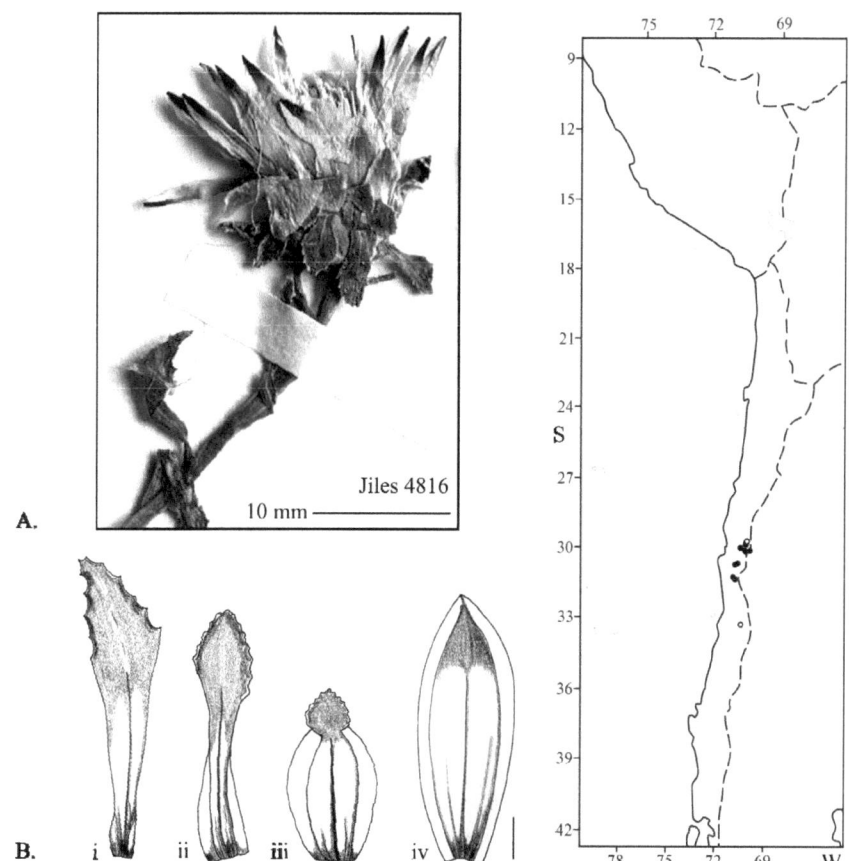

Figure 78: *Chaetanthera splendens. Jiles 4816.* **A.** Capitulum detail. **B.** Leaf & Bract detail, d.s. **i.** stem leaf; **ii.** OIB; **iii.** MIB; **iv.** IIB. Scale bar = 2mm.

Material seen. – **Argentina San Juan** Depto. Calingasta Manantiales, 3500 m., 02.1971, *Volponi & Zardini 181* (LP). – **Chile. IV Región de Coquimbo** Prov. de Elqui Llanos de Guanta, 2003 m, 11.1836, *Gay 387* (GH P). – Entre curvas 9 y 10 camino entre Guanta y Baños del Toro, 2700 m, 01.2003, *K. Arroyo* et al. *25081* (CONC). – Cordillera de Paihuano, La Cuchilla, 3000 m, 12.1942, *Gajardo s.n.* (CONC 57363, CONC 21122). – Prov. de Limarí Cordillera de Ovalle, Río Molles, 2700 m, 11.1952, *Jiles 2306* (CONC). – Cordillera de Ovalle, Morro Blanco, 2300 m, 10.1949, *Jiles 1567* (CONC LP). – Cordilleras de Combarbalá, Potrero Grande, Hacienda Ramadilla, 2700 m, 01.1966, *Jiles 4816* (CONC M). – Cordillera de Combarbalá, La Hierba Loca, 2600 m, 02.1962, *Jiles 4230* (CONC).

32. *Chaetanthera villosa* D. Don ex Taylor & Phillips Philos. Mag. Ann. Chem. 11: 391. 1832.

Typus CHILE "Herb. Gill." [Ascent to El Planchon, Andes of Mendoza. *Dr. Gillies*] Holotypus OXF (# 19)! Isotypus E! K!

= *Carmelita formosa* C. Gay in DC. Prodr. (DC.) 7 (1): 15. 1838. Typus CHILE. "Perenne in petrosis summarum Andium Chilensium ad Talcaregue febr. flor. legit cl. Gay" [Cl. Gay in litt. Decaisn. mss. cum icon. flor.] Isotypus P(*Gay 326*)! G! (microfiche)

Perennial monoecious herbs forming scattered, single-flowered rosettes with deep spreading root system. **Roots** thick (to 3 cm), woody, branching, rhizomatous, can be more than 50 cm long. **Stems** woody, flexible, short at substrate surface with densely whorled leaves, whole rosettes to 6 cm diam. Lower **stem leaves** 4–6 x 1.5–2.5 mm, loosely whorled, sessile (scale-like), linear, obtuse, indumentum short, sparsely lanate, mostly ventral or on margins. Upper stem leaves14–22 x 6–7 mm, densely whorled, sessile, linear-spathulate, obtuse (sometimes broadly acute in young plants). Midrib barely visible, leaves flat; indumentum densely straight villous (more on dorsal than ventral surface), silvery, hairs 3–4 mm L **Capitula** sessile, single, disk diameter usually > 2 cm. **Involucral bracts** imbricate, arranged in three types, initially foliaceous then reduced to entirely membranous. **Outer involucral bracts** 19–21 x 4–5 mm, as leaves but with short membranous alae to less than ½ height of bract; barely dilated at bases. **Middle involucral bracts** 24–29 mm L, membranous alate, ratio of lamina to alae decreases from ⅔ to nearly entirely alate; alae 0.5–1 mm W.; lamina component spathulate obtuse to broadly acute, densely pubescent, especially on dorsal surface. **Inner involucral bracts** 27–29 x 3–4 mm, entirely membranous alate, linear-lanceolate, apices acute, sparsely shortly pubescent. **Ray florets** 20–35, pistillate, yellow, apices and dorsal (and sometimes also ventral) surfaces dark reddish-brown; corolla 23–25 mm L, corolla tube 9–10 mm L, outer ligule shortly irregularly tridentate, 2.9–3.5 mm W., inner ligule bifid, 1.5–2 mm L **Disk florets** ca. 50, bisexual, yellow, distal part of corolla tube red-brown; corolla 16–17.5 mm L, corolla tube 13.5–15 mm L **Styles** [ray] 15 mm L, yellow, lobed but closed, obtuse. [disk] 15 mm L, yellow, stigma lobes green, obtuse to acute, closed, 0.5 mm L **Anthers** 9–9.5 (16.5–17.5 incl. filaments). **Achenes** brown, flattened pyriform, to 8 mm L; ray achenes sterile; carpopodium anular; pericarp pellucid, glabrous or rarely sparsely dotted with minute twin hairs to 25 µm L; testa epidermis surface of elongated parallel cells, margins entire, cells with O-formed sclerenchymatous thickenings. **Pappus** white, 2 rows, dehiscent; 13.5–15 mm L, 4–8 cells wide, cells 5–10 µm W.; no basal cilia; barbellate (5–6 barbs/ 100 µm), barbs 190 µm L, barb base medium adhered (505), free barb spreading. **Chromosome number:** 2n = 2x = 22 (Davies & Vosyka, 2002). *Davies & Grau 2002/058* (M).

Distribution and habitat. *C. villosa*'s distribution follows the upper reaches of the chain of active volcanoes from south east of Santiago (Chile) to San Carlos de Bariloche (Argentina) between latitudes 33°40' – 40°29'S. It is a colonizing species of volcanic scree and debris. It forms large

IX *Chaetanthera* Ruiz & Pav.

populations in Alpine/ Andean vegetation above and below the tree line. It is found at elevations between 1000–3000 m.a.s.l. and grows at lower altitudes further south.

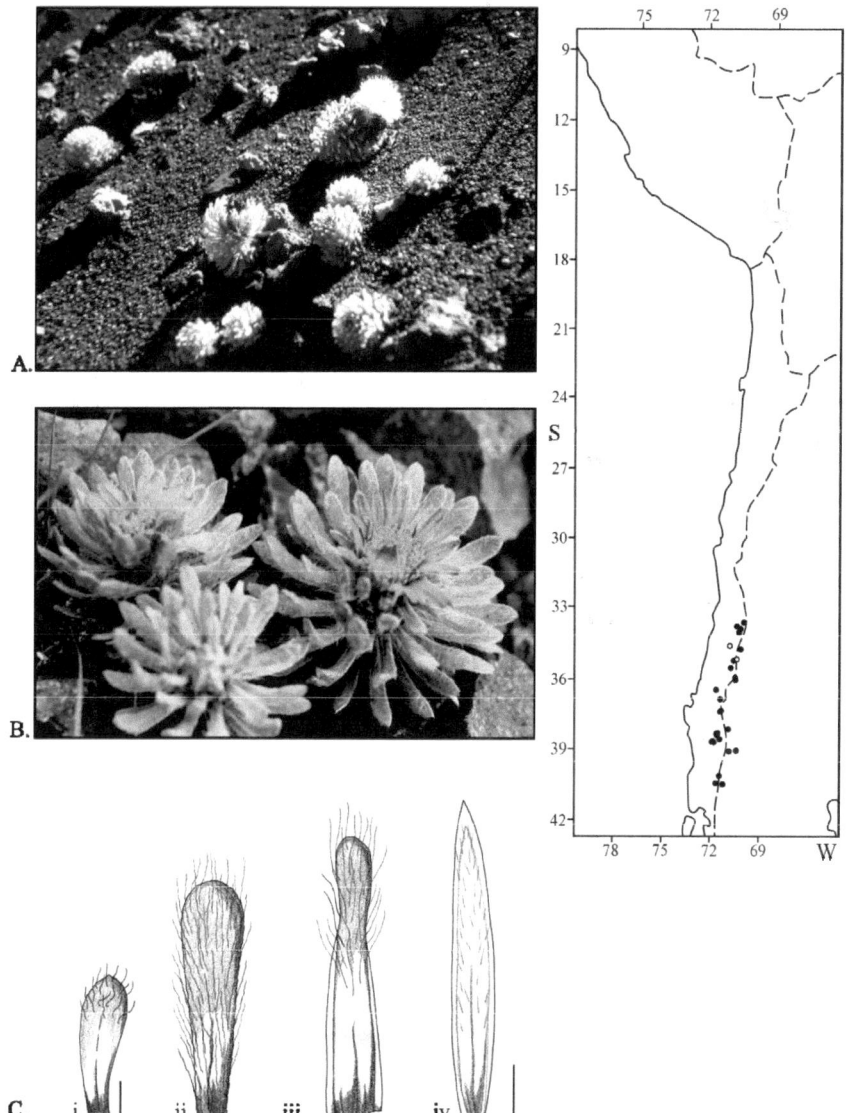

Figure 79: *Chaetanthera villosa*. **A.** Habit, photographed Lonquimay, 2002©Alison Davies. **B.** Capitulum detail, photographed Chillán, 1988©Jürke Grau. **C.** Leaf & Bract detail. *Davies & Grau 2002/46* **i.** Lower stem leaf; **ii.** Upper stem leaf; **iii.** MIB; **iv.** IIB. Scale bar i, ii – iv = 5 mm.

Differential diagnosis. Close to *C. spathulifolia*, *C. villosa* is distinguished by its linear-spathulate laminas, and membranous, lanceolate IIB. These two species have adjacent allopatric distributions, occuping the same ecological niche. It seems very susceptible to insect herbivory, often corollas are affected and many capitula were observed to contain larvae around the achenes.

Material seen. – **Argentina. Mendoza** St. Raphael et la vallee du Río Atuel, entre las Choicas de le col Tingui[ri]rica, 3000 m, 01.1897, *Wilczek 149* (G). – **Ñeuquen** Malargüe, Paso Pehuenche, Cerro detrás del Puesto de Gendarmeria, 01.1963, *Boelcke, Bacigalupo & Correa 10393* (LP). – Dept. Loncopue, chenque Pehuén ceito Butahuao, 2300 m, *Carique 1064* (CONC). – Norquineo (Pulmari), 7000 ft, 02.1926, *Comber 508* (E K). – Río Negro, Cerro Nireco, 1600 m, 03.1941, *Neumayer 492* (LP). – Cerro Bayo, 9000 ft, 01.1925, *Comber 338* (E K). – Estancia Meliquinn, Cerro Repollo, 2100 m, 02.1965, *Schajovskoy s.n.* (M). – Río Negro, Cerro Nireco [Laguna Blanca], 1600 m, 03.1941, *Neumayer 492* (LP). – Fuentes del Río Caleufú, 2100 m, 01.1896, *Roth s.n.* (LP). – Lago Villarino, 1896, *Roth s.n.* (LP). – Lago Laca [Lacar], 01.1898, *Spegazzini 1595* (LP).
Chile. Región Metropolitana de Santiago Prov. de Santiago Cordillera Laguna Yeso, 3000 m, 12.1956, *Lopez s.n.* (CONC). – RN Río Clarillo, Sector Las Cruces, 2870 m, 01.2000, *K. Arroyo et al. 20646* (CONC). – Nacimiento del Río Claro, 2930 m, 01.1978, *Niemeyer s.n.* (CONC). – Prov. de Cordillera Valle del Yeso, *Philippi s.n.* (BM). – Cordillera de Santiago, *Philippi s.n.* (F). – Valle del Yeso, 01.1866, *Herb. F. Philippi 1050a* (SGOx2). – **VI Región del Libertador G. B. O'Higgins** Prov. de Colchagua El Teniente, near Río Coya above E, 2900 – 3000 m, 01.1925, *Pennell 12326* (F GH K NY SGO). – **VII Región del Maule** Prov. de Curicó Baños del Azufre, 2100 m, 02.1959, *Barrientos 2068* (CONC). – Prov. de Talca Cord. de Talca, Str. del Descabezado [Chico or Grande?], 1877, *Williams s.n.* (LP). – Cord. del Maule, Westküste der Laguna südlich der Zufahrt, 2200m, 01.1981, *Grau 2912* (M). – **VIII Región del Biobío** Prov. de Ñuble Cordillera de San Carlos, 1800 m, 01.1925, *Barros s.n.* (CONC). – Volcán Chillán, Valle de las Nieblas, 1982, *Grau s.n.* (M). – Prov. del Biobío Cordillera de Chillán, 03.1860, *Pearse s.n.* (SGO). – Cord. Chillán, valle de las aguas calientes, 02.1862, *Philippi s.n.* (SGO). – Laguna de la Laja, 1400 m, 02.1989, *Niemeyer & Fernandez 8905* (CONC). – Volcán Antuco, Angostura, 1390 m, 01.1988, *Stuessy & Baeza 11065* (CONC). – Laguna de la Laja, Sierra Velluda, 1000 m, 02.1960, *Ricardi & Marticorena 5156* (CONC). – Laguna de la Laja, Volcán Antuco, 1000 m, 04.1965, *Ricardi 5269* (CONC). – Volcán Antuco, 02.1972, *Beckett, Cheese & Watson 5063* (SGO). – **IX Región de La Araucanía** Prov. de Malleco Volcán Lonquimay, 1500 m, 03.2002, *Davies & Grau 2002/058* (M). – Volcán Lonquimay, 1700 m, 03.1954, *Sparre & Constance 10893* (CONC). – Volcán Lonquimay, 01.1982, *Torres s.n.* (SGO). – Volcán Lonquimay, road to Lolco, 1500 m, 03.2002, *Davies & Grau 2002/46* (M). – Volcán Lonquimay, faldeos, 1500 m, 02.2002, *Stuessey, Baeza & Tremtzberger 18101* (CONC). – Escorial del Lonquimay, 1550 m, 01.1977, *Marticorena et al. 1333* (CONC). – Volcán Lonquimay, 1300 m, 02.2001, *K. Arroyo et al. 210671* (CONC). – Cordillera Lonquimay, 01.1920, *Hollermayer 415* (1 & 2) (W). – Prov. de Cautín Volcán Llaima, 1800 m, 01.1921, *Hollermayer 415* (CONC). – Refugio Volcán Llaima 2700 m, 01.1942, *Montero 4305* (CONC LP SGO). – Volcán Llaima, 1700 m, 03.1948, *Sparre 4857* (CONC, SGO). – Llaima, 1800 m, 04.1953, *Kunkel s.n.* (CONC). – Volcán Llaima, 1600 m, 03.1954, *Sparre & Constance 10638* (CONC). – Volcán Llaima, 1500 m, 02.1956, *Montero 4889* (CONC). – Refugio del Volcán Llaima, 1750 m, 02.1956, *Garaventa 5546* (CONC). – Volcán Llaima, 01.1972, *Beckett, Cheese & Watson 5000* (SGO). – Volcán Llaima, 1800 m, 03.1972, *Duek & Inostroza s.n.* (CONC). – Volcán Llaima, 2400 m, 02.1988, *Gardner & Knees 4248* (E). **S.l.d.** Chile. *Seibold 3083* (W).

X *Oriastrum* Poepp. & Endl.

1 Diagnosis

Oriastrum Poepp. & Endl. Nov. Gen. Sp. Pl. 3: 50. t. 257. 1842. – **Generitypus**: *Oriastrum pusillum* Poepp. & Endl. l.c. = *Chaetanthera* subgen. *Oriastrum* (Poepp. & Endl.) Cabrera, Revista Mus. La Plata, Secc. Bot. 1:96. 1937.

= *Aldunantea* J. Rémy, Fl. chil. 3: 320. 1849.
= *Egania* J. Rémy, Fl. chil. 3: 324. 1849. = *Chaetanthera* subgen. *Egania* (J.Rémy) Reiche, Fl. chil. 4: 355. 1905].

Etymology The name Oriastrum is taken from the greek όρειαστρον or όριαστρον meaning mountain (όρει – ori) star (αστρο – astro).

Plants lax to dense cushion-forming, herbaceous dwarf annuals or perennials. Perennials almost always with perennating buds at the base of the stems. **Stems** to 15 cm, trailing to erect, branching. **Leaves** laxly opposite to loosely or densely whorled below the capitulum, sometimes forming rosettes. Leaves sessile, linear, acute and mucronate, or indistinctly petiolate and spathulate with ovate to orbicular lamina, margins sometimes limbate. **Capitula** sessile, sometimes subtended by a minute scale; plants can be monoecious or gynodioecious. **Involucral bracts** are multiseriate, imbricate, and initially foliaceous then reduced to entirely membranous. **Outer involucral bracts** are foliaceous, lamina dorsally and ventrally pubescent; petiole or lower half dilated with membranous alae of ovate to rhombic outline; bracts become smaller along series; sometimes this series is absent. **Middle involucral bracts** half way to almost entirely alate, with diminuishing lamina component. **Inner involucral bracts** 1–2 (3) series, entirely membranous, outline narrowly oblanceolate to rhomboid, often dorsally pubescent. Apical appendages broadly acute to apiculate or rarely aristate, translucent to darkly coloured (e.g. red, black, green), internal anatomy densely packed with sclerified cells. **Ray florets** generally white or cream, often tinged with pink, or entirely pale yellow; one series of 6–25 florets (except for pistillate capitula with dimorphic pistillate ray florets). Ligule usually exerted beyond capitulum, always irregularly 2-3 dentate, 3–13 mm L **Disk florets** yellow, bisexual, 10–20 (rarely more than 30). **Styles** yellow, (2)3–7(8.5) mm L; stigmas of ray florets greenish-yellow, pale yellow or orange-yellow to red; stigmas shortly branched, obtuse; hairs long, rounded, echinate. **Anthers** caudate, ciliate; apical appendages acute, yellow. **Pollen** grains equatorially subrectangular, with microspines inconspicuous to conspicuous, grains medium sized (polar diameter x equatorial diameter) 23–49 μm x 18–36 μm; exine 5–7 μm thin, nexine elliptic, sexine divided into compact ectosexine and ramified columellate endosexine. **Achenes** brown, pyriform to flattened cylindric or turbinate; pericarp pellucid, glabrous or with indumentum of either long filiform hairs (~450 μm L), short oblate papillae (~35 μm L) or spherical to ampulliform twin hairs (10–50 μm L); carpopodium absent or very poorly developed; testa epidermis with long narrow parallel cells internally poorly structured or with sinuous pseudo-dendritic cells internally ribbed. **Pappus** 1–2 series, setae white, monomorphic, fused or united at base, dehiscent; barbellate, barbs spreading, barb base shortly to longly adhered, infrequently shortly barbed to ciliate.

2 Subgeneric division of *Oriastrum*

2.1 *Oriastrum* Poepp. & Endl. subgenus *Oriastrum*

= *Chaetanthera* subgenus *Oriastrum* (Poepp. & Endl.) Cabrera in Revista Mus. La Plata, Secc. Bot. 1: 96, 1937.
= *Oriastrum* section *Aldunatea* (J.Rémy) Reiche in Fl. Chile [Reiche] 4: 350, 352 – 353, 1904. ≡ *Aldunatea* J.Rémy in Fl. Chil. [Gay] 3: 320. t. 38 f. 1. 1849. – **Generitypus**: *Aldunatea chilensis* J.Rémy.
= *Tylloma* D. Don, in Trans. Linn. Soc. London 16: 238, 1830. [pro parte]
= *Oriastrum* section *Gnaphaliastrum* Reiche in Fl. Chile [Reiche] 4: 351, 353 – 355, 1904. nom. invalidum

Plants lanate herbaceous dwarf annuals or rarely facultative perennials, sometimes forming small ground-hugging cushions. **Stems** to 8 cm, trailing to erect, branching. **Leaves** laxly opposite to loosely or densely whorled below the capitulum, sometimes forming rosettes. Leaves indistinctly petiolate, narrowly to broadly spathulate with ovate to orbicular lamina, margins often limbate. **Capitula** sessile; plants monoecious. **Inner involucral bracts** 1 (–2) series, entirely membranous, outline narrowly oblanceolate to rhomboid, often dorsally pubescent. Apical appendages acute to apiculate, translucent to brightly coloured (e.g. red, black). **Ray florets** white or cream, or entirely pale yellow; one series of 8–14 florets. **Achene** brown, pyriform; pericarp pellucid, glabrous or indumentum of spherical to ampulliform twin hairs (10–50 µm L); testa epidermis with parallel or sinuous pseudo-dendritic tesselated cells, with internal ribbed strengthenings. **Pappus** free at base, dehiscent, setae 4–7 mm L, 1–2 (3) cells wide, cells 5–10 µm wide. Setae ciliate at base with barbs (90–140 µm L; *O. chilense* 50 µm L) in upper 2/3 of setae with 3–5 (6) barbs /100 µm, barbs always spreading, sometimes ciliate.

Five species, principally from the Chilean Andes. With the exception of the lowland *O. werdemanii* the species grow above 2000 m.a.s.l. up to about 4000 m. With the exception of *O. gnaphalioides* whose distribution has its northernmost limit around Antofagasta (24° 40'S), the species are found between Baños del Toro (29° 50'S) and the Río Clarillo, Cordillera de Santiago, (33° 50'S).

2.2 *Oriastrum* Poepp. & Endl. subgenus *Egania* (J.Rémy) nov. comb.

= *Chaetanthera* subgenus *Egania* (J.Rémy) Cabrera in Revista Mus. La Plata, Secc. Bot. 1: 96, 1937. = *Oriastrum* section *Egania* (J.Rémy) Reiche in Fl. Chile [Reiche] 4: 351, 355 – 358, 1904. ≡ *Egania* J.Rémy in Fl. Chil. [Gay] 3: 324, 1849. – **Generitypus** "non designatus" cfr. Farr et al. (1979) **Lectotypus hic loc. designatus** *Oriastrum acerosum*. J. Rémy in Fl. Chil. [Gay] 3: 325, 1849.

Plants lanate, herbaceous, dwarf cushion-forming perennials with perennating buds at the base of the stems; species monoecious or gynodioecious. **Stems** to 15 cm, trailing to erect, branching. **Leaves** decussate or laxly opposite to loosely or densely whorled below the capitulum, forming rosettes. Leaves sessile or indistinctly petiolate, linear or sometimes dilated to narrowly spathulate, obtuse to acute, and typically mucronate. **Capitula** sessile or shortly pedunculate, sometimes subtended by a minute scale. **Inner involucral bracts** 1–2 (3) series, entirely membranous, outline narrowly oblanceolate to rhomboid, often dorsally pubescent. Apical appendages broadly acute to apiculate or rarely mucronate, translucent and striped or darkly coloured (e.g. black, green). **Pappus** free at base, dehiscent, setae 2.5–6 mm L, setae width variable [1–2 (3) cells wide = *O. dioicum*, *O. polymallum*, *O. pulvinatum*); 4–8 cells wide = *O. acerosum*, *O. revolutum*; 12 cells wide = *O. cochlearifolium*], cells 5–10 μm wide (exceptions are *O. dioicum* = 15; *O. acerosum*, *O. pulvinatum* = 24). Setae ciliate at base with barbs (90–140 μm L; *O. cochlearifolium* = 230 μm L) in upper 2/3 of setae with 3–4 (5) barbs /100 μm, barbs always spreading. **Ray florets** white or cream, can be tinged with pink, or entirely pale yellow; one series of 6–25 florets (except for female capitula with dimorphic pistillate ray florets). **Achene**s brown, flattened cylindric to turbinate; pericarp pellucid, glabrous or indumentum of either long filiform hairs (~450 μm L) or short oblate papillae (~35 μm L); testa epidermis with long narrow parallel cells internally poorly structured.

Thirteen species and one variety distributed on the Altiplano from Peru (11°S) and Bolivia to the Altoandino in Chile and western Argentina, with its southern limit at about 35°S.

3   Key to the species

All *Oriastrum* taxa recognised in this monograph are keyed out here. As far as possible the characters used are suitable for both field and herbarium identifications. However, on herbarium specimens floret colour can be ambiguous, especially the difference between yellow and white, and often underground components are not well collected, if at all. **N.B.** Abbreviations used in figure captions and consepectus: IIB = Inner involucral bracts, MIB = Middle involucral bracts, OIB = Outer involucral bracts, RN = Reserva Nacional, v.s. = ventral surface, d.s. = dorsal surface.

1	Dwarf annual (or rarely biennial), basal bud clusters absent (*Oriastrum* subgenus *Oriastrum*)	2
1*	Dwarf perennial, basal bud clusters present (*Oriastrum* subgenus *Egania*)	6
2	Rays white	3
2*	Rays yellow	**4.** *O. pusillum*
3	Inner involucral bract apices entirely dark red to pink	**1.** *O. chilense*
3*	Inner involucral bract apices translucent ± dark maculae	4
4	Stems loose to lax in substrate, densely whorled leafy rosettes	**4.** *O. pusillum*
4*	Stems loose to lax on surface, not compact rosettes	5
5	Leaves with thickly limbate margins, linear to narrowly spathulate, leaves densely arranged and appressed on stems	**3.** *O. lycopodioides*
5*	Leaves limbate, spathulate, lamina cordate to orbicular, stems sparsely leafy to naked, leaves not appressed	6
6	Capitulum < 3 mm diameter, ray corolla < 6.5 mm long, occurs above 2000 m.a.s.l.	**2.** *O. gnaphalioides*
6*	Capitulum ≥ 4 mm diameter, ray corolla > 8.0 mm long, occurs below 1000 m.a.s.l.	**5.** *O. werdermannii*
7	Stems surrounded by compact, densely whorled leaves	8
7*	Stems with few leaves	13
8	IIB with entirely darkly maculate apices	9
8*	IIB with translucent or striped apices	10
9	Leaf laminas cordate, plant short compact rosette	**8.** *O. achenohirsutum*
9*	Leaf laminas linear, plant lax	**19.** *O. tontalensis*

X *Oriastrum* Poepp. & Endl.

10	Plant densely pubescent	11
10*	Plant lightly pubescent to nearly glabrous, wiry, stems with appressed leaves **12.** *O. famatinae*	

11	Leaves linear to slightly spathulate, short mucros	12
11*	Leaves linear with longly recurved (revolute) mucros	**15.** *O. revolutum*

12	Florets deep-set within the capitula	**13.** *O. polymallum*
12*	Florets, especially ray florets, conspicuously exerted from capitula	**14.** *O. pulvinatum*

13	Plant forms loose cushions, sometimes epiphytic, ray florets yellow	**6.** *O. abbreviatum*
13*	Plant not in cushions, ray florets usually white, white flushed with pink, or seldom creamy yellow	14

14	Leaves obtusely spathulate, succulent	**10.** *O. cochlearifolium*
14*	Leaves linear, apiculate	15

15	Plants densely pubescent	16
15*	Plants lighty pubescent or glabrescent	18

16	IIB darkly, often entirely maculate, or translucent, seldom striped, apices acuminate (27 – 35 °S Chile)	17
16*	IIB striped, apices longly aciculate (19 – 22 °S Tarapaca, Chile)	**18.** *O. tarapacensis*

17	Leaves and bracts narrowly linear-spathulate, apiculate, grey-green colour	**9.** *O. apiculatum*
17*	Leaves and bracts linear, mucronate, pale green, IIB apices variably white, striped or black-green **7.** *O. acerosum*	

18	Plant bright virid green, middle involucral bracts with maculate alae	**11.** *O. dioicum*
18*	Plant dull green, no maculate alae on MIB	**16.** *O. stuebelii*

4 Species descriptions

4.1 *Oriastrum* subgenus *Oriastrum*

1. *Oriastrum chilense* (J. Rémy) Wedd. Chlor. And. 1: 30. 1855.

≡ *Aldunatea chilensis* J. Rémy Fl. Chil. [Gay] 3: 322. t. 38. 1849. Typus CHILE "Se cria en el suelo espuestos al sol de las cordilleras de San José (provincia de Santiago) á una altura de 3,200 m". Holotypus P! Isotypus K! Isotypus probabilis GH! NY! W!

= *Tylloma pusillum* D. Don ex Taylor & Phillips Philos. Mag. Ann. Chem. 11: 391. 1832. Typus CHILE "Herb. Gill." [Las Hyades, and Valle del Yeso, Andes of Mendoza and Chili, *Gillies 17*] Holotypus OXF! Isotypus E! K!
- *Chaetanthera pusilla* (D.Don) Hook. & Arn. Companion Bot. Mag. 1: 106. 1835.

Annual monoecious dwarf cushion-forming herb. No stem buds. **Stems** to 6 cm, branching, sometimes very condensed (depending on substrate). **Leaves** 6–10 mm L, indistinctly petiolate, narrowly spathulate; lamina 3–4 x 2.5–3 mm, ovate acute or dilated, reflexed, conduplicate, mucronate (0.6 mm); margins thickened to limbate; indumentum of densely tangled lanate hairs (1.9–2.5 mm L) on petiole and lamina; decussate, connate. **Capitula** 7.5 mm L, disk diameter 3.5 mm; sometimes shortly pedunculate, leaves alternate (0–3) up stem, otherwise capitula sessile. **Involucral bracts** imbricate, arranged in two series (no **outer involucral bracts** with short membranous alae to less than ½ height of bract), initially foliaceous then reduced to entirely membranous. **Middle involucral bracts** as leaves, 5.9–6.5(11.7) mm L; petiole dilated with membranous alae, with ovate to rhombic outline, 0.6–1.0 mm W., ratio of lamina to alae decreases from ⅔ to nearly entirely alate; alae lanate on margins; lamina ovate, (0.6)1.9–2.5 x (0.6)1–1.2 mm, dorsally and ventrally lanate. **Inner involucral bracts** 8–8.6(11.7) x 1.5–2.2 mm, entirely membranous, dorsally hirtellous below apical appendage; Outline narrowly oblanceolate to rhomboid, with acute, conspicuously maculate apical appendages (dark red to pale orange-red), 1.9–2.5 mm L **Ray florets** (8)12–14, pistillate, white; corolla 7.4–8.9 mm L, corolla tube 3.4–3.7 mm L, outer ligule 1.5 mm W., inner ligule notched, 0.1–0.4 mm L **Disk florets** ca. 10, bisexual, yellow-orange; corolla 4.6–5.9 mm L, corolla tube 4–5.3 mm L **Styles** [ray] 4–5.6 mm L, bifid, obtuse. [disk] 4.3–5.6 mm L; stigma lobes 0.2–0.5 mm L, obtuse. **Anthers** 2.8–3.7 mm (4–5.3 mm L incl. filaments). **Achenes** brown, 2.5 mm L, pyriform; pericarp pellucid, coated in globular twin hairs (35–40 µm) or rarely glabrous [rays]; no carpopodium; testa epidermis with ribbed strengthenings, cells sinuous-dendritic in outline. **Pappus** white, 1–2 rows, dehiscent; (3.0)4.6–5.7 mm L, 1–2(3) cells wide, cells 5–10µm W.; basal cilia, sparsely barbellate (3–4 barbs/100 µm), barbs 50 µm L, barb base medium adhered (50%), free barb spreading.

X *Oriastrum* Poepp. & Endl.

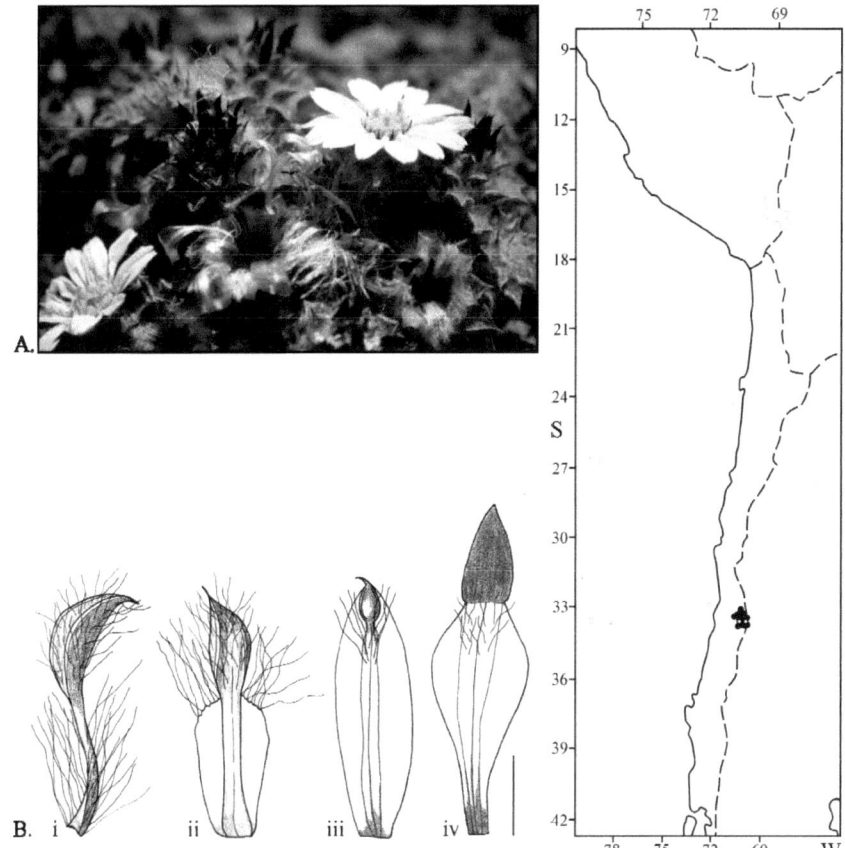

Figure 80: *Oriastrum chilense*. **A.** Habit, photographed Farellones, 1995©Jürke Grau. **B.** Leaf & Bract detail. **i.** Stem leaf d.s.; **ii.** MIB v.s.; **iii.** MIB d.s.; **iv.** IIB d.s. Scale bar = 2 mm.

Distribution and Habitat. *O. chilense* occurs in the Upper Andean zone (Altoandina) in the Eastern Santaginean Andes (33° 00' – 34° 00'S). It is found in loose steep scree or gravel banks, or on flat alluvial plains with flash flood activity between elevations of 2150 – 3800 m.a.s.l.

Differential diagnosis. *O. chilense* and *O. pusillum* are very similar in habit. *O. chilense* is distinguished by its pale orange to dark red-purple inner involucral bract apices, and the shortly barbellate pappus. The ray florets are always white. It is variable and can increase in size with increasing altitude. In loose scree it shows lax growth (as at Embalse del Yeso), when in firmer ground it grows as small compact clusters (as in Farellones).

Material seen. – **Argentina. Mendoza** Valle de Tunuyan, 1230 ft, I.1870, *Reed 95*? (K). – Los Penitientes near Puente del Inca F.E.T.A., 02.1931, *King 707* (LP). – **Chile. IV Región de Coquimbo** 1839, *Gay 947 & 948* (G). – **Región Metropolitana de Santiago** Prov. de Santiago Santiago, 4000-6000 ft., 22.1.1902, *Hastings 432* (NYx2). – Mina La Disputada, 2800 m, 04.1933, *Barros s.n.* (CONC). – Fierro Carrera, 2800 m, 01.1930 *Looser 1127* (CONC GH). – Fierro Carrera, 2800 m, 01.1930 *Garaventa 527* (CONC). – Fierro Carrera, 3800 m, 01.1930 *Montero 1039* (CONC). – Fierro Carrera, 2800 m, 01.1936 *Garaventa 1553* (CONC). – Perez Caldera, 2800 m, 01.1954 *Sparre 10596* (CONC). – cerca La Parva, 3000 m, 06.01.1979, *Muñoz & Meza 1399* (SGO). – La Parva, 3010 m, 04.1964 *Schlegel 5026* (CONC). – Farellones, Parva – Colorado, 2500 m, 12.1981, *Grau 3293* (MSB M). – camino Farellones, Cerro de los Andes, ca. 35 km NE of Santiago, 2920 m, 01.1981, *Schmid 1981-15* (NY). – Farellones – Valle Nevado, 01.1998, *Muñoz S. 3880* (SGO). – Valle Largo, 3000 m, 02.1950 *Barros s.n.* (CONC). – Valle San Francisco, 2800 m, 04.1933 *Grandjot 1089* (CONC M). – Laguna Los Franciscanos, 3500 m, 13.1990, *Solervicens s.n.* (SGO). – 40 km al noreste de Santiago, Cordillera de Las Condes, Cerro La Parva, al este del pueblo de La Parva, 3170 m, 02.2003, *K. Arroyo* et al. *25179* (CONC). – Complejo de Esqui "Valle Nevado", camino a Cerro Franciscano, 3360 m, 02.2003, *K. Arroyo* et al. *25180* (CONC). – Valle de Las Llaretas, 2800 m, 03.1956 *Schlegel 1072* (CONC). – Vegas Las Vacas, 2900 m, 03.1956 *Schlegel 1061* (CONC). – Camino entre Santiago y Farellones, curva 34, 2310 m, 10.2002, *K. Arroyo* et al. *25121, 25121a* (CONC). – Santuario El Arrayán, Fundo San Enrique, 01.1981, *Meza s.n.* (SGO). – 40 km al noreste de Santiago, Cordillera de Las Condes, 3500 m, 02.1921, *Jaffuel 509* (GH). – 50 km al noreste de Santiago, Cordillera Las Condes, 3000 m, 02.1921, *Jaffuel 691* (CONC GH). – San Gabriel, 2500 m, 01.1950 *Gunckel 21414* (CONC). – Cerro Abanico, 2200 m, 12.1951 *Gunckel 21403* (CONC). – Potrero Grande, 3200 m, 01.1967 *Zöllner 2883* (CONC). – Valle del Río Cepo, 3200 m, 02.1939 *Grandjot 3664* (CONC). – Valle de Los Paramillos, 3750 m, 02.1939 *Grandjot s.n.* (CONC). – Valle de Los Paramillos, 3500 m, 02.1951 *Barros s.n.* (CONC). – Prov. de Cordillera Valle del Yeso *Philippi s.n.* (W# 98173). – Cordilleras das Arañas, 12.1854, *Philippi s.n.* (W). – Cordill. de Santiago, *Philippi 719* (W). – Cordill. de Santiago, *Philippi s.n.* (CONC F Wx2). – Cordill. de Santiago, 1861, *Philippi s.n.* (K). – Cordill. de Santiago, 1856-1857, *Philippi s.n.* (G K W#1248). – Cordill. de St. Jago, *Philippi 827* (BM P). – Cordill. de Santiago, 1864, *Philippi s.n.* (NY). – Santiago, *Philippi s.n.* (E). – Cordillera Santiago, *Capt. Wilkes Exped. s.n.* (NY). – camino entre Laguna del Yeso y Baños del Plomo, 2500 m, 12.1992, *von Bohlen 1474* (SGO). – Embalse El Yeso, 01.1972, *Beckett, Cheese & Watson 4846* (SGO). – a lo largo Embalse El Yeso, 2520 m, 01.1993 *Taylor & Gereau 10921* (CONC). – Mountains above Río Colorado, 01.1902, *Hastings s.n.* (NY). – Lagunillas, 2700 m, 12.1950 *Barros s.n.* (CONC). – Lagunillas, 12.1971, *Beckett, Cheese & Watson 4491* (SGO). – Embalse El Yeso, above road at end of reservoir, 2500 m, 02.2002, *Davies & Grau 2002/008* (M). – Above San Gabriel, 30 km from turnoff at Romeral of road into the valley of the Río Yeso, at end of Embalse El Yeso, 2520 m, 01.1993, *Eggli & Leuenberger 2292* (B). – Above San Gabriel, 30 km from turnoff at Romeral of road into the valley of the Río Yeso, at end of Embalse El Yeso, 2520 m, 01.1993, *Eggli & Leuenberger 2291* (B). – Puente Alto, Cerro Las Tinajas, 2420 m, 03.1967 *Schlegel 5915* (CONC). – Puente Alto, Valle del Yeso, 2800 m, 05.1960 *Schlegel 2565* (CONC). – PN El Morado, frente a Santiago, 2320 m, 12.1991, *Teillier, Pauchard & P. García 2500* (SGO). – PN El Morado, frente a Santiago, 2150 m, 12.1991, *Teillier & González 2499* (SGO). – Cajón de Morales, 2400 m, 02.1963 *Ricardi* et al. *887* (CONC). – El Volcán, 2200 m, 01.1951 *Barros s.n.* (CONC). – RN Río Clarillo, Sector Los Cristales, 2640 m, 01.2000, *K. Arroyo 20746.5* (CONC). – Cordillera San José, 3600 m, 02.1950 *Soto s.n.* (CONC). – Cord. de Santiago, 1862, *Philippi s.n.* (G). – Cord. de Santiago, 1876, *Philippi s.n.* (G). – Cordillera de Santiago, 01.1895 *Johow s.n.* (CONC). – Santiago, *Burmeister s.n.* (CONC).
S.l.d. *Seibold 2990* (W). – Chili, *Gay s.n.* (K NYx2 W). – Chili, leg. *Gay, Sch.Bip.* (NY). – Andes of Chile, *Capt. Wilkes Exped. s.n.* (NY). – Chili, *Gay s.n.* (GH). – *Philippi F. s.n.* (BM). – *Gay 693* (MO).

X *Oriastrum* Poepp. & Endl.

2. *Oriastrum gnaphalioides* (J. Rémy) Wedd. Chlor. And. 1: 30. 1855.

≡ *Aldunatea gnaphalioides* J. Rémy in Fl. Chil. [Gay] 3: 323. 1849. Typus CHILE "Se cria en los cerros de las cordilleras de Doña Ana (provincia de Coquimbo), á una altura de 3,847 m" [...collines du cordilleres de Doña Ana, haute 3847 m, trés rare 11.1836, *Gay 399*] Holotypus P(141378 p.p.)!

= *Chondrochilus parvifolius* Phil. Linnaea, 33: 114. 1864-65. Typus CHILE "In prov. Coquimbo propre los Baños del Toro Volckmann" [Coquimbo, Baños del Toro, 01.1860, *Volckmann s.n.*] **Lectotypus hic loc. designatus** SGO 64979! Isotypus K! SGO 71766!
= *Tylloma gnaphalioides* Phil. Anales Univ. Chile 85: 843. 1894. Typus CHILE "In Andibus illapelinis loco dicto La Polcura januario 1888 lectum" [Andes Illapelinae, La Polcura, 01-02.1888] **Lectotypus hic loc. designatus** SGO 71308! Isotypus BM! SGO 45028!
= *Tylloma albiflorum* Phil. Anales Univ. Chile 85: 844. 1894. Typus CHILE "Habituat in Andibus Illapelinis loco dicto La Polcura" [And. Illapel, La Polcura, 01.1888] **Lectotypus hic loc. designatus** SGO 71311! Isotypus SGO 45029!
= *Tylloma minutum* Phil. Anales Mus. Nac. Santiago de Chile 2: 32. 1891. Typus CHILE "Encontrado en Pastos Largos a 4,000 m" [Pastos Largos, 01.1885, *F.Philppi s.n.*] **Lectotypus hic loc. designatus** SGO 71317! Isotypus SGO 45026!
= *Oriastrum leucocephalum* Phil. Anales Univ. Chile 87: 16. 1894. Typus CHILE "In editissimo monte Doña Ana legit Dr. Antonio Peralta" [Cerro de Doña Ana, *Dr. Peralta*] Holotypus SGO 64965!
= *Oriastrum gossypinum* Phil. Anales Univ. Chile 87: 19. 1894. Typus CHILE "In Andibus prope Huanta pariter novembri 1836 a cl. Gay lectum" [Prov. Coquimbo, in editillinum andium Doña Ana, 11.1836, *Gay 684*] Holotypus SGO 64975!
– *Oriastrum albiflorum* (Phil.) Reiche Anales Univ. Chile 115: 337. 1904, Fl. Chile [Reiche] 4: 354. 1905.
– *Chaetanthera gnaphalioides* (J. Rémy) I. M. Johnst. Physis (Buenos Aires) 9: 325. 1929. – *Chaetanthera minuta* (Phil.) Cabrera Revista Mus. La Plata, Secc. Bot. 1: 134, fig. 19. 1937.

Nomenclature notes The sheet #141378 from Paris has two labels – the holotype (*Gay 399*) and a second label *Gay 357*. It is not clear to which of the two plants on the sheet each label pertains.

Annual monoecious dwarf cushion-forming herb to 5 cm. No stem buds. **Stems** lax, branching, loosely lanate. **Leaves** 3–6 mm L, alternate, indistinctly petiolate, spathulate; lamina 1.5–2 x 0.8–1.5 mm, cordate to orbicular, grey-green; lamina margins pale limbate, apices mucronate; indumentum lanate, hairs simple, white, to 1.5 mm, shortly tomentose on lamina. **Capitula** cylindrical, disk diameter 1.5–2.5 mm. **Involucral bracts** imbricate, arranged in three types, initially foliaceous then reduced to entirely membranous. **Outer involucral bracts** 3.5–6 mm; as leaves but with short membranous alae to less than ½ height of bract; lamina 1–1.5 x 1 mm, shortly tomentose; alae inconspicuous, narrow, linear, 0.3–0.5 mm W.; distal dorsal surface and margins lanate. **Middle involucral bracts** 4–7.5 mm; lamina reduced, 1–1.5 x 0.8–1.3 mm, shortly tomentose; membranous alate, ratio of lamina to alae decreases from ⅔ to nearly entirely alate, alae ovate-linear, 1–2.2 mm W.; distal dorsal surface and margins lanate. **Inner involucral bracts** entirely membranous, 1 series; 5–8 mm L, obovate; apical appendages obtuse to broadly acute, pale translucent, 1.5 mm L **Ray florets** 8–14, pistillate, white to creamy yellow; corolla 4–6.5 mm L; corolla tube 2 mm L; outer ligule 1–1.2 mm W.; inner ligule reduced, bifid. **Disk florets** ca. 10, bisexual, yellow; corolla 2.5–3.7 mm; corolla tube 2–3.3 mm. **Styles** yellow [ray] 3.1–4.4 mm L [disk] 3.9 mm L, stigmas shortly lobed. **Anthers** 2.4 mm (3.6 mm incl. filaments). **Achenes** dark brown, pyriform, to 2 mm L; pericarp pellucid, glabrous or scattered with tiny (to 20 μm) globular to ampulliform twin hairs; no carpopodium; testa epidermis with ribbed strengthenings, cells linear-parallel. **Pappus** white, 1–2 rows, dehiscent; (2.5)3–4.3 mm L, 1–2(3) cells wide, cells 5–10 μm W.; basal cilia, sparsely barbellate (3–6 barbs /100 μm), barbs 90–140 μm L, barb base longly adhered (>60%), free barb spreading.

Distribution and Habitat. *O. gnaphalioides* is distributed in the high Chilean Andes from Antofagasta (Volcán Llullaillico) to Santiago (Cordillera del Plomo), and the Argentinean high Andes within the same range of latitudes (24°40'–33°20'S). It is found in steep dry screes, or level spots among rocks, rarely on rocky soil near streams at elevations between 2000–4300 m.a.s.l. It flowers at lower altitudes towards the south of its range.

Differential diagnosis. Divided by Cabrera (1937) into two species, *O. gnaphalioides* s.l. repesents a highly variable taxon, especially in terms of size. This variation in size is associated to the altitude at which the individuals grow (K. Arroyo pers. comm.), larger individuals (formerly *O. gnaphalioides* s.str.) being found at higher altitudes and smaller individuals at lower altitudes (formerly *Chaetanthera minuta*).

Material seen. – **Argentina.** Salta Depto. Pastos Grandes 4100 m, 02.1945, *Cabrera 8775* (LP). – **La Rioja** Cord. de la Brea, Quebrada del Cajón , 3600 m, 01.1949, *Krapovickas & Hunziker 5664* (LP). – **San Juan** Valle de la Cura, cerca junta arroyo Blanco, 4000 m, 01.1930 *Perez Moreau* 179 (LP). – Arroyo Tambillos, along trail from Paso de Valeriano, 4000 m, 01.1926, *Johnston 6094* (GH K LP). – Depto. Iglesia Valle del Río Blanco, Baños del Ballet, 02.1950, *Castellanos 71644* (W). – Cordillera de L'Espinacito, *Bodenbender s.n.* (G). – Depto. Caligasta Paso desde cuesta del Río Blanco hacia el Valle de los Patos Norte, 4000 m, 01.1991, *Kiesling, Peralta & Ulibarri 7768* (MO). – Manatiales, 3500 m, 01.1971, *Volponi & Zardini 217* (LP). – **Mendoza** La Penitiente, near Puente del Inca, 3000 m, 02.1931, *King 707* (BM LP). – **Chile. II Región de Antofagasta** Prov. de Antofagasta Volcán Llullaillico, 4000 m, 01.1994, *Arroyo* et al. *94033* (CONC). – Cerro del León, 4240 m, 02.1997, *Arancio & Squeo 10415* (CONC). – camino Salar Aguas Calientes a Cori, 3850 m, 01.1994, *Arroyo* et al. *94089* (CONC). – Salar Punta Negra, Río Frío, 3500 m, 01.2002, *K. Arroyo* et al. *25127* (CONC). – Quebrada Pereda, 3830 m, 12.1996, *Arancio & Squeo 10231* (CONC). – **III Región de Atacama** Prov. de Huasco Cordillera Laguna Grande, 3700 m, 01.1924, *Werdermann 252* (B BM CONC E F G GH K M). – Vallenar, vicinity of Laguna Valeriano, 4000 m, 01.1926, *Johnston 6074* (F GH K LP). – Quebrada Cantarito, entre Quebrada Marancel y Portezuelo de Cantarito, 3500 - 4300 m, 02.1981, *K. Arroyo 81625* (CONC). – Quebrada Vizcachas, emtre Quebrada Cantarito y Portezuelo Vizcachas, 3200 – 4000 m, 02.1981, *K. Arroyo 81573* (CONC). – Cordillera Laguna Grande, 3700m, 01.1924, *Werdermann 252* (CONC). – Quebrada Cantarito, entre el estremo oeste de la Laguna Grande y la Quebrada Marancel, 3100 – 3500 m, 01.1983, *Marticorena, K. Arroyo & Villagrán 83449* (CONC). – Vega de la Colgada en curso superior del Sancarrón, 4000 m, 01.1979, *Osorio s.n.* (SGO). – Camino entre Mina Pascua Lama y San Felix, ca. 5 Km al norte de la Mina, 3570 m, 01.2003, *K. Arroyo* et al. *25096* (CONC). – Prov. de Copiapó Laguna Santa Rosa, 3850 m, 02.1997, *Teillier 4188* (CONC). – camino Salar de Maricunga, Km 132, 4100 m, 01.1963, *Ricardi* et al. *603* (CONC). – Quebrada Salitral, 2 Km interior El Rodeo, 3570 m, 01.1973, *Marticorena* et al. *576* (CONC). – de Papiote a la Laguna Santa Rosa, km 108 depuis le cruisement de la route principale, 3730 m, 12.1993, *Charpin, Grenon & Lazare 23714* (G). – Quebrada de Papiote, above Pastillo, 2600 m, 11.1925, *Johnston 4885* (GH). – Copiapó-Laguna Sta Rosa, Schotterhang vor der letzten Kuppe am übergang zur Laguna, 3850 m, 12.2002, *Ehrhart 2002-187* (MSB). – Prov. de Chañaral camino Potrerillos a Salar Maricunga, 45 Km

X *Oriastrum* Poepp. & Endl.

al interior del Tranque La Ola, 4000 m, 02.1966, *Ricardi, Marticorena & Matthei 1625* (CONC). – **IV Región de Coquimbo** Pasto Largo, 01.1902, *Reiche s.n.* (SGO). – Prov. de Elqui *Gay 357* (P). – Camino entre Baños del Toro y puenta de la mina El Indio curva 27, 3250 m, 01.2003, *K. Arroyo* et al. *25085* (CONC). – Approx. 2 Km al oeste de Baños del Toro, 3500 m, 01.2003, *K. Arroyo* et al. *25086* (CONC). – Faldeos al sur de la entrada de la Mina del Indio, 3210 m, 01.2003, *K. Arroyo* et al. *25079* (CONC). – Camino a Indio Km 29, 3900 m, 01.1988, *Squeo 88050* (CONC MO). – Km 29, camino a Indio, 3900 m, 01.1988, *Squeo 88048* (CONC). – Baños del Toro, 3100 m, 01.1981, *K. Arroyo 81199* (CONC). – Baños del Toro, 3000 m, 01.1979, *Jiles 6475* (CONC). – Cordillera de Doña Ana, Chancha de Sky, 3350 m, 01.1988, *Squeo 88016* (MO). – Cordillera Doña Ana, Cerro E de Canchas de Ski, 3100 m, 01.1992, *Arancio 92094* (CONC). – 21,4 Km N de Juntas del Toro, 3100 m, 01.1993, *Stuessy & Ruiz 12781* (CONC). – Baños del Toro, 3200 m, 02.1963, *Ricardi* et al. *702* (CONC). – Cerro Oeste Canchas de Sky, 3350 m, 01.1988, *Squeo 88028* (CONC). – Baños del Toro, Quebrada Pastos, 3400 m, 05.02.1939, *Morrison 17269* (G GH K). – Cordillera de Paihuano, 3000 m, 12.1942, *Gajardo s.n.* (CONC). – Baños del Toro, 3500 m, 01.1948, *Wagenknecht 256* (CONC). – Canchas de Sky, 3350 m, 01.1988, *Squeo 88016* (CONC MO). – 16,5 Km N de Juntas del Toro, 2900 m, 01.1993, *Stuessy & Ruiz 12794* (CONC). – Entre Juntas y Embalse La Laguna, 2400 m, 01.1967, *Ricardi* et al. *1739* (CONC). – Baños del Toro, 3500 m, 12.1923, *Werdermann 207* (B BM CONC E G MO). – Prov. de Limarí Cerro Poncho, Río Torca (Teulahuan) 12.1890, *Geisse s.n.* (SGOx2). – Quebrada Larga, 3400 m, 02.1958, *Jiles 3390* (CONC). – Río Torca, 2400 m, 12.1961, *Jiles 4078* (CONC). – Cordillera de Ovalle, Portezuelo Pingo, 2000 m, 01.1972, *Jiles 5899* (CONC). – Cabreria, Morro Blanco, 2200 m, 10.1949, *Jiles 1564* (CONC). – Cordillera San Miguel, 2600 m, 01.1959, *Jiles 3602* (CONC). – Cordillera de Ovalle, Los Pingos, 3400 m, 01.1972, *Jiles 5880* (CONC). – Río Gordito, 3200 m, 01.1954, *Jiles 2505* (CONC). – Cordillera de Combarbalá, Laguna El Toro, 3450 m, 01.1966, *Jiles 4871* (CONC). – Cordillera de Ovalle, Serón, 2000 m, 11.1957, *Jiles 3312* (CONC). – Serón, 2000 m, 11.1957, *Jiles 3312* (CONC). – Río Flamencos, 2900 m, 01.1956, *Jiles 2966* (CONC). – Los Molles, 2500 m, 01.1972, *Zöllner 6293* (CONC). – Cordillera de Ovalle, Morro Blanco, 2000 m, 01.1949, *Jiles 1237* (CONC). – Cordillera San Miguel, 2600 m, 01.1959, *Jiles 3602* (CONC). – Cordillera de Palacios, 3200 m, 01.1954, *Jiles 2468* (CONC). – Hacienda Ramadilla, 2700 m, 01.1966, *Jiles 4817* (CONC). – El Derecho, 3300 m, 01.1966, *Jiles 4793* (CONC). – Río Flamencos, 2900 m, 01.1956, *Jiles 2966* (CONC). – Prov. de Choapa Cordillera de Illapel, 01.1860, *Volckmann s.n.* (SGO). – Ojetas valley near Cuncumén, 3550 m, 02.1984, *Zöllner 11968* (MO). – Illapel, Río Cenicero, 3000 m, 02.1962, *Jiles 4157* (CONC). – Cordillera de Illapel, Río Illapel, 3200 m, 02.1962, *Jiles 4158* (CONC Mx2). – Quebrada La Vega Escondida, 3 hrs by horse due east of Cuncumén, hacienda at fork of Río Tranquilla and Tencaan creek, SSW from Las Placetas, 2700 m, 11.1938, *Worth & Morrison 16579* (K). – Depto. Illapel, Cerro La Yerba Loca, 2 hrs by horse east of La Vega Escondida, 2800 m, 12.1938, *Morrison 16947* (G K MO). – Zona de Lagunas limite Chile-Argentina Mina Los Pelambres, 3500 m, 02.1999, *K. Arroyo* et al. *991486* (CONC). – Mina Los Pelambres, sector Hualtates, 3340 m, 02.1999, *K. Arroyo* et al. *991282* (CONC). – **V Región de Valparaíso** Prov. de Los Andes Baños del Inca, 01.1886, *Borchers s.n.* (BM SGOx2). – Prov. de Petorca Aconcagua, Junta de Piquenes, Río Sobrante, 02.1938-9, *Morrison 17301* (CONC G K). – **Región Metropolitana de Santiago** Prov. de Maipo Cerro Parva im Plomo-Massiv bei Santiago, 3000 – 3100 m, 02.1939, *Grandjot 3817* (GH).

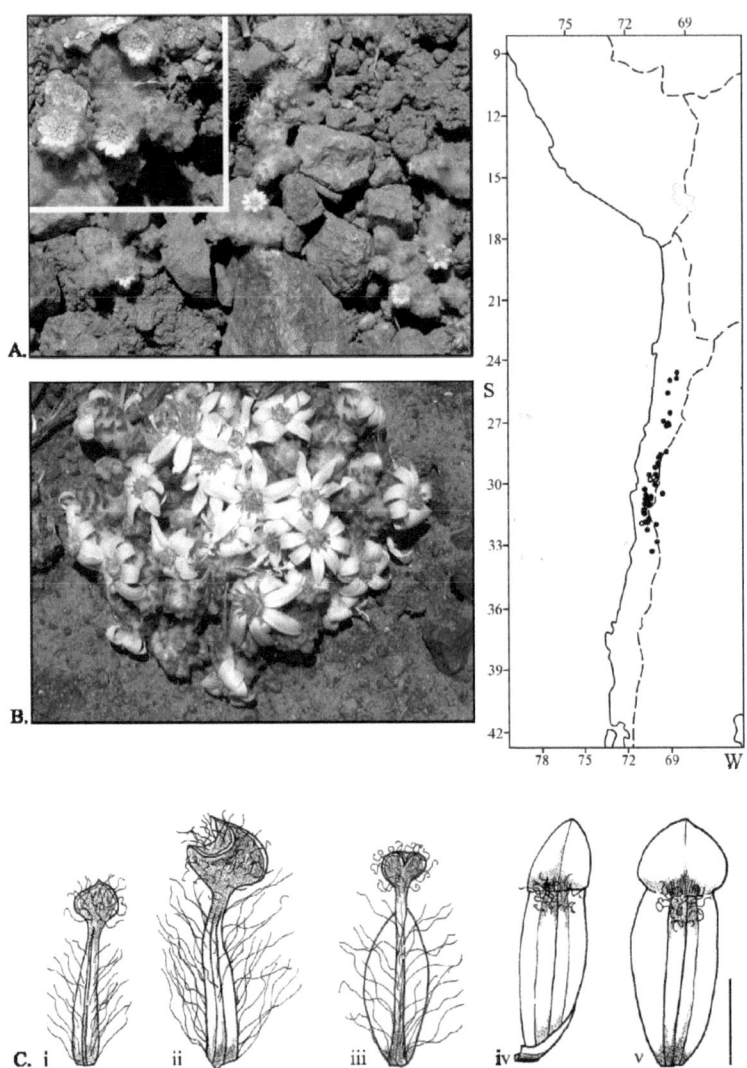

Figure 81: *Oriastrum gnaphalioides*. **A.** Habit "minuta" form, photographed Agua Negra, 2003©Iréne Till-Bottraud. **B.** Habit "gnaphalioides" form, photographed Agua Negra, 2003©Iréne Till-Bottraud. **C.** Leaf & Bract detail. *Ehrhart 2002-187* (i, iii – v), *Werdermann 207* (ii). **i.** Stem leaf d.s.; **ii.** Stem leaf v.s.; **iii.** MIB d.s.; **iv – v.** IIB d.s. Scale bar = 2.5 mm.

X *Oriastrum* Poepp. & Endl.

3.   *Oriastrum lycopodioides* (J. Rémy) Wedd. Chlor. And. 1: 30. 1855.

≡ *Aldunatea lycopodioides* J. Rémy Fl. Chil. [Gay] 3: 323. 1849. Typus CHILE "Se cria en las cordilleras de los Patos (provincia de Santiago), á una altura de 4,000 m" [Hautes cordilleras de los Patos, 01.1837, *Gay 468*] Holotypus P! Isotypus GH(photo)!

= *Oriastrum gayi* Phil. Anales Univ. Chile 87: 18. 1894. Typus CHILE "In andibus provinciae Coquimbo prope Hunata Novembri el 1836 legit cl. Gay" [Prov. Coquimbo, in collibus andium Huanta, 09.1836, Gay 738] Holotypus SGO 64956!

= *Oriastrum glabriusculum* Phil. Anales Univ. Chile 87: 16. 1894. Typus CHILE "Habitat in Andibus provinciae Santiago in Valle Largo Frid. Philippi" [Valle Largo, 02.1892, *F. Philippi*] Holotypus SGO 71305!

– *Oriastrum lycopodiodes* var. *glabriusculum* (Phil.) Reiche Anales Univ. Chile 115: 334. 1904, Fl. Chile [Reiche] 4: 353. 1905.

– *Chaetanthera lycopodioides* (J. Rémy) Cabrera in Pérez-Mor. Revista Centro Estud. Doct. Ci. Nat. 1: 58 Reprint page 12. 1935.

Nomenclature Notes Although Philippi mentioned having two examples of *Oriastrum gayi* Phil. in the Museo [de Historia Natural], only one still exists.

Perennial (growth intervals sometimes on old stems), monoecious dwarf cushion-forming herb. **Stems** to 6 cm L, laxly branching, densely covered in whorled leaves. **Leaves** 3–5 x 1 mm, slightly spathulate, lamina conduplicate, margins thickened to limbate, mucronate; leaves somewhat reflexed from stem, villous at bases. **Capitula** cylindrical, 10 mm high, sessile, disk diameter 4 mm. **Involucral bracts** imbricate, arranged in two series (no **outer involucral bracts** with short membranous alae to less than ½ height of bract), initially foliaceous then reduced to entirely membranous. **Middle involucral bracts** (3)5–8 mm L, as leaves but petiole dilated with membranous alae, ratio of lamina to alae decreases from ⅔ to nearly entirely alate; alae linear to ovate to rhombic outline, 0.5–1.3 mm W., lanate on margins; lamina 1.5–2.5 x 1.3–1.8 mm, somewhat reflexed, mucronate, conduplicate, dorsally and ventrally shortly pubescent. **Inner involucral bracts** entirely membranous, narrowly oblanceolate to rhomboid, dorsally hirtellous below apical appendages; apices diamond-shaped, entirely black-brown. **Ray florets** 7–14; pistillate, white often with pink dorsal surfaces (discoloured to brown on herbarium sheets); corolla 9.5–10 mm L, corolla tube 3.7–4 mm L, outer ligule 2.2–2.5 mm W., irregularly 3–(4)-dentate, inner ligule to 0.4 mm, bifid. **Disk florets** ca. 10, bisexual, orange-yellow; corolla 6.3 mm L, corolla tube 5.6 mm L, corolla teeth ± fibrillate margins. **Styles** yellow [ray] 6.5 mm L, upper third thickened but stigma lobes closed, can be dark red or yellow. [disk] 6.5 mm L, lobes 0.6 mm L, open. **Anthers** 4 mm L (10.7 mm L incl. filaments). **Achenes** brown, pericarp pellucid [ray] sterile, glabrous or sparsely covered with globular twin hairs (45 µm L). [disk] 2.5 cm L, fertile, pyriform, often densely covered in globular twin hairs (45 µm L); carpopodium absent; testa epidermis with ribbed strengthenings, cells parallel to sinuous tesselate. **Pappus** white, 1–2 rows, dehiscent; setae

5.7–7.4 mm L, 1–2(3) cells wide, cells 5–10 μm W.; basal cilia, sparsely barbellate (3–4 barbs/100 μm), barbs 90–140 μm L, barb base shortly adhered (<40%), free barb spreading.

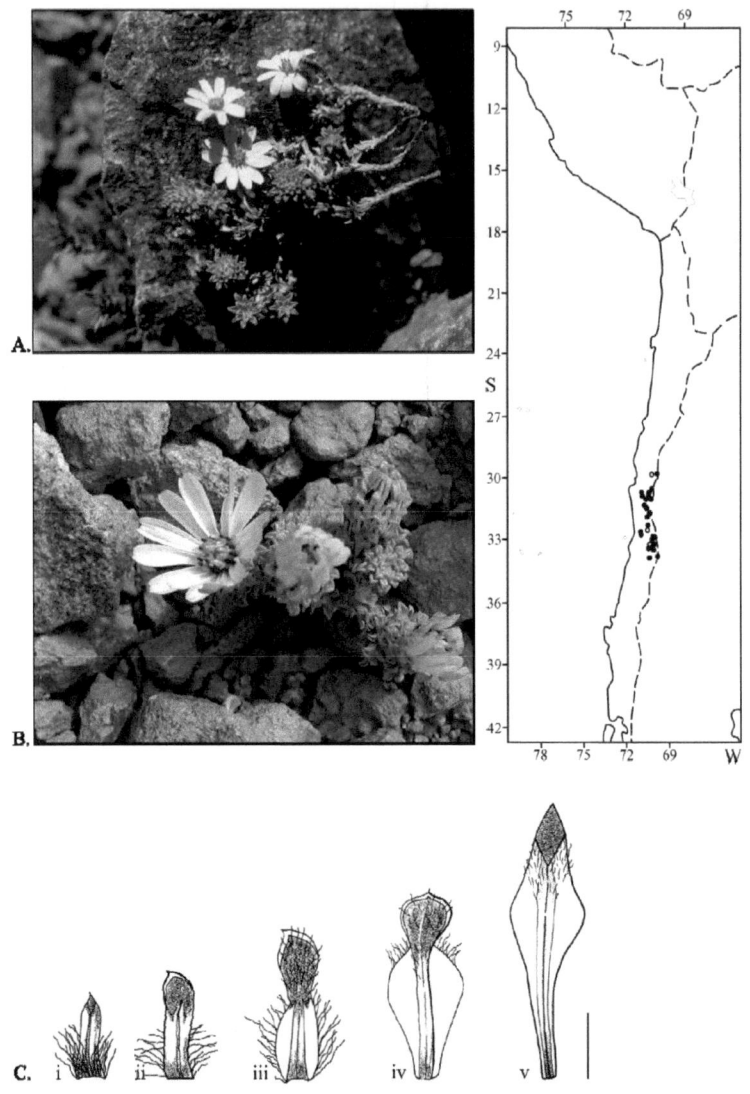

Figure 82: *Oriastrum lycopodioides*. **A.** Habit, photographed Laguna del Inca, 1991©Michel Grenon. **B.** Capitulum detail, photographed Valle Nevado, 2003© Iréne Till-Bottraud. **C.** Leaf & Bract detail. *Davies* et al. *2002/015*. **i.** lower stem leaf d.s.; **ii.** Upper stem leaf v.s.; **iii.** OIB v.s.; **iv.** MIB v.s.; **v.** IIB d.s. Scale bar = 2.5 mm

X *Oriastrum* Poepp. & Endl.

Distribution and Habitat. *O. lycopodioides* is distributed in the Chilean Andes from Baños del Toro to the RN Río Clarillo and San Juan and Mendoza in Argentina between the latitudes 29°50'–33° 50'S. The western bastion of *C. lycopodioides* is on Cerro Chache, stretching towards the Pacific. The altitude zone it occupies is somewhat lower in this locality. It forms a dominant part of the *Nassauvia pungens-Chaetanthera lycopodioides* vegetation association of the Upper Andean zone (Hoffman et al. 1997). Its habitat is open, exposed rocky soil, levels and slopes or screes at elevations between 200 –3900 m.a.s.l., although inhabits higher altitudes in the northern areas.

Differential diagnosis. *O. lycopodioides* is closely related to *O. pusillum* and *O. chilense*. As the epithet implies, it has superficial similarities to a "clubmoss", especially in the habit (erect, branching) and leaf arrangment (densely whorled around stem). Within *O. lycopodiodes* there is variation in density of indumentum, from densely to sparsely villous or nearly glabrous.

Material seen. – **Argentina. San Juan** Depto. Calingasta Los Morrillos, 3400 m, 31°46 70°22' 01.1997, *Kiesling, Moglia & Tombesi 8862* (SI). – Valle de los Patos N, nacimiento del Río Melchor, 3000 m, 01.1991, *Kiesling, Peralta & Ulibarri 7795* (SI). – Pachón, 3850 m, 02.1975, *Luti 5479* (SI). – Paso de La Guardia (camino a Pachon), 4000 m, 03.1975, Marceñido 1 (LP). – **Mendoza** Depto. Tunuyán Las Salinillas cerca del Cerro Marmolyn (Valle del Río Tunuyán, 02.1934, *Ruiz Leal 2067* (LP). – **Chile. IV Región de Coquimbo** Prov. de Elqui Baños del Toro, 3260 m, 02.1939, *Wagenknecht s.n.* (CONC). – Prov. de Limarí Río Torca, (Ovalle), 1889/90, *Geisse s.n.* (GH). – Quebrada Larga, 3200 m, 02.1958, *Jiles 3411* (CONC). – Río Molles, 3900 m, 02.1962, *Jiles 4129* (CONC). – Punta de Huanta, Río Pedregales, 3800 m, 02.1962, *Jiles 4326* (CONC). – Cordillera de Ovalle, Los Pingos, 3000 m, 01.1972, *Jiles 5881* (CONC). – Vega Negra, San Miguel, 3000 m, 01.1959, *Jiles 3643* (CONC). – Tulahuén, Cerro Loica, 2600 m, 12.1965, *Jiles 4732* (CONC). – Río Gordito, 3000 m, 01.1954, *Jiles 2504* (CONC). – Hacienda Ramadilla, 3200 m, 01.1966, *Jiles 4792* (CONC). – Prov. de Choapa Dept. Illapel, Cerrro La Yerba Loca, 2 hrs by horse east of La Vega Escondida, 2800 – 3000 m, 12.1938, *Morrison 16946* (CONC G GH K). – Cordillera de Illapel, Caletón Blanco, 3000 m, 02.1962, *Jiles 4239* (CONC). – Mina Los Pelambres, Zona de lagunas, límite Chile-Argentina, 3500 m, 02.1999, *K. Arroyo* et al. *991241* (CONC). – Mina Los Pelambres, Zona de lagunas, límite Chile-Argentina, 3580 m, 02.1999, *K. Arroyo* et al. *991375* (CONC). – Dept. de Illapel, Cajón de las Pelambres, Hacienda Cuncumén, 3300 m, 01.1932, *Looser 2150* (GH). – **V Región de Valparaíso** Prov. de Petorca Dept. Petorca, Cerro Chache, 5 hours by horse southeast of Patagua Mine, ca. 18 km east of La Ligua, southeast of Chache, 2200 m, 12.1938, *Morrison 17069* (G GH K). – Dept. Petorca, 5 km south of Junta de Piquenes, Río Sobrante, 3400 m, 02.1939, *Morrison 17300* (G GH K). – Prov. San Felipe de Aconcagua Los Caquis, 2000 m, 03.1964, *Zöllner 405* (CONC). – Prov. de Los Andes Portillo, Laguna del Inca, west shore above ski lift pylons, 2900 m, 02.2002, *Davies, Becker, A. & Becker, D. 2002/015* (M). – High Cordillera in Riecillo, near Río Blanco, 3000 m, 02.1914, *Zöllner 542* (GH). – Río Blanco, 3000 m, 01.1964, *Zöllner s.n.* (CONC). – Aliste, Río Blanco, 3000 m, 01.1972, *Zöllner 5564* (CONC). – **Región Metropolitana de Santiago** Prov. de Santiago Tal Los Paramillos/ Cepo Pass, 3750 m, *Grandjot 3747* (GH). – Cord. Río San Francisco, Fierro Carrera, 3200 m, 02.1925, *Werdermann 632* (BM CONC E F G GH K M). – Cordillera de Las Condes, 3200 m, 01.1934, *Grandjot s.n.* (CONC). – Complejo de Esqui "Valle Nevado", camino a Cerro Franciscano, 3390 m, 03.2003, *K. Arroyo* et al. *25183* (CONC). – Faldeos al oeste de Cerro La Parva, al este de La Parva, 3140 m, 01.2003, *K. Arroyo* et al. *25169* (CONC). – Vegas Las Vacas, 2900 m, 03.1956, *Schlegel 1063* (CONC). – Farellones, La Parva, 01.1972, *Beckett, Cheese & Watson 4855* (SGO). – Farellones, 02.1957, *Rassmusen s.n.* (CONC). – Farellones, camino a Casa de Piedra, 2500 m, 01.1957, *Garaventa 5361* (CONC). – Las Condes, Valle Largo, 3000 m, 01.1948, *Muñoz s.n.* (CONC). – San Gabriel, 3000 m, 12.1950, *Gunckel 21845* (CONC). – Volcán San José, 3600 m, 02.1947, *Muñoz s.n.* (CONC). – R. N. Río Clarillo, Sector Alto las Cruces, 2930 m, 12.2000, *K. Arroyo* et al. *206666* (CONC). – RN Río Clarillo, Sector Las Cruces, 2870 m, 01.2000, *K. Arroyo* et al. *20653* (CONC). – RN Río Clarillo, Sector Los Cristales, 2650 m, 01.2000, *K. Arroyo* et al. *20725* (CONC). – RN Río Clarillo, Sector Los Cristales, 2750 m, 02.2001, *K. Arroyo* et al. *210637* (CONC). – Prov. de Cordilleras cordilleras de Santiago, *Philippi 826* (BM P). – Valle del Yeso, *Philippi s.n.* (K W) - Cord. de Santiago ad limit nivis, 02.1854, *Philippi s.n.* (W). – Cord. de Santiago, *Philippi s.n.* (F NY Wx2). – Cord. de Santiago, 1876, *Philippi s.n.* (G). – Cordillera de Santiago, Caj. del Cepo, 3500 m, 02.1894 (M). – **S.l.d.** *Seibold 3088* (W). – *Gay s.n.* (F).

4. *Oriastrum pusillum* Poepp. & Endl. Nov. Gen. Sp. Pl. 3: 50, fig. 257. 1842.

Typus CHILE "Crescit in glareosis frigidissimis Andium chilensium (Cordillera de San Jago), locis nivi aeternae proximis. Januario floret." [Chile borealis in frigidissimus glareosis Cordillera de Santiago, 01.1828, *Poeppig s.n.*] Holotypus W! Isotypus P (fragment)! W (Herb. Reichenbach, fragment)!

- *Chaetanthera planiseta* Cabrera Revista Mus. La Plata, Secc. Bot. 1: 128 – 130, fig. 16. 1937.

Annual monoecious dwarf cusion-forming herb. **Stems** to 4 cm, branching, very compact. **Leaves** 6.5 mm L, indistinctly petiolate, narrowly spathulate, decussate, connate; lamina 3.1 x 1.9 mm, ovate acute, or dilated, margins thickened or even limbate; lamina reflexed, conduplicate, shortly mucronate; indumentum of lanate hairs (1.2–1.5 mm L) on petiole and lamina, densely tangled. **Capitula** sometimes shortly pedunculate, leaves alternate (0–3) up stem, otherwise capitula sessile; capitula 7 mm L, disk diameter 2.5–3.5 mm. **Involucral bracts** imbricate, arranged in two types (no **outer involucral bracts** with short membranous alae to less than ½ height of bract), initially foliaceous then reduced to entirely membranous. **Middle involucral bracts** indistinctly petiolate, 5.2–5.9 mm L; membranous alate, ratio of lamina to alae decreases from ⅔ to nearly entirely alate, alae with ovate to rhombic outline, 0.6–0.9 mm W.; dorsally pubescent, but on alae only on margins; lamina ovate, 1.2–2.5 x 0.6–1.9 mm, becoming reduced along series, dorsally and ventrally lanate. **Inner involucral bracts** 6.8–7.4 x 1.9–2.5 mm, entirely membranous; outline narrowly oblanceolate to rhomboid; apical appendages acute, conspicuously maculate (transparent or with black-brown stripe), 2.2–2.5 mm L; below macula dorsally hirtellous. **Ray florets** ± 9, pistillate, white, or yellow; corolla 4.9–5.1 mm L, corolla tube 2.2 mm L, outer ligule 1.5–1.9 mm W., inner ligule notched, 0.1 mm L **Disk florets** ca. 10, bisexual, yellow-orange; corolla 3.7 mm L, corolla tube 3 mm L **Style**s yellow [ray] 3.1–3.4 mm, bifid; stigma lobes, obtuse, closed. [disk] 3.7 mm L; stigma lobes 0.3 mm L, round, open. **Anthers** 2.5 mm (3.6 mm L incl. filaments). **Achenes** brown, pyriform, 1.9 mm L; pericarp pellucid, coated in globular twin hairs (35–40 µm), ray achenes sometimes glabrous; no carpopodium; testa epidermis with branched ribbed strengthenings, cells broadly sinuous tesselate. **Pappus** white, 1–2 rows, dehiscent; setae 2.3–3 mm L, 14 cells wide, cells 5–10µm W.; basal cilia present, sparsely barbellate (3–5 barbs/100 µm), barbs 90–140 µm L, barb base shortly adhered (<40%), free barb spreading to ciliate.

X *Oriastrum* Poepp. & Endl.

Figure 83: *Oriastrum pusillum*. **A.** Habit, white flowered form, photographed La Parva 03.2006©Michail Belov. **B.** Habit, yellow flowered form, photographed La Parva, 03.2006©Michail Belov. **C.** Leaf & Bract detail. **i.** Stem leaf v.s.; **ii.** MIB d.s.; **iii.** IIB d.s. Scale bar = 2mm. **D.** Pappus detail, showing the characteristic ciliate pappus barbs. S.E.M. image *Grandjot 3817*.

Distribution and Habitat. *O. pusillum* is a Chilean Andean endemic, and has been recorded from Baños del Toro (29°50'S), Junta de Piuquenes (32°15'S) and most commonly in the Andes east and south of Santiago (33°20' – 33°52'S). It is found in amongst rocks, scree, or on volcanic ash slopes at elevations between 2300 – 3400 m.a.s.l. The two northernmost collections were found at the highest altitudes.

Differential diagnosis. *O. pusillum* is similar to *O. chilense* but can be distinguished by its transparent or black-marked (like a fingernail) inner involucral bract tips, and its ciliate pappus setae. *O. pusillum* is polymorphic for flower colour and can be either white, yellow (see herbarium labels), or both within one population (K. Arrroyo, pers. comm.).

Material seen. – Chile. IV Región de Coquimbo Prov. de Elqui Baños del Toro, 3260 m, 12.1939, *Barros s.n.* (CONC). **– V Región de Valparaíso** Prov. de Petorca 5 km south of Junta de Piquenes, Río Sobrante, 3400 m, 02.1939, *Morrison 17299* (G K MO). **– Región Metropolitana de Santiago** Prov. Santiago Cerro Parva im Plomo-Massiv bei Santiago, 3000 – 3100 m, 02.1939 *Grandjot 3817* (M). – La Parva, 2750 m, 03.1992, *Gardner, Hoffman & Page 5131* (E#28558, #129169). – Farellones, camino a valle Nevado, antes portion valle Nevado, 2850 m, 01.1993, *Muñoz S. & Eggli 3209* (SGO). – Vegas Las Vacas, 2900 m, 03.1956, *Schlegel 1060* (CONC). – La Parva, entre 2700-3100 m, 01.1980, *Muñoz S. 1601* (SGO). – La Parva, entre 2700-3100 m, 01.1980, *Muñoz S. 1588* (SGO). – Farellones, camino a valle Nevado, 2300 m, 01.1993, *Muñoz S. & Eggli 3202* (SGO). – Farellones, camino a Casa de Piedra, 2500 m, 02.1957, *Garaventa 5362* (CONC). – Complejo de Esqui "Valle Nevado", laderas cerca del hotel, 3050 m, 01.2003, *K. Arroyo et al. 25170; 25171; 210674; 210675* (CONC). – Strasse nach Farellones, Umgebung de Skipisten, 02.1986, *Hellwig 3107* (G K). – El Colorado, 01.1993, *Muñoz S. & Eggli 3227* (SGO). – Potrero Grande, 3200 m, 01.1936, *Grandjot s.n.* (CONC). – Prov. de Cordillera Lagunillas, 2600 m, 01.1985, *Zöllner 12285* (MO). – Las Condes, *King s.n.* (E). – Valle del Yeso, *Reed s.n.* (Kx2). – Cordilleras de Santiago, 1856-1857, *Philippi s.n.* (K W). – Cordillera de Santiago, *Philippi 720* (W). – Puente Alto, Cerro Las Tinajas, 2420 m, 03.1967, *Schlegel 5915* (CONC F). – Mountains above Río Colorado, East of Santiago, about 5000 ft, 01.1902, *Hastings 633/s.n.* (NYx2). – Lagunillas, 2700 m, 12.1966, *Mooney 205* (CONC). – Lagunillas, 2700 m, 02.1958, *Brunner s.n.* (CONC). – Lagunillas, arriba del pueblo en las canchas de esqui, 2460 m, 01.2003, *K. Arroyo et al. 25120* (CONC). – R.N. Río Clarillo, Sector Los Cristales, 2870 m, 02.2001, *K. Arroyo et al. 210647* (CONC). – **S.l.d.** Andes du Chili, *Poeppig s.n.* (P). – Chile, *Seibold 3064* (W). – Chile borealis in glareosis frigidus Andium Decbr. *Poeppig s.n.* (W). – Chile, *Ball s.n.* (E).

X *Oriastrum* Poepp. & Endl.

5. *Oriastrum werdermannii* A.M.R.Davies sp. nov.

Typus CHILE "Chile. IV Región de Coquimbo, Prov. de Elqui, Rivadavia, 800 m, 11.1923, *Werdermann 96*" Holotypus CONC! Isotypus BM! E! F! G! GH!

Planta annua nana herbacea pulvinata ad 8 cm alta. Folia indistincte petiolata lamina orbiculata, marginibus porphyreo limbatis, dense lanata vel tomentosa antice mucronata. Capitula radiata, campanulata. Flores radii albidi, ad 7 mm, capitulis disctince exsertis. Achenia fusca, pyriformia, pericarpio pellucido, glabrescentia vel sparse pilis nanis globularibus vel ampulliformibus ornata. Pappus caducus; pappi setae 5–6 mm longae; basaliter ciliis ornatae, sparse barbellatae, barbis basaliter longe adhaerentibus parte superiore patentibus. Habitat in Chile.

Annual monoecious dwarf; forms small laxly branched cushions. No stem buds. **Stems** to 8 cm; shortly erect before branching from central node, lax, loosely lanate. **Leaves** on erect stem opposite, distant pairs to 6.5 mm L, leaves on branches alternate to whorled, to 10 mm L, indistinctly petiolate, spathulate, connate; lamina 2–3 x 3–4 mm, cordate to orbicular, grey-green; margins pale limbate, apices mucronate; indumentum lanate, hairs simple, white, to 3 mm on petiole, shortly tomentose on lamina. **Capitula** campanulate, disk diameter (4) 5–7.5 mm. **Involucral bracts** imbricate, arranged in two series (no **outer involucral bracts** with short membranous alae to less than ½ height of bract), initially foliaceous then reduced to entirely membranous. **Middle involucral bracts** 8–10 mm L; lamina 2–3 x 1–3 mm, reduces along series from orbicular cordate to lancolate acute, shortly lanate above alae; membranous alate, ratio of lamina to alae decreases from ⅔ to nearly entirely alate, alae ovate-linear, 1.5–1.7 mm W.; distal dorsal surface and upper margins lanate. **Inner involucral bracts** entirely membranous alate, 1(-2) series; 8–10 x 2–2.5 mm, obovate, apical appendages obtuse to broadly acute, pale translucent, sometimes rosy. **Ray florets** 8–14, pistillate, white to creamy sometimes tinged pink; corolla 9 mm L; corolla tube 4.5 mm L; outer ligule 2–2.5 mm W.; inner ligule bifid reduced to 0.5 mm L **Disk florets** ca. 12, bisexual, yellow; corolla 3.9 mm; corolla tube 3 mm. **Styles** [ray] 5 mm. [disk] 4.5 mm L, stigma lobes short, obtuse. **Achenes** brown, pyriform, to 2 mm L; pericarp pellucid, glabrous or densely scattered with tiny (to 20 µm) globular to ampulliform twin hairs; no carpopodium. **Pappus** white, 1–2 rows, dehiscent; setae 5–6 mm L; basal cilia present, sparsely barbellate, barbs 90–140 µm L, barb base longly adhered (>60%), free barb spreading.

Distribution and Habitat. *O. werdermannii* is locally endemic in the Chilean pre-cordilleran Andes around Rivadavia (ca 29°58'S; Coquimbo) below 1000 m.a.s.l. It is found in steep dry screes, on level spots among rocks, or in rocky soil near streams.

Differential diagnosis. Differs from *O. gnaphalioides* by having larger capitula, larger ray florets, growing at lower altitudes.

Material seen. – Chile. IV Región de Coquimbo Prov. de Elqui Rivadavia, 3 km east of the station, talus along road, 850 m, 10.1940, *Wagenknecht 18590* (CONC G GH). – Rivadavia, 800 m, 11.1923, *Werdermann 96* (BM CONC E F G GH). – Cerca de Rivadavia, 880 m, 10.1940, *Wagenknecht s.n.* (CONC).

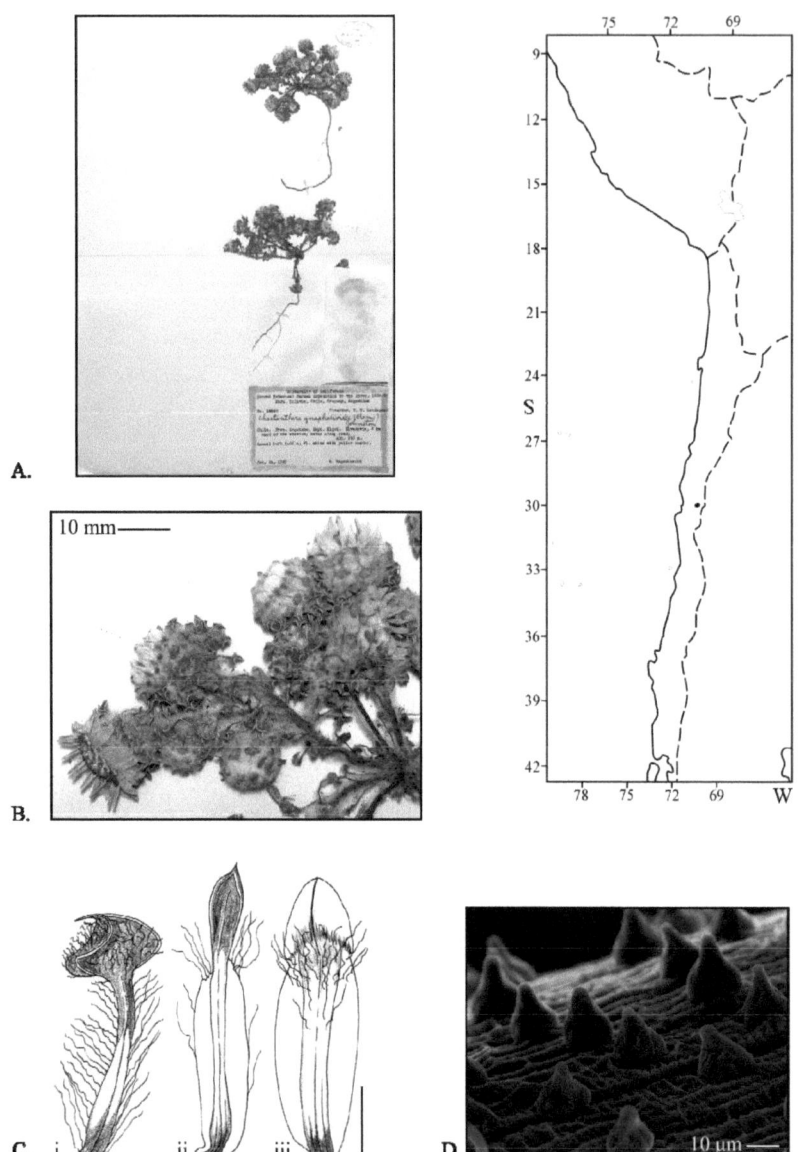

Figure 84: *Oriastrum werdermannii*. Holotype *Werdermann 96*. **A.** Herbarium sheet, habit. **B.** Capitulum detail. Scale bar = 10 mm. **C.** Leaf & Bract detail. **i.** Stem leaf v.s.; **ii.** MIB v.s.; **iii.** IIB d.s. Scale bar = 2 mm. **D.** S.E.M. Image of achene hairs *Wagenknecht 18590*.

X *Oriastrum* Poepp. & Endl.

4.2   *Oriastrum* subgenus *Egania*

6.   *Oriastrum abbreviatum* (Cabrera) A.M.R. Davies stat. nov.

≡ *Chaetanthera stuebelii* var. *abbreviata* Cabrera Revista Mus. La Plata, Secc. Bot. 1: 115, fig. 9. 1937. Typus ARGENTINA "Argentina, Tucumán, Cumbres Calchaquíes, Callejones, 4,200 m, 27.12.1913, *Castillón 3258*" Holotypus LIL 16305! Isotypus LP 66680!

= *Chaetanthera stuebelii* var. *argentina* Cabrera Revista Mus. La Plata, Secc. Bot. 1: 115, fig. 8. 1937. Typus ARGENTINA "Argentina, Salta, La Laguna Seca, Cerro Cajón, 4,280 m, 18.2.1914, *Rodriguez 1331*" Holotypus NY! Isotypus BA 25547(x2) F! K! LIL 76522 LP (photo) NY!

Perennial monoecious or rarely gynodioecious dwarf cushion-forming herb, sometimes epiphytic. **Stems** 4–5 cm high, with basal bud cluster; flexuous, ascending, branching; glabrescent and sparsely leafy in lower two thirds. **Leaves** 2–3.5 x 1 mm, linear obtuse, shortly mucronate, pale green, slightly succulent, decussate, connate, and appressed on upper stems; margins thickened, involute; glabrescent to lightly floccose, indumentum shortly curly on lower margins and upper ventral surfaces. **Capitula** cylindrical, 3–4 mm L; disk diameter 1–1.5 mm; sessile or shortly pedunculate, peduncle naked; capitula can be bisexual and female on same plant. **Involucral bracts** imbricate, arranged in two types (no **outer involucral bracts** with short membranous alae to less than ½ height of bract), initially foliaceous then reduced to entirely membranous.
Female capitula rare; capitula c. 3.5 mm L; disk diameter 1.5 mm. No **Outer involucral bracts**. **Middle involucral bracts** 3.1–4.9 x 1.5–2 mm; membranous alate, ratio of lamina to alae decreases from ⅔ to nearly entirely alate, alae ovate-linear, truncate, sometimes with distinctive black maculae on upper tips. **Inner involucral bracts** 5.6–6.2 x 1.2–1.5 mm, 2 series, lanceolate, entirely membranous alate; apical appendages shortly acute, brown-black, 2 mm L **Ray florets** ca. 11, yellow; corolla 3.4–3.7 mm L, corolla tube 2 mm L, outer ligule 0.5 mm W., irregularly 3-dentate, inner ligule broadly bifid, 0.2 mm L **Styles** [ray] 3.1 mm L, stigma lobes orange-brown, obtuse, 0.2 mm L **Achenes** brown, 2 mm L (immature); pericarp pellucid, glabrous. **Pappus** white, 1–2 series, setae 4 mm L, barbellate, dehiscent.
Bisexual capitula most common; capitula 3 mm L, disk diameter 1.1–1.5 mm. No **Outer involucral bracts**. **Middle involucral bracts** 2.8–5.5 x 0.9–2.5 mm; membranous alate, ratio of lamina to alae decreases from ⅔ to nearly entirely alate, alae ovate, truncate with distinctive black maculae on upper tips; indumentum on upper margins and ventral surface. **Inner involucral bracts** 4.5–7 x 1.2–1.7 mm, lanceolate, entirely membranous alate; apical appendages acute-obtuse, brown-black, 1.5–2.5 mm L **Ray florets** ca. 8, pistillate, yellow; corolla 4.2 mm L, corolla tube 1.5–1.9 mm L, outer ligule 0.8 mm W., irregularly 3-dentate, inner ligule 0.6 mm L, bifid. **Disk florets** 5–9, bisexual, yellow; corolla 2.2–3.7 mm L, corolla tube 1.2–3.1 mm L **Styles** [ray] red or orange-brown, 2.3 mm; stigma lobes 0.1–0.2 mm, obtuse; [disk] 1.6–3.4 mm L, lobes obtuse, 0.2

mm L, open. **Anthers** 1.8–2.2 mm. **Achenes** brown, narrowly turbinate; pericarp pellucid, glabrous, 0.6–1.2 mm L **Pappus** white, 1–2 series, setae 3.5 mm L, barbellate, dehiscent.

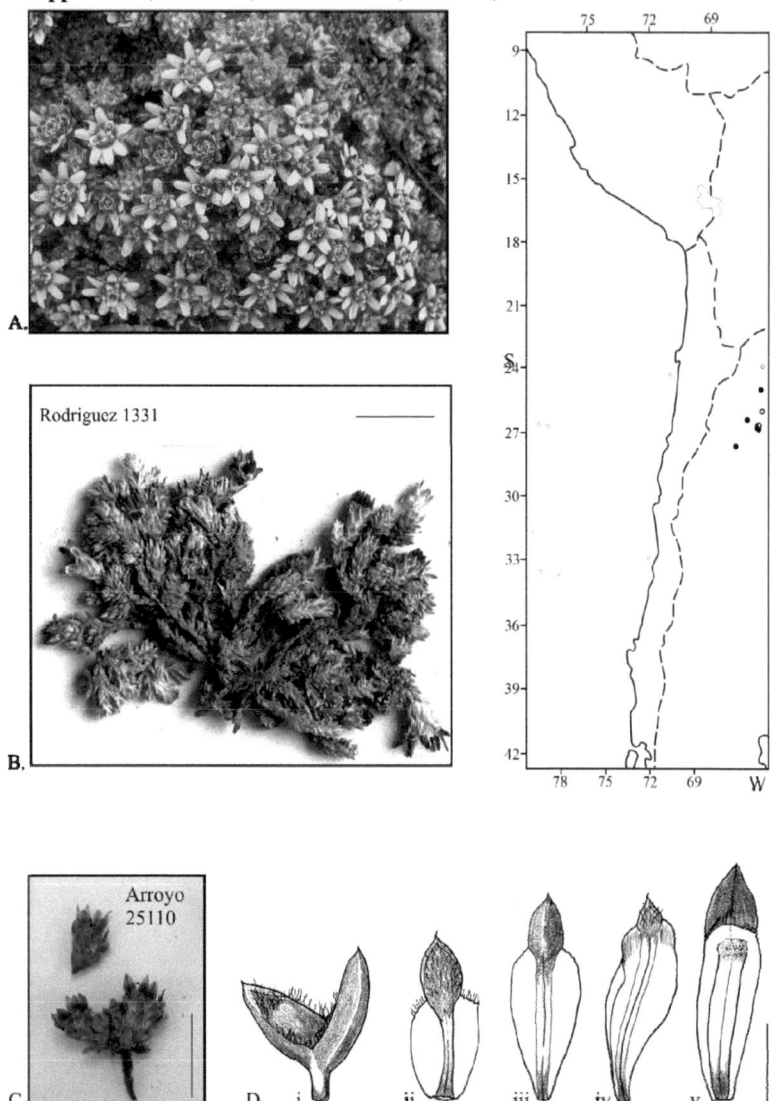

Figure 85: *Oriastrum abbreviatum*. **A**. Image taken 05.12.2008, Calchaquies, Argentina Projecto GLORIA ©Stephan Halloy. **B**. Habit *Rodriguez 1331* Scale bar = 10 mm. **C**. Capitulum detail, *Arroyo 25110* Scale bar = 10 mm. **D**. Leaf & Bract detail. **i**. pair of stem leaves; **ii – iv**. MIB v.s. & d.s.; **v**. IIB d.s. *Rodriguez 1331*, female capitulum. Scale bar = 2 mm.

X *Oriastrum* Poepp. & Endl.

Distribution and Habitat. Endemic to Argentina, *Oriastrum abbreviatum* is distributed from just north of the Nevado de Chañi, across the Valles Calchaquíes and into the Sierra del Aconquija at latitudes between 23°25'–27°40'S. It forms loose cushions, sometimes epiphytically, and is found at elevations between 3400–4450 m.a.s.l.

Differential diagnosis. Most easily confused with *O. famatinae*, *O. abbreviatum* belongs to a group of *Oriastrum* taxa with conspicuous pairs of linear stem leaves. It typically has only one series of IIB with shortly acute apices.

Material seen. – **Argentina. Prov. de Jujuy** Depto. Tumbaya Volcán, 3400 m, 01.1926, *Venturi 9257* (GH). – Depto. Tilcara Cerro Alto de Cima al oeste de Huacalera, 4450 m, 03.1967, *Werner 575* (LP). – **Prov. de Salta** Depto. Cachi Cerro de Cachi, 01.1897, *Spegazzini 1593* (LP). – Cerro Calchaquies, Cumbre del Cajón, 01.1913, *Castillon s.n.* (LP). – Depto. La Laguna Sierra del Cajón, 4250 m, 01.1914, *Rodriguez 1331* (F K NYx2). – Sierra del Cajón, 4100 m, 02.1914, *Rodriguez 1352* (LP). – **Prov. de Tucumán** Las Lagunas, 01.1935, *Castellanos s.n.* (LP). – Depto. Tafí Cerro de Las Animás, cerca de la Laguna Verde y Huesco, 4100 m, 01.1914, *Castillon 3263* (F). – Cerro Negrito, 4000 m., 02.1958, *Fabris 1426* (GH). – Cerro Negrito, subida desde El Infiernillo, Cumbres Calchaquíes, 4220 m, 02.2003, *K. Arroyo et al. 25110* (CONC). – Cerro Negrito, subida desde El Infiernillo, Cumbres Calchaquíes, 4220 m, 02.2003, *K. Arroyo et al. 25114* (CONC). – Cerro Negrito, subida desde El Infiernillo, Cumbres Calchaquíes, 3890 m, 02.2003, *K. Arroyo et al. 25109* (CONC). – Cerro Negrito, subida desde El Infiernillo, Cumbres Calchaquíes, 4000 m, 02.2003, *K. Arroyo et al. 25113* (CONC). – Cumbres Calchaquíes, 01.1907, *Lillo 5547* (LP). – **Prov. de Catamarca** Depto. Belén, faldeos al N. del Portezuelo del Río Blanco, arriba de Granadillas, 3600 m, 01.1952, *Sleumer & Vervoorst 2596* (LP). – El Cajón, Cerro Negro Ara, La Bolsa, 4400 m, 01.1914, *Castillon s.n.* (F, LP).

7. *Oriastrum acerosum* (J. Rémy) Phil. Anales Univ. Chile 87: 20. 1894.

≡ *Egania acerosa* J. Rémy in Fl. Chil. [Gay] 3: 325. 1849. Typus CHILE "Se cria en las cordilleras de Coquimbo, á Pasto Blanco, los Patos, á una altura de 3,000 m". [Cordilleras de Ovalle, Janv. 1837, *Gay 464*] Holotypus P! Isotypus probabilis [Coquimbo, *Gay s.n.*] G!

= *Egania pallida* Phil. Linnaea 28: 712. 1856. Typus CHILE "In saxosis Andium ad los Patos prov. Coquimbo legit cl. *Gay Herb Chile no. 745*" [Prov. Coquimbo, in saxosis andium Los Patos, *Gay 745*] Holotypus SGO 64957!
= *Oriastrum parviflorum* Phil. Anales Univ. Chile 87: 18. 1894. Typus CHILE "Habitat in monte ingenti Doña Ana prov. Coquimbo, locis Laguna seca Quebrada de Pastos etc. legit ubi *Frid. Philippi* Februario 1883" [Doña Ana, Febr. 1888, *Frid. Philippi s.n.*] Holotypus SGO 71269!
= *Oriastrum incanum* Phil. Anales Univ. Chile 87: 22. 1894. Typus CHILE "Ad originem fluminis Torca in prov. Coquimbo invenit onr. *Guill. Geisse*" [Nacimiento Río Torca, Teulahuen, Dicbr. 1890, *Geisse s.n.* - apical appendages 5 mm] Holotypus SGO 72401!
= *Oriastrum uncinatum* Phil. Anales Univ. Chile 87: 23. 1894. Typus CHILE "Habitat in provinciae Coquimbo Andibus loco Los Patos. *Cl. Gay*" [Prov. Coquimbo, in saxosis andium Los Patos, *Gay 745*] Holotypus SGO 64957!
= *Chaetanthera acerosa* var. *dasycarpa* Cabrera Revista Mus. La Plata, Secc. Bot. 1: 112. 1937. Typus CHILE "Chile, Coquimbo, Quebrada del Toro, 3,600 m, 1.1936, *Cabrera 3561*" Holotypus LP! Isotypus F! LPx2! NY!
− *Egania racemosa* Pritz. Icon. Bot. Index 2: 114. 1854-55. = *Chaetanthera acerosa* (orth. err.), illus. Icon. Weddell Chloris Andina fig. 9. 1855.
− *Chaetanthera acerosa* (J. Rémy) Benth. & Hook. f. Gen. Pl. 2(1): 496. 1873.
− *Chaetanthera acerosa* (J. Rémy) Griseb. Symb. Fl. Argent. : 214. 1879.
− *Oriastrum pallidum* (Phil.) Phil. Anales Univ. Chile 87: 20. 1894.
− *Oriastrum pusillum* var. *uncinatum* (Phil.) Reiche Anales Univ. Chile 115: 334. 1904; Fl. Chile [Reiche] 4: 353. 1905.
− *Chaetanthera acerosa* (J. Rémy) Hauman Anales Soc. Ci. Argent. 86: 316. 1918.
− *Chaetanthera pallida* (Phil.) Hauman Anales Soc. Ci. Argent. 86: 317. 1918.

Nomenclatural Notes *Gay 745* is the Typus for *Oriastrum uncinatum* Phil. and *Egania pallida* Phil. I.M. Johnston erroneously noted on I.M.J. 5955 (GH) that the Typus number of this collection was '945'.

Perennial monoecious dwarf cushion-forming herb. **Stems** to 5 cm (sometimes longer), with basal bud cluster; trailing and branching. **Leaves** 2–3.3 x 0.5–1.0(1.4) mm, sparse on stems, decussate, connate, linear or scale-like, mucronate, midrib ± visible; proximal margins and ventral surfaces pubescent, distal dorsal surfaces glabrous or hirtellous; leaves on flowering stems 7–8.5(14) x (0.6)0.8–1.2 (1.5) mm, opposite to alternate, linear, recurved, apiculate, mucronate; margins thickened to inrolled; proximal dorsal surface and ventral surface densely lanate (hairs 1.5–2.5 mm L). **Capitula** cylindrical, 10–13(16) mm L, pedunculate; disk diameter 2–4 mm. **Involucral bracts**

X *Oriastrum* Poepp. & Endl.

imbricate, arranged in three types, initially foliaceous then reduced to entirely membranous. **Outer involucral bracts** (6)8–12(14) x 0.8–2 mm; bracts as leaves but with short membranous alae to less than ½ height of bract, alae linear to narrowly ovate, (1.5)2–2.5(4) mm W.; indumentum on alate margins, and ventral lamina surfaces. **Middle involucral bracts** (7)8.5–11.5 x (1.7)2.5–3.5(4) mm, narrowly ovate to lanceolate; membranous alate, ratio of lamina to alae decreases from ⅔ to nearly entirely alate; indumentum on distal dorsal surface of alae and ventral lamina surface; innermost bracts sometimes with black maculae on distal dorsal alae margins. **Inner involucral bracts** (7.0)9–11(13) x (1.5)2.5–3.5(4) mm, narrowly ovate to lanceolate; entirely membranous alate; pubescent in dorsal distal third; apical appendages acute to shortly triangular, (0.9)2–3 mm L, distinctly black–green (in some examples striped or pale), mucronate. **Ray florets** 7–11, pistillate, white, sometimes tinged pink on dorsal apical appendages; corolla (6.5)8–9.5 mm L, corolla tube (2.5)3–4 mm L, outer ligule 1.2–2.2 mm W., inner ligule bifid, 0.1–0.4 mm. **Disk florets** < 10, bisexual, yellow; corolla 5–6.5 mm L, corolla tube 4.3–5.7 mm L **Styles** [ray] (4.3)6–8 mm L, yellow-green, lobes bifid, 0.25 mm L; [disk] yellow, 5–7 mm L, lobes 0.2–0.4 mm L **Anthers** 3–4 mm L (5.5–6.5) mm incl. filaments) **Achenes** brown, turbinate, 1.9–3.2 mm (immature); pericarp pellucid, densely hirsute (long simple hairs to 0.5 mm) or glabrous; no carpopodium; testa epidermis with no significant strengthenings, cells parallel. **Pappus** white, 1–2 rows, dehiscent; setae 5–6.5 mm L, 4–8 cell wide, cells 24μm W.; basal cilia present, sparsely barbellate, barbs 90–140 μm L, barb base longly adhered (>60%), free barb spreading.

Distribution and Habitat. *O. acerosum* it distributed from 27°10'–31°30'S. The distribution stops at the south end of the Cordillera del Guacho massif and does not continue on to the adjacent Cordillera de Espinazito. It occurs in dry rocky soil on slopes or screes, along the edges of quebradas at elevations between 2400–4300 m.a.s.l.

Differential diagnosis. *O. acerosum* has OIB that extend beyond the capitula, the linear leaves are a yellow-green to grey green in colour, and the yellow-green ray stigmas and its generally smaller stature distinguish it from the nearest species *O. apiculatum* (whose OIB do not extend beyond the capitula, the linear to spacellate leaves are blue-green often with reddish margins and it generally has a larger stature). *O. acerosum* is polymorphic. Material from the extreme points of the distribution range is disparate. The changes are gradual along the whole range, with the plants becoming smaller, the IIB apices more acute, smaller, and have more variable colouration (they vary from pale to marked or completely black-green) as the collections move north and gain altitude. *Marticorena et al. 83451B* and *K. Arroyo 81504* are anomalous.

Material seen. – **Argentina San Juan** Valle de La Cura cerca Pirca de los Ingenieros, 4000 m, 01.1930, *Perez Moreau 147* (LP). – **Chile. III Región de Atacama** Prov. de Copiapó Portezuelo Chinches, 3280 m, 01.1973, *Marticorena* et al. *570* (CONC). – Prov. de Huasco Quebrada Cantarito, 4300 m, 02.1981, *K. Arroyo 81615* (CONC). – Quebrada Cantarito, entre Quebrada Marancel y Portezuelo de Cantarito, 4300 m, 01.1983, *Marticorena, K. Arroyo & Villagrán 83471* (CONC). – Quebrada Cantarito, entre el estremo oeste de la Laguna Grande y la Quebrada Marancel, 3500 m, 01.1983, *Marticorena, K. Arroyo & Villagrán 83451B* (CONC). – entre Laguna Chica y Portezuelo Yerba Buena, 3950 m, 01.1983, *Marticorena, K. Arroyo & Villagrán 83591* (CONC). – entre Laguna Chica y Portezuelo Yerba Buena, 3950 m, 01.1983, *Marticorena, K. Arroyo & Villagrán 83588* (CONC). – Río Laguna Grande, entre Las Papas y Potrero de Toledo, 2400 m, 02.1981, *K. Arroyo 81504* (CONC). – Cordillera Laguna Chica, 4000 m, 01.1924, *Werdermann 256* (BM, CONCx2 E F G GH K M). – Quebrada Barriales, Mina de Pascua Lama, sector El Nevado, 3760 m, 01.2003, *K. Arroyo* et al. *25095* (CONC). – Vicinity of Laguna Chica, 3500 m, 01.1926, *Johnston 5955* (GH K). – Vicinity of Laguna Chica, 3800 m, 01.1926, *Johnston 5925* (GH). – **IV Región de Coquimbo** Prov. de Elqui Andes, "*Egania pallida*", 02.1888, *Philippi s.n.* (K W). – Portezuelo de Dona Aña, Baños del Toro, "*Egania acerosa*"

Philippi s.n. (K). – Vega La Colgada, en curso superior del Río Sancarrón, 4000 m, 01.1979, *Osario s.n.* (SGO). – camino a Tambo desde Pta Diamante, 4300 m, 01.1988, *Squeo 88041* (CONC MO). – Km 29, 3800 m, 02.1988, *Squeo 88092* (CONC). – Km 29, Camino a Indio, 3900 m, 01.1988, *Squeo 88053* (CONC). – Baños del Toro, 3600 m, 01.1936, *Carbera 3561* (F NY). – Baños del Toro, Quebrada Pastos, 3300-3600 m, 02.1939, *Morrison & Wagenknecht 17190* (CONC G MO). – Baños del Toro, 3200 m, 02.1963, *Ricardi* et al. *703* (CONC). – Baños El Toro, 3400 m, 01.1966, *Peña s.n.* (CONC). – Baños del Toro, 12.1971, *Beckett, Cheese & Watson 4640* (SGO). – Huanta, 2750 m, 01.1949, *Jiles 1215* (CONC). – Camino entre Baños del Toro y Mina El Indio, Km 29, 3700 m, 01.2003, *K. Arroyo* et al. *25087* (CONC). – Entrada a Mino El Indio, aproximadamente 2 km por debajo de Baños del Toro, 3210 m, 01.2003, *K. Arroyo* et al. *25077* (CONC). – Canchas de Sky, 3400 m, 02.1988, *Squeo 88156* (CONC). – Canchas de Sky, 3300 m, 02.1988, *Squeo 88127* (CONC MO). – Straße von Vicuna zum Embalse La Laguna, 2780 m, 11.2002, *Ehrhart* et al. *2002/062* (MSB). – camino a Embalse La Laguna, 3150 m, 02.1987, *Niemeyer 8704* (CONC). – 9 Km antes del Paso de Agua Negra, 4150 m, 01.1967, *Ricardi, Marticorena & Matthei 1761* (CONC). – Embalse de La Laguna, 3350 m, 02.1963, *Ricardi* et al. *718* (CONC). – entre Juntas y Embalse La Laguna, 3050 m, 01.1967, *Ricardi* et al. *1747* (CONC). – 11 Km antes del Paso de Agua Negra, 4050 m, 01.1967, *Ricardi* et al. *1768* (CONC). – <u>Prov. de Limarí</u> Coquimbo, *(Ball) Philippi s.n.* (E). – Quebrada Larga, 3000 m, 02.1958, *Jiles 3412* (CONC). – Río Mostazal, 3500 m, 02.1956, *Jiles 2975* (CONC). – San Miguel, 3500 m, 02.1972, *Jiles 5947* (CONC). – Los Pingos, 3000 m, 01.1972, *Jiles 5882* (CONC). – Río Torca, Las Galenas, 3000 m, 02.1961, *Jiles 3779* (CONC). – Gordito, 3000 m, 01.1954, *Jiles 2513* (CONC). – <u>Prov. de Choapa</u> Caleton Blanco, 3000 m, 02.1962, *Jiles 4239* (CONC). – Cerro Curimahuida, 10 km east of Matancilla, and 15 km northeast of Sanchez Mine, 2800 m, 11.1938, *Worth & Morrison 16660* (G GH K). **S.l.d.** *Gay s.n.* (F K).

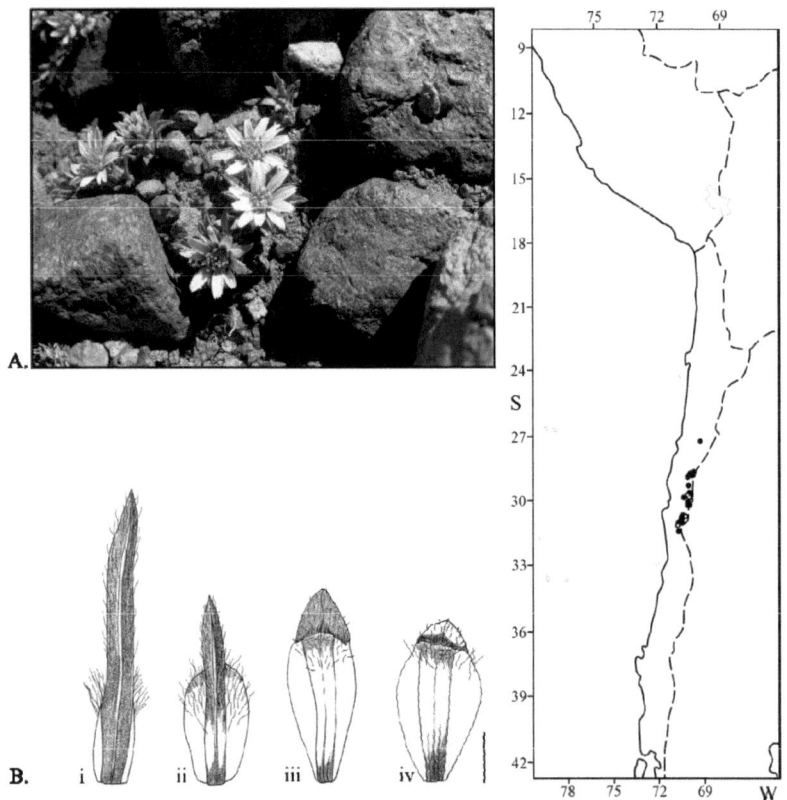

Figure 86: *Oriastrum acerosum*. **A.** Habit, photographed Paso Agua Negra, 4000 m, 2008©Mauricio Zuñiga. **B.** Leaf & Bract details *Cabrera 3561* **i.** Stem leaf d.s.; **ii.** MIB d.s.; **iii.** IIB d.s.; *Squeo 88053* **iv.** IIB d.s. Scale bar = 2 mm.

X *Oriastrum* Poepp. & Endl.

8. *Oriastrum achenohirsutum* (Tombesi) A.M.R. Davies comb. nov. & stat. nov.

≡ *Chaetanthera pulvinata* (Phil.) Hauman var. *acheno-hirsuta* Tombesi Hickenia 3 (29): 69 – 72, figs 1 L – O, 2. 2000. Typus ARGENTINA "Argentina. San Juan, Depto. Iglesia: Reserva de San Guillermo, Mina Las Carachas, 4050 m. s.m. 20.02.1981, *Nicora, Guaglianone & Ragonese 8217*" Holotypus SI!

= *Chaetanthera acheno-hirsuta* (Tombesi) Arroyo, A.M.R. Davies & Till-Bottraud Gayana Bot. 61(1): 27 – 31, 2004.

Perennial monoecious dwarf cushion-forming herb. **Stems** shortly lax to 8 cm, with bud clusters at base; wiry, distally lanate. **Leaves** 8–15 x 1–4 mm, narrowly spathulate to linear, lamina decurrent to petiole rendering spathulate part somewhat indistinct, opposite to rarely alternate, pale green, limbate, midrib visible, apex mucronate; indumentum to 2 mm, hairs shorter towards apex, dorsal and ventral surfaces densely lanate. **Capitula** sessile or shortly pedunculate, solitary, 1–3 per plant; small scale or bract subtending capitula, obovate, 2–4 x 1 mm, shortly mucronate; disk diameter 6–8 mm (fresh). **Involucral bracts** imbricate, arranged in three types, initially foliaceous then reduced to entirely membranous. **Outer involucral bracts** broadly spathulate, 13.5–11(9) mm L, lamina 3.5–2 mm W., somewhat conduplicate; as leaves but with short membranous alae to less than ½ height of bract, alae 0.5–0.8 mm W., linear oblong; bracts getting shorter along series; dorsal and ventral surfaces densely lanate except in areas with alae, which are glabrescent, indumentum shorter towards apex. **Middle involucral bracts** 10 x 1–2 mm; as reduced leaves, membranous alate, ratio of lamina to alae decreases from ⅔ to nearly entirely alate, alae oblong, 0.8–1 mm W. **Inner involucral bracts** 1 series, 9–13.5 x 2–3 mm, lanceolate; entirely membranous alate; apical appendages 3–3.5 mm L, broadly acute, green-black, hirtellous. **Ray florets** 12–25, pistillate, white; corolla 10–11 mm L, corolla tube 4–4.5 mm, outer ligule 1.5–2 mm W., inner ligule 0.5 mm, inconspicuous, bifid. **Disk florets** 27–60 (mean 39), bisexual, yellowish green; corolla 8 mm, corolla tube 7–7.5 mm. **Styles** [ray] 7–8 mm L, stigma shortly bifid, green. [disk] yellow, 7.5–8.5 mm L, stigma lobes 0.5 mm. **Anthers** 5–6 mm (7.5–8.5 mm including filaments). **Achenes** brown, turbinate, to 3.5 mm L (immature); pericarp pellucid, densely hirsute (long simple hairs). **Pappus** setae 7–8.5 mm L, white, barbellate.

Distribution and Habitat. *O. achenohirsutum* is found in the high Andes of Chile (Región de Atacama) and Argentina (San Juan) between 29°00'–30°00'S. It occurs in amongst rocks of scree slopes at elevations between 3600–4080 m.a.s.l.

Differential diagnosis. *O. achenohirsutum* is distinct from the nearest species *O. acerosum* because it has broadly spathulate, limbate leaf laminas. The IIB are in one series, with broadly acute green-black apices.

Material seen. – Argentina. San Juan Depto. Iglesia Mina Las Carachas, 4050 m, 02.1981, *Nicora, Guaglianone & Ragonese 8217* (SI). – al oeste del Cajon de la Brea, 4080 m, 01.1997, *Kiesling, Moglia & Tombesi 8831* (op. cit.). – Río Turbio, al oeste del Río de las Taguas, 4000 m, 03.1998, *Kiesling 9021b* (op. cit.). – Quebrada de la Ciénaga Colgada, 4080 m, 02.1983, *Pujalte 188* (op. cit.). – Cerro Torrecillas, 3950 m, 02.1983, *Pujalte 243* (op. cit.). – **Chile. III Región de Atacama** Prov. de Huasco Cuenca de Laguna Grande, Quebrada Cantarito, entre Quebrada Marancel y Portezuelo de Cantarito, 3500 – 4300 m, 02.1981, *K. Arroyo 81612* (CONC). – Cuenca de Laguna Grande, Quebrada Cantarito, entre Quebrada Marancel y Portezuelo de Cantarito, 3500 – 4300 m, 01.1983, *Marticorena, K. Arroyo & Villagrán 83473* (CONC). – Cuenca de Laguna Grande, Quebrada Vizcachas, entre Quebrada Cantarito y Portezuelo Vizcachas, 3200 – 4000 m, 01.1983, *Marticorena, K. Arroyo & Villagrán 83576* (CONC). – Cuenca de Laguna Chica, entre Laguna Chica y Portezuelo Yerba Buena, 3400 – 3950 m, 01.1983, *Marticorena, K. Arroyo & Villagrán 83590* (CONC). – Quebrada los Barriales, 4000 m, 01.1994, *Arancio, Squeo & León 94238* (CONC). – Quebrada Barriales, Mina de La Pascua, sector El Nevado, 3760 m, 01.2003, *K. Arroyo et al. 25093* (CONC).

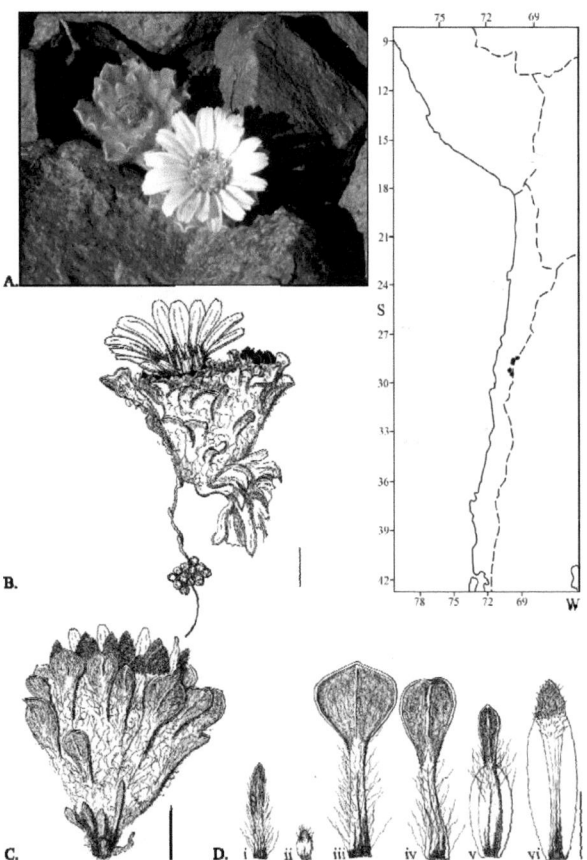

Figure 87: *Oriastrum achenohirsutum.* **A.** Habit, photographed Mina Pascua, 2003©Iréne Till-Bottraud. B – D. *K. Arroyo et al. 25093* **B.** Habit. **C.** Capitulum detail. **D.** Leaf & Bract detail. **i – ii.** Stem leaves; **iii – iv.** OIB d.s.; **v.** MIB d.s.; **vi.** IIB d.s. Scale bars (B – D) = 5 mm.

X *Oriastrum* Poepp. & Endl.

9. *Oriastrum apiculatum* (J. Rémy) A.M.R. Davies comb. nov.

≡ *Egania apiculata* J. Rémy Fl. Chil. [Gay] 3: 326. 1849. Typus CHILE "Se cria en los mismos lugares" [see *Egania acerosa* J. Rémy, "...en las cordilleras de Coquimbo, á Pasto Blanco, los Patos, á una altura de 3,000 m"] [Coquimbo, *Gay 361*] Holotypus P! Isotypus GH (photo)!

= *Chaetanthera lanigera* Phil. Anales Univ. Chile 87: 9. 1894. Typus CHILE "Cordillera del Peuco, 1886 *Cádiz*" Holotypus SGO 64969!
= *Oriastrum albicaule* Phil. Anales Univ. Chile 87: 20. 1894. Typus CHILE "In provinciae Santiago Andibus l.d. Cajon de la Yerba Loca invenit *Otto Philippi*, februario 1891" Holotypus SGO 64667!
= *Oriastrum nivale* Phil. Anales Univ. Chile 87: 21. 1894. Typus CHILE "Habitat in provinciae Santiago Andibus ad nivem perpetuam" Holotypus SGO 64961!
– *Chaetanthera apiculata* (J. Rémy) Benth. & Hook. f. Gen. Pl. 2(1): 496. 1873.
– *Chaetanthera apiculata* (J. Rémy) F. Meigen Bot. Jahrb. Syst. 18: 456. 1894.
– *Oriastrum albicaule* var. *nivale* (Phil.) Reiche Anales Univ. Chile 115: 337. 1904, Fl. Chile [Reiche] 4: 356. 1905.
– *Aldunantea sphacellata* Phil. ex.sched.

Nomenclature Notes Although there are no collections with good label identifications on them, and although this taxon clearly comes from further south than Coquimbo, there is one collection, *Gay 361*, which matches the description, and has the name *Egania apiculata* Rémy in Rémy's handwriting on it.

Perennial monoecious dwarf cushion-forming herb with deeply buried (to 30 cm) roots. **Stems** to 10 cm, with basal bud cluster; glabrous or lightly pubescent towards capitula. **Leaves** 7.5–9.5 x 1–1.5 mm, decussate, scale-like on lower stems, becoming linear to sphacellate, apices apiculate, mucro to 1 mm; green-grey; pubescent on dorsal and ventral surfaces, indumentum white, dense, 1–2 mm L; margins thickened or inrolled, often pinkish-red. **Capitula** cylindrical-campanulate, shortly pedunculate, peduncles with a few alternate leaves; disk diameter 3–4 mm. **Involucral bracts** imbricate, arranged in three types, initially foliaceous then reduced to entirely membranous. **Outer involucral bracts** 8.5–11.5 x 0.9–1.5 mm, as leaves but with short membranous alae to less than ½ height of bract, alae linear, 0.5–0.8 mm W., distal dorsal surface sometimes pink–brown maculate; indumentum on upper bract margins and dorsal surface of alae. **Middle involucral bracts** 9–12.5 x 2.5–3.5, ovate-lanceolate; membranous alate, ratio of lamina to alae decreases from ⅔ to nearly entirely alate; indumentum on distal bract margins and dorsal surface of alae; distal alae surface sometimes with pink–brown maculae. **Inner involucral bracts** 2 series, (8.5)11–13.5 x 2.5–3.7 mm, ovate-lanceolate, entirely membranous alate; apical appendages acute to acuminate, 2–3.5 mm L; sparsely pubescent. **Ray florets** (13)14–17, pistillate, white with bright pink dorsal surface; corolla 8.5–11.5 mm L, corolla tube 3.5–3.7 mm L, outer ligule 1.5–2 mm W., inner ligule shortly bifid, 0.1 mm L **Disk florets** 18–20, bisexual, yellow; corolla 6.5–7.5 mm L, corolla tube 5.5–6.5 mm L **Styles** [ray] 6–7.5 mm L, stigma lobes reddish brown, 0.3 mm L, obtuse, [disk] yellow, 6.3–7.4 mm L, stigma lobes obtuse. **Anthers** 5.2 mm L (6.5–7.5 mm L incl. filaments). **Achenes** brown-black, turbinate, 2–4(6) mm L (seed set poor, with only a few viable achenes per capitulum); pericarp pellucid, glabrous or longly pubescent (hairs simple, to 0.5 mm L). **Pappus** white, barbellate, setae 4.3–4.9 mm L, 1–2 rows, dehiscent.

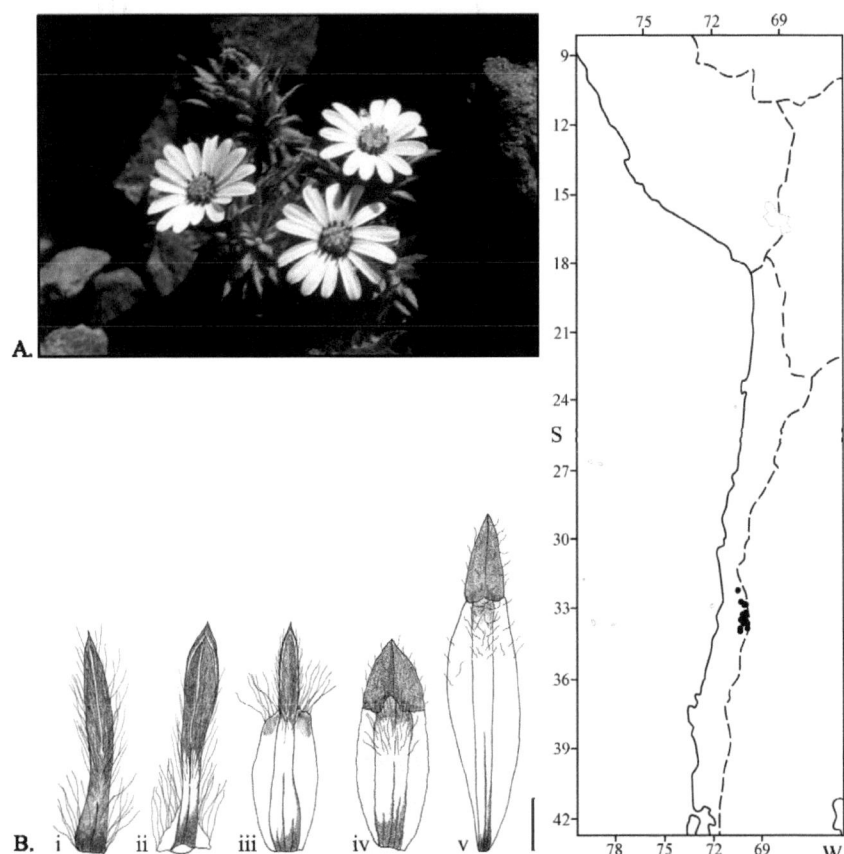

Figure 88: *Oriastrum apiculatum*. **A.** Habit, photographed Laguna del Inca, 1995©Jürke Grau. **B.** Leaf & Bract details. *Davies* et al. *2002/009* **i.** Stem leaf d.s.; **ii.** OIB v.s.; **iii.** MIB v.s.; **iv – v.** IIB d.s. Scale bar = 2 mm.

Distribution and Habitat. To date recorded only from Chile; the main distribution of *O. apiculatum* starts south of the Ventana de Horcones massif around the Laguna del Inca and continues along the west flank of the Andes to the Cord. de San José at latitudes between (32°13'S) 32°40'S–34°00'S (36°00'S). It occurs in steep scree and moraine, on dry slopes, above the snow line that are free of snow in summer at elevations between 2000–3600 m.a.s.l.

X *Oriastrum* Poepp. & Endl.

Differential diagnosis. Similar to *O. acerosum* and *O. dioicum*, *O. apiculatum* is bigger than both, and has a distinctive pink-red tinge to the otherwise grey-green vegetative parts. It has more ray florets, grows to the south of *O. acerosum*, and the leaves are linear to sphacellate. The apices of the involucral bracts are longer, and usually in 2 series instead of one.

Material seen. – **Chile. V Región de Valparaíso** Prov. de Petorca 5 km south of Junta de Piquenes, Río Sobrante, 3400 m, 02.1939, *Morrison 17299* (G GH F). – Prov. de Los Andes Los Maitenes, Río Colorado, 2500 m, 01.1980, *Zöllner 10651* (CONC). – Laguna del Inca, 3150 m, 01.1981, *K. Arroyo 81251* (CONC). – Laguna del Inca, am Ostufer, 3100 – 3200 m, 01.1981, *Grau 2960* (MSB M). – Portillo, Laguna del Inca, 3800 m, 02.1995, *Ehrhart & Grau 95/818* (MSB). – Portillo, Laguna del Inca, 1.5 hrs by foot along west shore, 3200 m, 02.2002, *Davies, Becker, A. & Becker, D. 2002/010* (MSB). – Portillo, Laguna del Inca, west shore past skilift pylons, 3200 m, 02.2002, *Davies, Becker, A. & Becker, D. 2002/009* (MSB). – Portillo, Laguna del Inca, west shore NW of upper ski lift pylons on scree, 3300 m, 02.2002, *Davies, Becker, A. & Becker, D. 2002/017* (MSB). – Portillo, 2800 m, a l'ouest du Grand Hotel Portillo, 01.1991, *Grenon 22746* (G). – Portillo, 2870 m, 03.1954, *Ricardi 2844* (CONC). – Llano de Juncalillo, 2700 m, 03.1954, *Ricardi 2924* (CONC). – Juncal, 10,000 ft, 01.1930, *Elloitt 633* (E K). – Between Caracules & Portillo F.C.T.C., 2800 m, 02.1931, *King 709* (BM). – **Región Metropolitana de Santiago** Prov. de Santiago Cord. de Santiago, *Philippi s.n.* (B W). – Cordillera de Santiago, 1856-1857, *Philippi s.n.* (G K P W). – Cord. de Santiago, 1861, *Philippi s.n.* (K). – Cordill. Santiago, *Philippi 728* (W). – Tal los Paramillos / Cepo pass, 3600 m, 02.1938, *C. Grandjot 3726* (GH). – Valle de Los Paramillos, 3500 m, 02.1950, *Barros s.n.* (CONC). – Valle del Cepo (bajo el Cerro Plomo), 3300 m, 02.1934, *C. & G Grandjot. s.n.* (M SGO). – Cord. Río San Francisco, Fierro Carrera, 3000 m, 02.1925, *Werdermann 627* (CONC E F G GH K M). – Cordillera de Las Condes, 3300 m, 02.1934, *Grandjot s.n.* (CONC). – Las Condes, 2500 m, 12.1951, *Castillo s.n.* (CONC). – Faldeos al oeste de Cerro La Parva, al este de La Parva, 3140 m, 02.2003, *K. Arroyo et al. 25174* (CONC). – La Parva, Sector Vega de Las Vacas, hacia portezuelo, 2900 m, 02.1999, *K. Arroyo et al. 991177* (CONC). – Complejo de Esqui "Valle Nevado", camino a Cerro Franciscano, 3310 m, 02.2003, *K. Arroyo et al. 25178* (CONC). – Vega Las Vacas, 2900 m, 03.1956, *Schlegel 1061* (CONC). – Cerro Ramon, 3000 m, 02.1948, *Castillo s.n.* (CONC). – Cerro San Ramon, 2700 m, 04.1959, *Schlegel 2482* (CONC). – Potrero Grande, 2000 m, 01.1967, *Zöllner 1463* (CONC). – Quebrada del Río Colorado, 2500 m, 02.1950, *Castillo s.n.* (CONC). – Straße von San José de Maipo nach Lagunillas oberhalb der Skihütten, 02.1985 *Hellwig 1019* (G). – above Laguna Negro, 11,000 ft, 02.1902, *Hastings 503* (NY). – PN El Morado, 2800 m, 01.1991, *Teillier, Pauchard & P. García 2496* (SGO). – Sector Agua Fría / Los Cristales, RN Río Clarillo, 2680 m, 02.2001, *K. Arroyo et al. 210635* (CONC). – Cajón de Morales, 3200 m, 03.1921, *Jaffuel 693* (CONC GH). – Cajón de Morales, 3200 m, 03.1921, *Jaffuel 416* (GH). – RN Río Clarillo, Sector Los Cristales, 2680 m, 02.2003, *K. Arroyo et al. 210640* (CONC). – RN Río Clarillo, Sector Los Cristales, 2650 m, 01.2000, *K. Arroyo et al. 20736* (CONC). – RN Río Clarillo, Quebrada Los Cipreses, Boquete Los Piquenes, 2700 m, 03.1992, *Solervicens s.n.* (SGO). – Mercedario, 3500 m, 01.1945, *Muñoz s.n.* (CONC). – **VII Región del Maule** Prov. de Linares Uspallata Pass, Las Calaveras, auf den Bergen, 36°S, 3100 m, 01.1903, *Buchtien s.n.* (E M). – Las Calveras, *Philippi s.n.* (BM). – **VIII Región de Bío Bío** Prov. de Ñuble Cordillere de Chillán, 1856-1857, *Philippi s.n.* (BM). **S.l.d.** *Seibold 2989* (W)

10. *Oriastrum cochlearifolium* A. Gray Proc. Amer. Acad. Arts 5: 144. 1861.

Typus PERU "Alpamarca, high Andes of Peru" [Peru. Depto. Junín Prov. Yauli Alpamarca, 1838-42, *Capt. Wilkes Exped. s.n.*] Holotypus GH! Isotypus K! US!

- *Chaetanthera cochlearifolia* (Gray) B. L Rob. Proc. Amer. Acad. Arts 49: 514. 1913.

Perennial monoecious dwarf cushion-forming herb. **Stems** lax with basal bud clusters; long to 6 cm, wiry, spreading. **Leaves** 2–10 x 1.5–3 mm, oblong-spathulate, obtuse; decussate, slightly succulent, ventrally and basally lanate-tomentose otherwise glabrescent; sessile, leaf margins entire, upper margins thickened and involute, distal, ventral surface concave. **Capitula** campanulate, sessile or shortly pedunculate, solitary, terminal, 10 x 15 mm; leaves densely whorled on peduncles. **Involucral bracts** imbricate, arranged in three types, initially foliaceous then reduced to entirely membranous. **Outer involucral bracts** 7–9.2(11) x 1.5–3.2(4) mm; sparsely longly pubescent (to 3 mm) in lower dorsal regions, mid-ventral regions and upper ventral surface, the latter curly tomentose; bracts as larger leaves but with short membranous alae to less than ½ height of bract. **Middle involucral bracts** 8–11 x 2.5–3.4 mm, bracts as leaves, glabrous except on upper margins; membranous alate, ratio of lamina to alae decreases from ⅔ to nearly entirely alate, alae rhomboid to ovate, 1–1.5 mm W. **Inner involucral bracts** 10.5–16 x 2–3.5 mm; entirely membranous alate; apical appendages 3.5–5 mm L, longly acute to acuminate, black-green. **Ray florets** ± 25, pistillate, white or yellow; corolla 7–8.5 mm L, corolla tube 3–4.5 mm L, outer corolla lip 1–1.5 mm W., inner lip to 0.5 mm, bifid. **Disk florets** ± 27, bisexual, yellow; corolla 6–7 mm L, corolla tube 4.2–5.2 mm L **Styles** 6–7 mm L; stigma lobes 0.4 mm, obtuse. **Anthers** 3.1 mm L (6.5 mm incl. filaments). **Achenes** brown, cylindric, 1.3–3.1 mm L; pericarp pellucid, glabrous. **Pappus** white, 1–2 rows, dehiscent; setae 7.5–9 mm L; 12 cells wide, cells 5–10 µm W.; longly ciliate at base, sparsely barbellate (3-4 barbs /100 µm), barbs 230 µm L, barb base longly adhered (>60%), free barb spreading.

Distribution and Habitat. *O. cochlearifolium* is endemic to Peru, and occurs on the eastern flank of the Cordillera Occidental in the upper valleys of the Río Mantara and Río Yauli between 11°00'–13°00'S. It is found on on rocky talus slopes, with Puna vegetation at elevations between 4650–5100 m.a.s.l.

Differential diagnosis. Morphologically, *O. cochlearifolium* is something of an anomaly. The pinkish-brown, slightly succulent, obtusely spathulate leaves and outer bracts are unique. The plants have 1-2 series of inner involucral bracts with longly acuminate, black-green apices, a little reminiscent of *O. tontalensis*. There is some conflict in the collection data as to the colour of the ray florets of this species. *Meza 225* recorded that the rays were white and the disks yellow; while *Dillon & Turner 1315* recorded that the rays were yellow.

X *Oriastrum* Poepp. & Endl.

Material seen. – Peru. Depto. Lima Prov. Huarochiri (La Oroya), Casapalca, 15,500 ft, 05.1922, *Macbride & Featherstone 845* (F GH). – Prov. Canta La Viuda, Km 165 carretera Lima-Cerro de Pasco, 5000 m, 08.1964, *Meza 225* (MO). – **Depto. Junín** Prov. Yauli Anticona Pass [Ticlio], ca. 140 km E of Lima on highway to La Oroya, c. 4890 m, 12.1973, *Dillon & Turner 1315* (F). – Anticona Pass [Ticlio], ca. 140 km E of Lima, 4800-4900 m, 12.1978, *Dillon & Turner 1477* (F). – An der Lima – Oroya Bahn, loc. von Buenaventuri bei Yauli, 4600 – 4700 m, 1906, *Weberbauer 357* (G). – Ticlio, 4900 m, 02.1974, *Tovar 7182* (USM). – Ticlio, 5000 m, 02.1974, *Tovar 7192* (USM). – **Depto. Huanaco** Prov. Don de Mayo Valle de Huallanca, 4850 m, 03.1983, *Tovar et al. 9921* (USM).

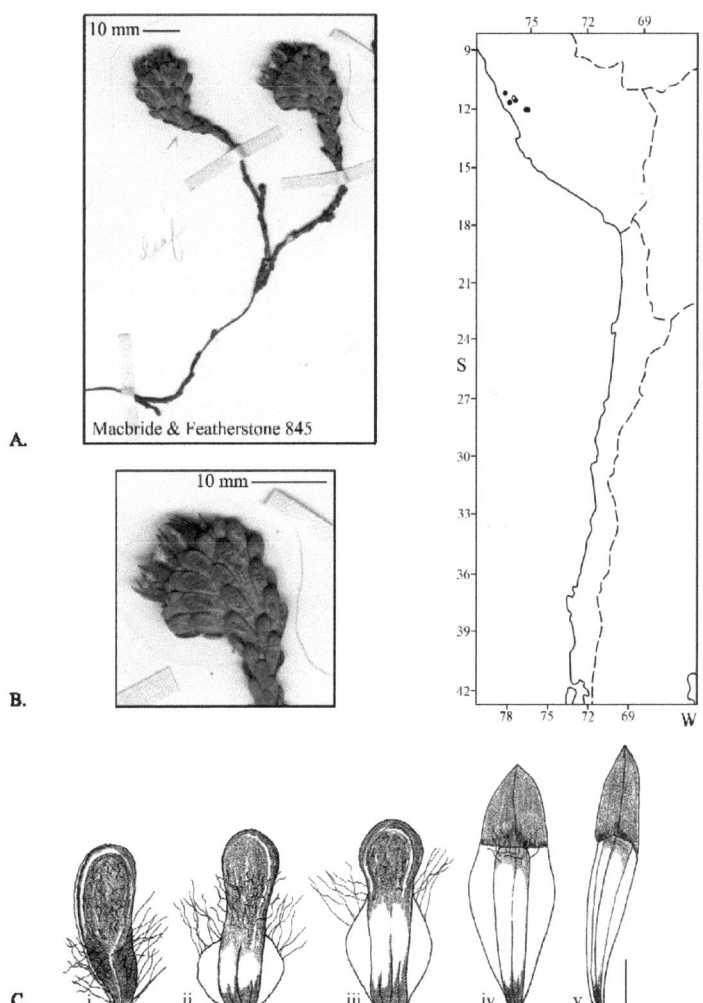

Figure 89: *Oriastrum cochlearifolium*. **A**. Habit. *Macbride & Featherstone 845*. **B**. Capitulum detail. *Macbride & Featherstone 845*. **C**. Leaf & Bract detail. *Dillon & Turner 1477*. **i**. Stem leaf v.s.; **ii**. OIB d.s.; **iii**. MIB v.s.; **iv – v**. IIB d.s. Scale bar = 2 mm.

11. *Oriastrum dioicum* (J. Rémy) Phil. Anales Univ. Chile 87: 21. 1894.

≡ *Egania dioica* J. Rémy Fl. Chil. [Gay] 3: 327, t. 26. 1849. Typus CHILE "Se cria en las cordilleras del valle de Coquimbo, cerca de Pasto Blanco, á una altura de 2,890 m. Florece en noviembre" [Coquimbo, plant ne l'elevant pas d'avantage croit entre les perrel a Pasto blanco, 2890 m, trés rare et solitaire, 09.1836, *Gay 391*] Holotypus P 141377 (p.p.)!

= *Oriastrum pentacaenoides* Phil. Anales Univ. Chile 87: 22.1894. Typus ARGENTINA "In via inter Mendoza et Sta. Rosa de los Andes lecta" [inter Mendoza et Sta Rose de los Andes, 1868/9, *Philippi s.n.*] **Lectotypus hic loc. designatus** SGO 64972! Isotypus SGO 43698!
− *Chaetanthera dioica* (J. Rémy) Benth. & Hook. f. Gen. Pl. 2(1): 496. 1873.
− *Chaetanthera dioica* (J. Rémy) B. L Rob. Proc. Amer. Acad. Arts 49: 514. 1913.
− *Chaetanthera pentacaenoides* (Phil.) Hauman Anales Soc. Ci. Argent. 86: 317 – 318, pl. 25, fig. 4. 1918.

Nomenclature Notes Although the Typus material of *Egania dioica* J. Rémy is quite a small, faded specimen, and comes from an extreme point in the range of the species, there is no doubt that it is the type of this taxon. This conclusion is based on the type, the description and the illustration which clearly show three floret forms (dimorphic ray florets, and disc florets), as well as the 'typical' maculate involucral bracts. Due to an oversight in Cabrera's revision (1937), almost all herbarium collections of this taxon are found under the epithet "*pentacaenoides*". Cabrera's "*C. dioica*" is described under *O. famatinae*.
The best preserved of two equally fit collections of *Oriastrum pentacaenoides* Phil. in SGO was selected as the Lectotype.

Perennial gynodioecious dwarf cushion-forming herb. **Stems** lax to 4–5 cm, with basal bud clusters; flexuous, creeping or ascending, shortly curly pubescent or glabrescent. **Leaves** 4.3 x 0.5 mm, bright green, decussate, connate; glabrescent, indumentum shortly curly on lower margins; margins thickened involute, midrib prominent. **Capitula** cylindrical, sessile or shortly pedunculate, no leaves on peduncle, urcinate, 8–10 mm L, disk diameter 3.5–4.5 mm; capitula bisexual and female on same plant, some individuals are entirely one or the other; bisexual capitula broader and bigger. **Involucral bracts** imbricate, arranged in three types, initially foliaceous then reduced to entirely membranous.
Female capitula 8 mm L, disk diameter 2.1 mm. **Outer involucral bracts** 4–4.3 x 0.6 mm; as leaves but with short membranous alae to less than ½ height of bract, alae triangular–obovate, 1.2 mm W., distally dorsally shortly curly pubescent. **Middle involucral bracts** 4.3–6.8 x 2.2–2.6 mm; membranous alate, ratio of lamina to alae decreases from ⅔ to nearly entirely alate, alae linear, truncate with distinctive black maculae on upper tips; indumentum dorsal, shortly curly pubescent where alae meet lamina. **Inner involucral bracts** 8.3–8.9 x 1.9–2.2 mm, 2 series, lanceolate, entirely membranous alate; apical appendages acute, brown-black, 2.8 mm L, mucronate; indumentum dorsal, shortly curly pubescent where alae meet lamina. **Ray florets** white, dimorphic. **Outer series:** corolla 5.8–6 mm L, corolla tube 2.8 mm L, outer ligule 0.5 mm W., irregularly 3-dentate, inner ligule 0.4 mm L, bifid. **Inner series:** corolla 4.4 mm L, corolla tube 3.2 mm L, outer ligule 0.3 mm W., irregularly 3-dentate, inner ligule broadly bifid, 0.4 mm L **Style** 5.4–5.6 mm L,

stigma lobes obtuse, 0.1–0.2 mm L, open. **Achenes** brown, narrowly turbinate; 2.1 mm L (immature); pericarp pellucid, glabrous. **Pappus** setae 4.8–5.2 mm L, white, 1–2 series, barbellate, dehiscent.
Bisexual capitula 10 mm L, disk diameter 2.4–3 mm. **Outer involucral bracts** 4.3 x 0.6–0.9 mm, as leaves but with short membranous alae to less than ½ height of bract, alae triangular–obovate, 0.4–0.7 mm W., dorsally shortly curly pubescent where alae meet lamina. **Middle involucral bracts** 4.3–6.8 x 2.7–3 mm; membranous alate, ratio of lamina to alae decreases from ⅔ to nearly entirely alate, alae linear, truncate with distinctive black maculae on upper tips; indumentum dorsal, shortly curly pubescent where alae meet lamina. **Inner involucral bracts** 8.6–9.9 x 2.2–2.8 mm, 2 series, lanceolate, entirely membranous alate; apical appendages acute, brown-black, 2.5–3 mm L, mucronate; indumentum dorsal, shortly curly pubescent where alae meet lamina. **Ray florets** ca. 10, pistillate, white; corolla 8.2 mm L, corolla tube 3.7 mm L, outer ligule 0.8 mm W., irregularly 3-dentate, inner ligule 0.6 mm L, bifid. **Disk florets** ca. 15, bisexual, yellow. Corolla 5.2 mm L, corolla tube 4 mm L **Styles** [ray] 6.8 mm L; [disk] 5.2 mm L, stigma lobes obtuse, 0.1–0.2 mm L, open. **Anthers** 3.4 mm (5.2 mm incl. filaments). **Achenes** brown, narrowly turbinate, 1.2–1.9 (2.1) mm L; pericarp pellucid, glabrous. **Pappus** white, 1–2 rows, dehiscent; setae 5–5.5 mm L; longly ciliate at base; setae 1–2(3) cells wide, cells 15 μm W.; basal cilia present, sparsely barbellate (3–4 barbs /100 μm), barbs 90–140 μm L, barb base longly adhered (>60%), free barb spreading. **Chromsome number** 2n = 2x = 20 (BAEZA & TORRES DIAZ 2006). *C. Torres s.n.* (CONC).

Distribution and Habitat. *O. dioicum* is found between the latitudes of 30°00'–34°00'S in the higher Andes of Chile and San Juan and Mendoza (Argentina) in the cordillera of Olivares, Doña Rosa, Espinazito, Ventana de Horcones and southwest of the Cordillera Chorilla (Chile) and Cordillera del Tigre (Argentina). Outliers to the east of the main Andean chain are recorded from Cerro El Tontal (Cordillera Ansilta) and Altos de Paramillos. It occurs among rocks, gravel or sandy soil on gentle to steep slopes (granite or volcanic) at elevations between 2500–4320 m.a.s.l.

Differential diagnosis. *O. dioicum* has bright green glabrescent linear leaves and multiple series of darkly maculate involucral bracts surrounding a relatively elongated cylindrical capitulum. There are several other gynodioecious taxa in *Oriastrum* (*O. abbreviatum, O. revolutum, O. polymallum*). *O. apiculatum* and *O. acerosum* also have maculate involucral bracts, but they are far fewer, and both species are distinctly more lanate/ hirsute, and the ray florets are always conspicuously exerted beyond the capitula.

Material seen. – **Argentina. San Juan** Depto. Igelsia 56 km W of junction Ruta 426 / Ruta 150 outside Las Flores, following Ruta 150 towards Paso del Agua Negra, 19 km above Gendarmeria Nacional, 3600 m, 01.1995, *Leuenberger, Arroyo-Leuenberger & Eggli 4423* (B). – Depto. Calingasta Cerro El Tontal, camino a la antena, 3770 m, 02.2003, *Arroyo* et al. *25242*. – Cerro El Tontal, al N de Barreal, camino a la antena, 3550-3750 m, 02.1990, *Kiesling, Ulibarri & Kravopickas 7354* (NY). – Depto. Barreal Paso de Espinacito, 01.1953, *Castellanos s.n.* (LP). – Cordillera de L'Espinazito, Los Patillos, au dessous de 3000 m, *Bodenbender s.n.* (G). – **Mendoza** Altos de Paramillo, entre Uspallata y Villavicencio, bajo casa de antena, 3180 m, 02.2003, *K. Arroyo* et al. *25099* (CONC). – Los Penitientes, nr. Puente del Inca, F.C.T.A., 3000 m, 02.1931, *King 708/s.n.* (BM LP). – in viciniis montis Aconcagua, Puente del Inca, 02.1903, *Malme 2906a* (G GH). – Puente del Inca, 03.1901, *Spegazzini 2532* (LP). – Puente del Inca, Las Banderas 3200 m, 12.1946, *Wall s.n.* (NY). – Puente del Inca, 3200 m, 12.1946, *Sparre 1572* (K). – somme[ts] du Río Tupungato, 3600 m, 01.1908, *Hauman 336* (G). – **Chile. Región de Coquimbo** Prov. de Elqui Coquimbo, Hautes cordilleres de los Patos, 01.1837, *Gay 467* (P). – Coquimbo, *Gay 359* (P#141377 en parte). – Coquimbo, in arenosis andium Los Patos, 1837, *Gay 744* (SGO 64973). – Prov. de Choapa Mina Los Pelambres, zona de lagunas, límite Chile-

Argentina, 3580 m, 02.1999, *K. Arroyo* et al. *991357* (CONC). – **Región Metropolitana de Santiago** Cerro San Francisco, 2500 m, 01, 1979, *K. Arroyo, Armesto & Uslar 7130* (CONC). – Camino de tierra entre La Parva y Cerro San Francisco, 3310 m, 01.2003, *K. Arroyo* et al. *25168* (CONC). – Faldeos al oeste de Cerro La Parva, al este de La Parva, 3350 m, 02.2003, *K. Arroyo* et al. *25173* (CONC). – Valle Nevado, 3315 m, *K. Arroyo* et al. *210673* (CONC).

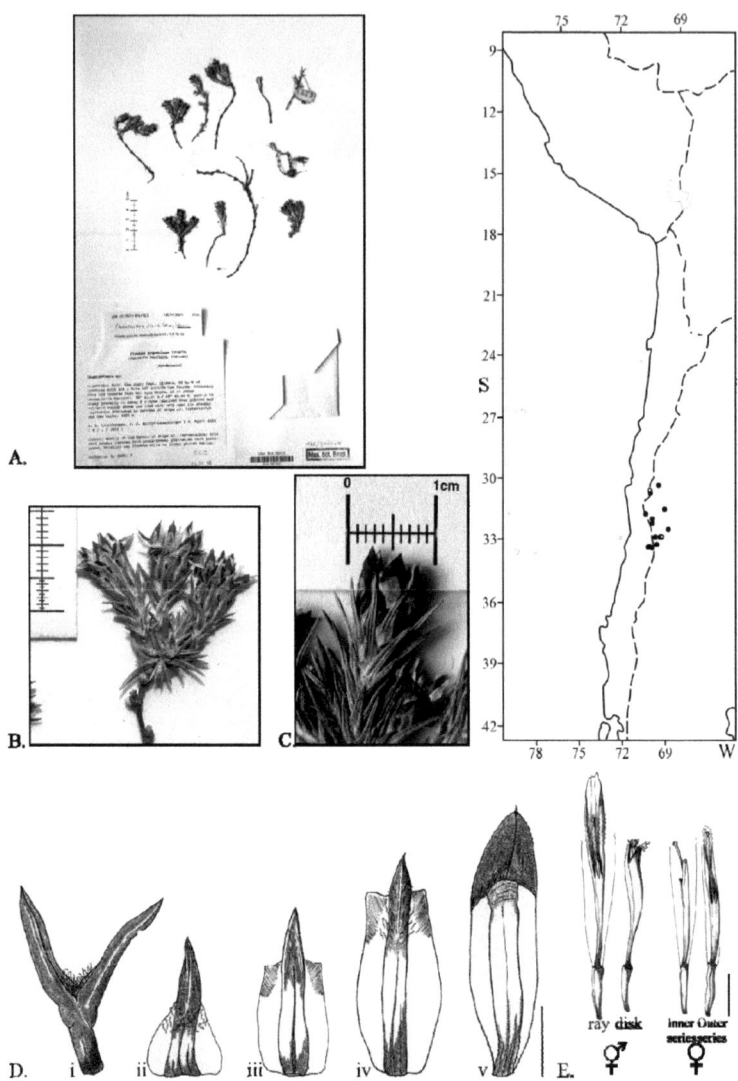

Figure 90: *Oriastrum dioicum. Leuenberger, Arroyo & Eggli 4423*. **A.** Herbarium sheet. **B.** Habit detail. **C.** Capitulum detail. **D.** Leaf & Bract detail. **i.** Pair of stem leaves; **ii.** OIB d.s.; **iii.** MIB v.s.; **iv.** MIB d.s.; **v.** IIB d.s. **E.** Floret detail for bisexual and female capitula. Scale bars D – E = 2 mm.

X *Oriastrum* Poepp. & Endl.

12. *Oriastrum famatinae* A.M.R. Davies sp. nov.

Typus ARGENTINA "Sierra La Famatina, ladera E. cerca del Paso Tocino, 3800 m, 01.1949, *Krapovikas & Hunziker 5342*" Holotypus LP!

Nomenclature Notes This taxon was formerly misidentified as *Chaetanthera dioica* sensu Cabrera. (see fig. 3 p. 105, 1937)

Planta perennis herbacea pulvinata. Caulis decumbens ad 10 cm altus, gemmis fasciculatis subterraneis ornatus glabrescens, trahens, ramosus. Folia linearia, 2.2–5.6 x 0.6 mm, dense breviter lanata, viridia, antice obtuse mucronata. Capitula sessilia, involucri bractaeis exterioribus destituta. Involucri bracteae interiores 1.5–2.8 mm, foliaceae ad membranaceae, antice breviter atrofusco obtusae et indistincte mucronatae. Flores radii albidi, flores disci aurei. Achenia fusca, anguste turbinata; pericarpium pellucidum, glabrum. Pappus caducus. Pappi setae 3–3.5 mm, sparsim barbellatae. Habitat in Argentina.

Perennial apparently monoecious dwarf cushion-forming herb. **Stems** to 10 cm, with basal bud cluster; glabrescent, wiry, flexuous, trailing and branching. **Leaves** 2.2–5.6 x 0.6 mm, linear, scale-like, obtuse, mucronate; bright green (live material), decussate, connate, densely arranged along stems, margins thickened to inrolled; lower margins and ventral surfaces pubescent, indumentum densely, shortly lanate. **Capitula** cylindrical, sessile, 2–3 mm L; disk diameter 1–2 mm. **Involucral bracts** imbricate, arranged in two series (no **outer involucral bracts** with short membranous alae to less than ½ height of bract), initially foliaceous then reduced to entirely membranous. **Middle involucral bracts** 4–5.2 x 0.6–2.5 mm, narrowly ovate to linear; membranous alate; ratio of lamina to alae decreases from ⅔ to nearly entirely alate. **Inner involucral bracts** 4.9–6.8 x 0.9–1.9 mm, entirely membranous, narrowly ovate to linear; apical appendages bluntly acute, 1.5–2.8 mm L, black–brown, very shortly mucronate. **Ray florets** ca. 10, pistillate, white; Corolla 4.4 mm L, corolla tube 2.4 mm L, outer ligule irregularly shortly tridentate, 0.5 mm W., inner ligule bifid, 0.4 mm. **Disk florets** ca. 8, bisexual, yellow; Corolla 3.6 mm L, corolla tube 2.8 mm L **Styles** [ray] 3.4 mm L [disk] 2.4 mm L; stigma lobes bifid, obtuse. **Anthers** 3.2 mm L Achenes brown, narrowly turbinate; pericarp pellucid, glabrous. **Pappus** setae 3–3.5 mm L, white, sparsely barbellate, 1–2 rows.

Distribution and Habitat. *O. famatinae* is an endemic of the Andes of Argentina, mainly found in the Sierra Famatina at latitudes between (25°50') 28°30'–29°00'S. It occurs in rocky habitats at elevations between 3340–4000 m.a.s.l.

Differential diagnosis. *O. famatinae*, with its lax, wiry, branching leafy stems, bright green leaves and white rays, is similar to *O. abbreviatum* (compact branched, sparsely leafy stems, dull green somewhat fleshy leaves and yellow rays). The former is found at slightly lower altitudes (ca 3650 m) than the latter (4060 m).

Material seen. – Argentina. Catamarca Reales Blancos, Cerro Aguas Blancas, 02.1930, *Castellanos s.n.* (LP). – **La Ríoja** Depto. Chilecito Sierra La Famatina, camino a La Mejicana, Estación 7, 4000 m, 02.1927, *Parodi 7883* (GH). – Sierra La Famatina, ladera E. cerca del Paso Tocino, 3800 m, 01.1949, *Krapovikas & Hunziker 5342* (LP). – entre la Mina Jarela y la allura del Esjurisu Santo Sierra Famatina, 07.1879, *Hieronymus & Niederlein 789* (G). – Quebrada Los Berros, Sierra Famatina, 3340 m, 02.2003, *K. Arroyo* et al. *25102* (CONC). – Quebrada Los Berros, Sierra Famatina, 3470 m, 02.2003, *K. Arroyo* et al. *25103* (CONC).

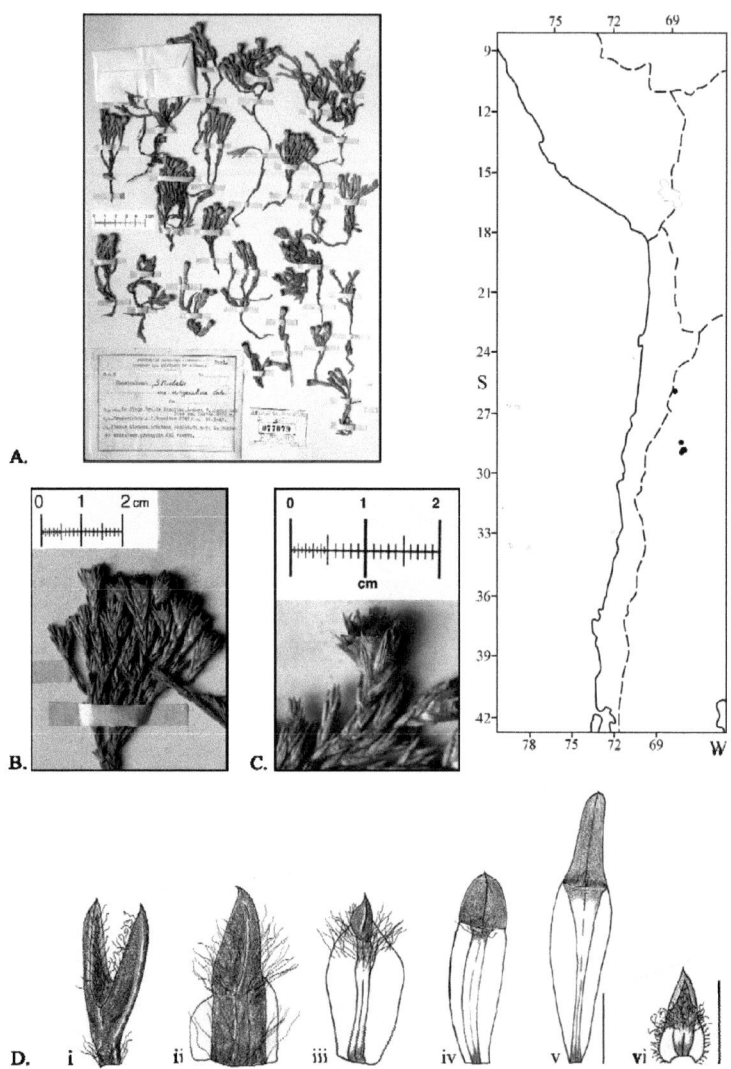

Figure 91: *Oriastrum famatinae*. Holotype *Krapovickas & Hunziker 5342*. **A.** Herbarium Sheet. **B.** Habit detail. **C.** Capitulum detail. **D.** Leaf & Bract detail. **i.** Pair of stem leaves; **ii – iii.** MIB d.s.; **iv – v.** IIB d.s.; **vi.** Stem leaf. *Krapovickas & Hunziker 5342* i – v; *Parodi 7883* vi. Scale bars = 2 mm.

X *Oriastrum* Poepp. & Endl.

13. *Oriastrum polymallum* Phil. Anales Univ. Chile 87: 17. 1894.

Typus CHILE "In monte altissimo Doña Ana 4,520 m. super mare invenit Fr. Philippi" [Portezuelo de Doña Ana, 07.02.1883, *F. Philippi s.n.*] **Lectotypus hic loc. designatus** SGO 64966! Isotypus SGO 43702!

= *Oriastrum sphaeroidale* Reiche Reiche, K. (1904) Anales Univ. Chile 115: 339, Reiche, K. (1905) Fl. Chile [Reiche] 4: 353 Typus CHILE "Cordilleras de Atacama" [Desierto de Atacama, aest. 1897, *Delaigne s.n.*] Holotypus SGO 64964!
= *Chaetanthera pulvinata* (Phil.) Hauman var. *acuminati-bracteata* Hauman Physis 3 (15): 419 – 420. 1917. Typus ARGENTINA [Cordilleras de San Juan, 1897, Burmeister s.n.]" Holotypus BRLU † (Lejoly, pers. comm.) Isotypus BA?? (fide Cabrera, 1937)
- *Chaetanthera pulvinata* var. *polymalla* (Phil.) Hicken Darwiniana 1: 41. 1922.
- *Chaetanthera sphaeroidalis* (Reiche) Hicken Darwiniana 1: 41. 1922.

Nomenclature Notes The Typus material of *Oriastrum polymallum* is not very good, and would appear to have arrived on Philippi's desk already partly eaten by larvae [Larvas de una mosca se habian comido todo el interior; una sola lígula de les habia escapado...] but there is a capitula dissection on the holotype which clearly designates the material under this name. On the isotype there are two pieces of plant, the right hand one of which is a greyer, smaller specimen. However, among the numerous collections observed during this current work, greyish collections occasionally occur (as opposed to the more typical reddish-pink). This colour difference is not obviously attributable to location or environmental differences. Reiche (1905) placed both *Oriastrum polymallum* Phil. and *Oriastrum sphaeroidale* Reiche under the heading of "Especias de clasificacion problemática", so he was obviously uncertain how to best treat this material.
After consultation with BRLU, where Hauman's holotypes are generally found, no material was found of *Chaetanthera pulvinata* (Phil.) Hauman var. *acuminati-bracteata* Hauman. The Latin diagnosis, unfortunately, could apply to any of the species in *Egania*. However, from the text description this would appear to be representative of *Chaetanthera sphaeroidalis*, with its distinctive globose lanate habit and reddish-pink colouration. Hauman writes "El tipo había sido mencionado para los alrededores del Aconcagua" referring to the "typical" species i.e. *C. pulvinata* var. *pulvinata*.

Perennial gynodioecious dwarf cushion-forming herb. **Stems** to 4 cm high, compact with basal bud clusters; leaves arranged in dense whorls. **Leaves** 4–8 x 1 mm, sclerophyllous, linear with somewhat dilated apices; prominent apical midrib becoming distinct towards tip, pale, with a blunt mucro; indumentum densely lanate-tomentose on both dorsal and ventral surfaces, hairs to 4 mm, white–grey except towards leaf apices where hairs are dark red distally; buds in leaf axils subtended by decussate leaves. **Capitula** of both female and bisexual plants are deeply recessed in the leaf rosettes. **Involucral bracts** imbricate, arranged in two series (no **inner involucral bracts** that are entirely membranous), initially foliaceous then reduced.
Female plants. **Outer involucral bracts** 9–13 x 1.5 mm; as leaves but with narrow membranous alae to less than ½ height of bract; apices dilated, with prominent pale mucro. **Middle involucral bracts** lamina as leaves, 8 x 1–2.5 mm, densely lanate on distal margins and dorsal surface only, lamina component reduced to acute tip, 1.5–2 mm L; membranous alate, ratio of lamina to alae

decreases from ⅔ to nearly entirely alate, alae 0.7–1.2 mm W., oblong to ovate or rarely truncate, lanate. No **inner involucral bracts**. **Ray florets** > 30, pistillate, white; Corolla 3.4–4.5 mm, corolla tube 3.1–4.2 mm L, outer ligule very reduced in length, 1 mm W., shallowly irregularly 3-dentate to truncate, inner ligule shortly bifid. **Styles** (4.2)5–6.5 mm L, stigmas shortly lobed; proportion of style to corolla is very variable–nearly dimorphic [4.5:5 or 4:6.5], but all florets had viable seed. **Achenes** brown, 3–3.5 mm L; narrowly turbinate; pericarp pellucid, glabrous. **Pappus** white, 1–2 series, barbellate, 6.5–8 mm, dehiscent.

Bisexual plants. **Outer involucral bracts** 7–8 x 1 mm; as leaves but with short membranous alae to less than ½ height of bract; apical appendages dilated, with prominent pale mucro. **Middle involucral bracts** 7–8.5 x 1–2.5 mm, lamina as leaves gradually reduced to prominent white, densely tomentose, recurved, obtuse tip, 1.5–3 mm L; apices exerted beyond capitulum, replacing ray floret function.; indumentum on dorsal and ventral surfaces; membranous alate, ratio of lamina to alae decreases from ⅔ to nearly entirely alate, alae 0.7–1.2 mm W., oblong to ovate or rarely truncate, lanate on distal margins and dorsal surface only; No **inner involucral bracts**. **Ray florets** < 10, pistillate, white, reduced; corolla 3.5 mm L, corolla tube 3 mm L; outer ligule reduced in length, 1 mm W., shallowly irregularly 3-dentate to truncate; inner ligule shortly bifid. **Disk florets** > 30, bisexual, yellow; corolla 5.5 mm L; corolla tube 4.5–5 mm L **Styles** [ray] 4.5 mm L, [disk] 5 mm L; stigmas shortly lobed. **Anthers** 4.5 mm L (5.5–6.5 incl. filaments). **Achenes** brown, 1–2 mm L (immature), narrowly turbinate; pericarp pellucid, glabrous; no carpopodium; testa epidermis with no significant strengthenings, cells parallel. **Pappus** white, 1–2 rows, dehiscent; 4.7–6 mm L; longly ciliate at base; setae 1–2(3) cells wide, cells 5–10 μm W.; basal cilia present, sparsely barbellate (3–4 barbs /100 μm), barbs 90–140 μm L, barb base longly adhered (>60%), free barb spreading.

Distribution and Habitat. *O. polymallum* is widely distributed between 21°00'–30°30'S, in Argentina (Catamarca, La Riója and San Juan), Bolivia and Chile (Tarapacá, Antofagasta, Atacama and Coquimbo). It is found among rocks, scree and moraine, around the permanent snow line, stable shale slopes, gravelly soil near rocks. It forms big populations, and many stems are detached and windblown (Brownless et al.). NAVARRO (1993) published the *Nototricho auricomae - Chaetantheretum sphaeroidalis* vegetation association that characterises gravelly gelifluxional slopes in high altitudes above 4800 m in southwest Bolivia. It is found at elevations between (1900) 4307–5500 m.a.s.l. The two collections [*Villagrán & K. Arroyo 4623; 4631*] from Cuesta El Salto in Copiapó occur at lower altitudes than usual for this species. Otherwise the lower end of the altitude range would lie around 2900 m.

Differential diagnosis. *O. polymallum* most closely resembles *O. pulvinatum*. *O. polymallum* has very densely compacted stems, forming a dwarf cushion, while *O. pulvinatum* has more loosely arranged stems. *O. polymallum* almost always has a very distinctive pink colouration and the capitula are embedded deep within the 'cushion', with pale MIB apices. It lacks the entirely alate IIB with coloured apical appendages typical of all other *Oriastrum* species. *O. pulvinatum* tends be more grey, and the bracts and capitula are visibly exerted. The IIB apices are yellow/black. *O. polymallum* is often referred to as "flor de puna" on herbarium sheets.

X *Oriastrum* Poepp. & Endl.

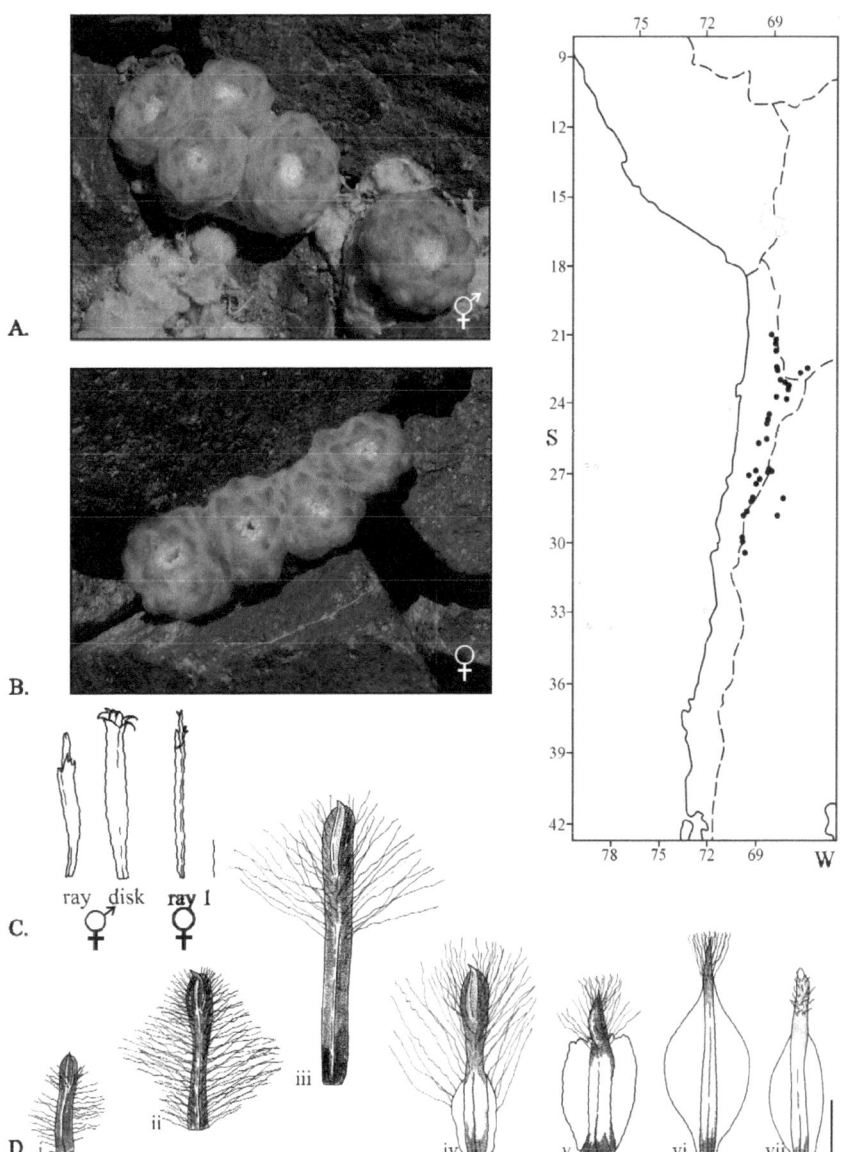

Figure 92: *Oriastrum polymallum*. **A.** Female plant. **B.** Bisexual plant. A & B Photographed Mina El Indio, 2003©Iréne Till-Bottraud. **C.** Floret detail. Scale bar = 1 mm. **D.** Leaf & Bract detail. *Brownless 554* (i – vi); *Werdermann 253* (vii) **i – iii.** Lower to upper stem leaves d.s.; **iv – v.** MIB d.s.; **vi.** Innermost MIB d.s. of female capitula; **vii.** Innermost MIB d.s. of bisexual capitula. Scale bar = 5 mm.

Material seen. – **Argentina. Catamarca** Depto. Tinogasta Tres Quebradas, en un valle (glacial?) del faldeo sud del Cerro de los Patos, 4400 m, 03.1951, *Vervoorst 3241* (K W). – **La Rioja** Depto. G. Sarmiento Cerro Rejas, entre Paso Pircas Negras y Laguna Brava, 11.1949, *Krapovikas & Hunkiker 5845* (LP). – Laguna Brava, 4000 m, 02.1949, *Krapovikas & Hunkiker 5848* (LP). – Depto. Chilecito Sierra La Famatina, quebrada Ensucijada, 3500-500? m, 03.1913, *Forsdorf 86* (LP). – Sierra La Famatina, Cuesta del Tocino, 02.02.1879, *Hieronymus & Niederlein 687* (LP). – **San Juan** Depto. Iglesias Cerro Alcaparrosa de Olivares, 5500 m, 1960, *Dawson s.n.* (LP). – **Bolivia. Depto. Potosí** Prov. Sud López Cordillera de López, cumbres del Cerro Laguna Colorada, 5200 m, 04.1990, *Navarro 498* (LPB). – Cerro Tapaquillcha, 4800 m, 04.1980, *Beck 185* (SI op. cit.). – altas laderas y vertientes entre el Volcán Apacheta y el Volcán Michina, a pocos kilómetros de la frontera con Chile, *Navarro* (op.cit.). – **Chile. I Región de Tarapacá** Prov. de Iquique Collaguasi, San Carlos, Cementerio, 4700 m, 01.1993, *Teillier 3056* (CONC). – Collahuasi, 4600 m, 01.1966, *Zöllner 929* (CONC). – Collahuasi, 4600 m, 01.1966, *Zöllner s.n.* (CONC). – **II Región de Antofagasta** Prov. de El Loa Volcán Öllague, 3700 m, 1945, *Perry s.n.* (CONC 21129; CONC 6986). – Cerro Carasilla, 4800 m, 1985, *K. Arroyo et al. 85356* (CONC). – frente Chuquicamata, 2760 m, 1936, *Perry s.n.* (CONC). – Volcan Tatio, 4100 m, 1961, *Ricardi et al. 479* (CONC). – Cerro Toco, ladera N, 4600 m, 04.1997, *K. Arroyo, Cavieres & Humaña 97042* (CONC). – entre Salar Aguas Calientes y Quebrada Quepiaco, 4800 m, 04.1997, *K. Arroyo, Cavieres & Humaña 97506* (CONC). – Cerro Losloyo, ladera O, 4600 m, 04.1997, *K. Arroyo, Cavieres & Humaña 97424* (CONC). – Cordon Ceja Alta, lado SE, 4600 m, 04.1997, *K. Arroyo, Cavieres & Humaña 97297* (CONC). – camino de Guaitiquina a El Laco, 4000-4500 m, 12.1978, *Krusell s.n.* (SGO). – Prov. de Antofagasta Antofagasta, 4700 m, 02.1950, *Castellanos 71648* (P W). – camino Monturaqui a Cerro Guanaqueros, 4300 m, 1983, *K. Arroyo & Villagran 831287* (CONC). – Volcán Llullaillaco, 4700 m, 01.1994, *K. Arroyo Grosjean, Messerli, Cuevas & Leonard 94017* (CONC). – Volcán Llullaillaco, 4350 m, 01.1994, *K. Arroyo, Grosjean, Messerli, Cuevas & Leonard 94003* (CONC). – S de Llullaillaco, 4800 m, 1993, *Baumann 220* (CONC). – inicio Quebrada Agua de La Piedra, 4300 m, 12.1996, *Arancio & Squeo 10255* (CONC). – **III Región de Atacama** Prov. de Chañaral camino entre Salares de Gorbea-La Isla, 4380 m, 01.1994, *K. Arroyo, Leonard & Cuevas 94141* (CONC). – camino Tinogasta, Cerro de La Mula Muerta, 4700 m, 03.1983, *Villagran & K. Arroyo 4526* (CONC). – Prov. de Copiapó Portezuelo Codocedo, 3600-3690 m, 01.1944, *Munóz P. 3992 & 3993* (SGO). – camino intern Tinogasta, Portezuelo Codoceo, 4400 m, 02.1966, *Ricardi et al. 1674* (CONC). – camino al Salar de Maricunga, Km 132, 4100 m, 01.1963, *Ricardi et al. 599* (CONC). – road to Nevados Tres Cruces, 12.1971, *Beckett, Cheese & Watson 4724* (SGO). – Paso fronterizo San Francisco, lado chileno, ca. 2 Km del paso, 4650 m, 02.2003, *K. Arroyo et al. 25117* (CONC). – Mula Muerta, Laguna Verde, 4500 m, 02.1973, *Zöllner 8404* (CONC). – Barranca Blanca y Ojo del Salado, 5000 m, *Siemben? s.n.* (CONC). – camino internacional Copiapó a Tinogasta, Cuesta El Salto, 1900 m, 03.1983, *Villagran & K. Arroyo 4623; 4631* (CONC 71315; CONC 71323). – camino Salar de Maricunga-Negro Francisco, 4000 m, 02.1958, *Behn F s.n.* (CONC). – Laguna del Negro Francisco, 4125 m, 03.1992, *Arancio 92229* (CONC). – Laguna del Negro Francisco, 3520 m, 02.1944, *Munóz P. 3984* (SGO). – southern slopes of Laguna del Negro Francisco, 4100 m, 03.1996, *Brownless, Gardner, Maxwell, & Rozzi 554* (E). – Portezuelo Peña Negra, 4270 m, 02.1975, *Niemeyer s.n.* (CONC). – Prov. de Huasco Quebrada Cantarito, entre Quebrada Marancal y Portezuelo de Cantarito, 4300 m, 02.1981, *K. Arroyo 81610* (CONC). – Quebrada Cantarito, entre Quebrada Marancal y Portezuelo de Cantarito, 4300 m, 01.1983, *Marticorena, K. Arroyo & Villagrán 83467* (CONC). – Quebrada Cantarito, entre Quebrada Marancal y Portezuelo de Cantarito, 4000 m, 01.1983, *Marticorena, K. Arroyo & Villagrán 83498* (CONC). – Cordillera Laguna Chica, 4300 m, 02.1924, *Werdermann 253* (B BM CONC E F G GH K M NY). – Vicinity of Laguna Chica, north of Portezuelo de Laguna Chica, 4200 m, 01.1926, *Johnston 5942* (GH). – **IV Región de Coquimbo** Prov. de Elqui Las Guatinas, camino a Tambo, Mina El Indio, 4360 m, 01.2003, *K. Arroyo et al. 25082* (CONC). – Cordillera Doña Ana, camino a Tambo desde Pta. Diamante, 4300 m, 01.1988, *Squeo 88030* (CONC MO). – Tortola, Valle del Elqui, 4500 m, 12.1948, *Castillo s.n.* (CONC).

X *Oriastrum* Poepp. & Endl.

14. *Oriastrum pulvinatum* Phil. Linnaea 33: 112. 1864-65.

Typus CHILE "In cordillera Doña Ana prov. Coquimbo legit orn. *Volckmann*" Holotypus SGO 64960! Isotypus K!

- *Chaetanthera pulvinata* (Phil.) Hauman Physis (Buenos Aires) 3: 420. 1917.

Nomenclatural Notes The only extant sheet on SGO has "*Oriastrum (Aldunatea) pulvinatum* Ph." inscribed in Philippi's hand without locality or collector information. The sheet in K has a Philippi label with the locality "Cord. Dona Ana" written on it.

Perennial gynodioecious dwarf herb, cushion forming. **Stems** to 4 (6) cm, with basal bud cluster; branched, densely white–grey lanate. **Leaves** 4–7 x 0.5–0.8 mm, linear to somewhat dilated at apex, remotely decussate, appressed on stem; distal margins thickened and distal midrib becoming prominent; indumentum long, lanate, hairs simple, 2.5–3 mm L, pubescent on dorsal and distal ventral surfaces. **Capitula** cylindrical, female and bisexual capitula similar in appearance: shortly pedunculate with densely whorled leaves subtending capitula. **Involucral bracts** imbricate, arranged in three types, initially foliaceous then reduced to entirely membranous.
Female capitula somewhat smaller and narrower than bisexual capitula. **Outer involucral bracts** as leaves, 7–9.5 x 1 mm, 3–4 series; alae oblong, to 0.5 mm W., pubescent only on distal dorsal surface and margins. **Middle involucral bracts** 7–7.5 x 2 mm, lamina reduced to 1–2 mm L apiculum; membranous alate, ratio of lamina to alae decreases from ⅔ to nearly entirely alate, alae oblong-ovate, distal margins ragged, pubescent only on distal dorsal surface and margins. **Inner involucral bracts** entirely membranous, ovate to linear-lanceolate, 8–10 x 2.5 mm, 2–3 series; apical appendages 1.5–2.5 mm L, acute, typically striped maculate; distal surface with sparse hairs. **Ray florets** white, pistillate, 2 series. **Outer series:** 11–15, corolla 6 mm; corolla tube 3.5–3.9 mm; outer ligule 1 mm W., irregularly 3-dentate, glabrous; inner ligule reduced to short bifid notches. **Inner series:** 5–15, corolla 3.6–3.9 mm; corolla tube 3–3.3 mm; outer ligule truncate or irregularly shallowly 3–dentate; inner ligule reduced to short bifid notches. **Styles** [outer series] 5.7–6.3 mm L [inner series] 5–5.5 mm; stigma lobes green-brown, short, < 1 mm, obtuse. **Achenes** [outer + inner series] to 3 mm, turbinate, glabrous or with dense coating of single celled papillae. **Pappus** ca. 6 mm L, white, 1 series, remotely barbellate.
Bisexual capitula somewhat larger and broader than the female capitula, otherwise bracts the same. **Outer involucral bracts** 7–9 x 1.5–2 mm; pubescent only on distal dorsal surface and margins. **Middle involucral bracts** 6–7 x 2 mm. **Inner involucral bracts** 7.7–11.5 x 1.8–3.5 mm, apical appendages 2.8–3.5 mm L **Ray florets** ca. 18–24; pistillate, white. Corolla 9–10 mm L, corolla tube 5.5, outer ligule 2 mm. **Disk florets** > 40; bisexual, creamy yellow; Corolla 6.3–7 mm L, corolla tube 5.5 mm L **Styles** [ray] 6.5–7 mm; [disk] 6–7 mm L, yellow; stigma lobes [ray] red-brown, short, obtuse. **Anthers** [disk] 3.5 mm L (6 mm incl. filaments) **Achenes** brown, narrowly turbinate or flattened, 2–3 mm L (immature); pericarp pellucid, glabrous or with dense coating of single celled papillae. **Pappus** white, 1–2 rows, dehiscent; 6–6.5 mm L; longly ciliate at base; setae 1–2(3)

cells wide, cells 24 μm W.; basal cilia present, sparsely barbellate (3–4 barbs /100 μm), barbs 90–140 μm L, barb base longly adhered (>60%), free barb spreading.

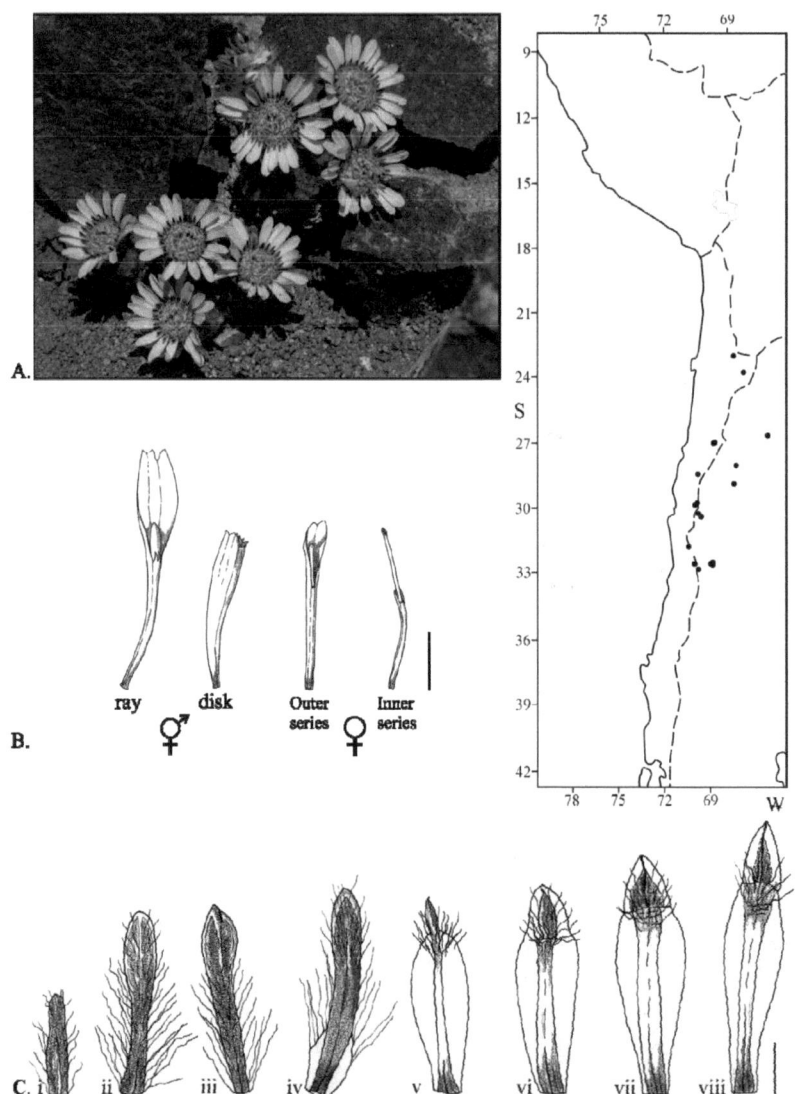

Figure 93: *Oriastrum pulvinatum*. **A.** Habit of bisexual plant, photographed Mina El Indio, 2003©Iréne Till-Bottraud. **B.** Floret detail from bisexual and female capitula. **C.** Leaf & bract detail. **i.** Lower stem leaf d.s.; **ii.** Upper Stem leaf v.s.; **iii.** upper stem leaf d.s.; **iv.** OIB v.s.; **v.** MIB d.s.; **vi – viii.** IIB d.s. Scale bar (B, C) = 2 mm.

X *Oriastrum* Poepp. & Endl.

Distribution and Habitat. *O. pulvinatum* is distributed in Chile (El Loa, Copiapó, Huasco, Elqui) and Argentina (Tucuman, Catamarca, La Rioja, San Juan and Mendoza) between 23°00'–33°00'S. It forms scattered populations in peri-glacial detritus of moraine, gentle to steep gravelly or sandy slopes (granite and volcanic) at elevations between 3000–5100 m.a.s.l.

Differential diagnosis. *O. pulvinatum* is often confused with lax examples of *O. polymallum*, but distinguished by the multiple series (2-3) of IIB with striped, acute apices. *O. tontalensis* has a similar habit, occurring in the southernmost part of the *O. pulvinatum* distribution, but is generally of larger stature and has distinct entirely maculate acute to acuminate IIB apices.

Material seen. – **Argentina. Prov. Salta** Depto. Caldera Quebrada del Río Potrero Castillo, a amba del Real de los Pastores, 3900 m, 15.3.1952, *Sleumer & Vervoorst 2931* (LP). – **Prov. Tucuman** Cerro Negrito, subida desde El Infiernillo, Cumbres Calchaquíes, 4220 m, 02.2003, *K. Arroyo et al. 25111* (CONC). – **Prov. Catamarca** Depto. Tinogasta Tres Quebradas, 4200 m, 03.1951, *Vervoorst 3239* (W). – **Prov. La Rioja** Sierra Famatina, Quebrada Los Berros, 5100 m, 02.2003, *K. Arroyo et al. 25104* (CONC). – **Prov. San Juan** Quebrada de la Concenta, Las Vianiatas, 2800 m, 03.01.1930, *Perez Moreau 125* (LP). – Valle del Cura, Pisca de los Ingenieros, 4000 m, 09.01.1930, *Perez Moreau 158* (LP). – Paso de Espinacito, 11.1953, *Castellanos s.n.* (LP). – Depto. Iglesia 56 km W of junction Ruta 426 / Ruta 150 outside Las Flores, following Ruta 150 towards Paso del Agua Negra, 19 km above Gendarmeria Nacional, 3600 m, 01.1995, *Leuenberger, Arroyo-Leuenberger & Eggli 4424* (B). – **Prov. Mendoza** Aconcagua, N-W face, Horcones valley, 3920 m, 02.1980, *Miehe 108/80/17.2* (LPB). – Aconcagua, N-W face, 4200 m, 02.1980, *Miehe 108, 109/80/17.2* (LPB). – Aconcagua, N-W facc, 4180 m, 02.1980, *Miehe 109/80/17.2* (LPB). – near Las Cuevas, valley below Tolosa, SW of glacier, 3810 m, 01.1980, *Miehe 193/80/4.5* (LPB). – Depto. Las Horas Cruz de los Paramillos, 3000 m, 01.1950, *Sleumer 374* (W). – Cord. del Tigre, Río Tambillitos, 3000 m, 12.1927, *King 338* (BM NY). – Cruz de Paramillos, Altos de Parmillo, entre Uspallata y Villavicencio, 3030 m, 02.2003, *K. Arroyo et al. 25100* (CONC). – **Chile. I Región de Tarapacá** San Nicolas, cerca de Collaguasi, 4500 m, 12.1984, "Cortomaltese" (image, www.chilebosque.cl). **II Región de Antofagasta** Prov. de El Loa Volcan Toco, 5000 m, 1954, *Aracena s.n.* (CONC). – Cerca frontera a Argentina, Paso Huaitiquina (Guaitiquina/ Huaytiquina) en el borde arenoso de un salar, 4300 m, 12.1986, *Beck 14147* (LPB). – **III Región de Atacama** Prov. de Huasco Vallenar, vicinity of Laguna Chica, north of Portezuelo de Laguna Chica, 4200 m, 01.1926, *Johnston 5943* (GH). – Quebrada Cantarito, entre Quebrada Marancal y Portezuelo de Cantarito, 4300 m, 01.1983, *Marticorena, K. Arroyo & Villagrán 83501* (CONC). – Quebrada Cantarito, entre Quebrada Marancal y Portezuelo de Cantarito, 4300 m, 02.1981, *K. Arroyo & Villagrán 81611* (CONC). – Prov de Copiapó camino intern. Tinogasta, Portezuelo Colorado, 4400 m, 02.1966, *Ricardi et al. 1651* (CONC). – camino intern. Tinogasta, Cuesta Los Colorados, 4250 m, 03.1983, *Villagrán & Arroyo 4554* (CONC). – **IV Región de Coquimbo** Prov. de Elqui Baños del Toro, Doña Ana, 4500 m, 01.1924, *Werdermann 233* (BM CONC E F G GH). – 11 Km antes del Paso de Agua Negra, 4050 m, 01.1967, *Ricardi et al. 1769* (CONC). – camino Tambo de Bifurcación Pta de Diamante, 4470 m, 01.1988, *Squeo 88031* (CONC). – Banos del Toro, 12.1971, *Beckett, Cheese & Watson 4654* (SGO). – camino entre Guatinas y la planta de la Mina El Indio, 4200 m, 01.2003, *K. Arroyo et al. 25083* (CONC). – Prov. de Choapa Mina Los Pelambres, Zona de lagunas, límite Chile-Argentina, 3500 m, 02.1999, *K. Arroyo et al. 991239* (CONC).

267

15. *Oriastrum revolutum* (Phil.) A.M.R. Davies comb. nov.

≡ *Egania revoluta* Phil. Anales Mus. Nac. Santiago de Chile 2: 34. 1891. Typus CHILE "Crece en Maricunga, etc. entre Aguas Calientes y Socaire; se llama 'flor de puna' Syntypus 1 [Aguas calientes – Socaire, 02.1885] SGO 43692! Syntypus 2 [Maricunga, 01.1885, *F. Philippi*] SGO 71270! K [Tarapaca]!

= *Egania appressa* Phil. Anales Mus. Nac. Santiago de Chile 2: 34. 1891. Typus CHILE "Prope Calalaste 3700 m. s.m. reperta" Holotypus [Calalaste, 01.1885] SGO 43687! Isotypus K [Tarapaca]!
- *Oriastrum dioicum* var. *revolutum* (Phil.) Reiche Anales Univ. Chile 115: 339. 1904, Fl. Chile [Reiche] 4: 358. 1905.
- *Chaetanthera revoluta* (Phil.) Cabrera Revista Mus. La Plata, Secc. Bot. 1: 106, fig. 4, 1937.

Perennial gynodioecious dwarf herb, forming cushions 4–5 cm above ground. **Stems** to 5 cm, with basal bud cluster; pubescent. **Leaves** 2–3(4) x 0.6–0.7 mm, linear with dilated apex to scale-like, densely decussate, connate; indumentum dense, long lanate (1.2–1.5 mm) on margins, lower dorsal and upper ventral surfaces, indumentum sparse on upper dorsal surface; apices with long (0.6 mm) reflexed mucro; margins ventrally thickened at apex; midrib prominent. **Capitula** sessile or shortly pedunculate, densely leafy on peduncle, cylindrical. Capitula bisexual and female on same plant. Some plants are entirely one or the other. **Involucral bracts** imbricate, arranged in three types, initially foliaceous then reduced to entirely membranous. **Outer involucral bracts** as leaves but with short membranous alae to less than ½ height of bract. **Middle involucral bracts** membranous alate, ratio of lamina to alae decreases from ⅔ to nearly entirely alate. **Inner involucral bracts** entirely membranous alate.
Female capitula cylindrical, 5–8 mm L, disk diameter 1.5–1.8 mm. **Outer involucral bracts** 3.5–5 x 0.6–0.7 mm; alae ovate–linear, 0.3–0.5 mm W., dorsally longly pubescent where alae meet lamina; lamina as for leaves. **Middle involucral bracts** 4.3–5.2 x 1–2 mm; alae outline linear-rhombic; indumentum dorsal, longly curly pubescent where alae meet lamina; lamina as for leaves. **Inner involucral bracts** 5.8–6.8 x 1.2–1.9 mm, 1–2 series, lanceolate; longly curly indumentum on distal dorsal surface; apical appendages acute, translucent with brown-black or ochre stripe, 1.2–2.5 mm L, shortly mucronate. **Ray florets** dimorphic **Outer series** ca. 10, pistillate, white; corolla 2.5–5.5 mm L, corolla tube 2.1–3 mm L, outer ligule 0.3–0.5 mm W., irregularly 0–3-dentate to entire, inner ligule inconspicuously shortly bifid, to 0.3 mm L **Inner series** ca. 10, pistillate, white; corolla 2.5–5 mm L, corolla tube 2–3.5 mm L, outer ligule to 0.5 mm W., irregularly 3-dentate, inner ligule inconspicuously shortly bifid. **Styles** [outer series] 3–4.5 mm L [inner series] 3.3–5 mm L; stigma lobes < 1 mm, obtuse. **Anthers** [disk] 3.7 mm (5.2 mm incl. filaments) **Achenes** [ray + disk] to 3.2 mm L (mature), turbinate; pericarp pellucid, glabrous or coated in 1-celled papillae. **Pappus** 3.2–3.5 mm L, white, 1 series, barbellate.
Bisexual capitula cylindrical, 4–5.5 mm L, disk diameter 1.5–2.1 mm. **Outer involucral bracts** 2.5–5 x 0.6–1.5 mm; alae linear-rhombic, 0.1–0.3 mm W., dorsally longly pubescent where alae

X *Oriastrum* Poepp. & Endl.

meet lamina; lamina as for leaves. **Middle involucral bracts** 3.5–5 x 1–2 mm; alae linear-rhombic; indumentum dorsal, longly curly pubescent where alae meet lamina; lamina as for leaves. **Inner involucral bracts** 4–6 (6.8) x 1–1.9 mm, 2 series, linear-lanceolate; indumentum on distal dorsal surface; apical appendages acute, with brown-black or ochre stripe, 1–2.0(2.7) mm L, mucronate. **Ray florets** ca. 12, pistillate, white; corolla 2.5–5.5 mm L, corolla tube 1.2–1.7(2.6) mm L, outer ligule 0.4–0.5 mm W., irregularly 0–3-dentate, inner ligule to 0.4 mm L, bifid. **Disk florets** ca. 14, bisexual, yellow; corolla 2.5–3.0(4.1) mm L, corolla tube 2–3 mm L **Styles** yellow [ray] 2–2.4(3.7) mm L [disk] 2.5–3.5 mm L; stigma lobes obtuse. **Anthers** [disk] 2–2.5 mm L (2.5–3.5 mm incl. filaments). **Achenes** [ray + disk] narrowly turbinate, 2.4–3 mm L; pericarp pellucid, glabrous or coated in 1-celled papillae; no carpopodium; testa epidermis with no significant strengthenings, cells parallel. **Pappus** white, 1–2 rows, dehiscent; 3–4 mm L; longly ciliate at base; setae 4–8 cells wide, cells 5–10 μm W.; basal cilia present, sparsely barbellate (3–5 barbs /100 μm), barbs 90–140 μm L, barb base medium adhered (50%), free barb spreading.

Distribution and Habitat. *O. revolutum* is recorded from Bolivia, Chile and Argentina between latitudes 21°14'–25°30'S. TOMBESI (2000) cites 2 records [*Nicora, Gómez-Sosa & Múlgura 8471* (SI) & *Pujalte 166* (SI)] from the Andes west of Iglesia in San Juan (Argentina). However, these would significantly extend the distribution of *O. revolutum* over 600 Km to the south. The identifications were not verified and may possibly be records of the morphologically close, novel variety *O. stuebelii* var. *cryptum*. It occurs in the tundra vegetation (*Stipa frigida, Pycnophyllum macropetalum*) surrounding the high altitude lakes on the plateau east of the Salar de Atacama at elevations between 3680–4800 m.a.s.l.

Differential diagnosis. *O. revolutum* has distinctive IIB apices (acute with reddish-brown stripe), and overall light green or brown-red colouration with dense lanate indumentum and "revolute" (reflexed) mucros on the laminas. Among the collections there is a diminuitive, precocious, more compact form. Its occurrence is not correlated to altitude, latitude, location, or phenology. Some populations of *O. revolutum* (Aguas Calientes, Nevados de Poquis, Cerro Losloyos) have both the diminuitive and normal form. The different forms could be a result of micro-climate.

O. revolutum shares traits with *O. tarapacensis*, which is distinguished by its much larger stature, especially by its aciculate leaves with long erect mucros.

O. revolutum, O. stuebelii var. *stuebelii* and *O. stuebelii* var. *cryptum* form a particularly tricky complex of taxa, whose key characters for identifying them are laid out in Table 14.

Species	*O. revolutum*	*O. stuebelii* var. *stuebelii*	*O. stuebelii* var. *cryptum*
Leaf colour	Reddish brown	Dark green	Bright green
Leaf shape	Linear	Linear	Linear dilated
Leaf Mucro	Long, reflexed	Short, inconspicuous	Short, hooked
Bract series	3	3	2: No OIB with alae to < ½ way up bract.
Bract apices (Colouration)	Pale with narrow reddish brown stripe	Black/brown, sometimes with pale edge	Black/brown, sometimes with pale edge
Bract apices (Shape)	acute	long acute-triangular	short acute
Location	Altiplano lakes east of the Salar de Atacama	Altiplano between Cord. Central & Oriental with some outliers in NW Argentina	Andean massifs east of main chain in NW Argentina.

Table 14: Key characters for identifying *O. revolutum, O. stuebelii* var. *stuebelii* and *O. stuebelii* var. *cryptum.*

Material seen. – **Argentina. Salta** Quebrada del Gallo, Poma, 4750 m, 12.02.1945, *Cabrera 8657* (LP). – **Jujuy** Depto. Rinconada Laguna Vilama, 4600 m, 03.1964, *Schwabe 984* (MA). – **Bolivia. Depto. Potosi** Prov. Sud-López Cerro Pabellón, abra entre Pabellon y Pabelloncito. 4700 m, 10.1989, *Salm 066* (LPB). – Laguna Verde 21 Km hacia Laguna Colorado, 4550 m, 04.2000, *Beck 27509* (LPB). – Cerro Tapaquillcha, 4600 m, 04.1980, *Liberman 156* (LPB). – Cerro Tapaquillcha, ladera Sur, 4680 m, 04.1980, *Liberman 194* (LPB). – **Chile. II Región de Antofagasta** Prov. de El Loa entre Laguna Miscante y Laguna Miñique, 4050 m, 05.1997, *Rodriguez & Ruiz 3679* (CONC). – Cerro Aucanquilcha, Quebrada del Inca, 4500 m, 1985, *K. Arroyo 85588* (CONC). – Altos de Cablor, 4400 m, 12.1995, *Villagran 8763* (CONC). – camino Volcán Becho – Zatio Linchon [Volcán Tatio?], *Navas 2148* (LP). – Cerro Toco, ladera N, 4600 m, 04.1997, *K. Arroyo, Cavieres & Humaña 97038* (CONC). – Cerro Toco, ladera N, 4600 m, 04.1997, *K. Arroyo, Cavieres & Humaña 97043* (CONC). – Cerro Toco, ladera N, 4320 m, 04.1997, *K. Arroyo, Cavieres & Humaña 97054* (CONC). – camino San Pedro de Atacama-Volcan Toco, 4600 m, 10.1958, *Ricardi & Marticorena 4830* (CONC LP). – faldeo O del Cerro Incahuasi, 4650 m, 04.1997, *K. Arroyo, Cavieres & Humaña 97516* (CONC). – Laguna ab 4300 m, Bolivia-Argentina-Chile border, Laguna de Tara, east of San Pedro de Atacama, 2000, *Richter s.n.* (?). – Cruce caminos Paso Jama y Salar de Tara, 4250 m, 12.1996, *Moreira 314* (CONC). – entre Salar Aguas Calientes y Quebrada Quepiaco, 4400 m, 04.1997, *K. Arroyo, Cavieres & Humaña 97479* (CONC). – Vegas de Aguas Amargas, 4450 m, 1992, *Arancio 92366* (CONC). – Cerro Nevados de Poquis, ladera SO, 4500 m, 04.1997, *K. Arroyo, Cavieres & Humaña 97384; 97376* (CONC). – Cerro Nevados de Poquis, ladera SO, 4350 m, 04.1997, *K. Arroyo, Cavieres & Humaña 97346* (CONC). – Al N Salar de Aguas Calientes, 4240 m, 01.1997, *Arancio 10731* (CONC). – entre Salar Aguas Calientes y Quebrada Quepiaco, 4600 m, 04.1997, *K. Arroyo, Cavieres & Humaña 97495* (CONC). – entre Salar Aguas Calientes y Quebrada Quepiaco, 4800 m, 04.1997, *K. Arroyo, Cavieres & Humaña 97505* (CONC). – Pampa Laguna Helada, 4300 m, 04.1997, *K. Arroyo, Cavieres & Humaña 97405* (CONC). – Cerro Losloyo, ladera SE, 4400 m, 04.1997, *K. Arroyo, Cavieres & Humaña 97335* (CONC). – Pampa Loyoques, 4350 m, 04.1997, *K. Arroyo, Cavieres & Humaña 97409* (CONC). – Cerro Losloyo, ladera O, 4400 m, 04.1997, *K. Arroyo, Cavieres & Humaña 97426* (CONC). – Cerro Curutu, lado S del Paso Jama, 4700 m, 04.1997, *K. Arroyo, Cavieres & Humaña 97267* (CONC). – San Pedro de Atacama, au llano de Pajonales, route du Paso de Jama, km 40, 4360 m, 12.1993, *Charpin, Grenon & Lazare AC 23521* (G). – route du Paso de Jama, depuis S. Pedro de Atacama, km 47, 4700 – 4800 m, 15.12.1993, *Charpin, Grenon & Lazare AC 23606* (G). – Cordon Ceja Alta, lado SE, 4600 m, 04.1997, *K. Arroyo, Cavieres & Humaña 97298* (CONC). – San Pedro de Atacama, Socaire, Cerro Casas, 4100 m, 04.1997, *Teillier 4045* (CONC). – Entre Laguna Miscante y Junta a Socaire, 4170 m, 05.1997, *Rodriguez & Ruiz 3691* (CONC). – Cerro Miñiques, 4500 m, 1993, *Baumann 190* (CONC). – Prov. de Antofagasta Cord. Volcan Llullaillaco, 4200 m, 02.1926, *Werdermann 1020* (B BM CONC E F G GH K LP M). – Volcan Llullaillaco, 4500 m, 01.1994, *K. Arroyo* et al. *94013* (CONC). – faldeos S de La Qda Llullaillaco, 3850 m, 01.1994, *K. Arroyo* et al. *94079* (CONC). – base Cerro de La Pena, 3680 m, 02.1997, *Arancio & Squeo 10407* (CONC). – Volcan Llullaillaco, 4450 m, 10.2002, *K. Arroyo* et al. *25126* (CONC). – Pampa Las Carretas Km 28,089, 4300 m, 02.1997, *Arancio & Squeo 10431* (CONC). – Prov. de Chañaral ladera oeste Cerro Los Patitos, 4500 m, 02.2001, *Latoore, Villagrán & Maldonado 199* (CONC).

X *Oriastrum* Poepp. & Endl.

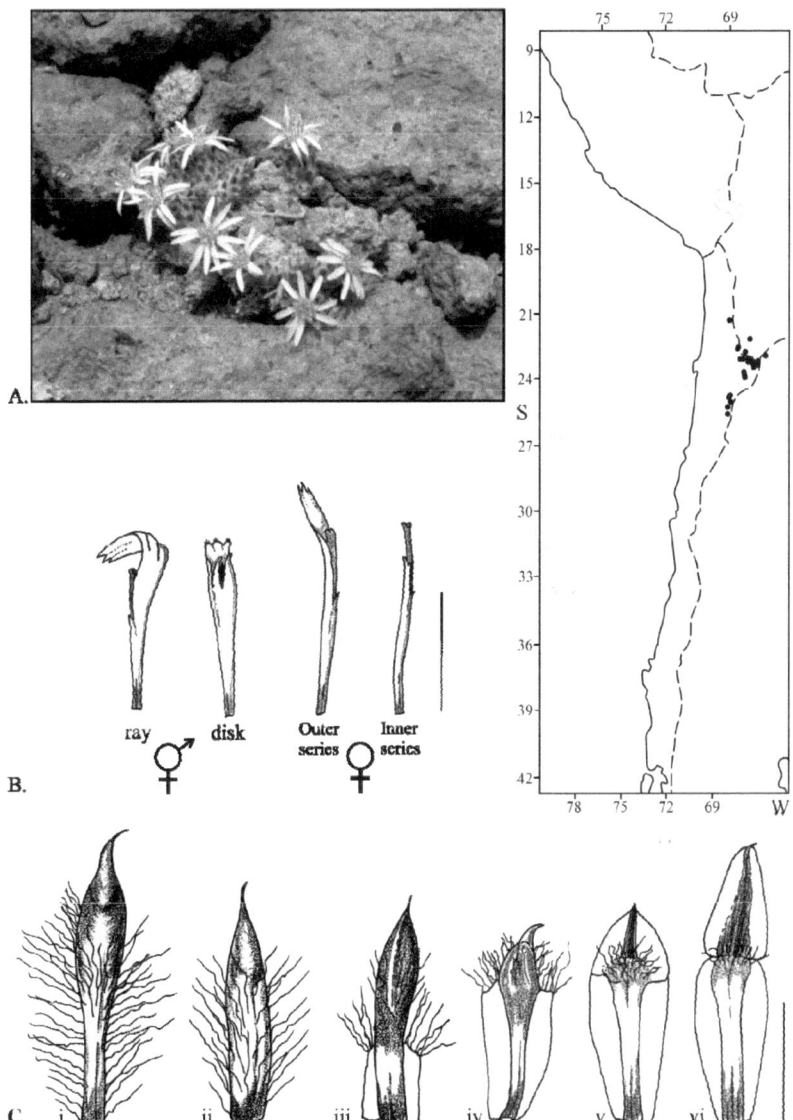

Figure 94: *Oriastrum revolutum*. **A.** Habit, photographed Paso de Jama, 1993©Michel Grenon. **B.** Floret detail from bisexual and female capitula. **C.** Leaf & bract detail. **i.** Stem leaf d.s.; **ii.** Stem leaf d.s.; **iii.** MIB d.s.; **iv.** MIB v.s.; **v – vi.** IIB d.s. B. & C. iii – vi *Charpin, Grenon & Lazarre 23606*. C. i – ii *Villagrán 8763*. Scale bar = 2 mm.

16. *Oriastrum stuebelii* (Hieron.) A.M.R. Davies comb. nov. var. *stuebelii*

≡ *Chaetanthera stuebelii* Hieron. Bot. Jahrb. Syst. 21: 368. 1895. Typus BOLIVIA "[Bolivia. Depto. La Paz Prov. Aroma/ Loayza] Crescit prope Sicasica inter Tomarapé et La Paz alt. 3800 m, ubi floret mense Octobri et Novembri. *Stuebel 15a*" Holotypus B† **Neotypus hic loc. designatus** Typus BOLIVIA Depto. Oruro, Prov. de Avaroa, de Challapata 80 Km hacia Potosi, 4050 m, 04.1992 *Beck 21142* LPB!

= *Chaetanthera boliviensis* J. Kost. Blumea 5: 673, fig. 5p-v. 1945. Typus BOLIVIA "Hab. an sonnigen, begrasten Erdhaengen von Choquecatachico, 4600 m alt., 10.1911, *Herzog 2339*" [Bolivia. Depto. Oruro Prov. Caranagas, Corque Cata chico] Holotypus Leiden!

Perennial gynodioecious dwarf herb, forming lax cushions. **Stems** with basal bud cluster; trailing, 4–10 cm, decumbent, glabrescent. **Leaves** (2)4–7 x 1–1.5 mm, linear, dilated at bases, sessile; dark green, decussate, nearly connate, indumentum long (1 mm) densely curly lanate hairs on margins, and lower dorsal and upper ventral surfaces; margins inrolled at apex, apices with short erect mucro. **Capitula** female or bisexual, sessile, solitary, (6)8–9 mm L, disk diameter 1.9–3 mm; campanulate. **Involucral bracts** imbricate, arranged in three series, initially foliaceous then reduced to entirely membranous. **Outer involucral bracts** 4.8–6.2 x 0.5–0.9 mm as leaves but with membranous alae to less than ½ height of bract; alae to 1 mm W., linear oblong in outline. **Middle involucral bracts** 4.4–4.8 x 1.2–1.6 mm; membranous alate, ratio of lamina to alae decreases from ⅔ to nearly entirely alate, alae linear-oblong, glabrous except for base of lamina. **Inner involucral bracts** entirely membranous, (5)6.8–10.8 x 1.6–2.8 mm, ovate-lanceolate; apical appendages 2.8–4.9 mm L, dark maculate sometimes with narrow pale margins, acute to long acute, shortly mucronate.

Female capitula with 2 series of ray florets; tend be narrower capitula than the bisexual ones. **Ray florets** 5–10, pistillate, white or white tinged with pink; **Outer series:** corolla (4) 6.7–7 mm L, corolla tube 2.5–3.2 mm L, outer ligule 0.6–0.9 mm W., inner ligule 0.6–0.8 mm L, longly bifid. **Inner series:** corolla 4.3 mm L, corolla tube 2.5 mm L **Styles** [Outer series] 4.5–5.4 mm L [Inner series] 5.4 mm L

Bisexual capitula Ray florets 5–10, pistillate, white or white tinged with pink; corolla 6.7–7 mm L, corolla tube ca. 3 mm L, outer ligule 0.6–0.9 mm W. with long tips, inner ligule 0.6–0.8 mm L, longly bifid. **Disk florets** 5–20, bisexual, yellow; corolla 4.6–5.2 mm L, corolla tube 2.5–4 mm L; **Styles** [ray] 4.5–5 mm L [disk] 4.2 mm L; stigma lobes obtuse, 0.2 mm L **Anthers** [disk] 3.7 mm (5.2 mm incl. filaments) **Achenes** brown, [ray + disk] (immature) (1.2) 4.8 mm L narrowly turbinate; pericarp pellucid, glabrous. **Pappus** 5–6 mm L, white, sparsely barbellate, 1–2 series.

X *Oriastrum* Poepp. & Endl.

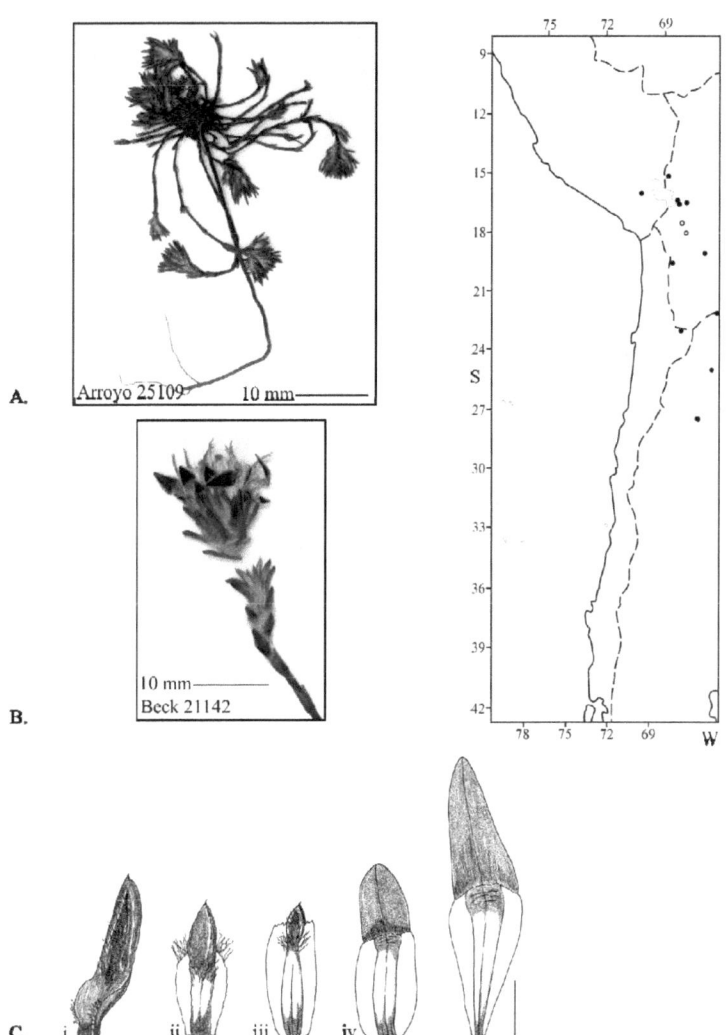

Figure 95: *Oriastrum stuebelii*. **A.** Habit. *Arroyo 25109*. **B.** Capitulum detail. *Beck 21142*. **C.** Leaf and Bract detail. *Stafford 873* **i.** One of stem leaf pair v.s. **ii.** MIB d.s. **iii.** MIB v.s. **iv. – v.** IIB d.s. Scale bar i – v. = 2 mm

Distribution and Habitat. *O. stuebelii* is distributed across the northern junction of the Cordillera Central and the Cordillera Oriental to the east of the Altiplano of Bolivia, just reaching southern Peru, northeast Chile (across the Altiplano) and NW Argentina between 16°00'–23°00'–27°40'S. It occurs in the high Andean Puna, on shrubby or *Lepidophyllum* slopes in loose sandy soil or stony/rocky ground ("cantizal") at elevations between 3800–5200 m.a.s.l.

Differential diagnosis. *O. stuebelii* is morphologically close to the more southerly distributed *O. acerosum* (23°45' – 31°30'S) and *O. dioicum* (30° – 34°S). The leaf and inner involucral bract colouration can give the impression of *O. dioicum*, but the distinctive maculate shoulders of the middle involucral bracts are lacking in *O. stuebelii*. *O. acerosum* has much smaller capitula, often exceeded by the outer involucral bracts and leaves, and shorter involucral bracts apices which are entirely coloured or pale, without paler margins.

Harder to distinguish are the three taxa *O. revolutum*, *O. stuebelii* var. *stuebelii* and *O. stuebelii* var. *cryptum*. Characters useful for this include leaf colour, shape, and mucro, colour and shape of the inner bract apices and the geographical location. These are laid out in Table 14, page 258. Genetic nrDNA showed a highly polymorphic *O. stuebelii* was closer to material representing *O. abbreviatum* and *O. tontalensis*.

Material seen. – **Argentina. Jujuy** Cerro Tuzgle, 5000 m, 02.1944, *Cabrera 8368* (LP). – **Salta** Rosario de Lerma, Nevado del Castillo, 3500 m, 01.1929, *Venturi 9267* (GH). – **Catamarca** Dept. Andagalá north of Andagalá, Cerro Negro, 4000 m, 01.1917, *Jörgensen 1325* (GH en parte, LP). – Andagalá, Río Potrero sup. La Overa, 4000 m, 03.1951, *Sleumer 1875* (W). – **Tucumán** Depto. Tafí Cumbres Calchaquíes, 01.1913, *Castillon s.n.* (LP 13151). – Cerro Negrito, subida desde El Infiernillo, Cumbres Calchaquíes, 3890m, 02.2003, *K.Arroyo 25109* (CONC). – Dept. Chichigasta Estancia Santa Rosa, habitat sobre las peñas, 4000 m, 01.1927, *Venturi 4812* (GH LP). – **Bolivia. Depto. Oruro** Prov. de Avaroa de Challapata 80 Km hacia Potosi, 4050 m, 04.1992 *Beck 21142* (LPB). – Prov. Atahullpa Coipasa, Cerro Villa Pucarini, 4700 m, 07.1982, *Menhofer 1524* (LPB). – **Depto. La Paz** 4811 m, *K.Arroyo 25200* (CONC). – Prov. de Murillo Nevado del Illimani, 4600 m, 07.1982, *Menhofer 1458* (F). – La Cumbre, ca. 16 km north of the puesto de transito Chuquiaguillo, vicinity of Laguna Estrelloni, 4600 m, 08.1986, *Solomon 15522* (LPB MO). – ca. 20 kms hacia el norte de La Paz, pie de los nevados Khala Huyo, 4700 m, 07.1988, *Beck 8482* (LPB). – Prov. de F. Tamayo, Estancia Olearia (Ulla Ulla), 4650 m, 02.1983, *B. Menhofer 2014* (LPB). – **Depto. Potosí** Prov. Sud López Cerro Pabellón, unos 14 km al sur de Laguna Colorada, 4620 m, 1991 – 1992, *Navarro* (in lit.) – **Chile. I Región de Tarapacá** Tarapacá *Arroyo 25204*. – **II Región de Antofagasta** Prov. de El Loa Cerro Toco, ladera Norte, 4320 m, 04.1997, *K. Arroyo, Cavieres & Humaña 97056* (CONC). – **Peru. Puno** San Antonio de Esquilache, 5200 m, 08.1937, *Stafford 873* (BM [8-3] F K).

X *Oriastrum* Poepp. & Endl.

17. *Oriastrum stuebelii* var. *cryptum* A.M.R.Davies var. nov.

Typus ARGENTINA "Argentina, Prov. de Catamarca, Dept. Andagalá, Andagalá, Río Potrero sup. Abra Grande, 3600 – 3900 m, 01.03.1951, *Sleumer 1876* Holotypus W!

A varietate typical differt foliis laete viridibus, linearibus, apice obtuso mucrone breve uncinato. Involucri bractaeis exterioribus destituta. Involucri bractaeis interioribus appendicibus acutis brevibus terminalibus (2.2–3.5 mm longis), capitulis brevioribus floribus radii 4.3–4.8 mm longis, floribus disci 3.2–4.4 mm longis. Pappus caducus. Pappi setis brevibus 2.6–3.4 mm longis. Habitat in Argentina.

Perennial gynodioecious dwarf herb, forming lax cushions. **Stems** short, 4–7 cm, with basal bud cluster; naked to sparsely leafy; glabrescent, indumentum sparsely floccose lanate. **Leaves** 4–5x 0.9–1.2 mm, linear, distally dilated, apex broadly acute with a short, hooked mucro; bright green, decussate, connate; indumentum floccose lanate (hairs to 1.5 mm L) on margins and distal ventral surface; margins inrolled. **Capitula** radiate. **Involucral bracts** imbricate, arranged in two series (no **outer involucral bracts** with short membranous alae to less than ½ height of bract), initially foliaceous then reduced to entirely membranous. **Middle involucral bracts** 4.3–6.8 x 2–3.1 mm, linear truncate to ovate; membranous alate, ratio of lamina to alae decreases from ⅔ to nearly entirely alate. **Inner involucral bracts** 5.9–8 mm x 1.9–2.8 mm, entirely membranous; apical appendages short (2.2–3.5 mm L) acute, midrib visible, entirely black-brown, sometimes with pale margin.
Female capitula rare, with ray-like florets in the disk region. **Ray florets** white, pistillate, dimorphic. **Outer series:** corolla 5.1–5.6 mm L, corolla tube 2.5–2.9 mm L **Inner series:** corolla 3.4–3.7 mm L, corolla tube 2.7–2.8 mm L **Styles** [Outer series] 3.4–4.6 mm L; [Inner series] 4.5–4.6 mm L
Bisexual capitula more common, cylindrical. **Ray florets** ca. 15, pistillate, white?; corolla 4.3–4.8 mm L, corolla tube 2–3.2 mm L **Disk florets** 15–21, bisexual, colour unknown; corolla 3.2–4.2 mm L, corolla tube 2.4–3.2mm L **Styles** [ray] 2.9 mm L [disk] 3–4mm L **Anthers** (incl. filaments) [ray + disk] 3–4 mm L **Achenes** [ray + disk] (immature) 1.2 mm L, sometimes with minute papillae.
Pappus setae short, 2.6–3.4 mm L, white, dehiscent, barbellate.

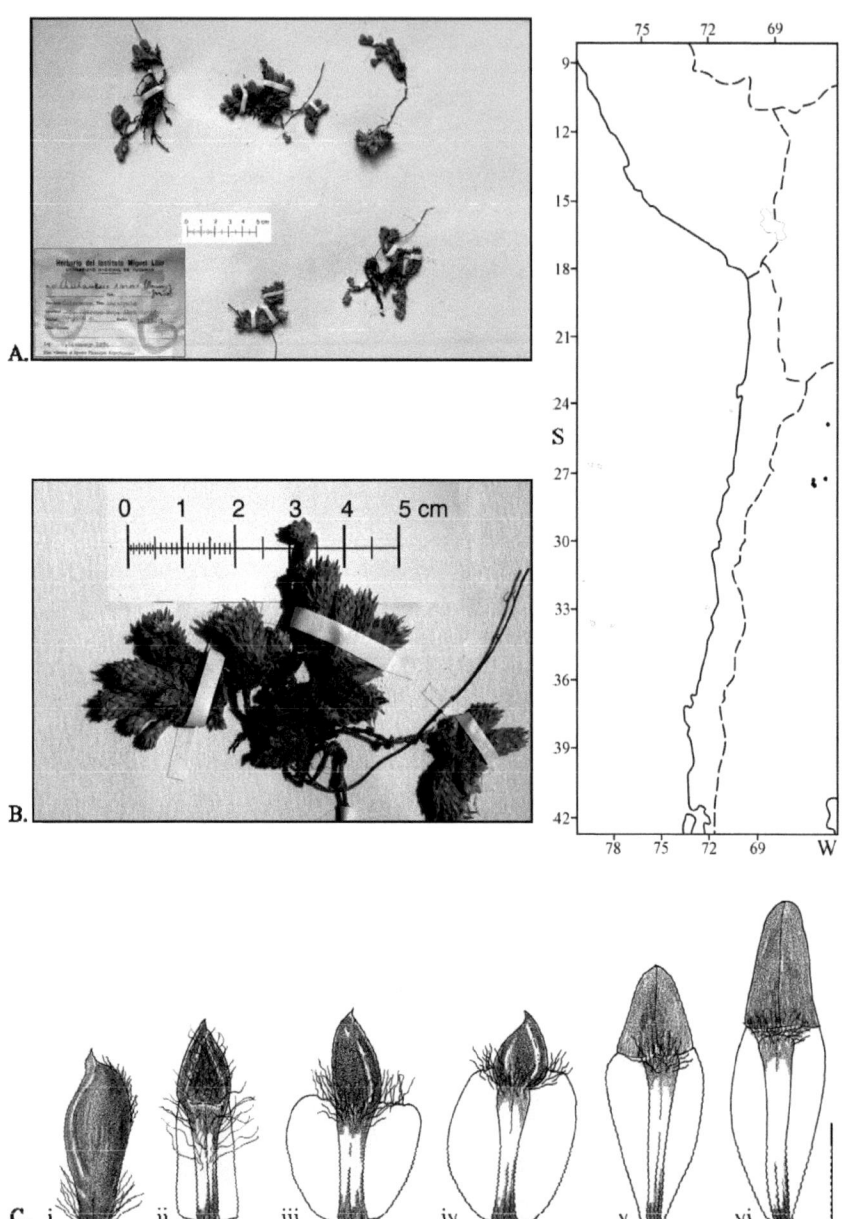

Figure 96: *Oriastrum stuebelii* var. *cryptum*. Holotype *Sleumer 1876*. **A.** Herbarium sheet, composite image. **B.** Habit. **C.** Leaf & Bract detail **i.** Stem leaf d.s. **ii – iv.** MIB v.s. & d.s. **v – vi.** IIB d.s. Scale bar i – vi = 2 mm.

X *Oriastrum* Poepp. & Endl.

Distribution and Habitat. *O. stuebelii* var. *cryptum* is represented by a few isolated collections from Salta and Catamarca (Argentina) between latitudes 25°00'–27°40'S. It occurs in among rocks, gravel or in rocky or sandy soil on gentle to steep slopes (granite or volcanic) with open, shrubby vegetation between 3500–5000 m.a.s.l. elevation.

Differential diagnosis. These collections are morphologically segregated from those numerically adjacent collections of *O. stuebelii* var. *stuebelii* (viz. *Sleumer 1875*, *Venturi 9267*, *Jørgensen 1325* pp). *O. stuebelii* var. *cryptum*, *O. stuebelii* var. *stuebelii* and *O. revolutum* are rather similar in size and stature. The two *O. stuebelii* varieties also share the same trailing habit and glabrescent stems. They are all differentiated on the basis of overall colour, leaf shape, leaf apices, bract apices and geographical location. These differences are outlined in Table 14, page 258.

Material seen. – **Argentina. Prov. de Salta** S of Abra Acra on road to Poma, 5000 m, 12.1978, *Hammel 5979* (MO). – Rosario de Lerma, Nevado del Castillo, 3500 m, 01.1929, *Venturi 9266* (GH). – **Prov. de Catamarca** Dept. Andagalá north of Andagalá, Cerro Negro, 4000 m, 20.01.1917. *Jörgensen 1325* (GH en parte). – Andagalá, Río Potrero sup. Abra Grande, 3600 – 3900 m, 01.03.1951, *Sleumer 1876* (W).

18. *Oriastrum tarapacensis* A.M.R. Davies sp. nov.

Typus CHILE "Chile, Prov. de Iquique, camino de Huara a Cancosa, Km 129, 4400 m, 02.1964, *Marticorena, Matthei & Quezada 348* Holotypus CONC!

Planta perennis herbacea decumbens laxe pulvinata. Caulis gemmis fasciculatis subterraneis ornatus, decumbens vel ascendens ad 12 cm altus, ramosissimus, dense villosus pilis ad 2.5 mm. Folia linearia 8 x 1.2 mm, erecta, antice aciculare acuminata. Capitula solitaria, sessilia vel breviter pedunculata, diametro 13 mm, infra foliis verticillatis ornata. Involucri bracteae imbricatae, dissimiles, foliaceae gradatim ad complete membranaceae; interiores 11–13 x 1.2–2 mm, lineari-lanceolatae, antice acutae vel acuminatae dorsaliter obscure univittatae. Achenia anguste turbinata, glabrescentia. Pappus caducus; pappi setae 6.5 mm, barbellatae. Habitat in Chili.

Perennial monoecious dwarf herb, forming lax cushions. **Stems** with basal bud cluster; decumbent to ascending to 12 cm, many branches 2.5–8 cm L; indumentum long, dense villous hairs (to 2.5 mm). **Leaves** 8 x 1.2 mm, linear, aciculate-acuminate apices, densely decussate, connate; upper margins inrolled, lower margins flat; indumentum dense on lower margins, distal ventral and proximal dorsal surfaces; midrib visible on distal dorsal surface. **Capitula** bisexual solitary, sessile or shortly pedunculate, leaves whorled below capitula; 13 mm L, disk diameter ca. 5 mm. **Involucral bracts** imbricate, arranged in three series, initially foliaceous then reduced to entirely membranous. **Outer involucral bracts** 8–9 x 1 mm, 2–3 series; as leaves but with short membranous alae (0.2–0.7 mm W.) to less than ½ height of bract. **Middle involucral bracts** 8.3–10.1 x 2.5–3 mm, 1–2 series, linear lanceolate; membranous alate, ratio of lamina to alae decreases from ⅔ to nearly entirely alate, alae margins pubescent where they meet the lamina. **Inner involucral bracts** entirely membranous alate, 11–13 x 1.2–2 mm, linear lanceolate; apical appendages acute to acuminate 2–4 x 1–1.8 mm, dark red-brown dorsal stripe. **Ray florets** ca. 10, pistillate, probably creamy-white; corolla 9.5–10 mm L, corolla tube 4.9 mm L, outer ligule 0.6–1.1 mm W., 0–3-dentate, apices minutely ciliate, inner ligule irregularly bifid, or notched (< 2 mm). **Disk florets** ca. 15, bisexual, yellow; corolla 6.1 mm L, corolla tube 5.2 mm L **Styles** [ray] 7.4 mm L, stigma lobes 0.25 mm L, obtuse; [disk] 4.3 mm L, stigma lobes < 0.2 mm L, obtuse. **Anthers** 4 mm (5.9 mm incl. filaments). **Achenes** brown, narrowly turbinate, 1.2 mm (immature); pericarp pellucid, glabrous. **Pappus** setae 6.5 mm L, white, dehiscent, barbellate.

Distribution and Habitat. *O. tarapacensis* is found at latitudes between 18°00'–20°00'S in Chile, north of Iquique on the Liparite plateau. It occurs at elevations between 4100–4600 m.a.s.l.

Differential diagnosis. Closely related to *O. revolutum*, *O. tarapacensis*, named after the Chilean region Tarapacá in which it is found, differs by having aciculate leaves that are more than twice the length of *O. revolutum* leaves. Densely hairy stems, elongated with solitary, ± sessile capitula further distinguish the species. It was not possible to determine whether *O. tarapacensis* was dioecious.

X *Oriastrum* Poepp. & Endl.

Material seen. – Chile. I Región de Tarapacá Tarapacá, Cerro Columbusca, Apaihete?, 4600 m, 03.1926, *Werdermann 1405* (B). – Prov. de Iquique camino de Huara a Cancosa, Km 129, 4400 m, 02.1964, *Marticorena, Matthei & Quezada 348* (CONC). – Prov de Parinacota camino de Putre a Chucuyo, Km 8, 4100 m, 02.1964, *Marticorena, Matthei & Quezada 184* (CONC). – cerca de Las Cuevas, 4250 m, 04.1984, *K. Arroyo 84934* (CONC). – camino Portezuelo de Chapiquiña, Km 111, 4100 m, 02.1964, *Marticorena, Matthei & Quezada 92* (CONC). – Portezuelo de Chapiquiña, 4350 m, 03.1961, *Ricardi* et al. *208* (CONC). – Portezuelo de Chapiquiña, 4400 m, 02.1964, *Marticorena* et al. *112* (CONC).
S.l.d. Sierra la Viuda, 15,200 ft, 1829, leg. *Poeppig fide Sch.Bip.* (P)

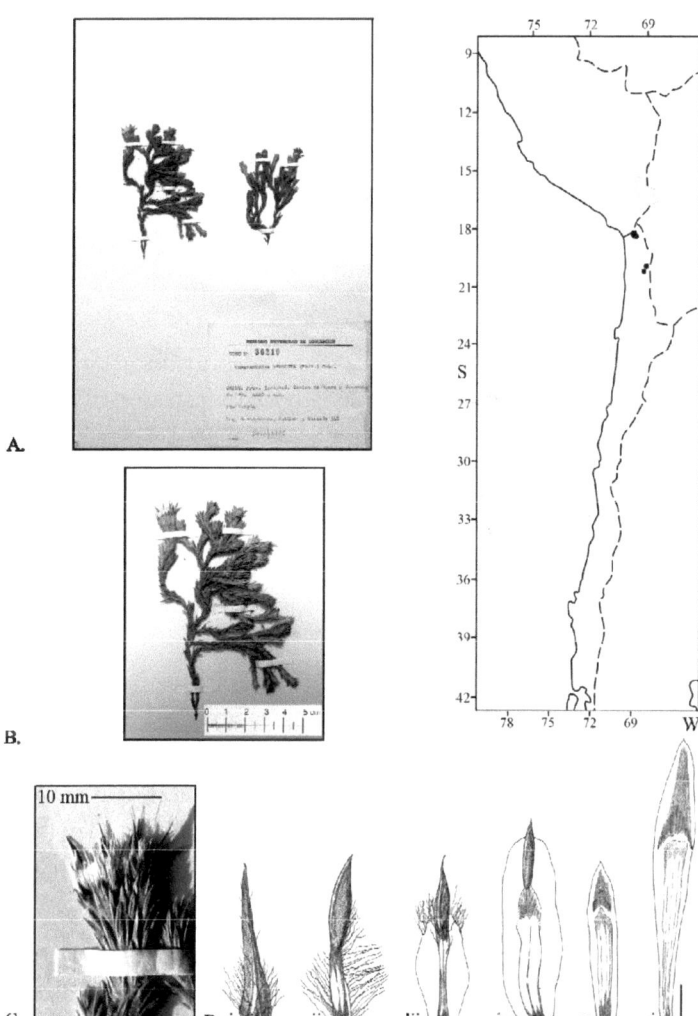

Figure 97: *Oriastrum tarapacensis*. Holotype *Marticorena, Matthei & Quezada 348*. **A.** Herbarium sheet. **B.** Habit of single plant. **C.** Capitulum detail. **D.** Leaf & bract detail **i.** Stem leaf d.s.; **ii.** OIB d.s.; **iii – iv.** MIB d.s.; **v – vi.** IIB d.s. Scale bar i – vi = 2 mm.

19. *Oriastrum tontalensis* A.M.R. Davies sp. nov.

Typus ARGENTINA "Argentina, Prov. San Juan, Dept. Calingasta, Cerro El Tontal, al N de Barreal, camino a la antena, 3550-3750 m, 05.02.1990, *Kiesling, Ulibarri & Kravopickas 7351*" Holotypus NY! Isotypus SI!

Planta perennis nana, pulvinos formans, dense crispato-lanosa, basaliter gemmis ornata. Caulis ad 10 cm altus, ramosus, decumbens. Folia 6-10 mm longa, linearia, sursum indistincte dilatata. Capitula campanulata, breve vel distincte pedunculata. Involucri bracteae imbricatae; exteriores foliis similares sed anguste membranaceo-marginatae, medianae 13-16 mm longae membranaceo-alatae aliis distaliter supra maculatae, interiores 2-3-seriatae ad 19 mm longae et 3 mm latae lineari-lanceolatae complete membranaceae; appendices terminales acutae ad longe acuminatae, obscure fuscae vel nigrae. Flores radii 20-30 feminei albi, corolla ad 13 mm longa. Flores disci ad 30 hermaphroditi aurei, corolla ad 8 mm longa. Pappi setae biseriatae deciduae 6-7 mm longae remote longe et patente barbellatae.

Perennial monoecious dwarf herb, laxly cushion-forming. **Stems** with basal bud cluster; to 10 cm, wiry, branched, decumbent on substrate; stems densely curly lanate. **Leaves** 6–10 x 1–1.5 mm, linear to somewhat dilated obtuse at apex, slightly succulent; distantly decussate up stem, buds in leaf axils; distal margins thickened, distal midrib becoming prominent to shortly mucronate; indumentum long, curly lanate, hairs simple, 1–2 mm L, pubescent on dorsal and distal ventral surfaces, hairs rarely exceeding the dorsal distal surface. **Capitula** campanulate, pedunculate, peduncles densely leafy but can vary in length from short peduncles with a rosette-like compacted appearance to elongated, densely whorled leafy peduncles. **Involucral bracts** imbricate, arranged in three series, initially foliaceous then reduced to entirely membranous. **Outer involucral bracts** (2)7–8 x 0.5–1 mm; dorsal and ventral surfaces covered with lanate (1–2 mm L) curly indumentum, becoming less dense towards apex; bracts as leaves but with inconspicuous, short membranous alae to less than ½ height of bract. **Middle involucral bracts** 2–4 series, 13–16 mm L; membranous alate, ratio of lamina to alae decreases from ⅔ to nearly entirely alate, alae distally dorsally maculate. **Inner involucral bracts** (12)16–19 x 3 mm, 2–3 series, linear-lanceolate, entirely membranous alate, lightly pubescent on dorsal apical region; apical appendages acute to longly acuminate, dark brown or black. **Ray florets** 20–30, pistillate, white; corolla 13 mm L; corolla tube 6–7 mm L; outer ligule 1.5 mm W., irregularly 3-dentate, glabrous; inner ligule reduced to short bifid appendages. **Disk florets** ca. 30, bisexual, yellow; corolla 8 mm L, corolla tube 6.5 mm L **Styles** [ray] 7 mm, shortly bifid. **Achenes** brown, narrowly turbinate, 1.5 mm L (immature); pericarp pellucid, glabrous. **Pappus** 2+ rows, white, dehiscent, setae 6–7 mm; barbellate, barb base shortly adhered (<40%), barbs infrequent, spreading.

X *Oriastrum* Poepp. & Endl.

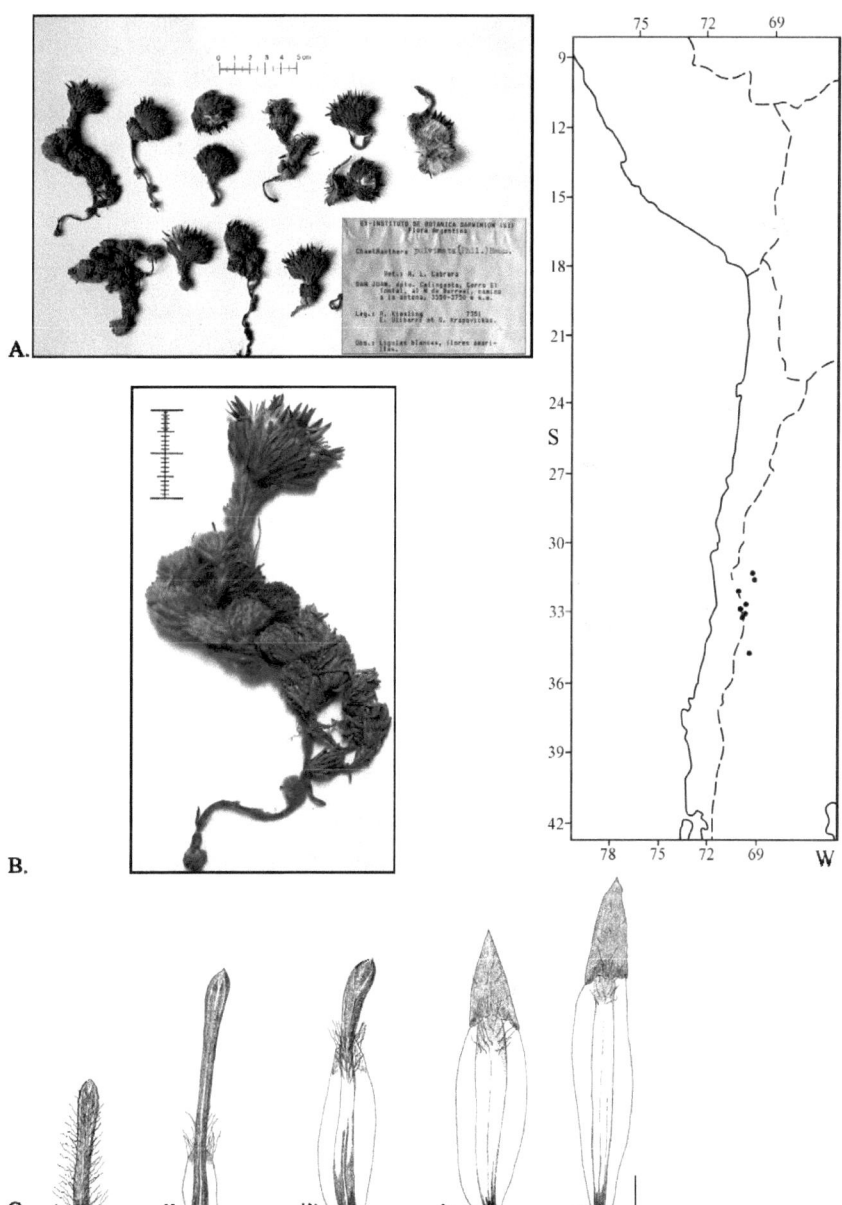

Figure 98: *Oriastrum tontalensis*. Holotype *Kiesling, Ulibarri & Krapovikas 7351*. **A.** Herbarium sheet. **B.** Habit of single plant. **C.** Leaf & bract detail. **i.** stem leaf v.s.; **ii.** OIB v.s.; **iii.** MIB d.s.; **iv – v.** IIB d.s. Scale bar i – v = 2 mm.

Distribution and Habitat. *O. tontalensis* is endemic to Argentina (San Juan & Mendoza), and recorded between latitudes 31°15'–34°45'S. It occurs in gravel, in a periglacial environment at elevations between 3000–4000 m.a.s.l.

Differential Diagnosis. *O. tontalensis* is distinguished from *O. pulvinatum* by its linear leaves and 2-3 rows of IIB with dark acute-apiculate apices.

Material seen. – **Argentina. Prov. San Juan** Cord. del Espinazito, La Colorada, 3000 m, *Bodenbender 28* (G). – Paso del Espinazito, 01.1953, *Castellanos s.n.* (LP). – Depto. Calingasta Cerro El Tontal, al N de Barreal, camino a la antena, 3550-3750 m, 02.1990, *Kiesling, Ulibarri & Kravopickas 7351* (NY SI). – Cerro El Tontal, camino a la antena, 3770 m, 02.2004, *Arroyo & Humaña 25240*. – Río del Palque, 02.1960, *Fabris & Marchioni 2379* (LP). – **Prov. Mendoza** Aconcagua, E-face, Relinchos valley, 4120 m, 03.1980, *Miehe 109/80/26.3* (LPB). – Quebrada de Navarro, west of Las Cuevas, 3680 m, 04.1980, *Miehe 193/80/5.4* (LPB). – Haute Cordillére de Mendoza, sources de río Tupungato, 3500 m, 01.1908, *Hauman 334* (G). – Río Plomo, Nevado de Plomo, SE glacier, 3910 m, 04.1980, *Miehe 92/80/5.2* (LPB). – Occidentale Cuchilla de la Tristeza, Cerro China Muerta, 02.1953, *Castellanos s.n.* (LP).

XI Summary of taxonomic revision

1 Changes in the nomenclature of *Chaetanthera* & *Oriastrum*

The re-evaluation of traditional characters and the discovery of novel character variation in the areas of architecture, anatomy, macro- and micro- morphology, and nrDNA information have necessitated significant changes at the generic and subgeneric levels of the two genera *Chaetanthera* and *Oriastrum*. Generic and subgeneric diagnoses are provided, as is a key to the species in each genus. Nomenclatural notes have been provided where necessary. Some issues were dealt with in prior publications by the author (PRUSKI & DAVIES 2004; DAVIES 2005). Latin descriptions of novel taxa are included.

In the light of the systematic analysis, the genus *Chaetanthera* has been revised. The lectotype of *Chaetanthera* is discussed and fifteen *Chaetanthera* names are lectotypified. The genus comprises thirty species. *Chaetanthera* subgenus *Chaetanthera* includes sixteen species, one variety and two hybrids distributed in western South America from Huancavelica (Peru) to Valdivia (Chile) and southwest Argentina (Neuquén). *Chaetanthera* subgenus *Tylloma* includes fourteen species distributed mainly in Chile, with sporadic collections from Argentina. One novel species (*C. pubescens* A.M.R. Davies), one novel variety (*C. glandulosa* var. *microphylla* A.M.R. Davies), and one new name (*C. frayjorgensis* A.M.R. Davies), and three new combinations: *C. albiflora* (Phil.) A.M.R. Davies, *C. depauperata* (Hook. & Arn.) A.M.R. Davies and *C. taltalensis* (Cabrera) A.M.R. Davies are presented. *C. kalinae* A.M.R. Davies was previously published (DAVIES 2006).

The genus *Oriastrum* has been re-instated. The generitype of *Oriastrum* is discussed and six *Oriastrum* names are lectotypified. The genus comprises eighteen species. *Oriastrum* subgenus *Oriastrum* includes five species, principally from the Chilean Andes. With the exception of the lowland *O. werdemanii* the species grow above 2000 m.a.s.l. up to about 4000 m. With the exception of *O. gnaphalioides* whose distribution has its northernmost limit around Antofagasta (24° 40'S), the species are found between Baños del Toro (29° 50'S) and the Río Clarillo, Cordillera de Santiago, (33° 50'S). *Oriastrum* subgenus *Egania* includes thirteen species and one variety distributed on the Altiplano from Peru (11°S) and Bolivia to the Altoandino in Chile and western Argentina, with its southern limit at about 35°S. Four novel species and one new variety are presented: *O. werdermannii* A.M.R. Davies, *O. famatinae* A.M.R. Davies, *O. tarapacensis* A.M.R. Davies, *O. tontalensis* A.M.R. Davies and *O. stuebelii* var. *cryptum* A.M.R. Davies. Five novel combinations are presented: *O. abbreviatum* (Cabrera) A.M.R. Davies, *O. achenohirsutum* (Tombesi) A.M.R. Davies, *O. apiculatum* (J.Rémy) A.M.R. Davies, *O. revolutum* (Phil.) A.M.R. Davies and *O. stuebelii* (Hieron.) A.M.R. Davies var. *stuebelii*.

2 Updates in the descriptive taxonomy of *Chaetanthera* & *Oriastrum* species

Traditional characters used in the description of the species - such as habit, capitula size, leaf shape, leaf length, leaf width, involucral bract series and achene features - have been re-evaluated, revised and updated. Ray ligule length has been rejected as a generic identifier but remains useful in the species descriptions. In addition to these descriptive and quantitative elements, novel variation in several characters was observed at the micro-morphological and anatomical level. These were sometimes important for differentiating species and therefore also included in the descriptions. Characters include:

- Life cycle, reproductive strategy, stem architecture, habit and stem details.
- Leaves: dimensions, outline, surfaces (e.g. colour, indumentum), arrangement.
- Capitula: shape, attachment.
- Involucral bracts: series, form; outer/ middle/ inner dimensions, descriptive features, anatomy.
- Ray and disk florets: number, sex, colour and measurements of corolla, corolla tube and ligules.
- Styles, stigmas, anthers: measurements, colour.
- Achenes: colour, shape, length, structure, pericarp features, pericarp indumentum, testa epidermis structure.
- Pappus: series, colour, dehiscence, barbs, cilia, barb size, attachment and frequency.
- Chromosome number.

Although pollen information was collated for some species in this study, the detailed publication by TELLERIA & KATINAS (2004) provides a better overview of the species. This information was not included in the species descriptions.

3 Additional information concerning *Chaetanthera* & *Oriastrum*

This study utilised more than 2000 herbarium collections and incorporates information gathered on field trips and also from anecdotal field evidence. Also, the availability of internet sources, particularly images and distribution/geographical data, greatly enriched the material. Thus the breadth and scope of information exceeds that of the previous revision.

The illustrations provided by CABRERA in his 1937 revision, and frequently re-used in later Floras and publications, are still adequate for the identification of many stable taxa (e.g. *C. microphylla*, *C. moenchioides*). They are not suitable for those groups of taxa that have been extensively revised, such as many species within *Oriastrum* subgenus *Egania*, or those species known to be polymorphic, e.g., *C. glabrata*, *C. chilensis*. The current revision takes advantage of the digital revolution and has, where possible, compiled and incorporated images of live material (reproduced with permission of the authors). These are supplemented by images of herbarium

XI Summary of taxonomic revision

material and original illustrations of important taxonomic details such as leaves, bracts and sometimes florets or other features necessary to identify the taxa.

Distribution and habitat information has been updated and presented as altitudinal and latitudinal ranges including pertinent geographical notes (e.g., mountain massifs, coastal plains). A map, as well as a complete geographical listing of the material seen, has also been made available for every species.

Differential diagnoses for all taxa have been compiled. In cases where species have been easily confused in the past, or where species boundaries remain unclear due to hybridisation or polymorphism, tabulated diagnoses are provided.

XII References

Ackerley, D.D. 2003. Community assembly, niche conservatism and adaptive Evolution in changing environments. *International Journal Plant Sciences* 164 (3 Suppl.): 165 – 184.

Adam, P. & Williams, G. 2001. Dioecy, self-compatibilty and vegetative reproduction in Australian subtropical rainforest trees and shrubs. *Cunninghamia* 7(1): 89 – 100.

Albarouki, E. & Peterson, A. 2007 Molecular and morphological characterisation of *Crataegus* L species (Rosaceae) in southern Syria. *Botanical Journal of the Linnean Society* 153: 255 – 263.Allan, R., Lindesay, J., & Parker, D. 1996. El Niño Southern Oscillation and Climatic Variability. Australia: CSIRO.

Alliende, M.C. & Hoffmann A.J. 1983. *Laretia acaulis*, a Cushion Plant of the Andes: Ethnobotanical Aspects and the Impact of Its Harvesting. *Mountain Research & Development*, 3 (1): 45 – 51

Anderberg, A.A. & Freire, S.E. 1990 *Luciliopsis perpusilla* Weddell is a species of *Chaetanthera* Ruiz & Pavon (Asteraceae, Mutisieae). *Taxon* 39(3): 430 – 432

Anderson, E. 1949. *Introgressive hybridisation*. Wiley, New York

Arnold, M.L, Bulger, M. R., Burke, J.M., Hempel, A.L & Williams, J.H. 1999. Natural hybridization: how low can you go and still be important? – Hybridization and Resistance to parasites. *Ecology*. 1 – 16.

Arroyo, M.T.K., Cavieres, LA., Peñaloza, A. & Arroyo-Kalin, M.A. 2003. Positive associations between the cushion plant *Azorella monantha* (Apiaceae) and alpine plant species in the Chilean Patagonian Andes. *Plant Ecology* 169: 121 – 129.

Arroyo, M.T.K., Chacon, P. & Cavieres, LA. 2006. Relationship between Seed Bank Expression, Adult Longevity and Aridity in Species of *Chaetanthera* (Asteraceae) in Central Chile. *Annals of Botany* 98: 591 – 600.

Arroyo, M.T.K., Davies, A.M.R. & Till-Bottraud, I. 2004. *Chaetanthera acheno-hirsuta* (Tombesi) Arroyo, Davies, A.M.R. & Till-Bottraud elevated to species, new for the Flora of Chile. *Gayana Botanica* 61(1): 27-31.

Arroyo, M.T.K., Muñoz, M.S., Henríqueza, C., Till-Bottraud, I. & Pérez, F. 2006. Erratic pollination, high selfing levels and their correlates and consequences in an altitudinally widespread above-tree-line species in the high Andes of Chile. *Acta Oecologia* 30: 248 – 257.

Arroyo, M.T.K., Squeo, F.A., Veit, H., Cavieres, L, Leon, P., & Belmonte, E. 1997. Flora and vegetation of the northern Chilean Andes. pp. 167 – 178. in González, R.C. (Ed.) *El Altiplano: Ciencia y conciencia en los Andes*. Universidad de Chile, Santiago.

Arroyo, M.T.K., Till-Bottraud, I., Torres, C., Henríquez, C.A. & Martínez, J. 2007. Display size preferences and foraging habits of high Andean butterflies pollinating *Chaetanthera lycopodioides* (Asteraceae) in the sub-nival of the central Chilean Andes. *Arctic, Antarctic and Alpine Research* 39: 347 – 352.

Atlas Geográfico de Chile para la educacion (Instituto Geografico Militar, 5a Ed., 1994, 1998)

Badano, E.I. & Cavieres, LA. 2006. Impacts of ecosystem engineers on community attributes: effects of cushion plants at different elevations of the Chilean Andes. *Diversity & Distributions* 16: 388 – 396.

Badano, E.I., Molina-Montenegro, M.A., Quiroz, C. & Cavieres, LA. 2002. Efectos de la planta en cojín *Oreopolus glacialis* (Rubiaceae) sobre la riqueza y diversidad de especies en una comunidad alto-andina de Chile central *Revista Chilena de Historia Natural* 75: 757-765.

Baeza Perry, C.M. & Schrader O. 2005b. Karyotype analysis in *Chaetanthera chilensis* (Willd.)DC. and *Chaetanthera ciliata* Ruiz & Pav. (Asteraceae) by double fluorescence in situ hybridization. *Caryologia* 58 (4): 332 – 338.

XII References

Baeza, C.M. & Schrader, O. 2005a. Karyotype analysis and detection of 5S and 18S/25S rDNA gene sequences in *Chaetanthera microphylla* (Cass.)Hook. & Arn. (Asteraceae). *Gayana Botanica* 62: 49 – 51.

Baeza, C.M. & Torres-Díaz, C. 2006. El cariotipo de *Chaetanthera pentacaenoides* (Phil.) Hauman (Asteraceae). *Gayana Botanica* 63(2): 180 – 182.

Balsamo, R.A., Bauer, A.M., Davis, S.D. & Rice, B.M. 2003. Leaf biomechanics, morphology, and anatomy of the deciduous mesophyte *Prunus serrulata* (Rosaceae) and the evergreen sclerophyllous shrub *Heteromeles arbutifolia*(Rosaceae). *American Journal of Botany.* 90: 72 – 77.

Barton, N.H. & Hewitt, G.M. 1989. Adaptation, speciation and hybrid zones. *Nature* 341, 497-503.

Bentham, G. & Hooker, J.D. 1873. *Genera Plantarum* 2(1): 495 – 497

Bremer, K. 1994. *Asteraceae – cladistics & classification.* Portland Oregon.

Brigg D. & Walters, S.M. *Plant variation & Evolution*

Cabrera A.L in Correa, M. N. 1971. *Flora Patagónica* Colección Científica del INTA, Parte VII Compositae. INTA, Buenos Aires.

Cabrera, A.L & Willink, A. 1973. Biogeografía de América Latina. Organización de los Estados Americanos (OEA), Serie de Biología, Monogr. No. 13, Washington, D.C. pp.117

Cabrera, A.L 1937 Revision del genero *Chaetanthera* (Compositae). *Revista Mus. La Plata Secc. Bot.* 1(3): 87 – 210. Lam. 1 – 4

Cabrera, A.L 1954 Compuestas sudamericanas nuevas o criticas II. *Not. Mus., Eva Peron, Bot.* 17(84): 71 – 80

Cabrera, A.L 1960 Notas sobre tipos de Compuestas Sudamericanas en Herbarios Europeos. III, Los Tipos de Ruiz y Pavon. *Bol. Soc. Argent. Bot.* 8(3 – 4): 195 – 215

Cabrera, A.L 1976. Regiones fitogeográficas argentinas. In Parodi, LR. (ed.), *Enciclopedia argentina de agricultura y jardinería*, 2nd edition. Vol. 2(1). Editorial Acmé, Buenos Aires. pp. 1 – 85.

Cabrera, A.L 1978. *Flora de Jujuy.* Colección Científica del INTA, Parte X Compositae. INTA, Buenos Aires.

Camp, W.H. 1940. The concept of genus. *Bulletin of the Torrey Botanical Club* 67: 349 – 389.

Candolle De, A.P. 1838. Prodromus VII. 1: 14; 29 – 32, 256.

Carlquist, S. 1974. *Island Biology.* Columbia University Press. New York, U.S.A.

Carlquist, S. 2008. www.SherwinCarlquist.com

Cassini, H. 1817 Dict. Sci. Nat. 8: 53. In *"Cassini on Compositae I"*. King, R. M. & H. W. Dawson, New York, 1975.

Cassini, H. 1826. *Opuscules Phytologiques* 2: 102 – 107

Cavieres, LA., Penloza, A. & Arroyo, M.T.K. 2000. Pisos altitudinales de vegetación en los Andes de Chile central (33°S). *Revista chilena historia natural* 73(2): 331 – 344.

Cavieres, LA., Arroyo, M.T.K., Posadas, P., Marticorena C., Matthei, O., Rodriguez, R., Squeo, F.A. & Arancio G. 2002. Identification of priority areas for conservation in an arid zone: application of parsimony anlysis of endemicity in the vascular flora of the Antofagasta region, northern Chile. *Biodiversity & Conservation* 11: 1301 – 1311.

Cavieres, LA., Quiroz, C.L, Molina-Montenegro, M.A., Muñoz, A.A. & Pauchard A. 2005. Nurse effect of the native cushion plant *Azorella monantha* on the invasive non-native *Taraxacum officinale* in the high-Andes of Central Chile. *Perspectives in Plant Ecology Evolution & Systematics* 7: 217 – 226.

Cepeda J., Squeo F., Cortés A., Oyarzún J. & Humberto Zavala Z. 2006 La Biota Del Humedal Tambo-Puquíos in Geoecología de los Aandes desérticos. *La Alta Montaña del Valle del Elqui.* Cepeda P., J. (ed) pp: 243-283. Ediciones Universidad de La Serena. La Serena. Chile.

Chaetanthera images: http://sajf.ujf-grenoble.fr/ Copyright © Irene Till-Bottraud.

Chaetanthera images: http://www.chileflora.com/ Copyright © 2006 Michail Belov.
Chaetanthera images: http://www.fundacionraphilippi.cl/ Copyright ©2009 Teresa María Eyzaguirre.
Chaetanthera images: www.Flickr.com
Cody LM. & Overton J. McC. 1996 Short-term evolution of reduced dispersal in island plant populations. *Journal of Ecology* 84: 53-61.
Cronquist, A. 1985. History of generic concepts in the Compositae. *Taxon* 34: 6 – 10.
Davies, A. & Facher, E. 2001. Achene hairs and their diversity in the genus *Chaetanthera* Ruiz et Pav. (Mutisieae, Asteraceae). *Sendtnera* 7: 13 – 33.
Davies, A.M.R. 2005. (1703) Proposal to reject the name *Euthrixia salsoloides* (Compositae). *Taxon* 54 (3): 838 – 839.
Davies, A.M.R. 2006. *Chaetanthera kalinae* (Mutisieae, Asteraceae) a new species from Chile. *Novon* 16 (1): 51 – 55.
Davies, A.M.R., Maxted, N. & Van der Maesen, LJ.G. 2007. A Natural Infrageneric Classification for *Cicer* L (Leuminosae, Cicereae). *Blumea* 52: 379 – 400.
Dillon, M.O. and Rundel, P.W. (1990). The botanical response of the Atacama and Peruvian Desert floras to the 1982-1983 El Niño event. In *Global Ecological Consequences of the 1982-1983 El Niño-Southern Oscillation* (Glynn, P.W., ed.). pp. 487-504. Elsevier Oceanography Series 52.
Dillon, S. & Hoffmann, A.E. 1997. Lomas Formations of the Atacama Desert: Northern Chile in Davis, S.D., Heywood, V.H., Herrera-MacBryde, O., Villa-Lobos, J. & Hamilton, A. (Eds.). *Centres of Plant Diversity: A Guide and Strategy for Their Conservation*. Volume 3: The Americas. IUCN Publications Unit, Cambridge, England http://www.nmnh.si.edu/botany/projects/cpd/
Dinerstein, E., Olson, D., Graham, D., Webster, A.L., Primm, S.A., Bookbinder, M.P., Ledec, G. 1995. *Conservation assessment of the terrestrial ecoregions of Latin America and the Caribbean*. Washington, D.C; Banco Mundial. 27: 129 pp.
Don D. 1830 Mr. D. Don's Descriptions of new Genera and Species of the Class Compositae. *Trans. Linn. Soc.* 16: 233 – 239, 257 – 259, 301.
Donoso C. & Landrum LR. 1979. *Nothofagus leoni* Espinosa, a natural hybrid between *Nothofagus obliqua* (Mirb.)Oerst. and *Nothofagus glauca* (Phil.) Krasser. *New Zealand Journal of Botany* 17: 353 – 360.
Duke, S.D., Kakefuda, G., Henson, C.A., Loeffler, N.L & van Hulle, N.M. 1986. Role of the testa epidermis in the leakage of intracellular substances from imbibing soybean seeds and its implications for seedling survival. *Physiologia Plantarum* 68 (4): 625 – 631.
Ebach, M., Williams, D. & Morrone, J. 2006. Paraphyly is bad taxonomy. *Taxon* 55: 831 – 832.
Ehrhart, C. 2000 Die Gattung *Calceolaria* (Scrophulariaceae) in Chile. *Bibliotheca Botanica* 153: 1 – 283.
Ehrhart, C. 2005. The Chilean *Calceolaria integrifolia* s.l. species complex (Scrophulariaceae). *Systematic Botany* 30 (2): 383 – 411.
Erdtman, P. 1960. *Svensk Bot. Tidskr.* 54: 561 – 564.
Estabrook, G.F. Nir L Gil-ad, Reznicek A.A 1996 Hypothesizing hybrids and parents using characters intermediacy, parental distance and equality. *Taxon* 45. 647 – 682..
Farr, E. R., Leussink, J. A. & Staffleu, F. A. 1979 *Index Nominum Genicorum* (plantarum) Vol 1 – 3 + suppl. Utrecht Publishers, Le Hague.
Farr, E.R. & Ziljstra, G. (eds) 1996 *Index Nominum Genicorum* (plantarum), http://rathbun.si.edu/botany/ing/ 9 February 1996.
Ferreyra R. 1953. Los especies peruanas del genero *Chaetanthera* (Compositae). *Publ. Mus. Nat. Hist. "Javier Prado"*, Ser. B. Bot. 6: 1 – 13 (Laminas 1 – 3)

XII References

Freire, S.E, Katinas, L & Sancho, G 2002. *Gochnatia* (Asteraceae, Mutisieae) and the *Gochnatia* complex: taxonomic implications from morphology. *Annals of the Missouri Botanical Garden* 89(4): 524 – 550.

Freire, S.E. & Katinas, L 1995. 6. Morphology and Ontongeny of the cypsela haris of Nassauviniiae (Asteraceae, Mutisieae). In D.J.N. Hind, C. Jeffrey and G.V. Pope (Editors). *Advances in Compositae Systematics* pp. 107 – 143. Royal Botanic Gardens Kew.

Funk, V.A. 1985. Cladistics and generic concepts in the Compositae. *Taxon* 34(1): 72 – 80.

Funk, V.A., Bayer, R.J., Keeley, S., Chan, R., Watson, L, Gemeinholzer, B., Schilling, E., Panero, J.L, Baldwin, B.G., Garcia-Jacas, N., Susanna, A. & Jansen, R.K. 2005. Everywhere but Antarctica: Using a supertree to understand the diversity and distribution of the Compositae. *Biol. Skr.* 55: 343 – 374.

Gay, C. 1854 *Atlas de la historia fisica y politica de Chile*. (E. Thundt y ca.) Paris.

Grau, J. 1980. Die Testa der Mutisieen und ihre systematische Bedeutung. *Mitteilungen der Botanischen Staatssammlung München* 16: 369 – 332.

Grau, J. 1987. Chromosomenzahlen chilenischer Mutisieen (Compositae). *Botanische Jahrbücher* 108: 229 – 237.

Grime, J.P. 1977. Evidence for the Existence of Three Primary Strategies in Plants and Its Relevance to Ecological and Evolutionary Theory. *The American Naturalist* 111(982):1169 – 1194.

Guillemin, J. B. A. 1833. Bulletin Bibliographique. *Arch. Bot. (Paris)* 2: 466 – 467

Gutterman, Y. 2002. Survival strategies of annual desert plants: Adaptations of desert organisms. Berlin, Heidelberg, New York, Springer.

Hamzeh, M, Sawchyn, C., Périnet, P. & Dayanandan, S. 2007. Asymmetrical natural hybridisation between *Populus deltoides* and *P. balsamifera* (Salicaceae). *Canadian Journal of Botany* 85: 1227 – 1232.

Hansen, H.V. 1991. Phylogenetic Studies in Compositae tribe Mutisieae. *Opera Botanica* 109: 5 – 50.

Harrison, R.G. 1993. Hybrid zones and the evolutionary process.

Hellwig, F.H. 1990. Die Gattung *Baccharis* L (Compositae-Asteraceae) in Chile. *Mitteilungen der Botanischen Staatssammlung München* 29: 1 – 456.

Hennig, W. 1950. *Grundzüge einer Theorie der phylogenetischen Systematik*. Deutscher Zentralverlag, Berlin.

Hennig, W. 1966. Phylogenetic Systematics. (Translated into English by Rainer Zangerl & D. Dwight Davis). University of Illinois Press, Urbana, Illinois, 263pp.

Hershkovitz, M.A., Arroyo, M.T.K., Bell, C. & Hinojosa, LF. 2006. Phylogeny of *Chaetanthera* (Asteraceae: Mutisieae) reveals both ancient and recent origins of high elevation lineages. *Molecular Phylogenetics and Evolution* 41: 594 – 605.

Heslop-Harrison J. 1953. *New concepts in flowering plant taxonomy*. Heinemann, London.

Hind, D.N. 2007. Compositae: tribe Mutisieae, In: Kadereit, J.W., Jeffrey, C. (eds). Families and Genera of Vascular Plants. Vol. VIII Flowering Plants. Eudicots. Asterales. Springer Verlag, Berlin pp. 90 – 112.

Hoffmann, A.E., Arroyo, M.K., Liberona, F., Muñoz, M. & Watson, J.M. 1998. Plantas altoandinas en la flora silvestre de Chile. Ediciones Fundación Claudio Gay, Santiago de Chile. 280 pp.

Hoffmann, A.E., Herrera-MacBryde, O. & S. Dillon. 1997. Mediterranean Region and La Campana National Park: Central Chile in Davis, S.D., Heywood, V.H., Herrera-MacBryde, O., Villa-Lobos, J. and Hamilton, A. (eds.). Centres of Plant Diversity: A Guide and Strategy for Their Conservation. Volume 3: The Americas. IUCN Publications Unit, Cambridge, England. http://www.nmnh.si.edu/botany/projects/cpd/

Holmgren M, Scheffer M, Ezcurra E, Gutiérrez JR, Mohren GMJ. 2001. El Niño effects on the dynamics of terrestrial ecosystems. *Trends in Ecology and Evolution* 16(2): 89 – 94.

Holmgren, M., Stapp, P., Dickman, C.R., Gracia, C., Graham, S., Gutiérrez, J.R., Hice, C., Jaksic, F., Kelt, D.A., Letnic, M., Lima, M., López B.C., Meserve, P.L., Milstead, W.B., Polis, G. A. †, Previtali M. A.,

Richter, M., Sabaté, S., & Squeo, F. A. 2006. A synthesis of ENSO effects on drylands in Australia, NorthAmerica and South America. *Advances in Geosciences*, 6, 69 – 72.

Hooker, W.J. & Arnott, G.A.W. 1835 Contributions towards a flora of South America *Comp. Bot. Mag.* 1: 104 – 106

Hörandl, E. 2007. Neglecting evolution is bad taxonomy. *Taxon* 56(1): 1 – 5.

Hornung-Leoni, C.T. & Sosa,V. 2008. Morphological phylogenetics of *Puya* subgenus *Puya* (Bromeliaceae) *Botanical Journal of the Linnean Society*, 156: 93 – 110.

Instituto Geográfico Militar Arengtina Mapas de las provincias Argentinas (http://logobali.netfirms.com/) 2002.

Instituto Geográfico Militar Chile (Ed.) 1994. Listado de Nombres Geográficos. Volume I A – M

Instituto Geográfico Militar Chile (Ed.) 1994. Listado de Nombres Geográficos. Volume II M – Z

Instituto Geográfico Nacional de Peru http://www.ignperu.gob.pe/

Jacquemyn, H. Honnya, O., Van Looy, K. & Breyne, P. 2006. Spatiotemporal structure of genetic variation of a spreading plant metapopulation on dynamic river banks along the Meuse River. *Heredity* 96: 471 – 478.

Jordan, R.S. 1991. Impact of ENSO events on the southeastern Pacific region with special reference to the interaction of fishing and climatic variability. In ENSO Teleconnections Linking Worldwide Climate Anomalies: Scientific Basis and Societal Impacts (Glantz, M. et al., eds). pp. 401-430. Cambridge University Press.

Katinas, L 2008a. The genus *Pachylaena* (Asteraceae, mutisieae). *Botanical Journal of the Linnaean Society* 157 (2): 373 – 380.

Katinas L, Pruski, J., Sancho, G. & Tellería, M.C. 2008b. The Subfamily Mutisioideae (Asteraceae) *Botanical Review* 74: 469 – 716.

Kim, H.-G., Loockerman, D.J. & Jansen, R.K. 2002. Systematic implications of ndhF sequence variation in the Mutisieae (Asteraceae). *Systematic Botany* 27: 597 – 609.

Kimball, R.T. & Crawford, D.J. 2004. Phylogeny of Coreopsideae (Asteraceae) using ITS sequences suggests lability in reproductive characters. *Molecular Phylogenetics and Evolution* 33 (1):127 – 139.

Klingenberg, L 2007. Monographie der südamerikanischen Gattungen *Haplopappus* Cass. und *Notopappus* L Klingenberg (Asteraceae - Astereae). *Biblioteca Botanica* 157: 1 – 331.

Klotz, S. & Kühn, I. 2002. Ökologische Strategietypen. *Schriftenreihe für Vegetationskunde* (Bundesamt für Naturschutz, Bonn) 38: 197 – 201.

Koerner, C., 1999. *Alpine Plant Life*. Springer, Berlin.

Koster, J. 1945 Plantae a Th. Herzogio in itinere euis Boliviensi collectae. *Blumea* 5(3): 670 – 674

Kuntze, O. 1898 *Revisio Genero Plantarum* 3(2): 137 – 138; 140 – 141.

Lessing, C. F. 1832 *Synopsis Generum Compositarum*. Sumtibus Dunckeri et Humboltii, Berlin.

Lexer, C., Joseph, J., van Loo, M., Prenner, G., Heinze, B., Chase, M. & Kirkup, D. 2009. The use of digital images-based morphometrics to study the phenotypic mosaic in taxa with porous genomes. *TAXON* 58(2): 349 – 364.

López, B.C. Rodríguez, R., Gracia C. A. Sabaté, S. 2006. Climatic signals in growth and its relation to ENSO events of two *Prosopis* species following a latitudinal gradient in South America *Global Change Biology* 12(5): 897 – 906.

Luebert, F., Wen, J. & Dillon, M.O. 2009 Systematic placement and biogeographical relationships of the monotypic genera *Gypothamnium* and *Oxyphyllum* (Asteraceae: Mutisioideae) from the Atacama Desert. *Botanical Journal of the Linnean Society* 159: 32 – 51.

XII References

Mabberley, D.J. 1981. Edward Nathaniel Bancroft's Obscure Botanical Publications and His Father's Plant Names *Taxon* 30 (1): 7 – 17.

Marticorena, C. & Quezada, M. 1974. Compuestas nuevas o interessantes para Chile. *Bol. Soc. Biol Concepcion* 48: 99 – 108.

Marticorena, C. & Quezada, M. 1985. Catalogo de la flora vascular de Chile. *Gayana Botanica* 42(1-2): 1 – 157.

Martorell, C. & Ezcurra, E. 2002. Rosette scrub occurrence and fog availability in arid mountains of Mexico. *Journal of Vegetation Science* 13: 651 – 662.

Miller, H. S. 1970. The herbarium of Aylmer Bourke Lambert. – TAXON 19: 489-656.

Mukherjee, S.K. & Nordenstam, B. 2004. Diversity of the carpopodial structures in some members of the Asteraceae and their taxonomic significance. *Compositae Newsletter* 41: 29 – 50.

Muñoz Pizarro, C. 1960. Las especies de plantas descritas por R. A. Philippi en el siglo XIX – estudio crítico en la identificación de sus tipos nomenclaturales. Ediciones de la Universidad de Chile.

Navarro, G. 1993. Vegetación de Bolivia: el Altiplano meridional. *Rivasgodaya* 7: 69 – 98.

Navas LE. 1979. *Flora de la Cuenca de Santiago*. Tomo III. Ediciones de la Universidad de Chile. Págs.187.

Oberprieler, C. 2001. Phylogenetic relationships in Anthemis L (Compositae Anthemideae) based on nr DNA ITS sequence variation. *Taxon* 50: 745 – 762.

Panero, J. L & Funk,V. A. 2008. The value of sampling anomalous taxa in phylogenetic studies: major clades of the Asteraceae revealed. *Molecular Phylogenetics and Evolution* 47: 757-782.

Parra, O. & Marticorena, C. 1972. Granos de polen de plantas Chilenas II Compositae - Mutisieae. *Gayana Botanica* 21: 1 – 107.

Philippi, F. 1881. Catalogus plantarum vascularum chilensium adhuc descriptarum. *Anales Univ. Chile* 29: 122 – 130, 138, 143

Poeppig, E.F. & Endlicher, S.F.L 1842. Nov. Gen. Sp. Pl. 3: 50 fig. 257.

Popper, K. 1934. *Logik der Forschung*.

Powell, A., Kyhos, D. & Raven, P. 1974. Chromosome numbers in Compositae. X. *American Journal of Botany* 61: 909 – 913.

Pruski, J. & Davies, A.M.R. 2004. On the Lectotypification of *Chaetanthera* Ruiz & Pav. (Compositae, Mutiseae). *Compositae Newsletter* 41: 54 – 57.

Raven P.H. 1973. Evolution of subalpine and alpine plant groups in New Zealand. *New Zealand Journal of Botany* 11: 177 – 200.

Reiche, K. 1904. Anales Univ. Chile 115: 317 – 339.

Reiche, K. 1905. *Estudios criticos sobre la Flora de Chile*. 4: 308 – 310; 330 – 359

Rémy, J. 1848. Historia Fisica y Politica de Chile, Botanica 3: 284 – 287; 300 – 328. t. 35 – 38.

Renner, S.S. & Hyosig Won 2001. Repeated Evolution of Dioecy from Monoecy in Siparunaceae (Laurales). *Systematic Biology* 50(5):700 – 712

Roque, N. & Hind, D.J.N. 2001. *Ianthopappus*, a new genus of the tribe Mutisieae (Compositae) *Novon* 11(1): 97 – 101.

Ruíz, H. & Pavón, J.A. 1794 Fl. Peruv. Prodr. p. 106, t. 23

Ruíz, H. & Pavón, J.A. 1798 Syst. Veg. Fl. Peruv. Chil. (1): 190 – 191

Shan F.; Yan G.; Plummer J.A. 2003. Cyto-evolution of *Boronia* genomes revealed by fluorescent in situ hybridization with rDNA probes *Genome* 46(3): 507 – 513.

Smith, P.M. 1994. University of Edinburgh Lecture notes "Taxonomy in the Service of Man"

Squeo, F. A., Tracol ,Y., López, D., Gutiérrez, J. R., Cordova, A. M. & Ehleringer, J. R. 2006. ENSO effects on primary productivity in Southern Atacama Desert. *Advances in Geosciences* 6: 273 – 277.

Squeo, F.A., Cepeda, P.J., Olivares, N.C. & Arroyo, M.T.K. 2006. Interacciones ecológicas en la alta Montaña del Valle del Elqui. in Geoecologia de los Andes desérticos. La Alta Montaña del Valle del Elqui. Cepeda P. J. (ed) pp. 69-103. Ediciones Universidad de La Serena. La Serena. Chile.

Squeo, F.A., G. Arancio & J.R. Gutiérrez 2001. *Libro Rojo* de la Flora Nativa de la Región de Coquimbo y de los Sitios Prioritarios para su Conservación. Ediciones de la Universidad de La Serena, La Serena. 388 pages.

Stevens, P.F. 2001 onwards. Angiosperm Phylogeny Website. Version 8, June 2007 [and more or less continuously updated since].

Stuessey, T.F., Tremetsberger, K., Müllner, A.N., Jankowicz, J., Guo, Y.-P., Baeza, C.M., Samuel, R.M. 2003. The melding of systematics and biogeography through investifgations at the populational level: examples from the genus *Hypochaeris* (Asteraceae). Basic Applied Ecology 4: 287 – 296.

Tauleigne-Gomes, C. & Lefèbvre, C. 2005. Natural hybridisation between two coastal endemic species of *Armeria* (Plumbaginaceae) from Portugal. 1. Populational in situ investigations *Plant Syst. Evol.* 250: 215 – 230

Taylor, C.M. & Munoz-Schick, M. 1994. The Botanical Works of Philippi, Father and Son, in Chile. *Annals of the Missouri Botanical Garden*, 81(4): 743 – 748.

Taylor, J.A. & Tulloch, D. 1985. El Niño effects on the dynamics and control of an island ecosystem in the Gulf of California. *Ecology* 78: 1884-1897.

Taylor, R. & Phillips, R. 1832. [An Abstract of the] Descriptive catalogue of the Compositae contained in the herbarium of Dr. Gillies, with some additions from other sources, by Mr. D. Don. *Philos. Mag. Ann. Chem.* 11: 387 - 392.

Teillier, S. 1999. Catálogo de las plantas vasculares del área altoandina de Salar de Coposa-cordón Collaguasi. Chile, Región de Tarapacá (I). *Chloris Chilensis*. 2(1). http://www.chlorischile.cl

Tellería, M.C. & Katinas, L 2004. A Comparative Palynologic Study of *Chaetanthera* (Asteraceae, Mutisieae) and Allied Genera. *Systematic Botany* 29 (3): 752 – 773.

Till-Bottraud, I., Giraud, T., Fournier, E., Torres, C., Vautrin, D., Solignac M, Genton B, & Arroyo, M.T.K. 2004. Isolation of seven polymorphic microsatellite loci, using an enrichment protocol, in the high Andean Asteraceous *Chaetanthera pusilla*. *Molecular Ecology Notes* 4: 462 – 464.

Tombesi, T.S. 2000. Novedades en *Chaetanthera* (Mutisieae, Asteraceae). *Hickenia* 3: 69-72.

Torres-Díaz, C., LA. Cavieres, C. Muñoz-Ramírez & Arroyo, M.T.K. 2007. Consecuencias de las variaciones microclimáticas sobre la visita de insectos polinizadores en dos especies de *Chaetanthera* (Asteraceae) en los Andes de Chile central. *Revista Chilena de Historia Natural* 80(4):

Turistel guia turistica de Chile (Turismo y Communicaciones, S. A. 2000)

Turner, B.L 1985. The Generic Concept in the Compositae: A Symposium - A Summing up. *Taxon* 34 (1): 85 – 88.

van der Pijl, L 1969. *Principles of dispersal in higher plants*. Springer

Werneck, T. *Die Raffiniertesten Denkspiele*. p. 24. Neff.

Young, K.R., León, B., Cano, A., & Herrera-MacBryde, O. 1997. Peruvian Puna: Peru in Davis, S.D., Heywood, V.H., Herrera-MacBryde, O., Villa-Lobos, J. and Hamilton, A. (eds.). *Centres of Plant Diversity: A Guide and Strategy for Their Conservation.* Volume 3: The Americas. IUCN Publications Unit, Cambridge, England. http://www.nmnh.si.edu/botany/projects/cpd/

Zander, R.H. 2007. Paraphyly and the species concept, a reply to Ebach et al. *Taxon* 56 (3): 642 – 644.

Zuloaga & Morrone. 1999. Cat. Pl. Vasc. Rep. Argent. 2: 143.

XIII Appendix 1

Table 15: Polymorphism & the El Nino effect. Leaf morph type of each *C. glabrata* collection, with longitude (degrees West), latitude (degrees South), altitude (m.a.s.l.), collection month and collection year (See Figure 35). Data ordered as follows: Leaf type; Collector; Collection number; Longitude (West); Latitude (South); Altitude (m.a.s.l.) ;Collection Month; Collection Year.

A	Johnston	5536	70.43	24.98	-	12	1925
A	Jiles	5424	70.43	25.02	700	-	1969
A	Quezada & Ruiz	207	70.43	25.02	500	-	1991
A	Rosas	1022	70.43	25.02	100	12	1987
A	Johnston	5598	70.42	25.03	-	12	1925
A	K. Arroyo et al.	25130	70.45	25.12	145	10	2000
A	Eggli & Leuenberger	1759	70.45	25.33	40	11	1991
A	Pisano & Bravo	220	70.43	25.38	100	9	1941
A	Grandjot	4400	70.48	25.40	20	-	1940
A	Rechinger & Rechinger	63504	70.48	25.40	-	11	1987
A	Werdermann	128	70.48	25.40	200	10	1925
A	Worth & Morrison	15805	70.48	25.40	75	-	-
A	Johnston	5115	70.43	25.45	-	11	1925
A	Rosas	1076	70.43	25.48	500	12	1987
A	Rodriguez	2638	70.67	26.17	15	10	1991
A	Behn	0	71.27	29.83	-	10	1948
A	Edding & Villagrán	0	70.80	30.08	700	12	1974
A	Behn K.	0	70.60	33.30	1000	10	1919
A	Ehrhart et al.	2002-003	70.60	33.30	1000	11	2002
A	Grau	2450	70.48	33.35	890	11	1980
A	Mahu	1519	70.47	33.35	800	11	1966
A	K. Arroyo et al.	25163	70.33	33.37	2190	12	2002
A	Gunckel	26667	70.72	33.38	800	10	1953
A	Looser	5554	70.48	33.42	1000	10	1948
A	Montero	290	70.38	33.58	1200	1	1928
A	Ehrhart et al.	2002-226	70.42	23.83	660	12	2002
A	Taylor	10763	70.43	25.20	-	-	1991
A	Ricardi	2521	70.42	25.23	200	-	1953
A	Werdermann	826	70.48	25.40	50	-	1925
A	Ricardi	2576	70.48	25.43	50	-	1953
A	Muñoz, Teillier & Meza	2838	70.63	26.13	-	10	1991
A	Marticorena & Weldt	605	70.67	32.97	1150	11	1970
A	Looser	66323	70.60	33.30	900	12	1948
A	Zöllner	5374	70.48	25.40	20	-	1970
A	Muñoz & Meza	2344A	70.83	28.00	-	10	1987
A	Elliott	293	70.37	33.25	1000	12	1927
A	Teillier, Rindel & García	2771B	70.50	25.05	-	9	1992
A	Guzman	0	70.60	33.30	900	10	1943
B	Biese	2263	70.50	26.07	200	2	1947
B	Kausel	3852	-	-	-	1	1954
B	Ricardi	2639	70.40	24.98	1300	-	1953
B	Hoffman	249	70.45	25.10	400	-	1988
B	Jaffuel	978	70.45	25.40	20	-	1930
B	Ricardi	2729	70.45	25.47	500	10	1953
B	Johnston	5793	70.62	26.03	-	12	1925
B	Pisano & Bravo	554	70.63	26.10	300	10	1941
B	Rosas	1135	70.58	26.17	500	12	1987
B	Teillier	4725	70.73	27.00	100	10	1999
B	Johnston	5035	70.35	27.33	370	11	1925
B	Ricardi et al.	666	70.47	27.67	700	2	1963
B	Pisano & Bravo	782	70.78	27.90	180	11	1941
B	Ricardi & Marticorena	4415	70.62	28.13	450	9	1957

B	Beckett, Cheese & Watson	4705	71.20	28.43	-	12	1971
B	Schlegel	5714	71.25	28.47	30	11	1966
B	Worth & Morrison	16227	71.25	28.47	20	10	1938
B	Cabrera	12688	70.90	29.03	-	9	1957
B	Rosas	1326	71.42	29.07	125	12	1987
B	Marticorena Rodriguez & Weldt	1819	71.40	29.10	110	10	1971
B	Ehrhart et al.	2002-316	71.30	29.43	-	11	2002
B	K. Arroyo et al.	25160	71.30	29.72	150	12	2002
B	Behn, F.	0	71.27	29.83	10	10	1948
B	Torres et al.	0	70.80	30.08	700	11	1974
B	Jiles	6415	70.63	30.37	2200	12	1976
B	Eggli & Leuenberger	1721	70.63	32.92	1140	11	1991
B	Gengler & Arriagada	156	70.68	32.95	1200	12	1995
B	Monero	136	70.90	33.08	700	10	1927
B	Caceres	0	70.60	33.18	900	10	1956
B	Uslar	121	70.30	33.30	2000	1	1980
B	Junge	0	70.47	33.35	885	12	1939
B	Mahu	2443	70.47	33.35	800	11	1965
B	Schlegel	5872	70.47	33.37	1010	2	1967
B	Gunckel	21029	70.72	33.38	700	12	1951
B	Grandjot	1007	70.43	33.42	1900	11	1932
B	Schlegel	943	70.37	33.48	800	11	1955
B	Soyka	0	70.32	33.70	1400	2	1950
B	Teillier	1007	71.18	28.08	-	10	1987
B	Marticorena & Matthei	210	71.17	29.43	500	10	1963
B	Montero	343	70.22	33.78	1500	12	1927
C	Rodriguez	2669	70.68	26.57	30	10	1991
C	Gigoux	16	70.85	27.07	-	-	1922
C	Muñoz & Johnson	1928	70.63	27.32	150	9	1941
C	Ricardi	1499	70.48	27.70	350	10	1965
C	Grau	2097	71.07	27.75	100	10	1980
C	Ricardi	2214	70.53	27.80	500	9	1952
C	Ehrhart et al.	2002-142	71.18	28.43	50	12	2002
C	Ehrhart & Grau	971197	71.40	28.87	250	10	1997
C	Ehrhart et al.	2002-118	71.40	28.87	250	12	2002
C	K. Arroyo et al.	25049	71.47	29.10	4	10	2002
C	Hannington	36	71.42	29.20	-	9	1987
C	Marticorena, Rodriguez & Weldt	1693	71.32	29.28	200	10	1971
C	Barros	3001	71.23	29.90	80	9	1928
C	Marticorena, Rodriguez & Weldt	1517	70.90	29.90	500	10	1971

Table 16: Mean annual Southern Oscillation Index (SOI) and the number of recorded leaf morph types from the north and south of the *C. glabrata* distribution for the years 1915 – 2005 (See Figure 37). The SOI means were calculated from data supplied by the Bureau of Meteorology (ftp://ftp.bom.gov.au/anon/home/ncc/www/sco/soi/soiplaintext.html), the National Climate Centre (http://www.bom.gov.au/climate/current/soihtm1.shtml) and the Climate Analysis Section of the Commonwealth Bureau of Meteorology; S.O.I. Archives, 1876 to present (2005). The data is listed as follows: Year; Leaf morph Typus & locality (A – North, B – North, C – North, A – South, B – South); Mean Annual SOI

Year					SOI	Year					SOI
1915	0	0	0	0	-0.34417	1926	0	0	0	0	-0.38
1916	0	0	0	0	0	1927	0	0	0	1	2	0
1917	0	0	0	0	0	1928	0	0	1	1	0	0
1918	0	0	0	0	0	1929	0	0	0	0	0
1919	0	0	0	1	0	-0.935	1930	0	1	0	0	0
1920	0	0	0	0	0	1931	0	0	0	0	0
1921	0	0	0	0	0	1932	0	0	0	1	-0.29
1922	0	0	1	0	0	1933	0	0	0	0	0
1923	0	0	0	0	-0.2575	1934	0	0	0	0	0
1924	0	2	0	0	0	1935	0	0	0	0	0
1925	5	2	0	0	-0.14083	1936	0	0	0	0	0

1937	0	0	0	0	0	0		1972	0	0	0	0	0	-0.735
1938	0	1	0	0	0	0		1973	0	0	0	0	0	0
1939	0	0	0	0	1	0		1974	1	1	0	0	0	0
1940	1	0	0	0	0	-1.385		1975	0	0	0	0	0	0
1941	1	2	1	0	0	-1.2825		1976	0	1	0	0	0	0
1942	0	0	0	0	0	0		1977	0	0	0	0	0	-0.99
1943	0	0	1	0	0	0		1978	0	0	0	0	0	-0.165
1944	0	0	0	0	0	-0.19417		1979	0	0	0	0	0	-0.19083
1945	0	0	0	0	0	0		1980	0	0	1	1	1	-0.30833
1946	0	0	0	0	0	-0.67083		1981	0	0	0	0	0	0
1947	0	1	0	0	0	0		1982	0	0	0	0	0	-1.305
1948	1	1	0	2	0	-0.11667		1983	0	0	0	0	0	-0.83333
1949	0	0	0	0	0	-0.11083		1984	0	0	0	0	0	-0.01083
1950	0	0	0	0	1	0		1985	0	0	0	0	0	0
1951	0	0	0	0	1	-0.06917		1986	0	0	0	0	0	-0.23833
1952	0	0	1	0	0	-0.22833		1987	3	3	2	0	0	-1.3075
1953	2	2	0	1	0	-0.68		1988	0	1	0	0	0	0
1954	0	0	0	1	0	0		1989	0	0	0	0	0	0
1955	0	0	0	0	1	0		1990	0	0	0	0	0	-0.21917
1956	0	0	0	0	1	0		1991	5	0	1	0	1	-0.87833
1957	0	2	0	0	0	-0.38917		1992	1	0	0	0	0	-1.03833
1958	0	0	0	0	0	-0.32		1993	0	0	0	0	0	-0.94667
1959	0	0	0	0	0	-0.00417		1994	0	0	0	0	0	-1.19333
1960	0	0	0	0	0	0		1995	0	0	0	0	1	-0.22667
1961	0	0	0	0	0	0		1996	0	0	0	0	0	0
1962	0	0	0	0	0	0		1997	0	0	1	0	0	-1.16667
1963	0	2	0	0	0	-0.195		1998	0	0	0	0	0	-0.10833
1964	0	0	0	0	0	0		1999	0	1	0	0	0	0
1965	0	0	1	0	1	-0.8425		2000	1	0	0	0	0	0
1966	0	1	0	1	0	-0.42417		2001	0	0	0	0	0	0
1967	0	0	0	1	0	0		2002	1	2	3	2	0	-0.61
1968	0	0	0	0	0	0		2003	0	0	0	0	0	-0.31417
1969	1	0	0	0	0	-0.53833		2004	0	0	0	0	0	-1.16
1970	1	0	0	1	0	0		2005	0	0	0	0	0	0
1971	0	2	2	0	0									

Table 17: Hybridisation. Identification (ID), Collectors and collection number, Longitude, Latitude, Altitude and Collection month of *C. albiflora* (ALBI), *C. linearis x albiflora* (ALBI X LINE) and *C. linearis* (LINE). Hybrid Zone 1 (Tulahuen - Vicuña), Hybrid Zone 2 (Canela Baja - Guanaqueros) Hybrid Zone 3 (Petorca - Salamanca) (See Figure 39) Data ordered as follows:ID; Collector; Collection; Number; Longitude (West); Latitude (South); Hybrid Zone; Altitude (m.a.s.l.); Collection Month.

ALBI	Jaffuel		2559	70.20	22.10	-	10	10
ALBI	Jaffuel		2565	70.20	22.10	-	10	10
ALBI	Worth & Morrison		15802	70.43	25.40	-	75	10
ALBI	Worth & Morrison		15842	70.43	25.40	-	150	10
ALBI	Johnston		5114	70.43	25.40	-	-	11
ALBI	Lopez		s.n.	70.48	25.40	-	20	-
ALBI	Werdermann		815	70.48	25.40	-	200	10
ALBI	K. Arroyo et al.		25129	70.42	25.42	-	-	
ALBI	Johnston		5637	70.47	25.42	-	-	12
ALBI	Johnston		5649	70.48	25.42	-	-	12
ALBI	Dillon & Trujillo		8050	70.43	25.43	-	890	11
ALBI	Teillier		685	70.75	26.15	-	350	10
ALBI	Marticorena, Rodriguez & Weldt		1870	70.78	27.17	-	70	10
ALBI	Zöllner		9282	70.77	27.28	-	265	9
ALBI	Geisse		s.n.	70.77	27.28	-	265	11
ALBI	Rose & Rose		19326	70.55	27.67	-	-	10
ALBI	Pisano & Bravo		794	70.78	27.90	-	180	11
ALBI	Muñoz S., Teillier & Meza		2937	70.93	28.13	-	-	11

ALBI	Rechinger & Rechinger	63365	71.08	28.13	-	-	11
ALBI	Rechinger & Rechinger	63366	71.08	28.13	-	-	11
ALBI	Ricardi et al.	1523	70.97	28.15	-	220	10
ALBI	Von Bohlen	1345	71.17	28.30	-	10	9
ALBI	Worth & Morrison	16271	70.67	28.33	-	500	10
ALBI	Jaffuel	1175	71.20	28.43	-	-	11
ALBI	Cabrera	12669	71.10	28.52	-	-	9
ALBI	Saavedra	474	70.75	28.57	-	585	12
ALBI	Rechinger & Rechinger	63321	70.27	28.58	-	-	11
ALBI	Barros	2442	70.77	28.58	-	450	9
ALBI	Ricardi	s.n.	70.78	28.78	-	1000	10
ALBI	Muñoz S., Meza & Barrera	1148	71.23	28.92	-	300	9
ALBI	Ricardi & Marticorena	4463	70.90	29.03	-	850	9
ALBI	Cabrera	12691	70.90	29.03	-	-	9
ALBI	Grau	2053	71.00	29.23	-	-	10
ALBI	Ricardi & Marticorena	4905	71.03	29.35	-	400	10
ALBI	Muñoz S., Teillier & Meza	2666	71.12	29.37	-	-	10
ALBI	Marticorena & Matthei	211	71.20	29.45	-	350	10
ALBI	Marticorena & Matthei	181	71.22	29.50	-	450	10
ALBI	Cabrera	12594	71.22	29.50	-	-	9
ALBI	Jiles	4010	71.28	29.62	-	50	11
ALBI	Marticorena et al.	1575	71.07	29.78	-	300	10
ALBI	Ehrhart et al.	2002/086	71.28	29.78	-	-	11
ALBI	K. Arroyo et al.	25048	71.30	29.82	-	110	10
ALBI	K. Arroyo et al.	25054	71.30	29.82	-	110	8
ALBI	Ricardi et al.	1812	71.27	29.83	-	10	1
ALBI X LINE	Jaffuel	1205	71.40	29.93	-	-	11
ALBI X LINE	Jaffuel	1206	71.40	29.93	-	-	11
ALBI X LINE	Jaffuel	2689	71.40	29.93	-	-	10
ALBI X LINE	Werdermann	115	71.35	29.97	-	100	11
ALBI	Marticorena et al.	9946	71.27	29.98	-	90	2
ALBI X LINE	Wagenknecht	272	71.37	29.98	-	75	1
ALBI X LINE	Jiles	536	71.50	30.83	-	350	2
ALBI X LINE	Jiles	4063	70.73	31.02	-	1000	12
ALBI X LINE	Jiles	1032	70.73	31.02	-	1300	10
ALBI X LINE	Jiles	5861	71.62	31.17	-	280	11
ALBI	K. Arroyo et al.	25025	70.98	31.18	-	1010	9
ALBI X LINE	Jiles	5002	71.48	31.30	-	100	10
ALBI	K. Arroyo et al.	25028	71.10	31.55	-	580	9
ALBI X LINE	Montero	7274	71.05	31.75	-	450	10
LINE	Ehrhart & Grau	94/295	70.97	31.77	-	-	11
LINE	Worth & Morrison	16470	71.38	31.83	-	450	11
ALBI	K. Arroyo et al.	25154	71.52	31.85	-	10	11
ALBI	K. Arroyo et al.	25012	71.52	31.85	-	950	9
LINE	Zöllner	9213	71.52	31.90	-	-	10
LINE	K. Arroyo et al.	992406	71.15	32.00	-	890	9
LINE	Marticorena & Matthei	509	70.57	32.07	-	1800	1
LINE	Rose & Rose	19380	71.42	32.33	-	-	10
LINE	Troncoso	s.n.	71.42	32.52	-	150	2
LINE	Bohm	s.n.	71.47	32.55	-	300	2
LINE	Johow	s.n.	71.47	32.55	-	300	2
LINE	Morrison	16873	71.23	32.67	-	1500	12
LINE	Morrison	16842	71.50	32.92	-	500	12
ALBI	Werdermann	1875	70.57	29.97	1	800	11
ALBI	Montero	11688	70.57	29.97	1	800	9
ALBI X LINE	Elliott	39	70.50	30.02	1	1300	10
ALBI X LINE	Gajardo	0	70.42	30.03	1	1100	12
ALBI	Ehrhart et al.	2002/079	70.72	30.07	1	880	11
ALBI X LINE	Elliott	97	70.72	30.07	1	-	10
ALBI X LINE	Ehrhart et al.	2002/083	70.72	30.07	1	850	11
ALBI X LINE	Ehrhart et al.	2002/078	70.72	30.07	1	880	11
ALBI	K. Arroyo et al.	25019	70.72	30.10	1	1050	9

ALBI X LINE	Rosas	1492	70.45	30.12	1	1310	12
ALBI X LINE	Pinto	18	70.48	30.12	1	1290	2
ALBI	Eggli & Leuenberger	3061	70.67	30.20	1	1650	10
ALBI	Wagenknecht	18489	70.53	30.47	1	1800	11
ALBI X LINE	Jiles	3259	70.62	30.73	1	1300	10
ALBI X LINE	Jiles	3253	70.68	30.75	1	1100	10
ALBI X LINE	Jiles	3598	70.55	30.83	1	3000	1
ALBI X LINE	Jiles	4064	70.70	30.98	1	1150	12
ALBI X LINE	Jiles	4722	70.70	31.03	1	2200	12
ALBI X LINE	Jiles	4723	70.70	31.03	1	2200	12
ALBI X LINE	Jiles	2447	70.57	31.10	1	1850	1
ALBI X LINE	K. Arroyo	81231	71.37	30.02	2	190	1
ALBI X LINE	Jiles	5837	71.37	30.10	2	50	10
ALBI	Grau	2542	71.42	30.20	2	60	11
ALBI X LINE	Muñoz	1891	71.42	30.20	2	60	10
ALBI X LINE	Rodriguez & Marticorena	1625	71.38	30.22	2	60	11
ALBI X LINE	Garaventa	5553	71.38	30.22	2	100	10
ALBI	Barros	4570	71.08	30.23	2	1000	9
ALBI	Jiles	3897	71.43	30.33	2	120	10
ALBI X LINE	Jiles	3098	71.05	30.42	2	1100	10
ALBI X LINE	Zöllner	10431	71.50	30.50	2		11
ALBI X LINE	Jiles	3048	71.40	30.58	2	250	10
ALBI	Ehrhart et al.	2002/040	71.63	30.65	2	-	11
ALBI	Muñoz P. & Coronel	1409	71.63	30.67	2	250	11
ALBI	Muñoz	B-105	71.63	30.67	2		9
ALBI X LINE	Werdermann	890	71.63	30.67	2	300	11
ALBI X LINE	Marticorena et al.	432	71.63	30.67	2	300	11
ALBI X LINE	Sparre	2958	71.63	30.67	2		10
ALBI	K. Arroyo et al.	25022	71.22	30.85	2	380	9
ALBI	K. Arroyo et al.	25067	71.22	30.85	2	380	9
ALBI X LINE	Marticorena & Matthei	365	71.30	30.90	2	630	10
ALBI	K. Arroyo et al.	25033	70.77	32.33	3	760	10
ALBI X LINE	K. Arroyo et al.	25034	70.77	32.33	3	760	10
ALBI	K. Arroyo et al.	25036	70.93	32.48	3	400	10
ALBI	K. Arroyo et al.	25037	70.88	32.53	3	500	10
ALBI	Zöllner	4416	70.72	32.75	3	630	10
LINE	King	417	70.33	32.83	3	2300	3
LINE	Mancilla	s.n.	70.60	32.83	3	820	9
LINE	Edwards	s.n.	70.98	32.83	3		1
LINE	Silva	s.n.	70.32	32.92	3	2000	2
LINE	Hutchinson	79	71.00	32.97	3	1200	12
LINE	Zöllner	322	71.00	32.97	3	2200	3
LINE	K. Arroyo et al.	209102	71.10	32.97	-	1100	12
LINE	K. Arroyo et al.	994030	71.03	32.98	-	1630	11
LINE	Zöllner	576	71.27	32.98	-	2100	3
LINE	Schlegel	3120	70.93	33.00	-	700	11
LINE	Muñoz S.	806	70.87	33.02	-	-	3
LINE	Zöllner	13026	71.25	33.03	-	1400	11
LINE	Jaffuel	751	71.63	33.03	-	150	10
LINE	Montero	1379	70.97	33.07	-	700	10
LINE	Looser	995	70.67	33.13	-	-	11
LINE	Villagran	4761	70.98	33.18	-	1600	12
LINE	Davies	s.n.	70.28	33.30	-	1800	11
LINE	Wall	s.n.	70.28	33.30	-	2000	1
LINE	Hellwig	3156	70.28	33.30	-	-	2
LINE	Davies & Grau	2002/001	70.30	33.30	-	1800	02
LINE	Davies & Grau	2002/001a	70.30	33.30	-	2200	02
LINE	Ehrhart et al.	2002/002	70.60	33.30	-	1000	11
LINE	Looser	5214	70.30	33.33	-	2000	01
LINE	Schlegel	1048	70.33	33.35	-	2000	3
LINE	K. Arroyo et al.	206159	70.33	33.35	-	2160	11
LINE	Junge	s.n.	70.47	33.35	-	800	12

LINE	Saa	s.n.	70.47	33.35	-	-	10	
LINE	K. Arroyo et al.	25164	70.33	33.37	-	2180	12	
LINE	Looser	666	70.83	33.38	-	-	11	
LINE	Espinosa	s.n.	70.83	33.38	-	-	1	
LINE	Barros	s.n.	70.48	33.42	-	1800	2	
LINE	Gunckel	23475	70.63	33.42	-	750	10	
LINE	Schlegel	3282	70.90	33.48	-	680	12	
LINE	Montero	293	70.38	33.58	-	1200	01	
LINE	Gunckel	26331	70.75	33.60	-	750	11	
LINE	Muñoz S., Teillier & Arrigada	3438	70.35	33.62	-	-	11	
LINE	Grau	2399	70.83	33.62	-	-	11	
LINE	K. Arroyo et al.	25165	70.47	33.73	-	1000	12	
LINE	Montero	342	70.22	33.78	-	1600	12	
LINE	Muñoz S. et al.	3708	70.23	33.78	-	-	1	
LINE	Marticorena & Weldt	522	70.72	33.92	-	450	11	
LINE	Mahu	4086	70.57	34.25	-	700	11	

Table 18: The number of collections recorded in each hybrid zone with a hybrid index scores of 0 – 3. Character scores: Ray colour: yellow (0) or white (1). Dorsal stripe on rays: absent (0) or present (1). Indumentum: present (0) or absent (1). (See Figure 41)

Hybrid Zone	Hybrid Index Score			
	0	1	2	3
1	0	9	4	6
2	0	12	3	3
3	6	1	1	3

Table 19: Data set for PCA and HYWIN of collections, characters and character scores (See Figure 43, 44, Table 6). [Character List: Number; Collector; Collection number; Latitude (S); Longitude (W); Leaf dentition: dentate(0), spinose (1); Leaf margins: serrate apex (0), entirely serrate (1); Outer Involucral Bract length (mm); Outer Involucral Bract width (mm); Inner Involucral Bract length (mm); Inner Involucral Bract width (mm); Leaf length average (mm); Leaf width average (mm); Pedicel length average (cm); Pubescent Leaves: no (0), yes (1); Number of leafy OIB series; Habit: roots (0), stolons (1); Pubescent Axils: no (0), yes (1); Pubescent Stems: no (0), yes (1); Capitula height average (mm); Capitula width average (mm); Pubescence > 3mm: no (0), yes (1); Pubescence sparse on leaves: no (0), yes (1); Outer Involucral Bract apical angle (degrees); Inner Involucral Bract apical angle (degrees); PCA Component 1 score; PCA Component 2 score]

1	ROSAS 1875	38.22	71.73 1 1 10.1 3.2 15.7 3.2 4.5 4.0 18.0 0 1 0 1 1 12.5 17.5 0 0 120 90 9.54 -4.74
2	GRAU s.n.	36.82	71.25 1 1 5.3 1.7 14.7 1.1 2.8 3.8 6.5 0 1 1 0 1 11.0 13.5 0 0 90 60 2.51 2.90
3	ROSAS 1888	37.18	72.30 0 0 8.5 1.9 15.7 1.1 2.3 1.5 7.5 1 2 0 1 1 11.5 13.0 0 0 90 60 -4.54 -1.95
4	JAFFUEL 1307	36.82	73.05 0 0 7.5 1.6 15.2 1.3 2.3 2.5 11.0 1 1 1 1 1 13.0 18.5 0 1 60 25 -1.41 3.10
5	WERDERMANN 1256	38.68	71.80 1 1 10.1 2.6 17.1 1.3 3.3 3.5 10.5 1 2 1 1 1 12.0 19.0 0 1 120 60 5.78 -1.33
6	MATTHEI & QUEZADA 786	34.35	71.73 0 0 5.5 1.5 9.0 1.5 3.0 2.3 10.0 1 1 1 1 0 10.0 12.0 0 0 30 90 -2.69 4.09
7	PENNELL 12836	37.85	72.60 0 0 6.9 1.7 13.3 0.8 3.5 1.3 16.0 1 1 1 1 1 13.0 16.0 0 1 60 35 -0.89 -0.97
8	GRAU s.n.	37.90	72.55 0 0 11.2 1.8 16.8 1.5 2.8 1.5 11.5 1 2 1 1 1 13.0 18.5 0 0 90 35 -5.31 3.24
9	WEST 4918	37.82	72.32 0 0 6.4 1.3 12.2 1.1 4.5 1.3 14.0 1 1 0 1 1 12.0 13.5 0 0 60 25 -1.88 -7.95
10	GAY 228	33.03	71.63 0 0 9.9 2.4 11.5 2.1 3.3 2.3 10.5 1 3 0 1 1 12.0 16.0 1 0 150 120 -0.20 -6.82
11	ELLIOTT 292	33.25	70.37 0 0 10.1 2.4 13.6 1.9 3.5 3.5 10.5 1 3 0 1 1 14.0 15.0 1 0 150 90 0.75 -6.41
12	O.KUNTZE s.n.	34.00	69.50 0 0 16.0 1.9 19.0 1.3 1.8 3.0 5.0 1 2 0 1 0 12.0 17.5 0 0 120 90 5.80 -0.33
13	POEPPIG s.n.	36.78	73.17 1 1 7.5 2.1 18.1 2.1 4.5 3.5 15.0 1 2 1 1 1 12.5 20.0 0 0 90 65 -8.90 1.76
14	JAFFUEL 1267	33.03	71.33 0 0 4.5 1.1 9.3 1.1 2.0 1.3 2.0 1 1 0 1 1 12.0 12.5 0 0 60 60 -6.63 -1.48

15	GOODSPEED 23322		
		33.10	71.65 0 0 8.0 1.6 13.9 1.1 3.0 2.0 7.5 1 3 0 1 1 8.5 9.5 0 0 60 60 -4.35 1.49
16	GRAU 2495	34.75	70.50 0 0 6.7 1.6 14.4 1.3 3.3 2.5 17.0 1 2 0 1 1 11.0 12.0 0 0 60 20 -0.61 0.956
17	ROSAS 1917	35.72	70.78 1 1 5.9 1.7 13.3 1.3 4.0 2.0 8.0 1 2 0 1 1 13.5 13.5 0 0 60 60 -1.54 4.48
18	F.PHILIPPI 359	40.25	73.10 0 0 5.0 1.6 13.9 1.1 3.3 3.3 21.0 1 1 1 1 1 15.0 20.0 1 0 60 25 3.26 -6.10
19	O.KUNTZE s.n.	34.00	69.50 0 0 10.1 2.9 16.8 2.1 4.0 2.5 7.0 1 3 0 1 0 15.0 20.0 0 0 120 65 -3.30 2.62
20	F.PHILIPPI 359a	40.25	73.10 0 0 8.5 1.3 14.7 0.8 4.0 2.0 17.5 1 2 1 1 1 12.0 17.5 1 0 60 35 -5.44 0.92
21	F.PHILIPPI s.n.	33.42	70.63 0 0 9.3 1.3 16.0 1.1 2.0 1.5 5.0 1 2 0 1 1 13.5 12.5 0 0 30 20 -6.33 -2.70
22	WERDERMANN 507		
		35.25	71.08 0 0 8.0 2.1 9.6 1.2 2.8 1.5 5.0 1 2 0 1 1 9.5 13.5 0 0 90 90 7.50 -1.85
23	WERDERMANN 1256		
		38.68	71.80 1 1 11.2 2.4 18.1 1.6 4.0 4.0 10.0 0 2 1 1 1 12.5 17.5 0 0 120 60 10.21 -1.46
24	GRAU s.n.	37.83	73.00 1 1 9.0 3.2 17.3 1.3 3.5 5.5 9.5 0 3 1 1 1 16.0 25.0 0 0 90 60 -5.10 4.02
25	GRAU 2998	36.93	72.92 0 0 5.0 1.6 13.3 0.9 3.0 2.5 11.0 1 1 1 1 1 11.0 12.5 0 0 60 25 1.28 -6.31
26	BERTERO 472	34.17	70.77 0 0 12.0 2.7 15.5 1.8 3.0 2.5 20.0 1 3 0 1 1 15.0 15.0 0 0 120 90 0.85 2.5
27	BERTERO 166	34.33	71.25 0 1 6.7 1.8 13.1 1.3 5.0 2.5 29.5 1 3 0 1 1 15.0 17.5 0 0 35 30 -4.62 2.49
28	MARKHAM 336	33.03	71.63 0 0 4.5 1.3 13.3 1.3 4.3 2.5 11.0 1 2 0 1 1 11.0 16.0 0 0 60 25 6.62 -2.31
30	COMBER 372	39.00	71.00 1 1 10.4 2.9 17.1 1.6 4.0 4.0 9.0 0 1 0 1 1 13.5 13.5 0 0 120 60 1.28 0.66
31	HUTCHINSON 241	36.78	73.17 0 0 8.0 2.1 17.3 1.6 2.3 3.5 7.0 1 1 1 1 1 15.0 17.5 0 1 90 35 10.55 0.82
32	HOLLERMAYER 160b		
		39.80	73.23 1 1 8.5 2.4 16.0 1.3 6.5 6.0 30.0 1 2 1 1 1 15.0 20.0 0 1 120 90 5.27 0.60
33	F.PHILIPPI s.n.	36.92	71.33 1 1 8.0 2.1 13.0 1.6 4.5 3.3 13.5 0 2 0 1 0 12.0 20.0 0 0 90 30 5.68 5.05
34	KUNKEL 461	37.78	72.47 1 1 7.5 1.6 16.0 1.2 3.8 3.5 21.5 0 1 1 1 1 13.5 22.5 0 0 60 25 -2.65 0.29
35	JUNGE 2728	36.90	73.00 0 0 6.7 1.9 14.1 1.6 1.8 1.3 9.0 1 2 1 1 1 12.5 17.5 0 1 60 60 -1.45 2.25
36	POEPPIG 110(62)	32.68	70.42 0 0 9.3 1.3 17.3 1.1 3.8 2.5 27.0 1 1 0 1 1 15.0 15.0 1 0 60 35 -7.22 2.19
37	GRANDJOT s.n.	35.02	70.80 0 0 8.0 1.3 11.5 0.8 2.5 1.5 7.0 1 2 0 1 1 12.5 12.5 0 0 25 25 8.48 0.14
38	GAY s.n.	-	- 1 1 13.3 2.9 17.3 1.3 4.8 2.0 27.0 1 2 1 1 1 15.0 20.0 0 1 90 35 -3.92 -6.68
39	CUMING 182	-	- 0 0 11.2 2.1 13.3 1.3 2.0 2.0 10.5 1 3 0 1 1 12.0 12.0 1 0 150 90 -2.44 -9.95
40	MONTERO 270	33.58	70.38 0 0 11.7 2.1 16.0 2.1 2.3 1.0 6.5 1 3 0 1 1 11.5 15.0 1 0 150 150 -5.12 -2.34
41	PIRION 86	34.83	70.58 0 0 8.5 1.6 12.0 1.3 3.0 2.5 8.0 1 2 0 1 1 11.0 11.0 0 0 90 90 -3.58 -3.24
42	MORRISON & WAGENKNECHT 17111		
		33.08	71.02 0 0 8.5 2.1 14.4 1.6 3.5 1.8 10.0 1 2 0 1 1 11.0 11.0 0 0 90 90 1.16 2.48
43	GARDNER, KNEES & EVORE 4590		
		35.92	70.72 1 1 7.0 1.9 11.2 1.3 3.0 3.0 10.0 0 1 0 1 0 10.5 14.0 0 0 60 30 5.683 3.85
44	MONTERO 3091	38.72	72.58 1 1 8.0 2.1 13.9 0.8 5.0 3.0 16.0 0 1 1 1 0 12.0 20.0 0 0 90 30 6.810 1.11
45	GRAU 2923	36.07	70.50 1 1 7.4 2.0 14.9 1.6 4.8 3.8 10.0 1 2 1 1 0 12.0 25.0 0 1 90 60 7.41 -0.19
46	POEPPIG 208	37.33	71.68 1 1 12.0 2.1 17.9 1.3 3.5 4.5 8.0 1 2 1 1 1 20.0 20.0 0 0 60 60 -4.549 -3.92
47	ARAVENA 33348	34.95	70.40 0 0 10.7 2.1 14.1 1.3 2.0 1.5 10.0 1 2 0 1 1 12.0 8.0 0 0 120 60 -6.754 -1.47
48	ROSAS 1956b	34.93	70.43 0 0 9.9 1.3 13.3 1.3 2.0 2.0 4.0 1 2 0 1 1 10.0 10.0 0 0 60 60 -0.77 4.14
49	OCHENSIS s.n.	-	- 0 0 8.8 1.3 14.7 1.3 3.5 3.5 20.0 1 1 1 1 1 15.0 20.0 1 0 30 25 -3.157 -0.3
50	JUNGE 2286	36.78	73.17 0 0 7.2 1.6 13.3 1.5 2.0 3.0 10.0 1 1 1 1 1 12.0 15.0 0 0 90 90 -2.746 1.13
51	LECHLER 488	40.30	73.10 0 0 8.5 1.9 11.7 1.3 3.0 3.0 9.0 1 1 1 1 1 12.0 20.0 1 0 60 60 -0.517 2.05
52	GAY 379	40.35	73.15 0 0 9.1 1.3 15.2 1.1 3.0 5.0 15.0 1 2 1 1 0 12.0 15.0 0 0 60 30 8.438 2.74
53	WERDERMANN 1256		
		38.68	71.80 1 1 7.2 2.6 16.5 1.3 4.5 4.5 12.5 0 1 1 1 1 14.0 25.0 0 0 90 35 -1.799 0.7640
54	LEUENBERGER ET AL 4039		
		35.73	70.78 0 0 9.1 1.6 15.2 1.3 4.5 2.0 7.5 1 2 0 1 1 13.0 15.0 0 1 60 30 -3.719 0.075
55	F.PHILIPPI s.n.	-	- 0 0 7.5 1.9 17.1 1.6 2.0 2.0 7.0 1 1 0 1 1 11.0 15.0 0 0 60 35 -1.238 1.616
56	PFISTER s.n.	36.78	72.75 0 0 6.4 1.8 13.9 1.6 2.5 2.5 10.0 1 1 1 1 1 15.0 20.0 0 0 90 30 12.06 1.602

57	BOEHNERT s.n.	38.72	71.72 1 1 10.1 2.4 18.1 1.1 3.8 4.0 10.0 0 1 1 0 0 18.0 30.0 0 0 90 60 -5.781 2.962
58	LECHLER 488	40.30	73.10 0 0 6.0 1.2 12.0 1.0 3.0 3.0 9.0 1 1 1 0 1 10.0 10.0 1 0 60 60 -2.791 3.091
59	F.PHILIPPI s.n.	40.25	73.10 0 0 7.2 1.6 12.3 1.3 3.0 3.0 17.5 1 1 1 1 1 13.0 18.0 1 0 60 30 1.116 3.650
61	GRAU s.n.	36.82	71.25 1 1 6.7 1.3 13.9 1.1 3.3 3.5 8.5 0 1 1 1 1 10.0 15.0 0 0 60 60 4.105 0.409
62	EHRHART & GRAU 95/919		
		36.63	71.87 1 1 9.1 1.6 17.3 1.3 3.5 3.5 9.5 0 2 0 1 1 13.0 17.5 0 0 90 35 -0.049 -1.397
63	GRAU s.n.	37.90	72.55 0 0 7.5 2.0 18.7 1.8 3.0 1.5 9.0 1 2 1 1 1 15.0 17.5 0 0 90 60 -5.595 -0.136
64	GRAU 2348	35.58	72.10 0 0 6.9 1.9 14.9 1.3 1.8 1.5 9.5 1 2 0 1 1 11.5 11.5 0 0 60 30 -4.771 0.6351
65	GRAU 2495	34.75	70.50 0 0 8.0 1.3 14.7 1.3 2.5 3.0 14.5 1 3 0 1 1 11.5 12.0 0 0 30 30 -6.004 0.7382
66	GRAU 2375	35.32	72.33 0 0 6.7 1.6 14.4 1.3 2.0 1.5 12.5 1 2 0 1 1 9.5 14.0 0 0 60 25 2.515 1.121
67	GRAU 3185	36.77	72.63 1 1 9.3 1.9 17.6 1.6 4.0 2.0 8.0 1 2 1 1 1 13.0 15.0 0 0 60 25 5.35773 1.450
69	GRAU 2923	36.07	70.50 1 1 8.0 2.1 15.7 1.9 3.5 4.3 10.5 1 1 1 1 1 14.0 22.5 0 0 60 60 -8.732 -0.29
70	GRAU 2736	37.50	72.28 0 0 7.2 1.3 10.7 1.1 2.0 1.5 2.5 1 2 0 1 1 8.5 12.0 0 0 60 60 -7.40815 0.5292
71	GRAU s.n.	36.58	73.13 0 0 7.2 1.3 11.0 1.3 2.8 1.5 3.3 1 2 1 1 1 10.0 9.0 0 0 60 60 -2.85323 1.894
72	GRAU 2991	36.77	72.67 0 0 6.0 1.5 15.0 1.3 4.3 3.5 13.0 1 1 1 1 1 10.0 11.5 0 0 90 60 10.137 -4.299
73	GRAU s.n.	37.83	73.00 1 1 11.2 3.2 18.1 2.7 3.3 5.5 11.5 0 2 1 1 1 12.5 17.5 0 0 90 90 0.23 -1.960
74	EHRHART & GRAU 95/577		
		35.30	71.52 0 0 9.3 2.1 14.7 1.3 3.3 2.0 11.5 1 2 0 1 0 15.0 20.0 0 0 120 35 -0.330 2.758
75	F.PHILIPPI 359a	40.25	73.10 0 0 8.0 1.6 17.3 1.1 4.8 4.5 20.0 1 2 1 1 1 10.5 17.5 1 0 60 25 -5.091 2.678
76	F.PHILIPPI 359	40.25	73.10 0 0 6.4 1.3 11.5 1.3 3.3 3.3 18.0 1 1 1 1 1 10.0 12.0 1 0 60 60 7.414 -3.301
77	F.PHILIPPI s.n.	35.43	71.67 1 1 8.9 2.6 16.1 2.3 3.0 4.0 8.0 0 2 0 0 1 14.0 18.0 0 0 90 90 -0.059 0.058
79	F.PHILIPPI s.n.	37.80	72.92 0 0 9.0 2.4 17.3 1.2 4.0 2.5 10.0 1 1 1 1 1 10.0 20.0 0 0 90 60 15.29 -0.54
80	F.PHILIPPI s.n.	36.92	71.33 1 1 12.7 1.6 16.8 1.9 3.3 7.0 13.0 0 0 3 1 0 0 20.0 30.0 0 0 1 90 60 13.80 4.64
81	O.PHILIPPI s.n.	39.33	71.80 1 1 9.6 2.1 17.7 2.1 8.5 4.0 36.0 0 1 1 1 0 15.0 20.0 0 1 60 35 -2.964 3.27
82	F.PHILIPPI s.n.	40.25	73.10 0 0 4.3 1.3 15.0 1.3 4.5 3.0 8.5 1 2 1 1 1 12.0 19.0 1 0 60 25 -3.299 -5.08
83	CUMING 182	-	- 0 0 10.6 1.7 13.6 1.4 3.3 2.5 8.5 1 3 0 1 1 13.0 12.0 1 0 150 60 -7.469 -0.76
84	F.PHILIPPI s.n.	34.24	70.72 0 0 6.4 1.5 11.7 1.0 2.0 1.2 1.5 1 2 0 0 1 10.0 10.0 0 0 90 60 -1.895 -4.79
85	F.PHILIPPI s.n.	33.42	70.63 0 0 11.2 1.9 14.4 1.9 3.5 3.5 8.5 1 2 0 1 1 11.0 12.5 1 0 150 50 -1.56 -8.22
86	O.KUNTZE s.n.	34.00	69.50 0 0 11.0 2.2 15.5 2.5 2.0 2.0 2.5 1 3 0 1 0 11.5 11.0 0 0 120 95 -5.683 -0.60
666	RICARDI 2831	36.30	71.65 0 0 9.3 1.6 10.7 1.3 3.5 1.8 9.0 1 2 0 1 1 10.0 12.5 0 0 90 30 -5.484 2.03
681	GRAU 2998	36.93	72.92 1 1 7.7 1.3 13.9 1.2 2.0 1.5 8.0 0 2 1 1 1 10.0 10.0 0 0 60 30 -1.532 2.37
682	GRAU 2998	36.93	72.92 0 0 6.7 1.2 15.4 0.9 3.8 2.5 13.5 1 1 1 1 1 10.0 12.0 0 0 60 25 -4.511 4.14
781	F.PHILIPPI s.n.	36.92	71.33 1 1 8.0 1.6 14.9 1.6 4.5 4.3 14.0 0 0 1 0 0 1 16.0 20.0 0 0 60 30 6.821 3.41
782	F.PHILIPPI s.n.	36.92	71.33 1 1 10.1 1.5 17.1 1.6 3.0 2.0 14.0 0 1 0 0 1 16.0 20.0 0 0 60 30 5.538 2.15

Table 20: HYWIN data table of frequency counts of top ranking 1000 hybrid triplets (P> 0.95), where Rank criterion = 1.0 * IN + 1.0 * (1-|EQ|) + 1.00 (equal weighted rank) and where Rank criterion = 0.1 * IN + 0.10 * (1-|EQ|) + 1.00 (parental distance weighted rank assuming introgression). (See Figures 45 and 46). HYB = Hybrid, WK HYB = Weak Hybrid, PAR = Parent, WK PAR = Weak Parent. Data is ordered as follows: ID Number; Equal Weighted hybrid frequency; Equal Weighted hybrid rank; Equal Weighted parent frequency; Equal Weighted parent rank; Equal Weighted posteriori category; Weighted hybrid frequency; Weighted hybrid rank; Weighted parent frequency; Weighted parent rank; Weighted posteriori category

1	8	249	30	134	WK PAR	17	59	0	0	HYB
2	19	9	9	14	WK HYB	8	7	0	0	HYB
3	18	27	8	141	WK HYB	3	262	0	0	HYB
4	17	84	5	468	WK HYB	7	50	0	0	HYB
5	69	12	1	905	WK HYB	18	5	0	0	HYB
6	1	564	22	56	WK PAR	7	28	0	0	HYB
7	4	208	18	27	WK PAR	6	71	0	0	HYB
8	28	5	0	0	HYB	13	35	0	0	HYB
9	0	0	41	1	PAR Z	4	355	33	22	PAR Z

10	0	0	39	14	PAR Z	2	229	63	119	PAR Z
11	3	508	16	391	WK PAR	4	126	0	0	HYB
12	0	0	15	70	PAR Z	7	122	0	0	HYB
13	29	152	5	658	WK HYB	16	27	0	0	HYB
14	0	0	76	7	PAR Z	1	409	76	1	PAR Z
15	1	438	14	11	WK PAR	3	240	0	0	HYB
16	4	281	14	37	WK PAR	5	261	0	0	HYB
17	42	42	3	687	WK HYB	11	41	0	0	HYB
18	-	-	-	-	VALD	-	-	-	-	VALD
19	4	460	31	103	WK PAR	12	44	0	0	HYB
20	-	-	-	-	VALD	-	-	-	-	VALD
21	0	0	22	35	PARENT	4	330	2	426	HYB
22	4	255	27	86	WK PAR	2	282	20	236	PAR Z
23	14	396	9	199	WK HYB	15	49	0	0	HYB
24	0	0	24	228	PAR Y	9	76	0	0	HYB
25	2	580	33	49	WK PAR	5	166	0	0	HYB
26	0	0	9	21	PAR Z	5	145	0	0	HYB
27	4	111	14	19	WK PAR	6	57	0	0	HYB
28	0	0	21	66	PAR Z	5	320	4	488	HYB
30	20	41	3	463	WK HYB	17	10	0	0	HYB
31	25	43	0	0	HYB	11	12	0	0	HYB
32	14	213	15	42	WK PAR	16	33	0	0	HYB
33	29	4	5	321	WK HYB	13	14	0	0	HYB
34	25	15	52	5	WK PAR	8	20	0	0	HYB
35	12	55	3	328	WK HYB	5	121	0	0	HYB
36	1	434	11	40	WK PAR	5	90	26	80	PAR Z
37	0	0	37	16	PAR Z	2	348	32	86	PAR Z
38	7	76	13	72	WK PAR	13	37	0	0	HYB
39	0	0	44	12	PAR Z	2	264	76	40	PAR Z
40	0	0	102	3	PAR Z	2	244	137	11	PAR Z
41	10	117	18	36	WK PAR	4	156	0	0	HYB
42	20	90	7	352	WK HYB	4	218	0	0	HYB
43	39	56	6	295	WK HYB	9	19	0	0	HYB
44	3	576	16	3	WK PAR	5	78	1	315	HYB
45	47	86	4	226	WK HYB	14	15	0	0	HYB
46	19	129	3	599	WK HYB	11	13	0	0	HYB
47	4	781	13	65	WK PAR	4	207	0	0	HYB
48	2	566	20	114	WK PAR	2	274	0	0	HYB
49	-	-	-	-	VALD	-	-	-	-	VALD
50	16	14	4	268	WK HYB	6	96	0	0	HYB
51	14	29	2	142	WK HYB	6	70	0	0	HYB
52	10	53	0	0	HYB	13	16	0	0	HYB
53	9	49	37	147	WK PAR	7	47	0	0	HYB
54	7	104	9	233	WK HYB	5	147	0	0	HYB
55	4	323	11	101	WK PAR	5	127	1	523	HYB
56	15	127	3	740	WK HYB	13	29	0	0	HYB
57	0	0	137	2	PAR Y	2	341	125	11	PAR Y
58	-	-	-	-	VALD	-	-	-	-	VALD
59	-	-	-	-	VALD	-	-	-	-	VALD
61	33	13	15	131	WK HYB	10	3	0	0	HYB
62	70	2	1	281	WK HYB	20	1	0	0	HYB
63	30	59	0	0	HYB	11	25	0	0	HYB
64	3	538	17	149	WK PAR	3	365	3	539	HYB
65	0	0	15	2	PAR Z	3	209	1	326	HYB
66	0	0	22	74	PAR Z	3	370	0	0	HYB
67	45	11	0	0	HYB	13	8	0	0	HYB
69	61	8	6	69	WK HYB	17	2	0	0	HYB
70	0	0	51	31	PAR Z	2	527	39	85	PAR Z
71	0	0	31	116	PAR Z	4	190	0	0	HYB
72	2	860	7	665	WK HYB	7	130	0	0	HYB
73	2	784	17	107	WK PAR	13	101	0	0	HYB
74	13	3	0	0	HYB	13	34	0	0	HYB

75/6	-	-	-	-	VALD	-	-	-	-	VALD
77	17	88	22	56	WK PAR	14	21	0	0	HYB
79	35	112	2	296	WK HYB	8	18	0	0	HYB
80	0	0	264	1	PAR Y	1	427	261	1	PAR Y
81	0	0	195	9	PAR Y	5	143	213	17	PAR Y
82	-	-	-	-	VALD	-	-	-	-	VALD
83	0	0	12	77	PAR Z	3	117	15	248	PAR Z
84	0	0	81	9	PAR Z	2	303	72	17	PAR Z
85	0	0	9	236	PAR Z	3	217	0	0	HYB
86	0	0	49	25	PAR Z	3	227	0	0	HYB
666	4	495	12	175	WK PAR	4	259	0	0	HYB
681	10	26	30	103	WK PAR	9	4	0	0	HYB
682	1	772	26	39	WK PAR	6	167	0	0	HYB
781	27	1	29	70	WK HYB	11	9	0	0	HYB
782	24	6	6	50	WK HYB	13	6	0	0	HYB

Table 21a: Data table giving mean altitude and floret measurements for species in *Oriastrum* and *Chaetanthera*. Data is ordered as follows: Species; Mean altitude; Ray L mm (RL); Ray tube L mm (RTL); Ligule Length; Ligule Width; Disk corolla length mm (DL); Disk tube length mm (DTL); RL/DL. Altitude Zones (m.a.s.l.): 0-1500 Matorral; 1800-2700 Subandino; 2700-3100 Andino; 3100-3500 Altoandino; >3500 Andino Superior

Species	Alt	RL	RTL	LL	LW	DL	DTL	RL/DL
O. werdermannii	600	9.0	4.5	4.5	2.3	3.9	3.0	2.3
O. chilense	2832	8.3	3.6	4.7	1.5	5.3	4.7	1.6
O. pusillum	2858	5.0	2.2	2.8	1.8	3.7	3.0	1.4
O. apiculatum	2963	10.0	3.6	6.4	1.8	7.0	6.0	1.4
O. lycopodioides	3022	9.8	3.9	5.9	2.4	6.3	5.6	1.5
O. gnaphalioides	3309	4.8	2	2.8	1.2	3.7	3.3	1.3
O. dioicum	3360	8.2	3.7	4.5	0.8	5.2	4.0	1.6
O. acerosum	3498	8.0	3.3	4.8	1.7	5.8	5.0	1.4
O. tontalensis	3635	13.0	6.5	6.5	1.5	8	6.5	1.6
O. famatinae	3650	4.4	2.4	2	0.5	3.6	2.8	1.2
O. abbreviatum	3982	4.2	1.8	2.5	0.8	2.9	2.2	1.4
O. achenohirsutum	3989	10.5	4.3	6.3	1.8	8.0	7.3	1.3
O. pulvinatum	4100	6.0	-	-	-	3.8	-	1.6
O. polymallum	4307	3.5	3.0	0.5	1.0	5.5	4.5	0.6
O. tarapacensis	4314	9.8	4.9	4.9	0.8	6.1	5.2	1.6
O. revolutum	4409	3.8	1.9	1.9	0.5	3.3	2.3	1.2
O. stuebelii	4588	7.0	3.0	4.0	0.7	4.9	3.2	1.4
O. cochlearifolium	4960	7.8	4.0	3.8	1.3	6.5	5.0	1.2
C. ciliata	154	13.0	3.5	9.5	2.0	7.3	6.0	1.8
C. x serrata	200	18.0	4.0	14.0	2.7	8.5	6.0	2.1
C. incana	285	9.5	3.3	6.2	2.8	4.9	3.5	1.9
C. taltalensis	310	3.7	2.2	1.5	0.3	2.8	2.5	1.3
C. frayjorgensis	437	17.5	4.3	13.3	4.0	9.4	7.7	1.9
C. albiflora	455	9.9	4.3	5.6	2.1	5.7	4.6	1.7
C. glabrata	585	18.0	4.3	13.8	3.0	9.7	8.0	1.8
C. multicaulis	600	10.0	3.0	7.0	2.6	6.0	5.0	1.7
C. microphylla	699	7.2	3.7	3.5	1.2	6.2	5.6	1.2
C. ramosissima	875	5.7	3.7	2.0	0.5	4.3	3.8	1.3
C. elegans	930	22.0	3.5	18.5	3.0	9.5	7.5	2.3
C. moenchioides	953	9.0	5.0	4.0	2.5	6.5	5.8	1.4
C. chilensis	1030	15.0	4.0	11.0	3.5	8.0	6.0	1.9
C. linearis	1289	9.6	3.7	5.9	1.5	6.2	5.2	1.5
C. depauperata	1863	4.6	3.0	1.6	0.6	4.0	3.2	1.2
C. villosa	1972	24.0	9.5	14.5	3.2	16.7	14.2	1.4
C. glandulosa	1973	11.3	5.0	6.3	1.7	9.0	6.8	1.3
C. flabellata	2121	10.0	4.3	5.7	2.2	8.0	6.4	1.3
C. euphrasioides	2370	5.9	2.8	3.1	1.0	4.4	2.7	1.3
C. kalinae	2700	14.5	6.5	8.0	2.6	9.5	8.0	1.5
C. perpusilla	2975	3.5	2.1	1.4	0.6	2.8	2.5	1.3
C. philippii	3098	16.8	6.9	9.8	2.3	11.8	10.8	1.4

C. splendens	3104	15.3	4.7	10.5	3.5	10.5	8.5	1.5
C. renifolia	3158	20.0	6.3	13.7	3.0	11.8	10.0	1.7
C. spathulata	3232	25.4	8.0	17.4	3.3	15.5	12.5	1.6
C. peruviana	3325	5.63	3.1	2.6	0.8	3.8	3.0	1.5

Table 21b: Data to calculate the probability (p) associated with a homoscedatic Student's paired t-Test, with a two-tailed distribution (two-sample, equal variance). Null hypothesis assumes both data sets have equal sample means (ray length: disk length ratio).

Genus:	RL:DL				p
Altitude zones	Min.	Mean	Max.	s.d.	
Oriastrum: andino-altoandino*	1.28	1.45	1.57	0.12	0.29
Oriastrum andino superior	0.64	1.32	1.63	0.30	
Chaetanthera: Mediterranean matorral	1.16	1.71	2.32	0.33	0.01
Chaetanthera: subandino-altoandino	1.15	1.42	1.70	0.18	
Chaetanthera & *Oriastrum*: Mediterranean matorral	0.64	1.39	1.70	0.21	0.0001
Chaetanthera & *Oriastrum*: subandino-andino superior	1.16	1.75	2.32	0.35	

* excluding the lowland *O. werdermannii*.

I want morebooks!

Buy your books fast and straightforward online - at one of world's fastest growing online book stores! Environmentally sound due to Print-on-Demand technologies.

Buy your books online at
www.morebooks.shop

Kaufen Sie Ihre Bücher schnell und unkompliziert online – auf einer der am schnellsten wachsenden Buchhandelsplattformen weltweit! Dank Print-On-Demand umwelt- und ressourcenschonend produziert.

Bücher schneller online kaufen
www.morebooks.shop

KS OmniScriptum Publishing
Brivibas gatve 197
LV-1039 Riga, Latvia
Telefax: +371 686 204 55

info@omniscriptum.com
www.omniscriptum.com

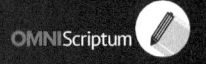

Printed by Books on Demand GmbH, Norderstedt / Germany